Molecular
Systematics
of Plants

Molecular Systematics of Plants

Pamela S. Soltis, Douglas E. Soltis,
and Jeff J. Doyle, Editors

Chapman
and Hall

New York

London

First published in 1992 by
Chapman and Hall
an imprint of
Routledge, Chapman & Hall, Inc.
29 West 35th Street
New York, NY 10001-2291

Published in Great Britain by
Chapman and Hall
2-6 Boundary Row
London SE1 8HN

Library of Congress Cataloging in Publication Data

Molecular systematics of plants / edited by Douglas E. Soltis, Pamela
  S. Soltis, and Jeff J. Doyle.
    p.   cm.
  Includes bibliographical references and index.
  ISBN 0-412-02231-1 ISBN 0-412-02241-9 (pb)
    1. Botany—Classification—Molecular aspects.   I. Soltis, Douglas
E.   II. Soltis, Pamela S.   III. Doyle, Jeff J.
QK95.6.M65   1991
581'.012—dc20   91-29233
                                          CIP

British Library of Congress Cataloguing in Publication Data also available

Contributors

Victor A. Albert
Department of Biology
University of North Carolina
Chapel Hill, NC 27599-3280

R. Appels
CSIRO Division of Plant Industry
P.O. Box 1600
Canberra ACT 2601, Australia

B. Baum
Biosystematics Research Center,
  Agriculture Canada
Ottawa, Ontario KIA 0C6, Canada

Anne Bruneau
L. H. Bailey Hortorium
Cornell University
Ithaca, NY 14853

Steven J. Brunsfeld
Department of Forest Resources
University of Idaho
Moscow, ID 83843

Mark W. Chase
Department of Biology
University of North Carolina
Chapel Hill, NC 27599-3280

Michael T. Clegg
Department of Botany and Plant Sciences
University of California
Riverside, CA 92521

Daniel J. Crawford
Department of Plant Biology
The Ohio State University
Columbus, OH 43210

John Doebley
Department of Plant Biology
University of Minnesota
St. Paul, MN 55108

Michael J. Donoghue
Department of Ecology and Evolutionary
  Biology
University of Arizona
Tucson, AZ 85721

Stephen R. Downie
Department of Biology
Indiana University
Bloomington, IN 47405

Jeff J. Doyle
L. H. Bailey Hortorium
Cornell University
Ithaca, NY 14853

James E. Eckenwalder
Department of Botany
University of Toronto
Toronto, Ontario M5S 3B2, Canada

R. Keith Hamby
Department of Biochemistry
Louisiana State University
Baton Rouge, LA 70803

Robert K. Jansen
Department of Ecology and Evolutionary
  Biology
University of Connecticut
Storrs, CT 06269

Sterling C. Keeley
Department of Ecology and Evolutionary
  Biology
University of Connecticut
Storrs, CT 06269

Ki-Joong Kim
Department of Ecology and Evolutionary
  Biology
University of Connecticut
Storrs, CT 06269

Marilyn Kobayashi
Department of Plant Biology
The Ohio State University
Columbus, OH 43210

Matt Lavin
Department of Biology
Montana State University
Bozeman, MT 59717

Helen J. Michaels
Department of Biology
Bowling Green State University
Bowling Green, OH 43402

Brook G. Milligan
Department of Botany
University of Texas
Austin, TX 78713

Brent Mishler
Department of Botany
Duke University
Durham, NC 27706

Jeffrey D. Palmer
Department of Biology
Indiana University
Bloomington, IN 47405

Loren Rieseberg
Graduate Program in Botany
Rancho Santa Ana Botanic Garden
Claremont, CA 91711

Kermit Ritland
Department of Botany
University of Toronto
Toronto, Ontario M5S 3B2, Canada

Michael J. Sanderson
L. H. Bailey Hortorium
Cornell University
Ithaca, NY 14853

James F. Smith
Department of Botany
University of Wisconsin
Madison, WI 53706

Douglas E. Soltis
Department of Botany
Washington State University
Pullman, WA 99164

Pamela S. Soltis
Department of Botany
Washington State University
Pullman, WA 99164

Kenneth J. Sytsma
Department of Botany
University of Wisconsin
Madison, WI 53706

Robert S. Wallace
Department of Ecology and Evolutionary
  Biology
University of Connecticut
Storrs, CT 06269

Linda E. Watson
Department of Botany and Microbiology
University of Oklahoma
Norman, OK 73019

Elizabeth A. Zimmer
National Museum of Natural History
Smithsonian Institution
4210 Silver Hill Rd.
Suitland, MD 20746

Gerard Zurawski
DNAX Research Institute
902 California Avenue
Palo Alto, CA 94304

Contents

Preface

The broad goals of systematics are phylogenetic reconstruction and the elucidation of the evolutionary processes that generate biological diversity. Over the past several decades, a diverse array of characters, including morphologic and cytologic features, flavonoid and protein variation, and now DNA restriction site and sequence variation, has been applied to the study of plant relationships and evolution. Recent advances in analytic techniques have improved our ability to reconstruct plant phylogeny. Regardless of the type of data used in phylogenetic reconstruction, however, in practice the evolutionary history of the characters is generally equated with the evolutionary history of the organisms, despite theoretical arguments against such inferences. Thus, current attempts to infer organismal phylogenies from gene phylogenies do not differ in principle from previous efforts at phylogenetic reconstruction, despite the attention that has recently been paid to the interpretation of gene trees versus species trees.

Rapid strides in molecular biology, including improvements in techniques of DNA isolation, advances in DNA sequencing, and the substitution of the polymerase chain reaction for cloning to amplify specific DNA sequences, have precipitated equally rapid changes in plant systematics. Not only have molecular techniques introduced new characters for systematic analysis, but the molecular characters themselves, so well suited for cladistic analysis, have done much to promote the use of phylogenetic methods in plant systematics.

The goals of this book are threefold: (1) to summarize the achievements of plant molecular systematics in its first decade; (2) to illustrate the potential of molecular characters for addressing a variety of phylogenetic and evolutionary questions; and (3) to suggest the appropriate technique(s) for a given level of systematic inquiry. To accomplish these goals, the book is divided into four sections, addressing types of molecular data, application of molecular data to evolutionary questions, specific examples of molecular phylogenetics, and issues in the analysis of molecular data, respectively. With both a review of previous studies in molecular systematics and suggestions for future research, the book is intended to be of value to fledgling and seasoned researchers alike.

The first five chapters illustrate the application of a diversity of molecular techniques and genomic constituents to questions in plant systematics. Clegg and Zurawski discuss the potential and limitations of chloroplast DNA sequence analysis for studies of plant phylogenetics. This chapter is particularly timely given the rate at which DNA sequences, especially *rbc*L, are being generated for phylogenetic inference. Downie and Palmer advocate the use of structural changes in the chloroplast genome (e.g., inversions, gene loss, intron loss, etc.) as major characters in phylogenetic reconstruction. The mitochondrial genome in plants has received little attention from systematists; its utility and potential in plant systematics are reviewed by Palmer. The chapters by Hamby and Zimmer and Appels and Baum address the role of nuclear ribosomal RNA genes (rDNA) in plant systematics. Hamby and Zimmer use sequences from the 18S–26S rDNA to examine relationships in the Poaceae and among groups of seed plants; Appels and Baum apply data from both the 18S–26S and 5S rDNA to studies of molecular and organismal evolution in the Triticeae (Poaceae).

Chapters 6 through 9 address questions of evolutionary processes in plants using molecular data. Recent studies have demonstrated levels of intraspecific chloroplast DNA (cpDNA) variation higher than previously believed on the basis of both early studies and the conservative nature of the chloroplast genome. The utility of this variation for addressing questions of microevolutionary processes is discussed by D. Soltis et al. Rieseberg and Brunsfeld illustrate the value of both cpDNA and rDNA markers in analysis of hybridization and introgression. The chapter by P. Soltis et al. uses similar logic and data to examine origins of polyploid species and various aspects of polyploid evolution. Molecular data have also been valuable in assessing relationships between crop plants and their wild relatives; the molecular systematics of domesticated plants are described by Doebley.

Chapters 10 through 14 represent what we consider to be model studies of phylogenetic relationships in plants using cpDNA. Each chapter uses molecular data to clarify relationships in a group of plants for which previous studies had yielded inadequate resolution. Doyle et al. illustrate the use of a variety of DNA characters, from structural mutations to restriction site mutations, at a diversity of taxonomic levels in the Leguminosae. Jansen et al. similarly present a variety of examples from the Asteraceae. Crawford et al. address the evolution of the annual versus perennial habit in *Coreopsis* (Asteraceae) by examining phylogenetic relationships using cpDNA. Speciational and biogeographic data in *Clarkia* and *Fuchsia* (Onagraceae) are discussed by Sytsma and Smith, who also address the congruence (or lack thereof) between morphologic and molecular data. This theme is echoed theoretically by Donoghue and Sanderson in Chapter 15. Chase and Palmer use a molecular phylogeny of the Oncidiinae (Orchidaceae) to infer directionality of changes in chromosome number and morphologic features in this highly variable group of orchids.

The final three chapters contribute a theoretical perspective to an otherwise largely empirical book. Donoghue and Sanderson explore the attributes of molecular and morphologic features for phylogenetic reconstruction. They conclude that, although both types of data may be plagued by a variety of factors, both molecular and morphologic data may provide valuable characters for phylogenetic analyses of plants and encourage their joint use by plant systematists. Albert et al. contend that the Dollo parsimony method of restriction site data analysis, often employed by plant molecular systematists, represents an unrealistic biological situation. They advocate that plant molecular systematists abandon the principle and practice of Dollo parsimony, and instead apply a system of character-state weighting that more accurately reflects biological reality. Aspects of molecular evolution and their effects on phylogenetic reconstruction are discussed by Ritland and Eckenwalder, who illustrate that polymorphism, differences in evolutionary rate (among taxa and among sequences), and hybridization, if undetected, may produce inaccurate phylogenies.

The reader will undoubtedly be struck by the diversity of analytic approaches and the diversity of opinions and perspectives presented by the authors. For example, several chapters (e.g., Jansen et al. and Sytsma and Smith) provide phylogenetic hypotheses based on the Dollo parsimony criterion, whereas Albert et al. strongly oppose its use. Some authors (e.g., Crawford et al., Sytsma and Smith, and Chase and Palmer) have interpreted patterns of morphologic, cytologic, and biogeographic data from phylogenetic hypotheses based on molecular data; Donoghue and Sanderson contend that greater resolution and more accurate phylogenies will result from the joint use of molecular and morphologic data in phylogenetic reconstruction. Furthermore, although most contributors prefer cladistic methods of analysis based on parsimony, Hamby and Zimmer, as well as Appels and Baum, also employ phenetic methods. This diversity of opinion should not be construed by the systematic community as a failure of molecular systematics. Instead, molecular systematics should be recognized as a vibrant field of inquiry with a dynamic tension among researchers. (Indeed, a tension—albeit generally friendly—was apparent as drafts of chapters were shared among authors prior to the completion of the book!) Only through dialogue will any field of scientific endeavor advance. We humbly hope that the diversity of techniques and perspectives presented in this book will contribute to the refinement of plant molecular systematics.

Finally, we wish to acknowledge some of the many people who have assisted with this book. Thanks to Jane L. Doyle for proofreading several of the manuscripts. Thanks also to Jerrold I. Davis, Michael W. Frohlich, Leslie D. Gottlieb, Kent E. Holsinger, Gerald H. Learn, and Jonathan F. Wendel for reviewing portions of the book. The following persons not only contributed chapters to the book but also reviewed portions of it: Michael T. Clegg, Daniel J. Crawford, John F. Doebley, Michael J. Donoghue, Robert K. Jansen, Brent D. Mishler, Loren H. Rieseberg, Michael J. Sanderson, and Kenneth J. Sytsma. Additional acknowledgments are listed in each chapter. We would like to thank the contributors for their cooperation in making what we hope will be a valuable reference for researchers in plant molecular systematics. Lastly, thanks to Greg Payne from Chapman and Hall for his encouragement and patience throughout this project.

Pamela S. Soltis, Douglas E. Soltis
Pullman, Washington, September 1990

Jeff J. Doyle
Ithaca, New York, September 1990

1

Chloroplast DNA and the Study of Plant Phylogeny: Present Status and Future Prospects

Michael T. Clegg and *Gerard Zurawski*

More than a quarter of a century has passed since the publication of Zuckerkandl and Pauling's (1965) seminal article on the evolutionary implications of macromolecular sequence data. Two fundamental suggestions were made in that article: first, Zuckerkandl and Pauling pointed out that molecular change (nucleotide or amino acid substitutions) might occur at a rate that is proportional to clock time (the molecular clock hypothesis); and second, they noted that the topology of evolutionary branchings (phylogenies) could be deduced from the pattern of molecular change. These two related suggestions form the theoretical underpinnings of the science of molecular evolution.

The field of molecular evolution developed rapidly into a significant area of research activity in the late 1960s, and this trend continued throughout the 1970s. Among the more noteworthy achievements were the use of cytochrome C amino acid sequence data for phylogenetic reconstruction by Fitch and Margoliash (1967), the estimation of the time of the human–chimpanzee–gorilla split by Sarich and Wilson (1967), and the development of the neutral theory of molecular evolution by Kimura (1968). Owing to the development of recombinant DNA technology and the invention of rapid DNA sequencing methods, the 1980s has seen an explosion in the use of molecular data for the study of evolutionary problems. The field of plant molecular evolution has participated in this expansion, especially where the chloroplast genome (cpDNA) is concerned.

The chloroplast genome is well suited for evolutionary and phylogenetic study, because cpDNA is a relatively abundant component of plant total DNA, thus facilitating extraction and analysis. A second advantage is an extensive background of molecular information on the chloroplast genome. For example, complete DNA sequences of three cpDNA genomes are known (the liverwort *Marchantia polymorpha,* Ohyama et al., 1986; tobacco, *Nicotiana tabacum,* Shinozaki et al., 1986; and rice, *Oryza sativa,* Hiratsuka et al., 1989). Knowledge

Supported in part by NSF grant BSR-8500206 and NIH grant GM45144.

of complete cpDNA sequences for three distantly related taxa is a great advantage because it facilitates the investigation of a variety of questions regarding (a) changes in gene content, (b) changes in genome structural organization, and (c) rates of gene evolution (see Clegg et al., 1990, for further discussion).

Another advantage of the chloroplast genome for evolutionary research is a conservative rate of nucleotide substitution. Conservative rates of cpDNA evolution have both a technical and a fundamental advantage: from a technical point of view, cloned cpDNA genes can be used as heterologous probes across virtually the entire plant kingdom. The universal applicability of cpDNA clones has led to a substantial economy of effort, because the initial molecular biology does not need to be repeated with each new taxon investigated. The fundamental advantage of a conservative rate of nucleotide substitution arises because cpDNA sequence change is appropriate to resolve plant phylogenetic relationships at deep levels of evolution. It is precisely at these levels that conventional methods for phylogenetic inference are most wanting because of problems of parallel and convergent character-state change. DNA sequence data also have the advantage of being independent of other biological characters, in the sense that no assumptions about relationships are necessary to infer phylogenies from sequence data. This means that molecular phylogenies can be contrasted to conventionally derived phylogenies and patterns of character evolution can be examined within the independent context of molecular trees. Owing to all these advantages, the comparative analysis of cpDNA sequence data has been a quickly expanding area.

Rapidly developing DNA sequencing and polymerase chain reaction (PCR) technologies present a host of questions and opportunities for the study of plant molecular evolution and systematics. Our goal in this chapter is to review some of these questions and opportunities. In addition, we take a prospective stance and discuss some issues that are likely to arise in the near future. One problem that will soon emerge is the management of what are for most plant evolutionary biologists relatively large data sets. A second issue, which has been the subject of much discussion, is the choice of appropriate methods of data analysis. A related problem concerns the statistical resolution of molecular data sets. The pace of technological innovation over the past decade has been breathtaking. What can we expect over the next few years? How large an impact will studies of fossil DNA have on evolutionary biology? In this chapter, we consider some of these questions and offer our opinions and guesses about the future.

Chloroplast Gene Evolution

In most respects, the molecular evolution of chloroplast genes mirrors that of nuclear genes. However, as noted above, chloroplast protein-coding genes evolve at a rate that is on average fivefold slower than plant nuclear genes (Wolfe et al., 1987, 1989). The average rates of synonymous substitution for cpDNA protein-coding genes vary approximately from 0.2 to 1.0×10^{-9} substitutions per site per year (Wolfe and Sharp, 1988). Ratios of synonymous to nonsynonymous substitution within genes are quite similar for nuclear and chloroplast protein-coding genes, indicating comparable patterns of selective constraint. It seems probable that the reduced rate of cpDNA gene evolution arises from a reduced mutation rate. According to population genetic theory, neutral gene substitutions are expected to occur at a rate equal to the mutation rate (Kimura and Ohta, 1973), and in those cases where careful statistical studies have been conducted (e.g., Li et al., 1985), synonymous mutations appear to be very weakly constrained, if constrained by selection at all. If we accept the conclusion that chloroplast mutation rates are reduced relative to nuclear gene rates, then we are left with the question of why the mutation rate is reduced. At this stage, we can only note that the DNAs are replicated in different compartments by different systems.

Noncoding regions of the chloroplast genome tend to evolve more rapidly than do coding regions (Wolfe and Sharp, 1988). Addition/deletion mutations accumulate in noncoding regions at a rate that is at least equal to nucleotide substitution (Golenberg et al., unpublished data), and this category of mutation accelerates the divergence of noncoding regions. It appears that many addition/deletion mutations are associated with short direct repeats (Zurawski et al., 1984) and probably arise from slipped-strand mispairing during replication. Owing to this association of addition/deletion mutation with direct repeats, it is probable that particular noncoding regions experience higher rates of these mutations because of local sequence features. It also seems probable that addition/deletion mutations may recur at specific sites, thus contributing to homoplasy in evolutionary studies (Golenberg et al., unpublished data).

Some chloroplast genes have introns, but unlike the introns of nuclear genes, cpDNA introns tend to have a high degree of secondary structure. Not surprisingly, the pattern of cpDNA intron evolution is clearly constrained by secondary structure. These constraints in mutational change reflect functional requirements associated with intron processing (Learn et al., unpublished data). Our limited studies indicate that the overall rate of chloroplast intron evolution is similar to overall rates of protein gene evolution (Learn et al., unpublished data; Zurawski and Clegg, 1984). There are a few cases of intron loss in cpDNA evolution. For example, rice has three fewer introns than found in the cpDNA of either *Marchantia polymorpha* or *Nicotiana tabacum* (Hiratsuka et al., 1989). The loss of an intron in the gene *rpl2* has been studied in 340 species representing 109 angiosperm families (Downie et al., 1991). The taxonomic distribution of intron presence versus absence indicates that the *rpl2* intron has been lost at least six times in angiosperm evolution.

Pseudogenes are also present in the chloroplast genome. In particular, the chloroplast ribosomal protein-coding gene *rpl23* is present as a pseudogene in many members of the Caryophyllales (Zurawski and Clegg, 1987). Interestingly, a portion of this same gene is present as a pseudogene in a region 3' to the gene *rbc*L in the rice chloroplast genome (Hiratsuka et al., 1989). There is no evidence for processed pseudogenes on the chloroplast genome, although they would be difficult to recognize because chloroplast mRNA is not polyadenylated and most cpDNA protein-coding genes lack introns. At this stage no attempts have been made to investigate rates of cpDNA pseudogene evolution to obtain a baseline estimate of the neutral mutation rate.

Two major factors that distinguish cpDNA evolution from nuclear gene evolution are the lack of clearly documented transposon activity associated with the chloroplast genome and the apparent lack of recombinational potential. Because biparental transmission is rare and intraspecific diversity is low, recombinational processes do not play a major role in cpDNA sequence evolution. One exception concerns the generation of major cpDNA inversions. The complete sequence of the rice genome has provided important insights into the mechanisms that generated a 20-kilobase pair (kb) inversion characteristic of the grass family. This inversion evidently arose as the result of nonhomologous recombination between different tRNA genes on two cpDNA molecules. Subsequent duplication and inversion events must then be posited to arrive at the contemporary structural arrangement (Hiratsuka et al., 1989).

From the perspective of using cpDNA gene sequence data for phylogenetic or evolutionary analyses, the salient fact is that cpDNA genes can be regarded as uniparentally transmitted molecules that collect mutations at a stochastically regular rate. Intracistronic recombination, which makes nuclear genes a mosaic of sequences with differing evolutionary histories, is not present for cpDNA genes. A consequence of this simplicity of transmission is that phylogenies derived from cpDNA sequences should be regarded as organellar (or gene) trees that may or may not be congruent with organismal history. However, it is reasonable to assume that the approximation to organismal history will improve as time increases, because the biases introduced by interspecific hybridization or intraspecific polymorphism will diminish with an increase in time scale.

Data Acquisition

Sequencing studies of cpDNA genes have utilized both the chemical cleavage reactions applied to end-labeled DNA fragments and dideoxy chain-termination reactions applied to DNA clones in double-stranded or single-stranded vectors. Both sequencing methods were laborious in that they demanded the preparation of specific gene fragments, the recloning of gene fragments, or the construction of directional gene deletions.

Early comparative sequencing studies uncovered the conservative nature of cpDNA evolution (Zurawski et al., 1984). This observation facilitated the development of sets of synthetic DNA primers which correspond to relatively conserved stretches of cpDNA genes. Such primer sets are now used with the dideoxy method to sequence specific genes rapidly from diverse plant species. Recently, these primers have been used with PCR technology to further simplify comparative cpDNA gene analysis. Primers corresponding to highly conserved sequences flanking the gene are used to amplify the intervening region from total plant DNA. The amplified cpDNA gene is then reamplified as single strands using each primer in separate PCRs. Sets of primers specific to each strand are then used with commercially available dideoxy sequencing kits and ^{35}S-labeled deoxynucleotides to generate sequences for both strands of the gene. These current methods, which are simple, rapid, and highly reliable, have enabled comparative cpDNA sequence analysis to become a routine tool for phylogenetic studies.

The area of DNA sequencing technologies is undergoing rapid change. It is expected that this change will accelerate with interest in the human genome sequencing project. Robot technologies are now available to execute many of the repetitive manipulations that are characteristic of present sequencing methods. Sequencing machines using either fluorescent primers or conventional radiolabeling have been developed, but cannot as yet match manual or robot-assisted methods for accuracy.

A further relevant developing technology is genomic sequencing (Church and Gilbert, 1984). This method utilizes chemical cleavage reactions on total DNA fragmented with various restriction enzymes. The reaction products are fractionated on sequencing gels and transferred to solid matrices. These matrices are then hybridized with radiolabeled specific gene fragments to reveal the gene sequence. One advantage of this method is that several gene sequences can be read from the same matrix by sequential rehybridization with new gene fragments. This method at present suffers from a loss of resolution resulting from the transfer process. However, the high dosage of cpDNA genes may make this a useful method for comparative studies examining several genes.

Chloroplast Gene Sequence Data Base

As already noted, complete DNA sequences of three chloroplast genomes are known. In addition, several chloroplast genes have been sequenced from a number of plant taxa. The gene *rbc*L that codes for the large subunit of ribulose-1,5-bisphosphate carboxylase/oxygenase (RUBISCO) has been sequenced from a moderately large number of plant taxa. As of this writing, our laboratories have sequenced approximately 40 *rbc*L genes (unpublished data), and we estimate that a much larger number of unpublished sequences of this gene are held by other labs. In addition, approximately 20 *rbc*L sequences have been published. Thus, a total of about 210 *rbc*L sequences are now available in the research community, and each sequence has been determined from a different plant species.

This large data base presents an unparalleled opportunity to study plant phylogeny. In addition, cpDNA sequence data are available from smaller numbers of taxa for the genes *atp*BE and *psb*A. Given present technologies, it is entirely reasonable to expect that *rbc*L sequence data can be obtained from representatives of all, or nearly all, of the approximately 400 flowering plant families. This opens up the very real possibility of constructing a molecular phylogeny of the angiosperms.

Chloroplast Gene Sequence Data and
Phylogenetic Reconstruction

There are three principal methods used for the reconstruction of phylogenies from molecular sequence data: the method of parsimony, distance matrix methods, and the method of maximum likelihood (MLE). Because much controversy has surrounded phylogenetic reconstruction in systematic biology, all three methods are often used in plant phylogenetic reconstruction. This is done to establish that each method is reasonably consistent for a given data set and to placate the proponents of each method.

Felsenstein (1988) has recently reviewed the properties of the three methods of phylogenetic reconstruction. From the point of view of statistical inference, it is important to realize that molecular phylogenies purport to show the evolutionary branching (topology) of various major biological groups. Usually each such group is represented by one or a very few sample taxa from which sequence data have been obtained. Clearly, there is an intragroup variance that needs to be determined to judge the reliability of the deduced branching pattern. The intragroup variance is a consequence of the random arrival of mutations in any lineage. Two lineages that have been separated for an identical (but reasonably long) period of time may have accumulated different numbers of mutations, and this variance can have the consequence that different sequences from the same groups may lead to different deduced topologies. This sampling variation is explicitly accounted for in MLE methods, because these methods are based on models of the mutational process. Resampling methods, such as bootstrap methods, can also be used to evaluate the intragroup variance of parsimony or distance-matrix derived trees (Felsenstein, 1988).

Statistical models of the mutational process require a number of assumptions about the relative occurrence of different categories of mutations, about the reversibility of the process, and about relative frequencies in the pool of introduced nucleotides. At first sight these assumptions seem to be a weakness of MLE methods, but, as argued by Felsenstein (1988), all methods involve either implicit or explicit assumptions. Because the assumptions of MLE methods are explicitly stated, they are open to statistical test. Ritland and Clegg (1987) have exploited this fact by subjecting model assumptions to likelihood ratio tests for cpDNA gene sequence data. They found that the assumption of reversibility is usually accepted, whereas assumptions about equality of nucleotide frequencies are usually rejected. They also found an approximately twofold excess of transition mutations for third codon positions and a slightly smaller excess for first and second codon positions for the cpDNA gene *rbc*L.

Ritland and Clegg (1987) also considered the statistical resolution of cpDNA data in attempting to decide among various competing phylogenies. They studied the relative branching order of the lineages leading to pea, spinach, and tobacco, which are representatives of three of the six major dicot subclasses (Caryophyllidae represented by spinach, Rosidae represented by pea, and Asteridae represented by tobacco; Cronquist, 1981). The analyses, based on complete DNA sequence data for the genes *rbc*L and *atp*B and representing nearly 3,000 base pairs of information, were unable to decide among the three competing phylogenies for these groups.

Recently, Giannasi et al. (1992) considered an expanded data base for *rbc*L which included sequences from the subclasses Dilleniidae, Magnoliidae, and the orders Polygonales and Plumbaginales, in addition to the pea, spinach, and tobacco sequences analyzed by Ritland and Clegg (1987). Giannasi et al. found that the additional taxa increased the statistical power of the MLE analyses and allowed a resolution of the topology of the various dicot subclasses. They also found that the orders Polygonales and Plumbaginales were distinct lineages that did not cluster with either the Caryophyllidae or the Dilleniidae, a result that is consistent with recent phenetic and cladistic analyses by Rodman et al. (1984).

Soltis et al. (1990) studied the broadly defined family Saxifragaceae. There is considerable debate about the boundaries of this family and about the evolutionary relationships of the Saxifragaceae with other major dicot groups. Eight *rbc*L sequences were obtained from species selected to span this diverse family. The resulting data were analyzed using MLE, parsimony, and UPGMA (unweighted pair group method with arithmetic averaging) algorithms. The various methods of analysis were in agreement and indicated that the Saxifragaceae are at least paraphyletic and probably polyphyletic. The work of Giannasi et al. (1992) and Soltis et al. (1990) strongly supports the contention that *rbc*L sequence data have great utility for resolving evolutionary relationships at the levels of family, order, and subclass.

In contrast, the studies of grass *rbc*L sequences by Doebley et al. (1990) suggest that the gene evolves too slowly to resolve relationships below the subfamily level. This work utilized ten sequences from three subfamilies (Pooideae, Panicoideae, and Bambusoideae) of the grass family (Gramineae). The data were subjected to a variety of phylogenetic analyses, including maximum likelihood, UPGMA, Wagner parsimony, and the least-squares method of Fitch-Margoliash. With the exception of UPGMA, all of the methods yielded consistent phylogenetic topologies, which differed from the traditional phylogenies proposed by grass systematists in the placement of rice. However, statistical tests failed to discriminate between the molecular phylogeny and the traditional phylogenies. In addition, the *rbc*L sequence data were unable to resolve the relationships among three taxa from the tribe Triticeae.

Chloroplast Gene Sequences from
Ancient Plant DNAs

The Clarkia fossil flora in northern Idaho has yielded an abundance of exceptionally well-preserved plant leaf compression fossils. More than a decade ago,
Giannasi and Niklas (1977) showed that intact flavonoid molecules could be
recovered from these materials. Based on this history, an attempt was made to
recover DNA by visiting the fossil site and extracting fossil leaf tissue. The
species or genus of the fossil leaves was identified and noted, the leaves were
photographed, and nucleic acid extraction was carried out *in situ,* immediately
after removing the samples from the fossil beds. Nucleic acid preparations were
made from 55 fossil leaf samples. Roughly 10% of the samples gave some
indication of nucleic acid present following ethidium bromide straining of crude
extracts. Several of these samples were then used as a source for PCR amplification of portions of the *rbc*L gene. A DNA fragment was amplified from a
Magnolia leaf compression fossil based on two 30mer primers that match the
coordinates 234–263 and 1,049–1,020 of the *Zea mays rbc*L gene. The DNA
sequence of the amplified fragment was determined and shown to be very similar
to that of other flowering plant *rbc*L genes (Golenberg et al., 1990).

To test further the fossil origin of the *rbc*L sequence, we determined the
complete DNA sequence of *rbc*L genes from two extant members of the Magnoliales (*Magnolia macrophyllum* and *Liriodendron tulipifera*) and one extant member of the Laurales (*Persea americana*). The sequence derived from the fossil
leaf material clusters with the other magnolioid sequences indicating a magnolioid
origin (Golenberg et al., 1990). Several other facts support the fossil origin of
the amplified sequence:

1. The environment of deposition was a cold anaerobic lakebed rich in organic
    sediments. Fossil fish recovered from the same bed show signs of toxic
    poisoning, but without subsequent microbiologic decay.

2. The leaves are exceptionally well preserved, and the leaf compression
    fossils retain intact cellular tissue with considerable ultrastructural preser-
    vation including cell walls, leaf phytoliths, and intracellular organelles.

3. Many organic molecules such as flavonoids and steroids have been ex-
    tracted from the leaf compression fossils and shown to be of fossil origin.

The Clarkia fossil flora is of Miocene origin and estimated to be 17–20 million
years old. This opens up the possibility of recovering DNA from a number of
fossil plant taxa. We are currently exploring this possibility by amplifying and
sequencing *rbc*L from additional fossil samples. A number of questions can be
addressed with these data including (1) determining the genetic relationships
between extant and extinct plant taxa, (2) estimating rates of genetic change based
on more accurate direct comparisons between fossil and extant genes, and (3)

establishing that the biogeographic relationships posited from morphologic considerations are consistent with genetic similarities. For example, the Clarkia fossil flora is believed to contain elements from southeastern China and elements from the southeastern United States. The fossil sequence data can be compared to sequence data from living representatives of these regions to validate these biogeographic inferences.

It seems probable that other well-preserved fossil floras may also yield good quality DNAs that can be studied. If this turns out to be the case, an exciting new window will have opened on gene evolution in plants. At this juncture, it is tempting to speculate that a field of molecular paleobotany will emerge to complement more traditional approaches to paleobotanical research.

Problems and Opportunities in
Plant Molecular Evolution

Up to this point, we have presented a brief review of current cpDNA research, and we have tried to use this review as a basis for evaluating research opportunities in the field of plant molecular evolution. Broadly speaking, research opportunities exist in the following three areas: (1) studies of the mechanisms of gene and genome evolution, (2) the investigation of plant phylogeny, and (3) the use of fossil DNAs to study rates of evolution and to characterize paleocommunities. What are the problems that confront the plant molecular evolutionist?

A major problem concerns the management and analysis of DNA sequence data. At the present time, data are reported in the primary literature and deposited in one or more of several data bases. Because nucleic acid sequence data are coded as long strings of simple characters, errors are easily introduced at any of several stages, including the stages of gel reading, data recording, and entry into computer files. There are no studies on error rates nor have any means of data cleaning been proposed. Nevertheless, it seems probable that some level of error exists in data bases. Error presents a particularly sensitive problem for evolutionary research where information about genetic relationships is derived from the pattern of mutational differences. There is a need to confront the problem of error and to establish its contribution to the total variance in phylogeny inference.

It may be desirable to establish sequence data bases dedicated to the collection of evolutionary data. A potential future problem is the sharing of sequence information with other members of the interested research community. At least two journals have incorporated sections where one can report nucleic acid sequence data with little or no comment (*Nucleic Acids Research* and *Plant Molecular Biology*). This is a valuable service to the evolutionary community because it provides a mechanism for the sharing of data. Both of the above mentioned journals also require that sequence data be deposited with a data base prior to publication. It would be particularly desirable for an evolutionary journal to establish similar policies. As the pace of evolutionary research in this field accelerates, it will be more and more difficult to rely on the molecular biology community to provide this service.

Perhaps the most vexing problem in the phylogenetic analysis of molecular data is the choice of an appropriate algorithm. There are many competing algorithms, and the proponents of some methods have adopted a very strong style of advocacy. The sometimes acrimonious debate that has ensued tends to obscure real issues of efficiency, precision, and computational speed. We have favored maximum-likelihood algorithms because of their well-established statistical properties; however, MLE algorithms are notoriously slow and demanding of computer time. At present, it is difficult to imagine the use of MLE algorithms on large data sets (50 taxa or more). The addition of bootstrap procedures to parsimony or UPGMA algorithms helps in the estimation of statistical error, but just compounds problems of computational inefficiency. Much work will have to be done to improve the computational efficiency of phylogeny estimation procedures. One solution is to combine simple and rapid, but biased, algorithms, with slow, but unbiased, methods. For instance, Ritland and Clegg (1987) suggested the use of UPGMA to obtain a first approximation of a phylogeny, followed by MLE as a basis for obtaining refined estimates.

Another problem that is just emerging with cpDNA gene sequence data is associated with variable rates of evolution. Recent work with the palm family indicates an eightfold reduction in the rate of molecular evolution in these long-lived plants relative to annual grass species (Wilson et al., 1990). Other workers have found hints of reduced rates in other tree species. Variable rates of evolution present serious problems for many estimation procedures. UPGMA assumes constant rates and is seriously compromised when this assumption is violated. Parsimony algorithms are also affected by variable evolutionary rates. Felsenstein (1978) noted that nonadjacent edges with higher numbers of changes will attract in parsimony algorithms. This can lead to an incorrect tree if some taxa evolve more rapidly than other taxa. Our preliminary calculations using *rbc*L sequences from representatives of the palms, the bromeliads, and the grasses, together with representatives of most dicot subclasses, gave highly discordant phylogenies for the monocots using parsimony methods (unpublished data). MLE analyses, on the other hand, yield reasonable phylogenies. Our working hypothesis is that the parsimony algorithm is seriously biased by unequal rates of evolution. Much more work will have to be done to establish the potential errors introduced by unequal rates of evolution.

Whether we can rely on many small research programs to yield the most rapid progress in molecular evolution is a debatable question. However, efforts to understand the history of life on earth, as recorded in DNA or RNA molecules, would certainly be greatly enhanced by some kind of national coordination. Problems of the maintenance of large data bases require some kind of national coordination. It is probable that the human genome project will lead to important advances in data management methods, but, because the focus is on the human genome, it is less likely that new advances in phylogenetic analysis will emerge from this project.

There is a present need to establish better linkage and coordination among the various research workers in the field of plant molecular evolution, and this need will become much more apparent over the next several years. A national repository for evolutionary data that also acts as a rapid clearinghouse for new analytic methods and that shares information on relevant new experimental techniques could greatly facilitate the assimilation and analysis of molecular information. Coordination at national or international levels would also prevent unnecessary duplication of effort and thereby enhance research efficiency.

Evolutionists and systematists have been handed a series of powerful new tools that offer a detailed view of biological history and of the genetic mechanisms that drive evolutionary change. These new tools have already yielded an unprecedented view of organismic relationships. Methods of data acquisition have vastly outstripped current systems of data management. Moreover, contemporary algorithms for phylogenetic and evolutionary analysis are not equal to the tasks presented by sequence data sets now in hand. Although most speculations about future trends seem comic when viewed retrospectively, it is nevertheless prudent to anticipate future problems. We believe that the issues for molecular evolutionary research in the 1990s will center on data sharing and data management, the development of better algorithms for the analysis of large data sets, and the coordination of efforts across many small research programs.

References

Clegg, M.T., Learn, G.H., and Golenberg, E.M. (1990) Molecular evolution of chloroplast DNA. In: *Evolution at the Molecular Level* (eds. R.K. Selander, A.G. Clark, and T.S. Whittam), Sinauer Associates, Sunderland, MA.

Church, G.M., and Gilbert, W. (1984) Genomic sequencing. *Proc. Natl. Acad. Sci. USA* **81,** 1991–1995.

Cronquist, A. (1981) *An Integrated System of Classification of Flowering Plants.* Columbia University Press, New York.

Doebley, J., Durbin, M.L., Golenberg, E.M., Clegg, M.T., and Ma, D.P. (1990) Evolutionary analysis of the large subunit of carboxylase (*rbcL*) nucleotide sequence among the grasses (Gramineae). *Evolution* **44,** 1097–1108.

Downie, S.R., Olmstead, R.G., Zurawski, G., Soltis, D.E., Soltis, P.S., Watson, J.C., and Palmer, J.D. (1991) Six independent losses of the chloroplast DNA *rpl2* intron in dicotyledons: molecular and phylogenetic implications. *Evolution,* in press.

Felsenstein, J. (1978) Cases in which parsimony or compatibility methods will be positively misleading. *Syst. Zool.* **27,** 401–410.

Felsenstein, J. (1988) Phylogenies from molecular sequences: inference and reliability. *Ann. Rev. Genet.* **22,** 521–565.

Fitch, W.M., and Margoliash, E. (1967) Construction of phylogenetic trees. *Science* **155**, 279–284.

Giannasi, D.E., and Niklas, K.J. (1977) Flavonoid and other chemical constituents of fossil Miocene *Celtis* and *Ulmus* (Succor Creek Flora). *Science* **197**, 765–767.

Giannasi, D.E., Zurawski, G., Learn, G.H., and Clegg, M.T. . (1992) Evolutionary relationships of the Caryophyllidae based on comparative *rbc*L sequences. *Syst. Bot.*, in press.

Golenberg, E.M., Giannasi, D.E., Clegg, M.T., Smiley, C.J., Durbin, M., Henderson, D., and Zurawski, G. (1990) Chloroplast DNA sequence from a Miocene *Magnolia* species. *Nature* **344**, 656–658.

Hiratsuka, J., Shimada, H., Whittier, R., Ishibashi, T., Sakamoto, M., Mori, M., Kondo, C., Honji, Y., Sun, C.-R., Meng, B.-Y., Li, Y.-Q., Kano, A., Nishizawa, Y., Hirai, A., Shinozaki, K., and Sugiura, M. (1989) The complete sequence of the rice (*Oryza sativa*) chloroplast genome: intermolecular recombination between distinct tRNA genes accounts for a major plastid DNA inversion during the evolution of the cereals. *Mol. Gen. Genet.* **217**, 185–194.

Kimura, M. (1968) Evolutionary rate at the molecular level. *Nature* **217**, 624–626.

Li, W.-H., Luo, C.-C., and Wu, C.-I. (1985) Evolution of DNA sequences. In: *Molecular Evolutionary Genetics* (ed. R.J. MacIntyre), Plenum Press, New York, pp. 1–94.

Ohyama, K., Fukuzawa, H., Kohchi, T., Shirai, H., Sano, T., Sano, S., Umesono, K., Shiki, Y., Takeuchi, M., Chang, Z., Aota, S., Inokuchi, H., and Ozeki, H. (1986) Chloroplast gene organization deduced from complete sequence of liverwort *Marchantia polymorpha* chloroplast DNA. *Nature* **322**, 572–574.

Ritland, K., and Clegg, M.T. (1987) Evolutionary analysis of plant DNA sequences. *Am. Nat.* **130**, S74–S100.

Rodman, J.E., Oliver, M.K., Nakamura, R.R., McClammer, J.U., Jr., and Bledsoe, A.H. (1984) A taxonomic analysis and revised classification of Centrospermae. *Syst. Bot.* **9**, 297–323.

Sarich, V.M., and Wilson, A.C. (1967) Immunological time scale for the hominid evolution. *Science* **150**, 1200–1203.

Shinozaki, K., Ohme, M., Tanaka, M., Wakasugi, T., Haysida, N., Matsubayashi, T., Zaita, N., Chungwongse, J., Obokata, J., Yamaguchi-Shinozaki, K., Ohto, C., Torazawa, K., Meng, B.Y., Sugita, M., Deno, H., Kamogashira, T., Yamada, K., Kusuda, J., Takaiwa, F., Kata, A., Tohdoh, N., Shimada, H., and Sugiura, M. (1986) The complete nucleotide sequence of the tobacco chloroplast genome. *Plant Mol. Biol. Rep.* **4**, 110–147.

Soltis, D.E., Soltis, P.S., Clegg, M.T., and Durbin, M.L. (1990) *rbc*L sequence divergence in the Saxifragaceae *sensu lato*. *Proc. Natl. Acad. Sci USA* **87**, 4640–4644.

Wilson, M.A., Gaut, B., and Clegg, M.T. (1990) Chloroplast DNA evolves slowly in the palm family (Arecaceae). *Mol. Biol. Evol.* **7**, 303–314.

Wolfe, K.H., and Sharp, P.M. (1988) Identification of functional open reading frames in chloroplast genomes. *Gene* **66**, 215–222.

Wolfe, K.H., Li, W.-H., and Sharp, P.M. (1987) Rates of nucleotide substitution vary greatly among plant mitochondrial, chloroplast, and nuclear DNAs. *Proc. Natl. Acad. Sci. USA* **84,** 9054–9058.

Wolfe, K.H., Sharp, P.M., and Li, W.-H. (1989) Rates of synonymous substitution in plant nuclear genes. *J. Mol. Evol.* **29,** 208–211.

Zuckerkandl, E., and Pauling, L. (1965) Evolutionary divergence and convergence in proteins. In: *Evolving Genes and Proteins,* (eds. V. Bryson and H.J. Vogel), Academic Press, New York.

Zurawski, G., and Clegg, M.T. (1984) The barley chloroplast DNA *atp*BE, *trn*M2, and *trn*v1 loci. *Nucl. Acids Res.* **12,** 2549–2559.

Zurawski, G., and Clegg, M.T. (1987) Evolution of higher-plant chloroplast DNA-encoded genes: implications for structure-function and phylogenetic studies. *Ann. Rev. Plant Physiol.* **38,** 391–418.

Zurawski, G., Clegg, M.T., and Brown, A.H.D. (1984) The nature of nucleotide sequence divergence between barley and maize chloroplast DNA. *Genetics* **106,** 735–749.

2

Use of Chloroplast DNA Rearrangements in Reconstructing Plant Phylogeny

Stephen R. Downie and *Jeffrey D. Palmer*

Reconstructing phylogenies among genera and at higher taxonomic levels always has been fraught with difficulties. Conventional plant classifications employ a diverse array of approaches (phytochemical, anatomic, morphologic, etc.) and often offer a synthesis of these data sets. Many of these traditional characters are susceptible to convergent evolution by natural selection; the ensuing homoplasy largely precludes robust phylogenies. Only recently have we been able to examine the genetic material itself to investigate phylogenetic relationships. Chloroplast DNA (cpDNA) variation has proven to be immensely valuable in reconstructing phylogenies at the species level, and the application of cpDNA comparisons at higher taxonomic levels is now being pursued actively.

One approach to extracting phylogenetic information from cpDNA is by analyzing the distribution of major structural rearrangements. Because of their infrequent occurrence, rearrangements usually can provide strong evidence of monophyly. In this chapter, we demonstrate the utility and significance of cpDNA rearrangements in reconstructing plant phylogeny. After an introduction to the salient features of the chloroplast chromosome, we briefly review the approaches used to detect and analyze rearrangements and discuss our current survey to detect and circumscribe cpDNA rearrangements among angiosperms. We will then examine the different classes of rearrangements and for each provide examples of their use in phylogenetic reconstruction. This chapter will deal exclusively with land plants. Algal genomes may have different structural dynamics than land plant genomes and are too poorly characterized to warrant discussion of their phylogenetic utility at the present time.

This research was supported by NSF Grant BSR-8996262 to JDP and a Natural Sciences and Engineering Research Council (NSERC) of Canada Postdoctoral Fellowship to SRD. We thank Richard G. Olmstead and Claude W. dePamphilis for critical reading of the manuscript and the many scientists who have generously provided us with DNAs to include in our survey of angiosperm cpDNA rearrangements.

The Chloroplast Chromosome

The chloroplast genomes of photosynthetic land plants are circular DNA molecules ranging in size from 120 to 217 kilobase pairs (kb) (Table 2.1). A list of land plants for which complete restriction endonuclease cleavage site maps of plastid genomes have become available since 1985 is presented in Table 2.1. This tabulation updates the previous compilation by Crouse et al. (1985). The chloroplast genome contains, with few exceptions, two duplicate regions in reverse orientation, known as the inverted repeat (IR). These repeated regions separate the remainder of the molecule into large single-copy (LSC) and small single-copy (SSC) regions (Fig. 2.1). The expansion or contraction of the IR into, or out of, adjacent single-copy regions, and changes in sequence complexity due to insertions or deletions of unique sequences are largely responsible for variation in size of the molecule. Several reviews of cpDNA structure, function, and evolution have been published recently (Whitfeld and Bottomley, 1983; Palmer, 1985a, 1985b, 1991; Zurawski and Clegg, 1987; Sugiura, 1989).

Recent studies of chloroplast genome evolution have revealed a high degree of conservation in size, structure, gene content, and linear order of genes among major lineages of land plants (Palmer, 1985b, 1991; Palmer and Stein, 1986). This conservative mode of cpDNA evolution suggests that any change in structure, arrangement, or content of the chloroplast genome may have significant phylogenetic implications.

Mutations in cpDNA are of two kinds: nucleotide substitutions (point mutations) and rearrangements. The detection of nucleotide substitutions through restriction site analysis or direct comparisons of homologous sequences currently is used widely in phylogenetic reconstruction (e.g., Palmer et al., 1988a, and other chapters in this volume). Major rearrangements of the chloroplast molecule include inversions, the insertion or deletion of genes and introns, and loss of one copy of the IR. Minor rearrangements consist of small insertions and deletions (1–1,000 bp). These events are much more common than major rearrangements and occur principally in noncoding intergenic spacer regions and introns (Palmer, 1985b). Because small length mutations have a tendency to cluster in "hot spot" regions (Kung et al., 1982; Palmer et al., 1988a), the assignment of exact homology for each mutation may be difficult. Owing to their homoplasious nature, they are difficult to use for systematic purposes, and are sometimes not included in phylogenetic analyses (Sytsma and Gottlieb, 1986; Palmer et al., 1988a), although in studies of closely related taxa they may provide useful information (Doebley et al., 1987a; Soltis et al., 1989, 1990).

Detection and Analysis of Rearrangements

Chloroplast DNA rearrangements most often are revealed using a heterologous filter hybridization approach in which cloned restriction fragments from one chloro-

Table 2.1. Land plant species for which plastid genome size (in kb) and complete restriction endonuclease maps are available.*

Taxon	Size	Reference
Angiosperms[a]		
Asteridae		
Apocynaceae		
Vinca minor	150	Palmer (unpublished)
Asteraceae		
16 tribes, 267 genera	148–151	Jansen et al. (this volume)
16 tribes, 57 genera, 57 spp.	148–151	Jansen et al. (1990)
Heliantheae (6 genera, 33 spp.)	151	Schilling and Jansen (1989)
Mutisieae (13 genera, 13 spp.)	151	Jansen and Palmer (1988)
Madiinae (5 genera, 26 spp.)	151	Baldwin et al. (1990)
Barnadesia caryophylla	151	Jansen and Palmer (1987a)
Carthamus tinctorius	151	Ma and Smith (1985)
Helianthus (7 spp.)	152	Heyraud et al. (1987)
Lactuca sativa	151	Jansen and Palmer (1987a)
Dipsacaceae		
Scabiosa sp.	154	Palmer (unpublished)
Orobanchaceae		
Epifagus virginiana[b]	71	dePamphilis and Palmer (1989)
Plantaginaceae		
Plantago sp.	144	Palmer (unpublished)
Solanaceae[c]		
Capsicum annuum	143	Gounaris et al. (1986)
Solanum tuberosum	155	Heinhorst et al. (1988)
Caryophyllidae		
Caryophyllaceae		
Cerastium arvense	147	Palmer (unpublished)
Chenopodiaceae		
Beta (11 spp.)	148	Kishima et al. (1987)
Phytolaccaceae		
Phytolacca heterotepela	153	Palmer (unpublished)
Polygonaceae		
Rumex sp.	158	Palmer (unpublished)
Dilleniidae		
Actinidiaceae		
Actinidia deliciosa	160	Hudson and Gardner (1988)
Begoniaceae		
Begonia sp.	170	Palmer (unpublished)
Clusiaceae		
Hypericum sp.	140	Palmer (unpublished)
Cucurbitaceae		
Cucumis (21 spp.)	150	Perl-Treves and Galun (1985)
Primulaceae		
Anagallis arvensis	148	Palmer (unpublished)
Primula sp.	149	Palmer (unpublished)
Salicaceae		
Populus (10 spp.),	155	Smith and Sytsma (1990)
Salix exigua		
Hamamelidae		
Urticaceae		
Pilea microphylla	150	Palmer (unpublished)

(continued)

Table 2.1.
(Continued)

Taxon	Size	Reference
Liliidae		
Alliaceae		
Allium cepa	145	Chase and Palmer (1989)
Asparagaceae		
Asparagus sprengeri	149	Chase and Palmer (1989)
Amaryllidaceae		
Narcissus X hybridus	157	Chase and Palmer (1989)
Narcissus pseudonarcissus	161	Hansmann (1987)
Dioscoreaceae		
Dioscorea bulbifera	152	Terauchi et al. (1989)
Orchidaceae		
Oncidium (9 spp.),	143	Chase and Palmer (1989)
Psychopsis sanderae,		
Rossioglossum schlieperianum,		
Trichocentrum (2 spp.)		
Commelinidae		
Poaceae		
Aegilops (10 spp.)	135	Ogihara and Tsunewaki (1988)
Avena (5 spp.)	135	Murai and Tsunewaki (1987)
Oryza sativa	135	Hiratsuka et al. (1989)
Secale (5 spp.)	136	Murai et al. (1989)
Sorghum bicolor	138	Dang and Pring (1986)
Triticum (10 spp.)	135	Ogihara and Tsunewaki (1988)
Zea (4 spp.),	135	Doebley et al. (1987b)
Tripsacum dactyloides		
Magnoliidae		
Aristolochiaceae		
Aristolochia durior	158	Palmer (unpublished)
Papaveraceae		
Eschscholzia californica	158	Palmer (unpublished)
Ranunculaceae		
Aquilegia sp.	151	Palmer (unpublished)
Delphinium sp.	149	Palmer (unpublished)
Ranunculus californica	157	Palmer (unpublished)
Rosidae		
Aceraceae		
Acer pseudoplatanus	142	Ngernprasirtsiri and Kobayashi (1990)
Apiaceae		
Coriandrum sativum	148	Palmer (unpublished)
Balsaminaceae		
Impatiens sp.	156	Palmer (unpublished)
Crassulaceae		
Sedum oreganum	146	Sundberg et al. (1990)
Fabaceae		
Lupinus polyphyllus	147	Palmer et al. (1987b)
Medicago sativa	126	Palmer et al. (1987b)
Trifolium subterraneum	142	Milligan et al. (1989)
Wisteria floribunda	130	Palmer et al. (1987b)
Geraniaceae		
Pelargonium X hortorum	217	Palmer et al. (1987a)

(continued)

Table 2.1.
(Continued)

Taxon	Size	Reference
Hippocastanaceae		
Aesculus californica	153	Palmer (unpublished)
Linaceae		
Linum grandiflorum	154	Palmer (unpublished)
Linum (8 spp.)	160, 173	Coates and Cullis (1987)
Onagraceae		
Clarkia (14 spp.)	170	Sytsma et al. (1990)
Fuchsia sp.	151	Palmer (unpublished)
Fuchsia (6 spp.)	150	Sytsma et al. (in press)
Oxalidaceae		
Oxalis oregana	152	Palmer (unpublished)
Rutaceae		
Citrus (7 spp.),	166	Green et al. (1986)
Poncirus trifoliata,		
Microcitrus sp.		
Saxifragaceae		
10 genera, 40 spp.	151	Soltis et al. (1990; unpublished)
Gymnosperms		
Ginkgoaceae		
Ginkgo biloba	158	Palmer and Stein (1986)
Pinaceae		
Pinus monticola	120	White (1990)
Pinus radiata	120	Strauss et al. (1988)
Pseudotsuga menziesii	120	Strauss et al. (1988)
Pteridophytes		
Adiantum capillus-veneris	153	Hasebe and Iwatsuki (1990)
Osmunda (3 spp.)	144	Stein et al. (1986)
Bryophytes		
Marchantia polymorpha	121	Ohyama et al. (1986)
Physcomitrella patens	122	Calie and Hughes (1987)

[*]Information not previously compiled by Crouse et al. (1985), who listed mapped genomes for 16 families, 29 genera, and 32 species of land plants, is presented. Unless otherwise stated (see text and other tables presented herein) all maps are colinear with that of *Nicotiana tabacum*.

[a]Restriction mapping data are currently being analyzed for 99 families (211 spp.) of angiosperms including 36 families from the Asteridae (Downie and Palmer, unpublished data). To date, 40 families (71 spp.) have been found whose cpDNA genomes are colinear with that of *Nicotiana tabacum*. Those taxa whose cpDNA genomes possess rearrangements are listed in Tables 2.2 and 2.3.

[b]*Epifagus virginiana* is nonphotosynthetic.

[c]A comprehensive comparative restriction mapping analysis of the Solanaceae (55 genera, 132 species) is currently underway (Olmstead and Palmer, unpublished data).

plast genome are hybridized to filter-bound fragments of a second. Defined segments of cloned restriction fragments obtained from mapped reference genomes, such as that of *Nicotiana tabacum,* are used as hybridization probes. Heterologous probes can be used effectively across widely divergent lineages of angiosperms because most cpDNAs are highly conserved in sequence and arrangement.

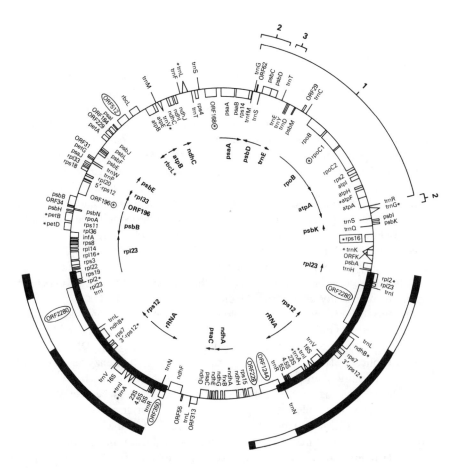

Figure 2.1. Physical and gene map of the 156-kb *Nicotiana tabacum* chloroplast genome, showing selected rearrangements found in *Oryza sativa* and *Marchantia polymorpha*. Genes transcribed clockwise are shown on the inside of the circle; those transcribed counterclockwise are on the outside. Arrows on the inside of the circle indicate sets of genes thought to constitute operons. The operon names are indicated in boldface. Circled gene names indicate genes present in *N. tabacum* and *M. polymorpha* but absent from *O. sativa*. The boxed gene name indicates a gene present in *N. tabacum* and *O. sativa* but absent in *M. polymorpha*. Asterisks denote genes that have the same intron(s) in all three sequenced genomes; circled asterisks denote genes that are split in the three genomes with two exceptions (ORF196 and *rpo*C1 are not split in *O. sativa*, see Table 2.3). The thickened parts of the circle represent the 25.3-kb IR of *N. tabacum*. The thick lines outside of the *N. tabacum* IR represent the extent of the *O. sativa* IR (20.8 kb), with regions deleted from *O. sativa* left open. The numbered brackets outside the circle indicate three overlapping inversions in *O. sativa* relative to *N. tabacum*. Gene nomenclature follows Hallick and Bottomley (1983), Hallick (1989), and Palmer (1991). *Nicotiana tabacum* cpDNA map is based on Shinozaki et al. (1986); *O. sativa* and *M. polymorpha* data are based on Hiratsuka et al. (1989) and Ohyama et al. (1986), respectively.

Currently, we are investigating chloroplast genome structure in angiosperms with emphasis on the subclass Asteridae. Our survey for major structural re-arrangements encompasses 88 species, representing 36 families of Asteridae (see Cronquist, 1981), and an additional 123 species comprising representatives from all subclasses of monocotyledons and dicotyledons. From an initial set of large, cloned restriction fragments (Sugiura et al., 1986), we have subcloned small fragments specific for many of the genes and introns found in the cpDNA of *N. tabacum*. To date we have used over 120 hybridization probes, ranging in size from 0.2 to 3.5 kb, which together comprise the entire *N. tabacum* chloroplast genome. These probes are smaller than the 5-to-15-kb probes commonly used in restriction site mapping studies. The use of small probes permits the detection of small rearrangement events that often are undetected when larger fragments are used as probes. Our methods are adapted primarily to survey, as rapidly and as economically as possible, large numbers of taxa for rearrangements. Once a mutation is detected and characterized at the molecular level it can, depending upon the availability of material, be circumscribed in related taxa.

Our survey for rearrangements employs a modification of the approach outlined in Palmer et al. (1988a) to accommodate both increased numbers of taxa and probes. To minimize the number of hybridization cycles, each cpDNA sample is run in triplicate. To accomplish this, triple-size restriction enzyme digests are prepared initially, then one-third of each digest is loaded on each gel. Double-sided blotting yields six identical filters, which greatly reduces the amount of time necessary to complete the required number of hybridizations. To lessen the size and cost of such an undertaking, çare is taken so that the bromophenol blue dye marker in each of the gels migrates no more than 6 cm. In this way, four 20-cm-wide filters can be placed on a single sheet of standard 20 × 25-cm x-ray film. Resolution is sufficient to detect gene and intron losses, inversions, and changes in size of the IR, but often it is difficult to map inversions precisely and to measure fragment sizes accurately. When this becomes necessary, follow-up gels can be run in which the dye marker migration is 12–20 cm.

Rearrangements can be detected by arranging the autoradiograms according to the order in the chloroplast genome of the hybridization probes and by observing both fragment number and size as one "walks" along the chloroplast chromosome from one hybridization probe to the next. Any anomaly in the number of fragments detected, their size, or the intensity of hybridization may be indicative of a mutation. The construction of restriction site maps aids in diagnosing certain types of rearrangements. The detection of specific rearrangements is described in the appropriate sections below.

DNA sequence analysis is necessary when filter hybridization data are ambigu-ous; it can be used to corroborate the results of the hybridization experiments. In some instances, extreme base sequence divergence or shuffling of sequences by rearrangement may preclude a significant level of cross-hybridization. The puta-tive absence of genes or introns, or portions thereof, can be confirmed by sequencing the region of the suspected absence. DNA sequence analysis may also be necessary to ascertain homologous mutations via correct sequence alignment and can be used to determine the presence and extent of length mutations within coding sequences.

The use of the polymerase chain reaction (PCR) in systematic studies is rapidly gaining in popularity (Arnheim et al., 1990). Although comparative DNA sequence data are often routinely obtained using PCR, the use of PCR in surveying for structural rearrangements has been limited. Once a rearrangement has been found by filter hybridizations, PCR is well suited for rapid screens of large numbers of DNAs. The use of PCR to diagnose inversions and the presence or absence of genes and introns is described in the appropriate sections below.

The Phylogenetic Utility of
Chloroplast DNA Rearrangements

Compared to the large amount of phylogenetic data now available from the analysis of restriction site variation, relatively little effort has been made to survey and identify major rearrangements among land-plant cpDNAs, even though several identified rearrangements have been quite useful in phylogenetic reconstruction. To illustrate the phylogenetic utility of cpDNA rearrangements and the value of searches to find them, examples of each of the major classes of structural rearrangements are given below. Other rearrangements that will be discussed include the expansion or contraction of the IR, and the occurrence of small length mutations within evolutionarily constrained regions. We elaborate on how we can exploit this variation as phylogenetic information and offer insight into how these structural rearrangements are detected.

Inversions

An inversion occurs when any segment of the chloroplast chromosome has been rotated 180° relative to the regions on either side. The gene order exemplified by *N. tabacum* (Fig. 2.1) is similar to the ancestral vascular plant gene order, because it is found, with few exceptions, in all other examined angiosperms, ferns, and *Ginkgo biloba* (Palmer, 1985a, 1985b; Palmer and Stein, 1986; Palmer et al., 1988a). In most altered genomes, the order of genes can be derived from the ancestral form by one to a few inversions. For example, three inversions characterize the chloroplast genome of the monocot *Oryza sativa* relative to *N. tabacum* (Hiratsuka et al., 1989; Sugiura, 1989). Similar gene arrangements have been documented in *Triticum aestivum* (Quigley and Weil, 1985; Howe et al., 1988) and *Zea mays* (Palmer and Thompson, 1982), suggesting that these rearrangements predate the divergence of the grasses from other monocots. All other monocots examined exhibit the consensus gene order found in *N. tabacum* (de Heij et al., 1982; Palmer et al., 1988a; Chase and Palmer, 1989; Downie and

Palmer, unpublished data). The chloroplast genomes of the liverwort *Marchantia polymorpha* and the moss *Physcomitrella patens* differ from that of *N. tabacum* by only one 30-kb inversion (Calie and Hughes, 1987; Ohyama et al., 1988), despite over 400 million years of evolutionary divergence (Stewart, 1983). The polarity of this 30-kb inversion is unknown. A summary of known inversions in land plants is presented in Table 2.2. Relatively few taxa, most notably in the conifers, Geraniaceae, Fabaceae, Campanulaceae, and Lobeliaceae, contain multiple rearrangements.

An example of the utility of a cpDNA inversion in phylogenetic reconstruction has been documented recently. Chloroplast DNAs from subtribe Barnadesiinae (tribe Mutisieae) of the Asteraceae have the typical gene order found in most land plants, whereas all other Asteraceae share a derived 22-kb inversion (Jansen and Palmer, 1987a, 1987b). This finding and congruent results obtained through a phylogenetic analysis of restriction site mutations (Jansen and Palmer, 1988) and nucleotide substitutions (Jansen et al., Chapter 11, this volume) demonstrate that the Mutisieae is not monophyletic (because its three other subtribes possess the inversion) and that the Barnadesiinae should be considered the sister group to the remainder of the family (see also Jansen et al., Chapter 11, this volume). Consequently, it is now possible to root unambiguously phylogenetic trees using Barnadesiinae as the outgroup in cladistic analyses of the family.

Inversions can be detected in the following ways. The hybridization of two nonadjacent restriction fragments from a genome lacking an inversion to the same two fragments in another genome indicates that an inversion has occurred (see Figs. 2 and 3 in Jansen and Palmer, 1987a). Conversely, two adjacent fragments in an uninverted genome that have become separated by an inversion will hybridize to different fragments in the inverted genome. The hybridization of small probes to those fragments containing the presumed inversion endpoints can provide a more precise localization of both the endpoints and size of the inverted segment. Once restriction maps are constructed, gene mapping and sequencing studies can confirm the difference in gene order and the direction of transcription (via the differential hybridization of 5' and 3' gene probes). Additional taxa can be surveyed for the inversion by performing filter hybridization using cloned restriction fragments that contain the inversion endpoints. The PCR technique can also be used to detect inversions. Primers synthesized for two conserved sequences closely flanking a known inversion endpoint will yield a small PCR product after amplification when applied to a species possessing the same inversion. However, if an inversion is not present, the size of the PCR product will be much larger, if a product is produced at all.

Gene/Intron Loss

The complete sequences of the chloroplast genomes of *N. tabacum* (Shinozaki et al., 1986), *M. polymorpha* (Ohyama et al., 1986), and *O. sativa* (Hiratsuka et al., 1989) provide invaluable information on chloroplast gene content and organization. Comparisons of homologous sequences between *N. tabacum* and the liverwort *M. polymorpha* (Wolfe and Sharp, 1988) and between *N. tabacum* and *O. sativa* (Sugiura, 1989) reveal a high degree of conservatism in gene content. Of the 20 distinct introns previously demonstrated or tentatively identified in *N. tabacum* cpDNA (Shinozaki et al., 1986), 18 are present also in *M. polymorpha* (Ohyama et al., 1986) and 17 in *O. sativa* (Hiratsuka et al., 1989). Among these three sequenced chloroplast genomes, there are no known examples of gene or intron gains. Moreover, our observations and those of many other laboratories indicate that genes and introns have been gained rarely, if at all, during land-plant evolution. Consequently, our discussion will deal only with the loss of these sequences.

Any disruption in gene integrity will result in a loss of function. In some instances, gene losses are viewed more appropriately as gene transfers because some genes lost from the chloroplast genome have been found in the nucleus (Baldauf and Palmer, 1990; Gantt and Palmer, unpublished data). Nucleotide substitutions and length mutations occur readily in intron sequences (Ritland and Clegg, 1987; Zurawski and Clegg, 1987) and are of little systematic value. However, the loss of entire intron sequences is a relatively rare event and therefore phylogenetically informative. Palmer (1991) reviews the evolutionary processes and mechanisms responsible for the loss/gain of genes and introns in cpDNA.

The stability in gene/intron content among land-plant cpDNAs can make their absence valuable as a systematic marker at a number of taxonomic levels. For example, the genes *rpo*A, *rpl*22, and *tuf*A are absent from the chloroplast genomes of *Pelargonium*, the Fabaceae, and land plants, respectively (Table 2.3). The intron in the gene *rpl*2 is absent from all members examined of the Caryophyllales (Zurawski et al., 1984; Downie et al., 1991). Furthermore, filter hybridization surveys of more than 300 chloroplast genomes show that this intron is absent also from members of the Convolvulaceae, *Cuscuta, Menyanthaceae,* two genera within the Geraniaceae (*Sarcocaulon* and *Monsonia*), Saxifragaceae *s.s.,* and *Drosera filiformis* (Table 2.3; Downie et al., 1991). This intron loss can be considered to have occurred independently in at least six different lineages of dicots.

The presence or absence of a particular gene or intron may be assayed by hybridization using a probe specific to that gene or intron. Subsequent sequencing of the region in question can confirm its presence or absence, its fragmentation, or its change in position. Currently, in our laboratory, we are using the PCR technique to survey for the presence or absence of introns in some tRNA genes (Kuhsel et al., 1990). Primers are synthesized for conserved sequences flanking the region of interest and the intervening sequence is amplified by PCR. Comparing the size of the resultant PCR product to a sequence of known length on an agarose or polyacrylamide gel can indicate the presence or absence of a specific gene or intron (Bruzdzinski and Gelehrter, 1989). In general, deletions or inser-

Table 2.2. Summary of known inversions in land plant cpDNA (relative to vascular plant consensus gene order, as exemplified by Nicotiana tabacum).

Taxon	Size (or number)[a]	Reference
Angiosperms		
Asteraceae		
(all subtribes except Barnadesiinae)	22	Jansen and Palmer (1987a)
Lactuca sativa	4	Downie, Knox, Jansen, and Palmer (unpublished)
Cactaceae		
Pereskia sacharosa	(1)	Wallace (unpublished)
Campanulaceae (4 spp.)	(several)	Downie and Palmer (unpublished)
Chenopodiaceae		
Atriplex (60 spp.)	(1)	Palmer (unpublished)
Fabaceae		
Fabaceae	50	Palmer and Thompson (1982)
Phaseolinae	78	Palmer et. al. (1988b) Bruneau et al. (1990)
Robinieae (except Sesbania)	30	Lavin (unpublished)
Trifolium subterraneum	(ca. 8)	Milligan et al. (1989)
Vicia faba	(2 or 3)	Palmer et al. (1987b)
Pisum (4 spp.)	(ca. 8)	Palmer et al. (1988b)
Pisum humile	4	Palmer et al. (1985)
Medicago lupulina	11	Johnson and Palmer (unpublished)
Medicago (3 spp.)	62	Johnson and Palmer (unpublished)
Medicago arabica	(1)	Johnson and Palmer (unpublished)
Medicago tornata	(1)	Johnson and Palmer (unpublished)
Geraniaceae		
Erodium chamaedryoides	(1 or 2)	Calie and Palmer (unpublished)
Geranium grandiflorum	(several)	Calie and Palmer (unpublished)
Pelargonium X hortorum	(ca. 6)	Palmer et al. (1987a)
Sarcocaulon vanderietiae	(several)	Calie and Palmer (unpublished)
Lobeliaceae		
Lobelia (27 spp.), Sclerotheca jayorum	(2)	Knox, Downie, and Palmer (unpublished)
Lobelia erinus, L. fervens	(5)	Knox, Downie, and Palmer (unpublished)
Lobelia cardinalis, L. holstii, Monopsis lutea	(3)	Knox, Downie, and Palmer (unpublished)
Oleaceae		
Jasminum (2 spp.)	(2)	Downie and Palmer (unpublished)
Onagraceae		
Oenothera spp.	50	Herrmann et al. (1983) Sytsma (unpublished)
Orobanchaceae		
Conopholis americana	(1)	Downie and Palmer (unpublished)
Poaceae		
Triticum, Oryza, Zea	(3)[b]	Howe et al. (1988) Quigley and Weil (1985) Palmer and Thompson (1982) Hiratsuka et al. (1989)

(continued)

Table 2.2.
(Continued)

Taxon	Size (or number)[a]	Reference
Ranunculaceae		
Adonis aestivalis	(1)	Hoot and Palmer (unpublished)
Anemone (40 spp.), *Hepatica, Knowltonia, Pulsatilla*	(4)	Hoot and Palmer (unpublished)
Anemone (3 spp.), *Clematis* (2 spp.)	(6)	Hoot and Palmer (unpublished)
Scrophulariaceae		
Striga asiatica	(3)	Downie and Palmer (unpublished)
Gymnosperms		
Conifers	(several)	Strauss et al. (1988) Raubeson and Jansen (unpublished)
Pseudotsuga menziesii	45	Strauss et al. (1988)
Pteridophytes		
Adiantum capillus-veneris	(2)	Hasebe and Iwatsuki (1990)
Bryophytes		
Marchantia polymorpha	30	Ohyama et al. (1988)
Physcomitrella patens	30	Calie and Hughes (1987)

[a]Estimated sizes in kb provided. If more than one inversion is present, or if inversion is not well characterized, the number of postulated inversions is presented in parentheses.

[b]These three inversions are 28 kb, 10 kb, and approximately 1 kb (see Fig. 1).

tions of moderate size (less than 2 kb) can be detected in this way. Furthermore, amplified intron or gene-sequence products can be readily isolated and subsequently cloned and/or sequenced.

A summary of genes and introns known (by DNA sequencing) or suspected (by filter hybridization) to be lost from the cpDNAs of various land plants is presented in Table 2.3. Our results suggest that the unidentified open reading frames (ORFs) are most amenable to loss. In several situations it appears that similar losses have occurred in parallel. The ribosomal protein genes and introns are also sometimes lost, whereas no photosynthetic gene is known to have been lost from any cpDNA of a photosynthetic land plant.

Loss of the Inverted Repeat

One of the most intriguing rearrangements is the loss of one copy of the IR. Although the presence of the IR may confer a certain stability upon the cpDNA molecule, making it less prone to rearrangement (Palmer and Thompson, 1982; Strauss et al., 1988), the absence of one copy suggests that it is not fundamental to genome function. With few exceptions, all angiosperm cpDNAs possess a large IR, usually approximately 20 to 30 kb in size, that encodes a duplicate set of ribosomal RNA genes (Fig. 2.1). The deletion of one entire segment of this duplicated sequence is a significant mutation, which, when considered in a phylogenetic context, can define monophyletic groups. This rare deletion has

Table 2.3.   Summary of known cpDNA gene and intron losses in land plants.

Gene/Intron[a]	Taxon[b]	Reference
many*[c]	*Epifagus virginiana*	dePamphilis and Palmer (1989)
	Conopholis americana	Downie and Palmer (unpublished)
*tuf*A*	land plants	Baldauf and Palmer (1990)
*rpo*A*	*Pelargonium* (40 spp.)	Calie and Palmer (unpublished)
*rpl*20	*Sarcocaulon* (2 spp.)	Downie and Palmer (unpublished)
*rpl*22*	Fabaceae (3 subfamilies)	Palmer and Doyle (unpublished)
		Spielmann et al. (1988)
*rps*7	*Podophyllum peltatum*	Downie and Palmer (unpublished)
*rps*16	Fabaceae (5/9),	Downie and Palmer (unpublished)
	Linum grandiflorum,	
	Malpighia coccigera,	
	Passiflora sp.,	
	Polygala lindheimeri,	
	Populus deltoides,	
	Salix amygdaloides,	
	Securidaca diversifolia,	
	Turnera ulmifolia,	
	Viola (2 spp.)	
*clp*P	*Geranium* (2 spp.),	Downie and Palmer (unpublished)
(orf196)	*Jasminum* (2 spp.),	
	Linum grandiflorum,	
	Lobelia holstii,	
	Lonicera subsessilis,	
	Monopsis lutea,	
	Monsonia (2 spp.),	
	Oenothera missouriensis,	
	Sarcocaulon (2 spp.)	
*ndh*F	*Hebestigma cubense*	Downie and Palmer (unpublished)
*zfp*A*	*Oryza sativa*	Hiratsuka et al. (1989)
(orf512)	*Bambusa* sp., *Zea mays,*	Downie and Palmer (unpublished)
	Campanulaceae (3/4),	
	Geraniaceae (4/8),	
	Lobeliaceae (4/5),	
	Oleaceae (2/3)	
orf184*	*Pisum sativum*	Sasaki et al. (1989)
	Fabaceae (11/16)	Downie and Wolfe (unpublished)
orf228*	*Oryza sativa*	Hiratsuka et al. (1989)
orf1244*	*Oryza sativa*	Hiratsuka et al. (1989)
	Bambusa sp., *Zea mays,*	Downie and Palmer (unpublished)
	Campanulaceae (3/4),	
	Convolvulaceae (3/4),	
	Lobeliaceae (4/5),	
	Cuscuta sp.,	
	Linum grandiflorum,	
	Pisum sativum	
orf2280*	*Oryza sativa*	Hiratsuka et al. (1989)
	Bambusa sp., *Zea mays,*	Downie and Palmer (unpublished)
	Campanulaceae (2/2),	
	Geraniaceae (3/6)	

(continued)

Table 2.3.
(Continued)

Gene/Intron[a]	Taxon[b]	Reference
*rpo*C1 intron*	*Oryza sativa*	Hiratsuka et al. (1989)
rpl2 intron*	Caryophyllales	Zurawski et al. (1984)
		Downie et al. (1991)
	Convolvulaceae (4/5),	Downie et al. (1991)
	Menyanthaceae (4/5),	
	Saxifragaceae (24/50),	
	Cuscuta sp.,	
	Drosera filiformis,	
	Monsonia (2 spp.),	
	Sarcocaulon (2 spp.)	
rpl16 intron*	Geraniaceae (5/44),	Downie et al. (unpublished)
	Limonium gmelinii	
*trn*I intron	*Campanula garganica*	Downie and Palmer (unpublished)
orf196 introns	*Oryza sativa*	Hiratsuka et al. (1989)
1 and 2*	*Zea mays, Bambusa sp.*	Downie and Palmer (unpublished)

[a]Gene/intron absence postulated only on the basis of filter hybridizations. Asterisks denote those genes/introns whose absence has been confirmed by DNA sequencing in at least one of the taxa.

[b]Numbers in parentheses indicate # of genera/ # of species exhibiting loss.

[c]*Epifagus virginiana* and *Conopholis americana* are nonphotosynthetic, parasitic plants. Most photosynthetic genes, NADH dehydrogenase genes, and ORFs are missing.

now been found in four independent lineages of vascular plants (Table 2.4). Particularly notable from a phylogenetic standpoint is the absence of one copy of this repeat from six tribes and the putatively allied genus *Wisteria* within the subfamily Papilionoideae (Lavin et al., 1990; Doyle et al., Chapter 10, this volume), and from the conifers, including Taxaceae (Lidholm et al., 1988; Strauss et al., 1988; Raubeson and Jansen, unpublished data). The strategy used to detect the presence or absence of the IR involves hybridization assays using small probes homologous with the conserved ends of the IR and single-copy regions (Palmer et al., 1988a; Lavin et al., 1990).

Expansion/Contraction of the Inverted Repeat

The expansion or contraction of the IR into, or out of, the two single-copy regions significantly influences the variability in size of the chloroplast genome. At one extreme is the 217-kb *Pelargonium X hortorum* cpDNA possessing a greatly enlarged IR of 76 kb, almost three times the size found in most angiosperms (Palmer et al., 1987a). Consequently, many protein genes that are present only once in most other plants are duplicated in *P. X hortorum*. At the other extreme in angiosperms with the IR is *Coriandrum sativum* with an IR less than half the normal size (Palmer, 1985b; Downie, unpublished data). In *C. sativum*, the gene *rpl2*, which normally is located near the terminus of the IR (Fig. 2.1), is a single-copy gene some 10 kb away from the end of the repeat. In *Oryza sativa* the IR segments have expanded into single-copy regions; however, a series

Table 2.4. Land plant cpDNAs possessing only one inverted repeat segment.

Taxon	Reference
Conifers	Strauss et al. (1988)
	Lidholm et al. (1988)
Fabaceae	
Papilionoideae	Lavin et al. (1990)
(6 tribes and *Wisteria*)	Palmer et al. (1987b)
Geraniaceae	
Erodium and *Sarcocaulon*	Calie and Palmer (unpublished)
Orobanchaceae	
Conopholis americana	Downie and Palmer (unpublished)

of deletions within the repeat makes the IR smaller than that found in *N. tabacum* (Hiratsuka et al., 1989). Variation in IR size is common but has not yet been used in phylogenetic analyses. Any length mutation that occurs within the IR undoubtedly will affect its size, thus making homologous size variants difficult to assess.

Length Mutation in Constrained Regions

Although small length mutations occur predominantly in noncoding DNA, they are also occasionally found within genes and other evolutionarily constrained portions of the genome. In order not to disrupt the reading frame in protein-coding genes, only insertions or deletions of just 3 bp, or in multiples of three, are permitted (e.g., Blasko et al., 1988). Small insertions or deletions within coding regions that are conserved evolutionarily may be considered as phylogenetic characters independent of nucleotide substitutions (Meyer et al., 1986; Morden and Golden, 1989).

Use of Rearrangements in Phylogenetic Reconstruction

The paucity of major structural rearrangements found to date within the chloroplast genomes of vascular plants suggests that they occur rarely during cpDNA evolution. However, once a rearrangement is found, characterized, and its distribution circumscribed in related taxa, its presence can make a profound phylogenetic statement. These unique characters are prominent and powerful systematic markers that offer the following advantages to phylogenetic reconstruction: (1) assessing the homology of the rearrangement usually is straightforward; (2) the polarity of each structural mutation is ascertained readily when it is compared to an outgroup; and (3) once a particular rearrangement is identified, it can be surveyed rapidly in other taxa through simple filter hybridization or PCR assays.

Although shared structural mutations can provide strong evidence of common ancestry, it is apparent now that similar rearrangements can occur independently, such as the loss of the *rpl2* intron. However, because the intron is absent in otherwise distantly related clades and present in their immediate, respective outgroups, the assessment of homology can be made confidently, and the loss can be considered to have occurred independently in at least six different lineages of dicots.

As in any systematic endeavor, the selection of representative specimens is critical to the outcome of the analysis. Because of the conservative nature of cpDNA evolution, specifically as it relates to its generally invariant order and content of genes, relatively few species are necessary to represent most taxa at the generic level and above. However, as previous studies attest, rearrangements can identify major dichotomies within clades at any level, so unless the group in question is well represented, some rearrangements may go undetected. Incorporating additional specimens into the analysis also is necessary when doubts arise concerning monophyly.

Extensively rearranged genomes are encountered rarely in land plants, and have so far been well characterized only in *Pisum* (Palmer et al., 1988b), *Trifolium subterraneum* (Palmer et al., 1987b; Milligan et al., 1989), *Pelargonium X hortorum* (Palmer et al., 1987a), and conifers (Strauss et al., 1988). The processes that contributed to the formation of these rearranged chloroplast genomes are not clear (Palmer, 1991) but may have involved either some major alteration of the IR (its loss or manyfold expansion) (Palmer 1985a, 1985b; Palmer et al., 1987a; Strauss et al., 1988) or the occurrence of dispersed, recombinogenic repeat elements (Howe, 1985; Palmer et al., 1987a; Blasko et al., 1988; Bowman et al., 1988; Milligan et al., 1989). Extensive genome rearrangement makes it almost impossible to align restriction fragment maps and largely precludes phylogenetic analyses of comparative restriction site variation. Furthermore, determining the nature and polarity for each mutation, providing they can be delimited, would be an arduous task. Fortunately, most altered genomes can be explained by a few discrete inversions (Palmer, 1985b; Downie and Palmer, unpublished data).

The analysis of major structural rearrangements is a complementary approach to comparative sequencing for studying the higher-level relationships among angiosperms. As phylogenetic distance increases among taxa, comparative restriction site mapping is plagued by excessive homoplasy and length mutation. Since many genes are conserved more than the genome as a whole, the direct comparison of homologous coding sequences (such as *rbcL*) is more appropriate for studying higher levels of plant phylogeny. Accordingly, in collaboration with others from our laboratory, we are in the process of sequencing *rbcL* from representatives of the Asteridae and outgroups from the Rosidae to acquire complementary information to that obtained from the rearrangement study described herein. Sequencing provides a large number of phylogenetically informative characters, whereas fewer cpDNA rearrangements are expected simply due to the rarity of these events. However, once found, these mutations should be considered to be more powerful characters than individual nucleotide substitutions, as data sets constituting the latter are inevitably afflicted with certain levels of homoplasy. Moreover, comparative sequence data may not resolve relatively ancient and compressed evolutionary radiations, whereas each rearrangement has the potential to resolve with confidence a particular branching point in a phylogeny (Palmer et al., 1988a). The integration of rearrangement data with other cpDNA-derived data (such as restriction site mutations and nucleotide substitutions) in phylogenetic analyses is an issue that has not yet been seriously explored.

Conclusions

The distribution of major structural rearrangements has the potential to illuminate the deeper branches of plant evolution and in doing so to define monophyletic groups. We have initiated a large-scale survey to detect and circumscribe major structural rearrangements in the chloroplast genomes of angiosperms, with special reference to the subclass Asteridae. Even though rearrangements alone are unlikely to provide a comprehensive framework of plant relationships, simply because of the small number of phylogenetically informative characters they represent, when used in conjunction with other molecular and traditional approaches, they have the power to help resolve many questions of plant phylogeny.

References

Arnheim, N., White, T., and Rainey, W.E. (1990) Application of PCR: organismal and population biology. *Bioscience* **40,** 174–182.

Baldauf, S., and Palmer, J.D. (1990) Evolutionary transfer of the chloroplast *tuf*A gene to the nucleus. *Nature* **344,** 262–265.

Baldwin, B.G., Kyhos, D.W., and Dvořák, J. (1990) Chloroplast DNA evolution and adaptive radiation in the Hawaiian silversword alliance (Asteraceae-Madiinae). *Ann. Missouri Bot. Gard.* **77,** 96–109.

Blasko, K., Kaplan, S.A., Higgins, K.G., Wolfson, R., and Sears, B.B. (1988) Variation in copy number of a 24-base pair tandem repeat in the chloroplast DNA of *Oenothera hookeri* strain Johansen. *Curr. Genet.* **14,** 287–292.

Bowman, C.M., Barker, R.F., and Dyer, T.A. (1988) In wheat ctDNA, segments of ribosomal protein genes are dispersed repeats, probably conserved by nonreciprocal recombination. *Curr. Genet.* **14,** 127–136.

Bruzdzinski, C.J., and Gelehrter, T.D. (1989) Determination of exon-intron structure: a novel application of the polymerase chain reaction technique. *DNA* **8,** 691–696.

Bruneau, A., Doyle, J.J., and Palmer, J.D. (1990) A chloroplast DNA inversion as a subtribal character in the Phaseoleae (Leguminosae). *Syst. Bot.* **15,** 378–386.

Calie, P.J., and Hughes, K.W. (1987) The consensus land plant chloroplast gene order is present, with two alterations, in the moss *Physcomitrella patens*. *Mol. Gen. Genet.* **208,** 335–341.

Chase, M.W., and Palmer, J.D. (1989) Chloroplast DNA systematics of lilioid monocots: resources, feasibility, and an example from the Orchidaceae. *Amer. J. Bot.* **76,** 1720–1730.

Coates, D., and Cullis, C.A. (1987) Chloroplast DNA variability among *Linum* species. *Amer. J. Bot.* **74,** 260–268.

Cronquist, A. (1981) *An Integrated System of Classification of Flowering Plants,* Columbia University Press, New York.

Crouse, E.J., Schmitt, J.M., and Bohnert, H. (1985) Chloroplast and cyanobacterial genomes, genes and RNAs: a compilation. *Plant Mol. Biol. Reporter* **3,** 43–89.

Dang, L.A., and Pring, D.R. (1986) A physical map of the sorghum chloroplast genome. *Plant Mol. Biol.* **6,** 119–123.

de Heij, H.T., Lustig, H., Moeskops, D.M., Bovenberg, W.A., Bisanz, C., and Groot, G.S.P. (1983) Chloroplast DNAs of *Spinacia, Petunia,* and *Spirodela* have a similar gene organization. *Curr. Genet.* **7,** 1–6.

dePamphilis, C.W., and Palmer, J.D. (1989) Evolution and function of plastid DNA: a review with special reference to nonphotosynthetic plants. In: *Physiology, Biochemistry, and Genetics of Nongreen Plastids* (eds. C.D. Boyer, J.C. Shannon, and R.C. Hardison), American Society of Plant Physiologists, pp. 182–202.

Doebley, J.F., Ma, D.P., and Renfroe, W.T. (1987a) Insertion/deletion mutations in the *Zea* chloroplast genome. *Curr. Genet.* **11,** 617–624.

Doebley, J., Renfroe, W., and Blanton, A. (1987b) Restriction site variation in the *Zea* chloroplast genome. *Genetics* **117,** 139–147.

Downie, S.R., Olmstead, R.G., Zurawski, G., Soltis, D.E., Soltis, P.S., Watson, J.C., and Palmer, J.D. (1991) Six independent losses of the chloroplast DNA *rpl2* intron in dicotyledons: molecular and phylogenetic implications. *Evolution,* in press.

Gounaris, I., Michalowski, C.B., Bohnert, H.J., and Price, C.A. (1986) Restriction and gene maps of plastid DNA from *Capsicum annuum. Curr. Genet.* **11,** 7–16.

Green, R.M., Vardi, A., and Galun, E. (1986) The plastome of *Citrus.* Physical map, variation among *Citrus* cultivars and species and comparison with related genera. *Theor. Appl. Genet.* **72,** 170–177.

Hallick, R.B. (1989) Proposals for the naming of chloroplast genes. II. Update to the nomenclature of genes for thylakoid membrane polypeptides. *Plant Mol. Biol. Reporter* **7,** 266–275.

Hallick, R.B., and Bottomley, W. (1983) Proposals for the naming of chloroplast genes. *Plant Mol. Biol. Reporter* **1,** 38–43.

Hansmann, P. (1987) Daffodil chromoplast DNA: comparison with chloroplast DNA, physical map, and gene localization. *Z. Naturforsch.* **42c,** 118–122.

Hasebe, M., and Iwatsuki, K. (1990) Chloroplast DNA from *Adiantum capillus-veneris* L., a fern species (Adiantaceae); clone bank, physical map and unusual gene localization in comparison with angiosperm chloroplast DNA. *Curr. Genet.* **17,** 359–364.

Heinhorst, S., Gannon, G.C., Galun, E., Kenschaft, L., and Weissbach, A. (1988) Clone bank and physical and genetic map of potato chloroplast DNA. *Theor. Appl. Genet.* **75**, 244–251.

Herrmann, R.G., Westhoff, P., Alt, J., Winter, P., Tittgen, J., Bisanz, C., Sears, B.B., Nelson, N., Hurt, E., Hauska, G., Viebrock, A., and Sebald, W. (1983) Identification and characterization of genes for polypeptides of thylakoid membrane. In: *Structure and Function of Plant Genomes* (eds. O. Ciferri and L. Dure), Plenum Press, New York, pp. 143–153.

Heyraud, F., Serror, P., Kuntz, M., Steinmetz, A., and Heizmann, P. (1987) Physical map and gene localization on sunflower (*Helianthus annuus*) chloroplast DNA: evidence for an inversion of a 23.5-kbp segment in the large single copy region. *Plant Mol. Biol.* **9**, 485–496.

Hiratsuka, J., Shimada, H., Whittier, R., Ishibashi, T., Sakamoto, M., Mori, M., Kondo, C., Honji, Y., Sun, C.-R., Meng, B.-Y., Li, Y.-Q., Kanno, A., Nishizawa, Y., Hirai, A., Shinozaki, K., and Sugiura, M. (1989) The complete sequence of the rice (*Oryza sativa*) chloroplast genome: intermolecular recombination between distinct tRNA genes accounts for a major plastid DNA inversion during the evolution of the cereals. *Mol. Gen. Genet.* **217**, 185–194.

Howe, C.J. (1985) The endpoints of an inversion in wheat chloroplast DNA are associated with short repeated sequences containing homology to *att*-lambda. *Curr. Genet.* **10**, 139–145.

Howe, C.J., Barker, R.F., Bowman, C.M., and Dyer, M. (1988) Common features of three inversions in wheat chloroplast DNA. *Curr. Genet.* **13**, 343–349.

Hudson, K.R., and Gardner, R.C. (1988) Organisation of the chloroplast genome of kiwifruit (*Actinidia deliciosa*). *Curr. Genet.* **13**, 339–342.

Jansen, R.K., and Palmer, J.D. (1987a) Chloroplast DNA from lettuce and *Barnadesia* (Asteraceae): structure, gene localization, and characterization of a large inversion. *Curr. Genet.* **11**, 553–564.

Jansen, R.K., and Palmer, J.D. (1987b) A chloroplast DNA inversion marks an ancient evolutionary split in the sunflower family (Asteraceae). *Proc. Natl. Acad. Sci. USA* **84**, 5818–5822.

Jansen, R.K., and Palmer, J.D. (1988) Phylogenetic implications of chloroplast DNA restriction site variation in the Mutisieae (Asteraceae). *Amer. J. Bot.* **75**, 753–766.

Jansen, R.K., Holsinger, K.E., Michaels, H.J., and Palmer, J.D. (1990) Phylogenetic analysis of chloroplast DNA restriction site data at higher taxonomic levels: an example from the Asteraceae. *Evolution* **44**, 2089–2105.

Kishima, Y., Mikami, T., Hirai, A., Sugiura, M., and Kinoshita, T. (1987) *Beta* chloroplast genomes: analysis of fraction I protein and chloroplast DNA variation. *Theor. Appl. Genet.* **73**, 330–336.

Kuhsel, M.G., Strickland, R., and Palmer, J.D. (1990) An ancient Group I intron shared by eubacteria and chloroplasts. *Science* **250**, 1570–1573.

Kung, S.D., Zhu, Y.S., and Shen, G.F. (1982) *Nicotiana* chloroplast genome III. Chloroplast DNA evolution. *Theor. Appl. Genet.* **61**, 73–79.

Lavin, M., Doyle, J.J., and Palmer, J.D. (1990) Evolutionary significance of the loss of chloroplast-DNA inverted repeat in the Leguminosae subfamily Papilionoideae. *Evolution* **44**, 390–402.

Lidholm, J., Szmidt, A.E., Hällgren, J.E., and Gustafsson P. (1988) The chloroplast genomes of conifers lack one of the rRNA-encoding inverted repeats. *Mol. Gen. Genet.* **212**, 6–10.

Ma, C., and Smith, M.A. (1985) Construction and mapping of safflower chloroplast DNA recombinants and location of selected gene markers. *Theor. Appl. Genet.* **70**, 620–627.

Meyer, T.E., Cusanovich, M.A., and Kamen, M.D. (1986) Evidence against use of bacterial amino acid sequence data for construction of all-inclusive phylogenetic trees. *Proc. Natl. Acad. Sci. USA* **83**, 217–220.

Milligan, B.G., Hampton, J.N., and Palmer, J.D. (1989) Dispersed repeats and structural reorganization in subclover chloroplast DNA. *Mol. Biol. Evol.* **6**, 355–368.

Morden, C.W., and Golden, S.S. (1989) *psb*A genes indicate common ancestry of prochlorophytes and chloroplasts. *Nature* **337**, 382–385.

Murai, K., and Tsunewaki, K. (1987) Chloroplast genome evolution in the genus *Avena*. *Genetics* **116**, 613–621.

Murai, K., Naiyu, X., and Tsunewaki, K. (1989) Studies on the origin of crop species by restriction endonuclease analysis of organellar DNA. III. Chloroplast DNA variation and interspecific relationships in the genus *Secale*. *Jap. J. Genet.* **64**, 35–47.

Ngernprasirtsiri, J., and Kobayashi, H. (1990) Application of an efficient strategy with a phage λ vector for constructing a physical map of the amyloplast genome of sycamore (*Acer pseudoplatanus*). *Arch. Biochem. Biophys.* **276**, 172–179.

Ogihara, Y., and Tsunewaki, K. (1988) Diversity and evolution of chloroplast DNA in *Triticum* and *Aegilops* as revealed by restriction fragment analysis. *Theor. Appl. Genet.* **76**, 321–332.

Ohyama, K., Fukuzawa, H., Kohchi, T., Shirai, H., Sano, T., Sano, S., Umesono, K., Shiki, Y., Takeuchi, M., Chang, Z., Aota, S., Inokuchi, H., and Ozeki, H. (1986) Chloroplast gene organization deduced from complete sequence of liverwort *Marchantia polymorpha* chloroplast DNA. *Nature* **322**, 572–574.

Ohyama, K., Kohchi, T., Sano, T., and Yamada, Y. (1988) Newly identified groups of genes in chloroplasts. *TIBS* **13**, 19–22.

Palmer, J.D. (1985a) Evolution of chloroplast and mitochondrial DNA in plants and algae. In: *Molecular Evolutionary Genetics* (ed. R.J. MacIntyre), Plenum Press, New York, pp. 131–240.

Palmer, J.D. (1985b) Comparative organization of chloroplast genomes. *Ann. Rev. Genet.* **19**, 325–354.

Palmer, J.D. (1991) Plastid chromosomes: structure and evolution. In: *Cell Culture and Somatic Cell Genetics in Plants*, Vol. 7, *The Molecular Biology of Plastids* (eds. L. Bogorad and I.K. Vasil), Academic Press, New York, pp. 5–53.

Palmer, J.D., and Stein, D.B. (1986) Conservation of chloroplast genome structure among vascular plants. *Curr. Genet.* **10**, 823–833.

Palmer, J.D., and Thompson, W.F. (1982) Chloroplast DNA rearrangements are more frequent when a large inverted repeat sequence is lost. *Cell* **29**, 537–550.

Palmer, J.D., Jorgensen, R.A., and Thompson, W.F. (1985) Chloroplast DNA variation and evolution in *Pisum*: patterns of change and phylogenetic analysis. *Genetics* **109**, 195–213.

Palmer, J.D., Nugent, J.M., and Herbon, L.A. (1987a) Unusual structure of geranium chloroplast DNA: a triple-sized inverted repeat, extensive gene duplications, multiple inversions, and two repeat families. *Proc. Natl. Acad. Sci. USA* **84**, 769–773.

Palmer, J.D., Osorio, B., Aldrich, J., and Thompson, W.F. (1987b) Chloroplast DNA evolution among legumes: loss of a large inverted repeat occurred prior to other sequence rearrangements. *Curr. Genet.* **11**, 275–286.

Palmer, J.D., Jansen, R.K., Michaels, H.J., Chase, M.W., and Manhart, J.R. (1988a) Chloroplast DNA variation and plant phylogeny. *Ann. Missouri Bot. Gard.* **75**, 1180–1206.

Palmer, J.D., Osorio, B., and Thompson, W.F. (1988b) Evolutionary significance of inversions in legume chloroplast DNAs. *Curr. Genet.* **14**, 65–74.

Perl-Treves, R., and Galun, E. (1985) The *Cucumis* plastome: physical map, intrageneric variation and phylogenetic relationships. *Theor. Appl. Genet.* **71**, 417–429.

Quigley, F., and Weil, J.H. (1985) Organization and sequence of five tRNA genes and of an unidentified reading frame in the wheat chloroplast genome: evidence for gene rearrangements during the evolution of chloroplast genomes. *Curr. Genet.* **9**, 495–503.

Ritland, K., and Clegg, M.T. (1987) Evolutionary analysis of plant DNA sequences. *Am. Nat.* **130**, S74–S100.

Sasaki, Y., Nagano, Y., Morioka, S., Ishikawa, H., and Matsuno, R. (1989) A chloroplast gene encoding a protein with one zinc finger. *Nucleic Acids* Res. **17**, 6217–6227.

Schilling, E.E., and Jansen, R.K. (1989) Restriction fragment analysis of chloroplast DNA and the systematics of *Viguiera* and related genera (Asteraceae: Heliantheae). *Amer. J. Bot.* **76**, 1769–1778.

Shinozaki, K., Ohme, M., Tanaka, M., Wakasugi, T., Hayashida, N., Matsubayashi, T., Zaita, N., Chunwongse, J., Obokata, J., Yamaguchi-Shinozaki, K., Ohto, C., Torazawa, K., Meng, B.-Y., Sugita, M., Deno, H., Kamogashira, T., Yamada, K., Kusuda, J., Takaiwa, F., Kato, A., Tohdoh, N., Shimada, H., and Sugiura, M. (1986) The complete nucleotide sequence of the tobacco chloroplast genome: its gene organization and expression. *EMBO J.* **5**, 2043–2049.

Smith, R.L., and Sytsma, K.J. (1990) Evolution of *Populus nigra* L. (sect. *Aigeiros*): introgressive hybridization and the chloroplast contribution of *Populus alba* L. (sect. *Populus*). *Amer. J. Bot.* **77**, 1176–1187.

Soltis, D.E., Soltis, P.S., Ranker, T.A., and Ness, B.D. (1989) Chloroplast DNA variation in a wild plant, *Tolmiea menziesii*. *Genetics* **121**, 819–826.

Soltis, D.E., Soltis, P.S., and Bothel, K.D. (1990) Chloroplast DNA evidence for the origins of the monotypic *Bensoniella* and *Conimitella* (Saxifragaceae). *Syst. Bot.* **15**, 349–362.

Spielmann, A., Roux, E., Allmen, J. von, and Stutz, E. (1988) The soybean chloroplast genome: complete sequence of the *rps*19 gene, including flanking parts containing exon2 of *rpl*2 (upstream), but not *rpl*22 (downstream). *Nucleic Acids Res.* **16**, 1199.

Stein, D.B., Palmer, J.D., and Thompson, W.F. (1986) Structural evolution and flip-flop recombination of chloroplast DNA in the fern genus *Osmunda*. *Curr. Genet.* **10**, 835–841.

Stewart, W.N. (1983) *Paleobotany and the Evolution of Plants*, Cambridge University Press, Cambridge, UK.

Strauss, S.H., Palmer, J.D., Howe, G.T., and Doerksen, A.H. (1988) Chloroplast genomes of two conifers lack a large inverted repeat and are extensively rearranged. *Proc. Natl. Acad. Sci. USA* **85**, 3898–3902.

Sugiura, M. (1989) The chloroplast chromosomes in land plants. *Ann. Rev. Cell Biol.* **5**, 51–70.

Sugiura, M., Shinozaki, K., Zaita, N., Kusuda, M., and Kumano, M. (1986) Clone bank of the tobacco (*Nicotiana tabacum*) chloroplast genome as a set of overlapping restriction endonuclease fragments: mapping of eleven ribosomal protein genes. *Plant Sci.* **44**, 211–216.

Sundberg, S.D., Denton, M.F., and Rehner, S.A. (1990) Structural map of *Sedum oreganum* (Crassulaceae) chloroplast DNA. *Biochem. Syst. Ecol.* **18**, 409–411.

Sytsma, K.J., and Gottlieb, L.D. (1986) Chloroplast DNA evolution and phylogenetic relationships in *Clarkia* sect. *Peripetasma* (Onagraceae). *Evolution* **40**, 1248–1261.

Sytsma, K.J., Smith, J.F., and Gottlieb, L.D. (1990) Phylogenetics in *Clarkia* (Onagraceae): restriction site mapping of chloroplast DNA. *Syst. Bot.* **15**, 280–295.

Sytsma, K.J., Smith, J.F., and Berry, P.E. (1991) Biogeography and evolution of morphology, breeding systems, flavonoids, and chloroplast DNA in the four Old World species of *Fuchsia* (Onagraceae). *Syst. Bot.* **16**, 257–269.

Terauchi, R., Terachi, T., and Tsunewaki, K. (1989) Physical map of chloroplast DNA of aerial yam, *Dioscorea bulbifera* L. *Theor. Appl. Genet.* **78**, 1–10.

White, E.E. (1990) Chloroplast DNA in *Pinus monticola*. 1. Physical map. *Theor. Appl. Genet.* **79**, 119–124.

Whitfeld, P.R., and Bottomley, W. (1983) Organization and structure of chloroplast genes. *Ann. Rev. Plant Physiol.* **34**, 279–310.

Wolfe, K.H., and Sharp, P.M. (1988) Identification of functional open reading frames in chloroplast genomes. *Gene* **66**, 215–222.

Zurawski, G., and Clegg, M.T. (1987) Evolution of higher-plant chloroplast DNA-encoded genes: implications for structure-function and phylogenetic studies. *Ann. Rev. Plant Physiol.* **38**, 391–418.

Zurawski, G., Bottomley, W., and Whitfeld, P.R. (1984) Junctions of the large single copy region of the inverted repeats in *Spinacia oleracea* and *Nicotiana debneyi* chloroplast DNA: sequence of the genes for tRNA[HIS] and the ribosomal proteins S19 and L2. *Nucleic Acids Res.* **12**, 6547–6558.

3

Mitochondrial DNA in Plant Systematics: Applications and Limitations

Jeffrey D. Palmer

In the last decade, mitochondrial DNA (mtDNA) analysis has had a major impact on the study of phylogeny and population genetics in animals (Avise et al., 1987; Moritz et al., 1987; Harrison, 1989). In plants, however, chloroplast DNA (cpDNA) has been the molecule of choice for molecular phylogenetic studies (Palmer et al., 1988; see also many other chapters in this volume). Studies of nuclear DNA, principally of the nuclear ribosomal RNA genes, have also been important in these efforts (Hamby and Zimmer, chapter 4, and Appels and Baum, Chapter 5, this volume). It is the purpose of this chapter to review the factors that have contributed to the relative neglect of mtDNA for phylogenetic reconstruction in plants and to assess the applications of mtDNA analysis for future systematic studies. I first review information on the structure and evolution of plant mtDNA, with emphasis on the genus *Brassica,* and then discuss the application of mtDNA data to phylogenetic studies.

Overview of Plant Mitochondrial DNA Structure and Evolution

The reader should consult several recent reviews for detailed information on the molecular biology and evolution of plant mtDNA and for original literature citations for the results reviewed below (Newton, 1988; Gray, 1989; Levings and Brown, 1989; Lonsdale, 1989, 1991; Palmer, 1990, 1991). It is important to realize that plant mtDNAs are rather poorly studied compared to their counterparts in animals, fungi, and certain protists, many of which have been completely sequenced. Furthermore, among land plants, the only published information on mtDNA is for angiosperms, whereas, among algae, only the genome of

I am grateful to my collaborators—Chris Makaroff, Jackie Nugent, Magid Shirzadegan, Laura Herbon, Ingrid Apel, and Clark Shields—in the studies of crucifer mitochondrial genomes described in this chapter. I also thank Stephen Downie and Bob Price for critical reading of the manuscript and acknowledge financial support for this research from the NIH (GM-35087).

Chlamydomonas reinhardtii has been well characterized. Accordingly, the term "plant mtDNA" will be used in this chapter to refer to angiosperm genomes.

Like all other characterized mitochondrial genomes, plant mtDNA encodes approximately 5% of the proteins found in the mitochondrion. Most (12–15) of these mitochondrially encoded proteins are subunits of four respiratory chain enzyme complexes—cytochrome c oxidase, cytochrome bc_1 complex, NADH dehydrogenase, and ATP synthetase. In addition, plant mtDNA encodes at least three ribosomal protein genes, one or more proteins of unknown function, all three mitochondrial rRNAs, and some, but not all, mitochondrial tRNAs.

The structure and evolutionary dynamics of plant mtDNA are completely unlike those of any other characterized genome. Most aspects of genome structure, that is, genome size, configuration, and gene order, change extremely rapidly in plant mitochondria, whereas the primary sequence of the genome is exceptionally slow to change. The profound differences between mtDNAs of plants and other organisms are highlighted below.

1. *Plant mtDNAs are abnormally large and variable in size.* Even the smallest angiosperm mtDNAs are greater than 200 kb in size, several times larger than all examined animal mtDNAs and most fungal and protistan mtDNAs. Most angiosperm mtDNAs are 300–600 kb in size, and, in two families (Cucurbitaceae and Malvaceae), genomes exceeding 2,000 kb have been reported. Recent work suggests that mtDNAs of other vascular plants (ferns and horsetails) are also quite large (Palmer, Soltis, and Soltis, unpublished data).

2. *Many foreign sequences are found in plant mitochondrial genomes.* Chloroplast DNA sequences of all kinds, some as large as 12 kb in length, are found integrated in all examined plant mtDNAs. These cpDNA sequences can persist in the mitochondrial genome for many millions of years and generally comprise 5–10% of the genome. Most of these foreign sequences seem to be genetically inert and functionless, although it has recently been shown that several tRNA genes of chloroplast origin are functional within the plant mitochondrial genome. Only a few nuclear sequences have been reported in plant mtDNA, but no concerted effort has been made to search for such sequences (as has been done for cpDNA).

3. *Large duplications are readily created and lost.* With but one known exception, all examined plant mtDNAs contain at least one large (1–14 kb) repeated sequence (usually a duplication, sometimes a triplication). However, compared to cpDNA, which contains the same large inverted duplication in almost all land plants (Downie and Palmer, Chapter 2, this volume), the large repeats in mtDNA are ephemeral. Most mitochondrial duplications are restricted to a few closely related species or even populations of plants. Accordingly, there is no pattern to the sequences (including genes) that are duplicated in the mitochondrial genomes of different plants.

4. *Recombination between repeats creates a complex, multipartite genome structure.* All of the large repeats found in plant mtDNAs appear to be engaged

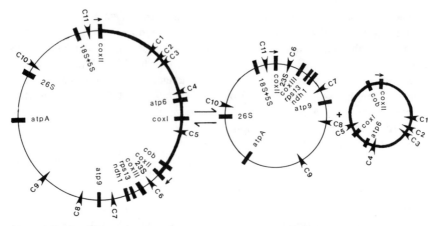

Figure 3.1. Tricircular structure of the *Brassica campestris* mitochondrial genome. Recombination between the two copies of a 2-kb repeat (indicated by arrows and containing the *cox*II gene) interconverts the 218-kb master chromosome with the subgenomic circles of 135 kb (thin line) and 83 kb (thick line) (Palmer and Shields, 1984). Mitochondrial genes are indicated by solid boxes (Makaroff and Palmer, 1987). Integrated cpDNA sequences are indicated by arrows and are labeled C1–C11 (Nugent and Palmer, 1988).

in high-frequency inter- and intramolecular recombination, such that nearly equi-molar proportions of interconvertible molecules are found. Most of the repeats are arranged as direct repeats, such that in the simple case of a molecule bearing a single two-copy repeat family, a single master circular chromosome (one carrying the entire sequence content of the genome) and two subgenomic circles will be present (e.g., see Fig. 3.1). In some genomes, multiple pairs and trios of repeats are found, resulting in an extremely complex multipartite organization of interconverting small and large circles.

5. *Unstable extrachromosomal plasmids are commonly found in plant mito-chondria.* These small (1–11 kb) plasmids exist as both circular and linear molecules and are of unknown origin, function, and selective value. They are sporadic in distribution even among closely related taxa, and sometimes are transmitted differently than the main mitochondrial genome in sexual crosses.

6. *Short dispersed repeats are common in plant mtDNA.* Animal mtDNAs lack short dispersed repeats, and these elements are generally rare in other mitochondrial genomes and in cpDNA. However, all plant mtDNAs that have been examined contain many short dispersed repeats of 50–1,000 bp scattered throughout the genome.

7. *Plant mtDNAs rearrange very rapidly.* No two examined species of flow-ering plants have the same mitochondrial gene order. Even closely related species differ by one or a few large inversions, whereas the genomes of more distantly related species are virtually randomized with respect to sequence arrangement (e.g., see Fig. 3.2). Most of the described rearrangements appear to have occurred

as inversions via recombination between pairs of oppositely oriented short dispersed repeats.

8. *Plant mtDNAs change very slowly in nucleotide sequence*. Rates of nucleotide substitutions, estimated either directly via sequencing of short regions of the genome or indirectly via restriction mapping of whole genomes, are 3–4 times lower in plant mtDNA than in cpDNA, 12 times lower than in plant nuclear DNA, and 40–100 times lower than in animal mtDNA.

Contrasts in Mitochondrial and Chloroplast Genome Evolution: The *Brassica* Example

The strikingly different ways in which plant mtDNA and cpDNA evolve are best illustrated by studies in which both genomes have been completely mapped and their structures and sequences compared within the same group of closely related species. The most comprehensive such comparisons have been made for the crucifer genus *Brassica* and will be reviewed below. Similar conclusions have been reached from more limited studies in which the organelle genomes of fertile and cytoplasmic male sterile (CMS) cultivars of *Zea mays* (Fauron and Havlik, 1989) and *Helianthus annuus* (Siculella and Palmer, 1988) were compared.

Palmer et al. (1983a) examined the evolution and phylogeny of cpDNA from a total of 22 cultivars representing seven agriculturally important species of *Brassica* plus the closely related *Raphanus sativa* (radish). All *Brassica* cpDNAs have the same gene order (indeed, are colinear with the tobacco-like genome representing the ancestral genome of vascular plants, see Downie and Palmer, chapter 2, this volume) and are essentially identical in size (151 kb). Aside from a few deletions/insertions of 50–400 bp, all of the mutations detected upon comparison of the 22 cpDNAs with 28 restriction enzymes were restriction site mutations, whose underlying cause is usually single nucleotide substitution. The estimated percent sequence divergence ranged from 0.3% to 2.6% for interspecific comparisons (considering diploid species only) and from 0.0% to 0.2% for intraspecific comparisons. Figure 3.3 shows a maternal phylogeny for the five diploid species and the one cytoplasmically distinct amphidiploid (*B. napus*) based on a parsimony analysis of the cpDNA restriction site mutations. This phylogeny clearly indicates that *R. sativa* is derived from within *Brassica*, a result confirmed by more recent studies by Warwick (personal communication). Therefore, in the rest of this chapter, *R. sativa* will be referred to as a *Brassica*.

Mitochondrial DNAs from this same set of *Brassica* taxa are highly diverse structurally (Fig. 3.3; Palmer and Herbon, 1988). Five of the six cytoplasmically distinct species have tripartite genomes as the result of recombination between a pair of large direct repeats. However, different sequences are duplicated and recombining in different lineages within the genus; the sizes of subgenomic circles also vary among species as the result of inversions that change the separation of repeats. The ancestral situation within the genome is inferred to be the presence

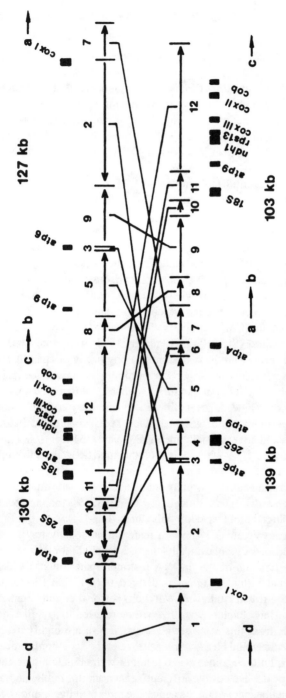

Figure 3.2. Rearrangements in mtDNAs of fertile and CMS *Raphanus sativa*. To facilitate comparison, the circular maps of the master chromosome are shown linearized (at a site internal to the 10-kb recombination repeat). The horizontal arrows numbered from 1 to 12 indicate the positions and orientations of blocks of cross-hybridizing sequences whose arrangement has been conserved (within blocks, but not between them) in the genomes of CMS (top line) and fertile (bottom line) *R. sativa*. The crossing lines connect homologous sequence blocks. A large region unique to the CMS genome is indicated by A. Positions of genes are marked with solid rectangles. The outermost horizontal arrows indicate the position and relative orientation of the two copies of the 10-kb repeat present in each genome. Letters denote the unique sequences flanking the repeats, and numbers between these arrows indicate the size of subgenomic circles resulting from recombination across the repeats. Modified from Makaroff and Palmer (1988).

40

of a single *cox*II gene and a 10-kb duplication ("recombination repeat" of Fig. 3.3), one copy of which was lost independently in two lineages. Concomitant with one loss of the 10-kb repeat was the duplication of a 2-kb element (containing the *cox*II gene) in the common ancestor of *B. campestris, B. oleracea,* and *B. napus* (Figs. 3.1 and 3.3). *Brassica hirta* lacks any large duplicated sequences and is the only plant known to have a largely homogeneous, unicircular genome (Palmer and Herbon, 1987). Two *Brassica* species also contain a linear, viral-like 11.3-kb mitochondrial plasmid of unknown function that is phylogenetically labile, highly variable in abundance when present (Fig. 3.4), and transmitted through both egg and pollen (Palmer et al., 1983b; Erickson et al., 1989).

The size of the mitochondrial genome varies from 208 to 257 kb within the eight examined *Brassica* species (Fig. 3.3). Approximately 10 kb of this size range results from differences in repeat-sequence content and the remainder from differences in sequence complexity. The origin of these species-specific unique sequences is unknown; however, they are not of chloroplast origin: all *Brassica* mtDNAs examined have the same 11 distinct regions of integrated cpDNA sequences, summing to 12–14 kb (Fig. 3.1; Nugent and Palmer, 1988). Mitochondrial DNA variability in cpDNA composition is, however, apparent upon comparison of different genera within the Brassicaceae (Nugent and Palmer, 1988).

The two most closely related *Brassica* species compared, *B. campestris* and *B. oleracea,* differ in mitochondrial gene order by three large inversions (Fig. 3.3; Palmer and Herbon, 1988). Most genome pairs differ by at least 10 rearrangements, far too many for elucidation of the nature of individual events with any certainty (Fig. 3.2). As many as 50 individual repeat elements, ranging in size from 40–700 bp and distributed among approximately 10 repeat families, are scattered fairly randomly throughout the various *Brassica* mitochondrial genomes (Shirzadegan and Palmer, unpublished data). Sequencing across the endpoints of several defined inversions reveals that the rearrangements are generated by recombination between these short dispersed repeats (Makaroff et al., 1989; Shirzadegan, Makaroff, and Palmer, unpublished data).

In spite of their huge diversity in structure and arrangement, the *Brassica* mitochondrial genomes are virtually identical at the nucleotide sequence level. Not a single variant restriction site was found in a limited study of intraspecific mtDNA diversity (Palmer, 1988), and interspecific divergence values ranged from 0.07% to 0.80% (Palmer and Herbon, 1988), four times lower than for the corresponding cpDNAs (Palmer et al., 1983a).

Comparison of mtDNA and cpDNA restriction patterns vividly illustrates the remarkably different evolutionary tempos and modes of the two plant organellar genomes (Fig. 3.4). The cpDNA patterns are extremely similar, showing only a few small deletions/insertions and one clear example of a restriction site mutation. The mtDNA patterns, on the other hand, are quite dissimilar. In fact, none of the differences in *Sal*I fragment patterns shown in Fig. 3.4 results from the point mutational loss or gain of restriction sites. Rather, these differences derive entirely

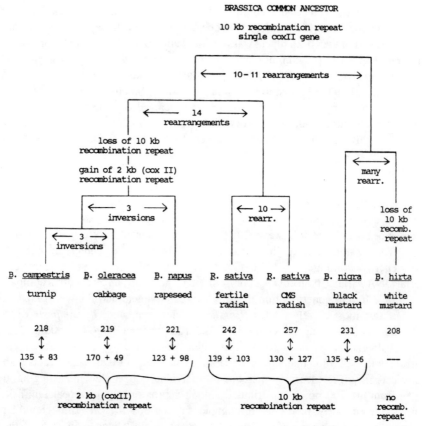

Figure 3.3. Phylogenetic history of mtDNA rearrangements in *Brassica* and *Raphanus*. *Top:* Cytoplasmic phylogeny for six cytoplasmically distinct species (five diploids and one amphidiploid) based on cpDNA restriction site mutations (Palmer et al., 1983a; Makaroff and Palmer, 1988). This phylogeny is cladistically derived and is not intended to convey divergence times. Numbers of mtDNA rearrangements are given relative to the reference genome, *B. campestris*, except for the 10 rearrangements that have been shown (Makaroff and Palmer, 1988) to distinguish the mtDNAs of CMS and fertile radish. *Bottom:* Sizes of master chromosomes and (where present) subgenomic circles resulting from high-frequency recombination at the indicated recombination repeats. Reprinted with permission from *Journal of Molecular Evolution* (Palmer and Herbon, 1988).

from structural changes, including deletions/insertions (of 0.1–12 kb), duplications, and multiple inversions (Palmer and Herbon, 1986, 1987, 1988).

Use of Mitochondrial DNA Sequence Variation in Systematics

The study of restriction site variability in cpDNA of plants (Palmer et al., 1988; and many chapters in this volume) and in mtDNA of animals (Avise et al.,

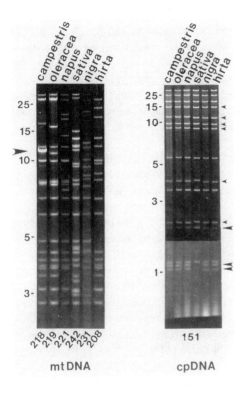

Figure 3.4. Contrasting patterns of restriction fragment variability in mtDNA (left) and cpDNA (right). DNAs from the five species of *Brassica* and one of *Raphanus* indicated were digested with the restriction enzymes *Sal*I and *Sac*I, respectively, and the resulting fragments were separated by size on 0.7% agarose gels. Sizes marked at the left of each panel are in kilobase pairs. Sizes marked below the panels indicate the sizes of the respective genomes in kilobase pairs. The small arrows indicate fragments that differ slightly in mobility (and therefore size) among the cpDNAs. The medium arrows indicate cpDNA fragments of 1.1 and 1.05 kb that are converted to one of 2.2 kb in *Raphanus sativa* by the loss of a restriction site. The large arrow indicates an 11.3-kb linear plasmid found in mitochondria of *B. campestris* and, in a 100-fold lower amount, in *B. napus*.

1987; Moritz et al., 1987; Harrison, 1989) has had a revolutionary impact on phylogenetic reconstruction at the population, interspecific, and, sometimes, intergeneric levels. However, for the following reasons, I see little, if any, value in applying this approach to plant mtDNA. First, plant mtDNA, which changes in sequence three to four times slower than cpDNA, is too limited in restriction site variability to be useful at the interspecific level, at which most plant systematists work. Second, at the higher taxonomic levels where useful amounts of mtDNA restriction site variation might theoretically be expected, this variation will be essentially impossible to use because of the confounding effects of high rates of rearrangement in plant mtDNA. Even a few inversions seriously hamper one's ability to compare restriction maps and determine homology of restriction sites, whereas the literally dozens of rearrangements that distinguish most plant mitochondrial genomes (Figs. 3.2 and 3.3; Fauron and Havlik, 1989) render such analysis impossible. A further practical complication is that plant mtDNA contains many cpDNA sequences. The much greater abundance of cpDNA compared to mtDNA (Bendich, 1987) in the preparations of total leaf DNA commonly used for restriction site analysis means that many mtDNA probes will actually hybridize more strongly to cpDNA than to mtDNA.

The direct sequence analysis of specific mitochondrial genes should obviate problems caused by the rapid rearrangement of mtDNA and the abundance of cpDNA sequences in the genome. However, the rate of synonymous (silent) nucleotide substitutions is three to four times lower in mtDNA than in cpDNA (specifically its single-copy regions, in which, for example, the *rbc*L gene is found). Thus, mitochondrial gene sequencing is probably most valuable at greater time depths, of at least 100 myr, than those to which chloroplast genes are now actively being applied (see Clegg and Zurawski, Chapter 1, and Jansen et al., Chapter 11, this volume).

The choice of mitochondrial genes for comparative sequencing is rather limited in plants. Several genes (*atp*8, *atp*9, *nad*3, all three ribosomal protein genes) are too short (200–400 bp) to provide sufficient information. Several others (*nad*1, *nad*4, *nad*5, *cox*II) are of adequate length but contain one or more introns. The intron sequences are subject to too much internal deletion and insertion (including complete loss of the intron; see the next section) to be useful for comparative analysis; also, their presence in the middle of coding regions provides major barriers to the efficient accumulation of coding-sequence information. The *atp*6 gene is known to sustain substantial size variation within its coding regions; this, combined with its moderate, average size (850 bp), makes it a poor choice for sequencing for phylogenetic purposes. At present, therefore, the best candidates among protein genes are the uninterrupted (at least as so far surveyed) genes *cox*I (1590 bp), *nad*2 (1545 bp), *atp*A (1525 bp), *cob* (1180 bp), and *cox*III (795 bp). In addition, the two large ribosomal RNA genes (18S, ca. 1900 bp, and 26S, ca. 3400 bp) also lack introns and should be good choices for sequencing projects. Thus far, although several of these candidate genes have been sequenced in several monocots and dicots, no explicit phylogenies have been constructed to allow evaluation of the reliability of any plant mitochondrial gene for phylogenetic reconstruction.

Plant mitochondrial genes are known to be duplicated fairly often. Their duplication as part of large recombining repeats (see previous two sections), whose sequences remain identical by copy correction, should pose no problem for phylogenetic studies. However, several instances are also known of partially duplicated mitochondrial pseudogenes, which are sometimes part of chimeric genes (Lonsdale, 1989). These could inadvertently be studied instead of the real gene, particularly in studies in which a gene amplified by the polymerase chain-reaction product is cloned and analyzed.

Chlamydomonas reinhardtii and *Prototheca wickerhamii* are the only algae for which any mitochondrial gene-sequence data are available. The mitochondrial genome of *C. reinhardtii* is unlike those of land plants in all respects, including size (15.8 kb), conformation (linear), gene content (only half the number of protein genes), gene structure (its rRNA genes are fragmented into short dispersed pieces), and gene sequence (its rRNA genes are very different from those of plants in sequence and, in fact, appear more similar to those of animals by formal phylogenetic analysis) (Gray and Boer, 1988; Gray et al., 1989). The mitochondrial genome of *C. reinhardtii* is missing *atp*A and *cox*III, two of the five good candidates for phylogenetic sequencing among plant mitochondrial protein genes. The highly divergent structure and sequence of its rRNA genes may also make these very difficult to relate to those of land plants in phylogenetic reconstruction. All that is known about the mitochondrial genome of *P. wickerhamii* is that its small subunit rRNA gene, unlike that of *C. reinhardtii,* is closely related in sequence to those of land plants (Wolff and Kück, 1990). It is clearly desirable to have information on more green algal mitochondrial genomes to know if *C. reinhardtii* is atypical or not. If it is atypical, as the *P. wickerhamii* data would suggest, then perhaps mitochondrial genes will be as useful as chloroplast genes (Turner et al., 1989; Morden and Golden, 1989; Baldauf and Palmer, 1990; Douglas et al., 1990) for elucidating phylogenetic relationships among land plants and algae.

Use of Mitochondrial DNA Rearrangements in Systematics

Large rearrangements produce readily detectable restriction fragment pattern differences (e.g., Fig. 3.4). However, determining the nature of individual rearrangements (in particular, inversions) in situations where two genomes differ by more than one or two events (as is usually the case even between closely related species) requires a large amount of work and is often impossible due to the complex and overlapping nature of the changes. Indeed, the latter is the case for most of the rearrangements detected among *Brassica* species (Fig. 3.2–3.4) and for all those that distinguish two cytoplasms of *Z. mays* (Fauron and Havlik, 1989). Thus, the use of rearrangements as discrete characters for phylogenetic analysis seems largely impossible or unjustifiable for mtDNA, whereas single inversions in cpDNA have served as valuable phylogenetic characters (Downie and Palmer, Chapter 2, and Jansen et al., Chapter 11, this volume).

Although mtDNA rearrangements cannot readily be used in a rigorous phylogenetic analysis, they are still of limited utility for grouping closely related genomes (and organisms). Much effort has been expended in this respect, principally among cultivated cereals (e.g., Weissinger et al., 1983; Terachi and Tsunewaki, 1986; Graur et al., 1989) and, to a lesser extent, members of the Solanaceae (Bland et al., 1985; McClean and Hanson, 1986). In some cases (e.g., Terachi and Tsunewaki, 1986; McClean and Hanson, 1986), phenograms have been constructed based on the numbers of shared restriction fragments. I consider these efforts potentially dangerous given the complex mode of plant mtDNA evolution, which violates most of the assumptions on which these distance formulations are based.

The use of rearranged mtDNAs as crude genetic markers may have a special application in conifers. Certain conifers transmit their mitochondrial and chloroplast genomes through opposite sexual partners (Neale and Sederoff, 1989) and thus provide an exceptional situation for studying paternal and maternal genetic lineages within a single species. Opposite inheritance of organellar genomes has also been reported for *Chlamydomonas reinhardtii* (Boynton et al., 1987).

Whereas inversions and large duplications have a limited applicability at lower taxonomic levels, another class of structural change, gene and intron losses, has limited application for reconstructing higher level phylogeny. The limitation here is the rarity of the losses. However, it is this rarity that also makes them useful at deep phylogenetic levels. Mitochondrial DNA contains only 20–25 genes for proteins and 20–30 genes for stable RNAs. Southern hybridization experiments suggest the possible absence of three different genes from specific lineages (Stern and Palmer, 1984; Bland et al., 1986; Wahleithner and Wolstenholme, 1988), and recent studies strongly suggest gene relocation to the nucleus in the case of the *cox*II gene in certain legumes (Nugent and Palmer, unpublished data). Three of the eight introns described in plant mitochondrial genes are missing from at least one lineage of plants. For example, the *cox*II genes of most monocots and dicots have a single homologous intron, whereas some dicot *cox*II genes completely lack the intron (reviewed in Lonsdale, 1989). Our lab is currently surveying widely among angiosperms for the presence or absence of these introns using the approaches described for chloroplast introns by Downie and Palmer (Chapter 2, this volume).

Conclusions

Compared to cpDNA, the current genome of choice for systematic studies, mtDNA is a much less versatile molecule. Mitochondrial DNA is much harder to purify than cpDNA and is much less abundant in leaves; mtDNAs are relatively poorly characterized: few mitochondrial genes have been sequenced, and relatively few mtDNA clone banks are available. Most importantly, the distinctive evolutionary dynamics of plant mtDNA—high rates of rearrangements and low rates of point mutations—make the molecule essentially worthless for the restriction site-based reconstructions of intrafamilial phylogeny for which cpDNA is so well suited. However, the frequency of rearrangements and ease with which rearranged genomes are distinguished by restriction pattern analysis can make the genome useful for rapid surveys and classifications of genome types at lower taxonomic levels. At higher levels, the occasional losses of mitochondrial genes and introns may also serve as useful markers of phylogeny. The low rate of mtDNA substitutions suggests that comparative sequencing efforts will be most rewarding at higher phylogenetic levels than those for which cpDNA (specifically, the *rbc*L gene) is now being applied. These conclusions all apply specifically to mtDNA of land plants. What little is known about algal mtDNA suggests that it may have very different structural and evolutionary dynamics (Gray, 1989;

Palmer, 1991). Therefore its suitability for phylogenetic studies may be correspondingly different and remains largely to be investigated.

References

Avise, J.C., Arnold, J., Ball, R.M., Bermingham, E., Lamb, T., Neigel, J. E., Reeb, C.A., and Saunders, N.C. (1987) Intraspecific phylogeography: the mitochondrial DNA bridge between population genetics and systematics. *Ann. Rev. Ecol. Syst.* **18**, 489–522.

Baldauf, S.L., and Palmer, J.D. (1990) Evolutionary transfer of the chloroplast *tuf*A gene to the nucleus. *Nature* **344**, 262–265.

Bendich, A.J. (1987) Why do chloroplasts and mitochondria contain so many copies of their genomes? *Bioessays* **6**, 279–282.

Bland, M.M., Matzinger, D.F., and Levings, C.S. III (1985) Comparison of the mitochondrial genome of *Nicotiana tabacum* with its progenitor species. *Theor. Appl. Genet.* **69**, 535–541.

Bland, M.M., Levings, C.S. III, and Matzinger, D.F. (1986) The tobacco mitochondrial ATPase subunit 9 gene is closely linked to an open reading frame for a ribosomal protein. *Mol. Gen. Genet.* **204**, 8–16.

Boynton, J.E., Harris, E.H., Burkhart, B.D., Lamerson, P.M., and Gillham, N.W. (1987) Transmission of mitochondrial and chloroplast genomes in crosses of *Chlamydomonas*. *Proc. Natl. Acad Sci. USA* **84**, 2391–2395.

Douglas, S.E., Durnford, D.G., and Morden, C.W. (1990) Nucleotide sequence of the gene for the large subunit of ribulose-1,5-bisphosphate carboxylase/oxygenase from *Cryptomonas* sp.: evidence supporting the polyphyletic origin of plastids. *J. Phycol.* **26**, 500–508.

Erickson, L., Kemble, R., and Swanson, E. (1989) The *Brassica* mitochondrial plasmid can be sexually transmitted. Pollen transfer of a cytoplasmic genetic element. *Mol. Gen. Genet.* **218**, 419–422.

Fauron, C., and Havlik, M. (1989) The maize mitochondrial genome of the normal type and the cytoplasmic male sterile type T have very different organization. *Curr. Genet.* **15**, 149–154.

Graur, D., Bogher, M., and Breiman, A. (1989) Restriction endonuclease profiles of mitochondrial DNA and the origin of the B genome of bread wheat, *Triticum aestivum*. *Heredity* **62**, 335–342.

Gray, M.W. (1989) Origin and evolution of mitochondrial DNA. *Ann. Rev. Cell Biol.* **5**, 25–50.

Gray, M.W., and Boer, P.H. (1988) Organization and expression of algal (*Chlamydomonas reinhardtii*) mitochondrial DNA. *Phil. Trans. R. Soc. Lond. [Biol.]* **319**, 135–147.

Gray, M.W., Cedergren, R., Abel, Y., and Sankoff, D. (1989) On the evolutionary origin of the plant mitochondrion and its genome. *Proc. Natl. Acad. Sci. USA* **86**, 2267–2271.

Harrison, R.G. (1989) Animal mitochondrial DNA as a genetic marker in population and evolutionary biology. *Trends Ecol. Evol.* **4,** 6–11.

Levings, C.S. III, and Brown, G.G. (1989) Molecular biology of plant mitochondria. *Cell* **56,** 171–179.

Lonsdale, D.M. (1989) The plant mitochondrial genome. In: *The Biochemistry of Plants* (eds. P.K. Stumpf and E.E. Conn), Vol. 15, Molecular Biology, Academic Press, New York, pp. 230–295.

Lonsdale, D.M. (1991) Structure and expression of mitochondrial genomes. In: *Plant Gene Research,* Vol. 6, Organelles (ed. R.G. Hermann), Springer-Verlag, Wien, in press.

Makaroff, C.A., and Palmer, J.D. (1987) Extensive mitochondrial specific transcription of the *Brassica campestris* mitochondrial genome. *Nucleic Acids Res.* **15,** 5141–5156.

Makaroff, C.A., and Palmer, J.D. (1988) Mitochondrial DNA rearrangements and transcriptional alterations in the male-sterile cytoplasm of Ogura radish. *Mol. Cell Biol.* **8,** 1474–1480.

Makaroff, C.A., Apel, I.J., and Palmer, J.D. (1989) The *atp*6 coding region has been disrupted and a novel reading frame generated in the mitochondrial genome of cytoplasmic male-sterile radish. *J. Biol. Chem.* **264,** 11706–11713.

McClean, P.E., and Hanson, M.R. (1986) Mitochondrial DNA sequence divergence among *Lycopersicon* and related *Solanum* species. *Genetics* **112,** 649–667.

Morden, C.W., and Golden, S.S. (1989) *psb*A genes indicate common ancestry of prochlorophytes and chloroplasts. *Nature* **337,** 382–385.

Moritz, C., Dowling, T.E., and Brown, W.M. (1987) Evolution of animal mitochondrial DNA: relevance for population biology and systematics. *Ann. Rev. Ecol. Syst.* **18,** 269–292.

Neale, D.B., and Sederoff, R.R. (1989) Paternal inheritance of chloroplast DNA and maternal inheritance of mitochondrial DNA in loblolly pine. *Theor. Appl. Genet.* **77,** 212–216.

Newton, K.J. (1988) Plant mitochondrial genomes: organization, expression and variation. *Ann. Rev. Plant Physiol.* **39,** 503–532.

Nugent, J.M., and Palmer, J.D. (1988) Location, identity, amount, and serial entry of chloroplast DNA sequences in crucifer mitochondrial DNAs. *Curr. Genet.* **14,** 501–509.

Palmer, J.D. (1988) Intraspecific variation and multicircularity in *Brassica* mitochondrial DNAs. *Genetics* **118,** 341–351.

Palmer, J.D. (1990) Contrasting modes and tempos of genome evolution in land plant organelles. *Trends Genet.,* **6,** 115–120.

Palmer, J.D. (1991) Comparison of chloroplast and mitochondrial genome evolution in plants. In: *Plant Gene Research,* Vol. 6, Organelles (ed. R.G. Hermann), Springer-Verlag, Wien, in press.

Palmer, J.D., and Herbon, L.A. (1986) Tricircular mitochondrial genomes of *Brassica* and *Raphanus*: reversal of repeat configurations by inversion. *Nucleic Acids Res.* **14,** 9755–9764.

Palmer, J.D., and Herbon, L.A. (1987) Unicircular structure of the *Brassica hirta* mitochondrial genome. *Curr. Genet.* **11,** 565–570.

Palmer, J.D., and Herbon, L.A. (1988) Plant mitochondrial DNA evolves rapidly in structure, but slowly in sequence. *J. Mol. Evol.* **28,** 87–97.

Palmer, J.D., and Shields, C.R. (1984) Tripartite structure of the *Brassica campestris* mitochondrial genome. *Nature* **307,** 437–440.

Palmer, J.D., Shields, C.R., Cohen, D.B., and Orton, T.J. (1983a) Chloroplast DNA evolution and the origin of amphidiploid *Brassica* species. *Theor. Appl. Genet.* **65,** 181–189.

Palmer, J.D., Shields, C.R., Cohen, D.B., and Orton, T.J. (1983b) An unusual mitochondrial DNA plasmid in the genus *Brassica*. *Nature* **301,** 725–728.

Palmer, J.D., Jansen, R.K., Michaels, H., Chase, M.W., and Manhart, J. (1988) Phylogenetic analysis of chloroplast DNA variation. *Ann. Missouri Bot. Gard.* **75,** 1180–1218.

Siculella, L., and Palmer, J.D. (1988) Physical and gene organization of mitochondrial DNA in fertile and male sterile sunflower. CMS associated alterations in structure and transcription of the *atp*A gene. *Nucleic Acids Res.* **16,** 3787–3799.

Stern, D.B., and Palmer, J.D. (1984) Recombination sequences in plant mitochondrial genomes: diversity and homologies to known mitochondrial genes. *Nucleic Acids Res.* **12,** 6141–6157.

Terachi, T., and Tsunewaki, K. (1986) The molecular basis of genetic diversity among cytoplasms of *Triticum* and *Aegilops*. 5. Mitochondrial genome diversity among *Aegilops* species having identical chloroplast genomes. *Theor. Appl. Genet.* **73,** 175–181.

Turner, S., Burger-Wiersma, T., Giovannoni, S.J., Mur, L.R., and Pace, N.R. (1989) The relationship of a prochlorophyte *Prochlorothrix hollandica* to green chloroplasts. *Nature* **337,** 380–382.

Wahleithner, J.A., and Wolstenholme, D.R. (1988) Ribosomal protein s14 genes in broad bean mitochondrial DNA. *Nucleic Acids Res.* **16,** 6897–6913.

Weissinger, A.K., Timothy, D.H., Levings, C.S. III, and Goodman, M.M. (1983) Patterns of mitochondrial DNA variation in indigenous maize races of Latin America. *Genetics* **104,** 365–379.

Wolff, G., and Kück, U. (1990) The structural analysis of the mitochondrial SSUrRNA implies a close phylogenetic relationship between mitochondria from plants and from the heterotrophic alga *Prototheca wickerhamii*. *Curr. Genet.* **17,** 347–351.

4

Ribosomal RNA as a Phylogenetic Tool in Plant Systematics

R. Keith Hamby and *Elizabeth A. Zimmer*

The traditional classification of plants into respective classes, orders, families, genera, and species has until recently been based on shared morphologic, cytologic, biochemical, and ecologic traits. The development of techniques in molecular biology including those for molecular hybridization, cloning, restriction endonuclease digestions, and protein and nucleic acid sequencing have provided many new tools for the investigation of phylogenetic relationships. At the molecular level, the most fundamental comparison possible is of the primary nucleotide sequences of homologous genes in different populations or species.

Within their genomes all organisms have DNA sequences that code for ribosomal RNAs (rRNAs), essential components of cellular protein synthesis. In plants, ribosomal DNA (rDNA) is found in nuclear, mitochondrial, and chloroplast genomes. The ubiquity of rRNA throughout nature and the development of techniques for the rapid determination of the primary nucleotide sequence of rRNA molecules make rRNA a good tool for inferring evolutionary relationships. Not all regions of the rDNA evolve at the same rate, so, even though some regions are useful for comparisons at or below the genus level, other regions are only useful at the family level or above. In this chapter, we report on the use of nucleotide sequence information from nuclear rDNA sequences in elucidating the evolutionary history of land plants.

Until recently, the greatest use of rRNA sequences had been in the investigations of bacterial evolution. Woese (1987) used a parsimony analysis of complete 16S rRNA sequences to propose three main lines of descent in nature: eubacteria, archaebacteria, and eukaryotes. Other analyses of the same data support the

We would like to thank C. Bult, R. L. Chapman, M. Donoghue, J. A. Doyle, M. Kallersjo, Y. Suh, and two anonymous reviewers for comments that led to improvements in our discussion of the angiosperm RNA sequencing project. This work was supported in part by NSF Grant DEB-BSR-86-15212 and LEQSF Contract 86-LBR-(048)-08.

archaebacteria tree (Gouy and Li, 1989a). Lake (1988) disputes this interpretation of the rRNA sequence data, suggesting that the archaebacteria are paraphyletic.

Aside from work in this lab (Hamby and Zimmer, 1988; Zimmer et al., 1989; Knaak et al., 1990) and that of our collaborators (Buchheim et al., 1990; Chapman and Avery, 1989; Kantz et al., 1990; Zechman et al., 1990), there is little use of comparative rRNA sequences in plant evolutionary studies. Nickrent and Franchina (1989) used nuclear 18S rRNA sequences to define the relationships within the parasitic flowering plants of the order Santalales. Wolfe et al. (1989) have compared published 18S and 26S sequences to calibrate the divergence of monocot and dicot lineages. Previous studies using chloroplast or nuclear rRNA sequences to study taxonomic or phylogenetic relationships are summarized in Table 4.1.

Function, Structure, and Evolution of Ribosomal RNA

Ribosomal RNA *Function*

The main function of rRNAs is in protein synthesis. It was previously thought that rRNAs served primarily as a scaffolding for ribosomal proteins, but recent evidence suggests that rRNA molecules are the basic functional element of the ribosome and that the proteins serve to mediate interactions between mRNA, tRNA, and rRNA (reviewed by Gerbi, 1985; and Dahlberg, 1989). Most detailed studies of ribosome action are based on the bacterium *Escherichia coli*, but the results are generally valid for higher taxa as well. The 70S *E. coli* ribosome consists of a 30S subunit and a 50S subunit, which come together in the presence of mRNA and other cofactors. The 16S rRNA (analogous to the plant 18S rRNA) is part of the 30S subunit, the 5S and 23S (analogous to the plant 26S rRNA) combine with various proteins to make up the 50S subunit. In plants and other eukaryotes, the large subunit of the ribosome also contains a 5.8S rRNA molecule.

During translation initiation, sequences near the 3' end of the 16S rRNA base pair with the Shine-Dalgarno sequence upstream of the initiation codon in bacterial mRNA. Interference with this base pairing interaction by mutation in the 16S rRNA molecule leads to significant reductions in the level of protein synthesis (Jacob et al., 1987; Hui et al., 1988). Ribosome activity can be restored by a compensatory mutation in the Shine-Dalgarno sequence of the target mRNA (Hui et al., 1988). Base pairing between mRNA and the same region of the 16S rRNA molecule also may be responsible for maintaining the correct reading frame during elongation (Trifonov, 1987; Weiss et al., 1987, 1988). In addition, translation termination at the stop codons appears to rely upon specific RNA–RNA interactions between the 16S rRNA and mRNA (Murgola et al., 1988).

The proper association of the small and large subunits is also dependent to

Table 4.1. A chronological summary of phylogenetic and taxonomic studies using nuclear or chloroplast rrna sequences.

Investigators	Groups	Subunit	Comments
Kumazaki et al. (1983)	Protists	Nuclear 5S	Green algae share common ancestor with vascular plants.
McCarroll et al. (1983)	Eukaryotes	Nuclear 18S	*Dictyostelium* represents earliest divergence of eukaryotes.
Hori et al. (1985)	Plants	Nuclear 5S	*Cycas* is a gymnosperm. Land plants are most closely related to charophyte algae.
Woese (1987)	Bacteria	16S	There are three primary lines of descent: archaebacteria, eubacteria, and eukaryotes.
Hori and Osawa (1987)	Prokaryotes and eukaryotes	5S and Nuclear 5S	Red algae are the most primitive eukaryotes. Archaebacteria and eukaryotes split off after eubacteria.
Vossbrinck et al. (1987)	Eukaryotes	Nuclear 18S	Microsporidia diverged early in eukaryotic evolution.
Lake (1988, 1989)	Bacteria	16S	Evolutionary parsimony analysis says archaebacteria are paraphyletic.
Edman et al. (1988), Stringer et al. (1989)	Protozoa and fungi	Nuclear 16S	*Pneumocystis carinii* is a fungus.
Field et al. (1988), Raff et al. (1989)	Animals	Nuclear 18S	Cniderians are separate from other animal lineages. Coelomates are monophyletic.
Nairn and Ferl (1988)	Eukaryotes	Nuclear 18S	Angiosperms are monophyletic.
Gouy and Li (1989a)	Bacteria	16S and 23S	Neighbor joining and maximum parsimony analysis support Woese (1987) above.
Gouy and Li (1989b)	Eukaryotes	Nuclear 18S and 26S	Fungi diverged first from the common ancestor of plants and animals.
Wolfe et al. (1989)	Angiosperms	Nuclear 18S and 26S, Chloroplast 16S.	Monocots and dicots diverged from one another 200 million years ago.
Perasso et al. (1989)	Algae	Nuclear 26S	Rhodophytes, chromophytes, and chlorophytes are each monophyletic groups. Plants are closest to chlorophytes.
Turner et al. (1989)	Prokaryotes	16S	Prochlorophytes are holophyletic with cyanobacteria and chloroplasts, but not progenitors of chloroplasts.
Watanabe et al. (1989)	Protozoa and fungi	Nuclear 5S	*Pneumocystis carinii* is closer to Zygomycota fungi than to Ascomycota or Basidiomycota.

(continued)

Table 4.1.
(Continued)

Investigators	Groups	Subunit	Comments
Schleifer and Ludwig (1989)	Bacteria	23S	23S rRNA trees support the 16S rRNA trees as well as those based on EF TU and subunit of ATPase.
Hillis and Dixon (1989)	Vertebrates	Nuclear 28S	Coelacanths belong among the tetrapods. Weak support for a bird–mammal relationship.
Sogin et al. (1989)	Eukaryotes	Nuclear 18S	Earliest eukaryotes are microsporidia and diplomonads. Fungi, plants, and animals diverged relatively recently.
Larson and Wilson (1989)		Nuclear 18S, 28S	Relationships among 7 families of salamanders and 3 orders of amphibians.

some degree on sequences within the 16S rRNA molecule, although no particular sequence dependence has been identified within the 23S rRNA molecule (Dahlberg, 1989). Methylation of two consecutive adenine residues near the 3' end of 16S rRNA is required for correct association of the subunits. The stem structure immediately upstream of the stem-loop containing the methylated adenines is also important in the formation of an active ribosome, as is the sequence around position 790 (of 1,542 bases in the 16S molecule).

The activities within the ribosome decoding site, which consists of the amino-acyl (A) site and the peptidyl (P) site, are dependent on the tertiary structure of the 16S rRNA. Several different regions of the 16S rRNA secondary structure are brought together by three-dimensional folding to line the cleft of the 30S subunit which has been shown to be only a few angstroms from the codon-anticodon site. Transfer RNA protection experiments indicate that the tRNAs interact with specific 16S rRNA nucleotides in this cleft region (Noller et al., 1987). Footprinting experiments have implicated specific nucleotides within the 16S rRNA as sites of action for antibiotic agents known to cause miscoding; resistance to the antibiotic is associated with modifications of the rRNA sequence (Moazed and Noller, 1987). Recently, Moazed and Noller (1989) have identified sequences within the 23S rRNA that make up parts of the A and P sites on the 50S subunit. They have also described the E site, the site where the deacylated tRNA resides before it dissociates from the ribosome completely, and have shown that the CCA conserved nucleotides at the end of all tRNA molecules interact with the 23S rRNA at the A, P, and E sites.

The peptidyl transferase activity of the ribosome catalyzes the formation of the peptide bond between the growing protein and the new amino acid. This activity can be significantly disrupted by base modifications in domain V of the 23S rRNA. The action of antibiotics known to inhibit transferase activity also map to

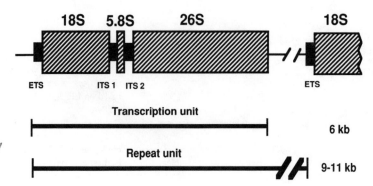

Figure 4.1. A typical plant rDNA repeat unit shown to scale. ETS is the external transcribed spacer, ITS1 and 2 are the internal transcribed spacers. The hatched boxes are the coding units.

this domain. Finally, specific nucleotides in the 23S rRNA have been shown to be involved with translocation of the peptidyl tRNA from the A site to the P site (Dahlberg, 1989).

Nuclear Ribosomal Gene Organization

The nuclear genes that code for rRNA are reiterated thousands of times within the typical plant genome (Appels and Honeycutt, 1986). In fact, they can comprise as much as 10% of the total plant DNA (Hemleben et al., 1988). Ribosomal DNA is arranged in tandem repeats in one or a few chromosomal loci. Only among closely related species are the chromosomal locations similar.

Each repeat unit consists of a transcribed region separated from the next repeat unit by an intergenic spacer (IGS). Figure 4.1 shows that, beginning from the 5′ end, the transcribed region consists of an external transcribed spacer (ETS), the 18S gene, an internal transcribed spacer (ITS1), the 5.8S gene, a second ITS (ITS2), and the 26S gene. Transcription by RNA polymerase I is thought to end immediately after the 26S gene, although in some animal systems transcription is known to continue on through most of the IGS and end just before transcription of the next repeat unit begins (DeWinter and Moss, 1986; Labhart and Reeder, 1986). In wheat, most transcripts end at or near the 3′ end of the 26S gene, but some transcription proceeds through the intergenic regions as in *Xenopus* and *Mus* (Vincentz and Flavell, 1989). Presumably, the 3′ trailer is rapidly discarded to yield the precursor rRNA molecule. This 45S precursor is enzymatically cleaved and trimmed to produce the three mature rRNA molecules.

There is another cytoplasmic rRNA molecule, the 5S rRNA, which is transcribed by RNA polymerase III. In prokaryotes and some lower eukaryotes, the 5S gene is linked to the other rDNA, but, in higher eukaryotes, the 5S genes lie in independent unlinked arrays (Appels and Honeycutt, 1986). In maize, for example, rDNA arrays are on the short arm of chromosome 6 (McClintock, 1934;

Givens and Phillips, 1976; Phillips, 1978), whereas the 5S rDNA repeats have been localized to the long arm of chromosome 2 (Steffensen and Patterson, 1979; Mascia et al., 1981).

important
✓ TP

Evolution of the Nuclear Ribosomal DNA

The most remarkable feature of rDNA is the overall sequence homogeneity among members of the gene family. If all parts of the genome were evolving independently, comparisons of nucleotide sequences between members of the same gene family within a species would show about the same level of similarity as comparisons of the same gene between two closely related species, as long as the duplication events creating the gene family preceded the divergence of the two species. Studies consistently show that this is not the case for rDNA (Arnheim, 1983). Brown et al. (1972) first demonstrated by hybridization experiments that within *Xenopus laevis* the several hundred rDNA repeats were essentially identical at both the coding and the intergenic regions, but that when the rDNAs of *X. laevis* were hybridized to those of *X. borealis* (misidentified as *X. mulleri* in the original reference) a much lower level of overall similarity was found. Whereas the coding regions were still highly conserved, the IGS regions were sharply divergent, although within each species the IGS was conserved. This same motif of conserved coding regions and nonconserved intergenic spacers with species-specific mutations has been identified in the rDNA of all species studied (Dover and Flavell, 1984). The process by which this pattern of intraspecific homogeneity and interspecific heterogeneity is maintained has been called horizontal evolution (Brown et al., 1972) and coincidental evolution (Hood et al., 1975), but it is now usually termed concerted evolution (Zimmer et al., 1980).

The mechanisms of concerted evolution are primarily unequal crossing over or unequal exchange, and gene conversion (Dover, 1982; Arnheim, 1983). To achieve overall homogeneity, one or both of these processes (and possibly others) must take place within each individual locus, between rDNA loci on homologous chromosomes, and between rDNA loci on nonhomologous chromosomes. There is evidence supporting the occurrence of all three types of exchange in yeast (Petes, 1980; Klein and Petes, 1981; Fogel et al., 1978). Computer modeling studies and analytic treatments have shown that the processes of unequal exchange and gene conversion alone or together can eventually lead to the fixation of a mutant gene within a population, even with only one or a few original copies of the mutant (Smith, 1974, 1976; Ohta, 1983, 1984; Birky and Skavaril, 1976).

Theoretically a gene conversion can proceed in either direction when a heteroduplex is recognized, that is, the mutant may be converted to wild type or vice versa. However, if there is even a small bias in one direction or the other, the rate of concerted evolution can be significantly increased (Nagylaki and Petes, 1982). Dover (1982) coined the term molecular drive to describe the process of gene family homogenization and fixation due to unequal crossing over and biased gene conversion. Transposition may also play an important role in molecular drive, but it has not yet been demonstrated as a mechanism in the concerted evolution of rDNA families.

Nuclear Ribosomal DNA *Copy Number*

The copy number of rDNA repeat units is highly variable in plants (Appels and Honeycutt, 1986), as well as animals (Long and Dawid, 1980). In plants, the variation exists at the interspecific and intraspecific levels as well as between individuals of the same population (Rogers and Bendich, 1987). Within a species, rDNA copy number can have a fourfold level of variation (Jorgensen and Cluster, 1988). In inbred lines of maize, rDNA copy number has been shown to have a 10-fold range (Rivin et al., 1986). Within a population of wild barley, a sixfold range in copy number was detected between different individuals, and, within a large population of broad bean, the copy number ranged from 500 to 44,000 per individual, and the copy number was found to vary in different tissues (Rogers and Bendich, 1987). Experiments in *Drosophila* have shown that there is a minimum level of rDNA required, and possession of genes in excess of those required has no discernible effect on phenotype (Shermoen and Kiefer, 1975; Tartof, 1975). An overabundance of rDNA might be one way for the cell to insure that, at critical times during development or in cases of stress, there is sufficient cellular machinery for protein synthesis.

There is evidence that there is a large excess of rDNA within the plant nuclear genome; structural studies in maize and DNAse digestion experiments in wheat have shown that a large amount of rDNA lies within the heterochromatic (i.e., nontranscribed) region of the chromosome (Flavell, 1986). Those rRNA genes that are transcribed lie in the nucleolar organizer region (NOR) of the chromosome. The genes within the NOR are methylated to a lesser degree than those in the heterochromatin; the loss of methyl groups from cytosine residues in animal genes has been associated with gene activation (Razin and Riggs, 1980). In wheat, the relative size of the NOR at a chromosomal locus, and hence the activity of that NOR, is proportional to the fraction of the rRNA genes without methylated cytosines (Flavell et al., 1983). Deletion of the NOR with the high activity results in a decrease in the methylation at the other NORs and a concomitant increase in rDNA expression at the other NORs (Flavell, 1986). Inactivity of hypermethylated rDNA has been demonstrated recently in maize (Jupe and Zimmer, 1989).

Copy number variation is due to the mechanisms of concerted evolution, primarily unequal crossing over (Flavell, 1986), although gene conversion can also contribute to the variation in copy number of members of a gene family (Li et al., 1985).

135 = 1800 bp
135 = 3400 bp
2L S =

Nuclear Ribosomal DNA *Length Variation* Not sure what this has to do w/ anything.

Restriction site analysis shows that there is no measurable variation in the lengths of the coding regions of the rDNA repeat units of plants (Jorgensen and Cluster, 1988). Sequencing the soybean, maize, and rice 18S genes has shown these cistrons to be 1,807 bp, 1,809 bp, and 1,812 bp in length, respectively (Eckenrode et al., 1985; Messing et al., 1984; Takaiwa et al., 1984). Among plants, only the 26S gene of rice has been completely sequenced, and it is 3,376 bp in length (Takaiwa et al., 1985a). The lengths of the 5.8S genes of rice and broad bean are 163 bp (Takaiwa et al., 1985b; Tanaka et al., 1980). No plant ribosomal genes are known to have intervening sequences (IVS) within the coding regions, so the lengths of the mature RNAs are the same as those of the coding regions, although some species of insect and protozoa do have an IVS within a subset of their 25S genes (Appels and Honeycutt, 1986), and recently an IVS was identified within the 18S gene of *Pneumocystis carinii* (Edman et al., 1988). In *Drosophila*, the genes with the intervening sequences are not expressed (Long and Dawid, 1980), but in *Tetrahymena* the precursor rRNA acts as a catalyst for splicing out the IVS to form the mature rRNA (Cech, 1983).

In the rDNA of rice and cucumber, ITS1 is 194 and 229 bp and ITS2 is 233 and 245 bp, respectively (Hemleben et al., 1988). No ITS length variation was detected within species of broad bean or pea, but comparisons between different legume genera showed some slight variation in length (Jorgensen and Cluster, 1988).

The length of the intergenic spacer ranges from 1 to 8 kb in plants thus far examined (Jorgensen and Cluster, 1988). The IGS heterogeneity accounts for the interspecific range of 8 to 15 kb in repeat unit length (Hemleben et al., 1988). The IGS may also show considerable length variation within populations of one species, within individuals of a population, and even within individual chromosomal loci (Schaal and Learn, 1988).

Intraspecific variation in IGS length is caused by the presence of varying numbers of subrepeats in the middle region of the IGS. In most plant species, the subrepeats range from 100–200 bp. In species of wheat, barley, and broad bean, the subrepeats are 130 bp, 115 bp, and 325 bp (consisting of two copies of a 155-bp repeat and an unrelated 14-bp fragment), respectively (Appels and Dvorak, 1982; Saghai-Maroof et al., 1984; Yakura et al., 1984). Samples of wheat have shown heterogeneity for IGS length between individuals of a population, each variant differing from the others by a multiple of 130 bp (Appels and Dvorak, 1982). In broad bean, individual plants can exhibit as many as 20 different size classes of IGS each differing by a multiple of 325 bp. The broad bean only has one chromosomal locus for rDNA, so the heterogeneity must occur among neighboring repeat units (Rogers et al., 1986). Not all species show length heterogeneity, however. Soybean and *Lisianthius skinneri* have shown no variation within their rDNA for repeat unit size (Doyle and Beachy, 1985; Sytsma and Schaal, 1985). The mechanism for the variation in IGS length presumably is unequal crossing over within an individual repeat unit.

Nuclear Ribosomal DNA Sequence Variation

Within the coding regions of the small (16S-like) and large (26S-like) rRNAs are stretches of nucleotides conserved across all species examined from bacteria to yeast, plants, and animals (Gerbi, 1985). Other regions of the small and large rRNA primary sequence are conserved only between related species, whereas a certain fraction of the rRNA is not conserved to any significant extent. In some of the areas where the primary sequence is divergent, computer modeling and chemical probing have shown that the secondary structures of the rRNA molecules are conserved. Both the small and large rRNA molecules have areas of base-paired nucleotides that form stems; at the ends of these stems lie single-stranded loops. It is believed that this core secondary structure is maintained through selection by the stringent requirements of protein synthesis (Gerbi, 1985). In the double-stranded stems, there may be compensatory mutations that restore base pairing after one nucleotide of the pair mutates (Wheeler and Honeycutt, 1988).

Comparisons of the small and large rRNA molecules of bacteria and various eukaryotes have revealed that the differences in length (e.g., 2,500 for the *E. coli* 23S and 3,300 for rice 26S) can be accounted for by the insertion of so-called expansion segments within the bacterial sequences (Clark et al., 1984). The expansion segments are usually found in the same location in the rRNA of eukaryotes, but their lengths and sequences are not conserved (Gerbi, 1985).

The 5.8S sequences are conserved at the same level as the other coding regions: sequencing has shown that there is a difference of only 1 bp between pea and broad bean and a difference of 2 bp between pea and lupine (Jorgensen and Cluster, 1988). The sequences of the internal transcribed spacers are much more divergent. Comparisons of ITS1 of pea and broad bean showed one region of 16% to 18% difference and the remainder at 55% difference. The second ITS was constructed similarly with two regions of different levels of conservation (Jorgensen and Cluster, 1988). The two levels of conservation could reflect the presence of processing signals within the ITS regions, perhaps for the posttranscriptional modifications.

The intergenic spacer is by far the most divergent part of the rDNA, making it useful for microevolutionary phylogenetic comparisons. The sequences of the subrepeats within the IGS are substantially conserved within a species, although not necessarily identical. Sequencing the broad bean subrepeats indicated that only five or fewer of the 325 nucleotides were not conserved through all copies of the subrepeat (Yakura et al., 1984). Interspecifically, there is generally little conservation of subrepeat structure, although some similarity has been detected between wheat and maize subrepeats (Schaal and Learn, 1988). It is possible that the subrepeats function as hot spots for recombination or possibly as enhancers of transcription (Rogers and Bendich, 1987). In *Xenopus*, the subrepeats within the IGS have been shown to possess enhancer activity: they increase the level of transcription from downstream coding regions irrespective of their orientation (Reeder, 1984).

The region downstream from the subrepeats, which contains the ribosomal gene promoter, shows little interspecific conservation; only short stretches can be found among closely related species. Sequence comparisons from different taxa have shown that there does not seem to be a consensus sequence analogous to the TATA box of genes transcribed by RNA polymerase II. In animal systems, it has been shown that RNA polymerase I (pol I) of one species generally is incapable of transcribing the rDNA from another species (Grummt et al., 1982). This stands in stark contrast to RNA polymerase II transcription, in which yeast can faithfully transcribe mammalian genes. The lack of sequence conservation and the species-specific nature of pol I transcription indicate that the promoter region of the rDNA IGS has been evolving rapidly and that pol I must be coevolving at a similar rate (Flavell, 1986).

Chloroplast Ribosomal DNA Organization

The typical plant chloroplast genome exists in multiple copies of a circular molecule of approximately 150 kb (Whitfeld and Bottomley, 1983). In all angiosperm plant families but one examined to date, the chloroplast genome contains a pair of inverted repeats approximately 22–25 kb long. One copy of the rDNA repeat unit lies within each of the inverted repeats. In certain members of the legume family (Fabaceae), one copy of the inverted repeat has been lost and the other copy retained in its entirety, so that these plants contain only one copy of the rDNA locus (Palmer, 1985). Within the gymnosperms, the inverted-repeat structure is missing in the conifers, but present in *Ginkgo* (Strauss et al., 1988; Palmer, 1985).

The rDNA repeat unit of the chloroplast genome is more similar to that of prokaryotes than to that of eukaryotes, reflecting the presumed symbiotic origin of chloroplasts from a cyanobacteria-like ancestor. The repeat unit consists of a 16S gene, a transcribed spacer, a 23S gene, and a linked 5S gene; it does not contain a 5.8S gene. Appended to the 23S gene of the chloroplasts of higher plants is a coding region for a 4.5S rDNA, which also functions as part of the chloroplast ribosome. The 16S, 23S, and 4.5S genes are transcribed as one unit and later cleaved to form the mature rRNAs. The 5S gene is thought to be independently transcribed from a promoter lying between the 4.5S and 5S genes (Palmer, 1985).

The chloroplast rDNA shows the same patterns of conservation of sequence seen in the nuclear genes. The coding regions are the most invariant, followed by the transcribed spacers and the intergenic spacers. In general, the chloroplast genome is thought to be evolving at a slower rate than the nuclear genome

(Palmer, 1985). In the chloroplast genomes of tobacco and maize, the genes coding for the 16S rRNA are 1,486 bp and 1,490 bp long, respectively (Tohdoh and Sugiura, 1982; Schwartz and Koessel, 1980), and the genes coding for the 23S rRNA are 2,804 and 2,881 bp, respectively (Takaiwa and Sugiura, 1982; Edward and Koessel, 1981). The 4.5S genes of tobacco and maize chloroplast rDNA are 101 and 95 bp, respectively (Takaiwa and Sugiura, 1982; Edwards et al., 1981).

Experimental Procedure

Here we present a brief description of our experimental procedure for direct rRNA sequencing. We have described it in detail elsewhere (Hamby et al., 1988). We isolate total RNA from fresh, quick frozen, preferably young, plant tissue with either a guanidinium isothiocyanate buffer or a hot borate buffer procedure. Yields are in the range of 40 to 400 μg per gram of fresh tissue. Variation in yield occurs with the age of the plant material, the degree of cellularization of the starting tissue, and the degree of contaminating secondary metabolites. Alignment of published rRNA small and large subunit sequences has identified regions absolutely conserved throughout evolution (Gerbi, 1985). We synthesize oligonucleotide primers 18–20 bases long that will anneal specifically to these conserved regions and use the primers with reverse transcriptase and dideoxynucleotide triphosphates to make complementary DNAs. We then separate the DNA strands by electrophoresis on polyacrylamide gels to determine the rRNA nucleotide sequence. We have used primers from five different regions of the 18S gene (18E, 18G, 18H, 18J, and 18L) and from three different regions of the 26S gene (26C, 26D, and 26F) (see Hamby et al., 1988, for sequences of the primers). The relative location of each primer is shown in Fig. 4.2. Each primer yields about 225 bases of sequence data; our total number of sites determined is approximately 1,700 per species. We have found a few regions that we cannot sequence through, presumably due to secondary structure interference with the reverse transcriptase. These regions are eliminated from the comparisons used to produce evolutionary trees.

We have shown previously that it is possible to sequence chloroplast rRNAs directly by exactly the same procedure as above, that is, from the total RNA (Hamby et al., 1987). We aligned published sequences for chloroplast 16S and 23S rRNAs, identified the conserved sites and synthesized oligonucleotide primers complementary to these regions, making sure at the same time that these primers would not anneal to 18S or 26S nuclear rRNA. We are able to read about the same number of nucleotides per gel, and the autoradiographs are of the same overall quality.

The primary nucleotide sequences are aligned using the University of Wisconsin Genetics Computer Group package of programs (Devereux et al., 1984). We first GAP each new sequence against a common sequence (usually soybean or

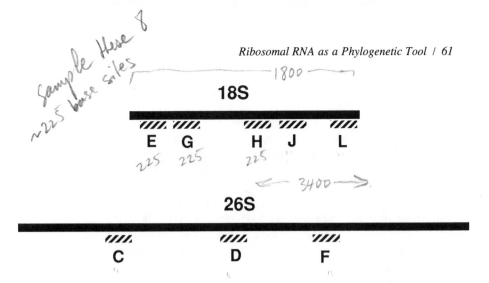

Figure 4.2. A schematic showing the location and the relative length of the regions sequenced in this study. The letters E, G, H, etc. refer to the primers used, e.g., 18E, 18G, 18H, etc. For complete details of the primer location and sequence, see Hamby et al. (1988).

rice); this program makes optimal pairwise alignments with the Needleman-Wunsch algorithm, inserting gaps into either sequence as necessary. We then take the resultant sequence (including any gaps), import it into LINEUP, and fine tune the alignments by visual inspection. LINEUP may display up to 31 sequences simultaneously.

We maintain a file for each species-primer combination (60 taxa times 8 primers equals 480 separate files) on the local VAX computer for archival purposes. Our files are organized into separate subdirectories, one for each different primer. Each file is named in the same manner: six or fewer letters to describe the species name followed by a three-character extension which names the primer. For example, we have a file called soy.18J in a subdirectory named 18J, soy.18L in a subdirectory named 18L, etc. The consistent use of this naming protocol makes file retrieval simple. After we GAP a sequence, we write the results to a new file in which we use up to nine characters to name the species and primer followed by the extension .GAP, e.g., soy18g.gap. This is the file that we import into LINEUP. At the end of the editing session, LINEUP renames the individual sequences by replacing the previous extension with .FRG, so it is important to have all descriptive information before the extension. Otherwise, the next time one calls up LINEUP, the program may not retrieve the correct files.

The complete aligned sequences, including the invariant positions, then are used as input into Swofford's PAUP 3.0 (1989), which calculates phylogenetic relationships based on the principle of parsimony. Appended to the 1,701 nucleotides for each species are 13 more characters representing the presence or absence of gaps, presumably due to insertion or deletion events, at particular sites within the sequences. Within the 1,701 aligned positions, the gaps are scored as missing so that they are not given double weight. Because our data set is so large (60 taxa, 1,714 positions), we are limited to using the heuristic search methods, although we may use the branch-and-bound option or the exhaustive-search option on subsets of our data.

The program Hennig86 (Farris, 1986), another parsimony program, which runs on the IBM PC, has been used to compare to PAUP. PAUP has a utility to convert data sets from its format to the format of Hennig86.

We also use the MacClade program package of Maddison and Maddison (1990) to aid in data analysis. We edit our aligned sequences with the MacClade editor and sometimes skip the LINEUP step and go directly to the MacClade editor to add new species to the alignments. The creators of MacClade have thoughtfully consulted with Swofford to create a common data format recognized by both programs (the NEXUS format). Our main use of MacClade has been for interactive rearrangement of phylogenetic trees and recalculation of tree parameters according to the new arrangement and for data management. Because we have a test version of MacClade, we present no results here based exclusively on MacClade; all of our results are from PAUP.

Results and Discussion

Plant Ribosomal RNA *Sequence Evolution below the Family Level*

We initially tested the ability of rRNA sequences to resolve relationships below the family level by comparing 18S and 26S sequences of members of the grass family, Poaceae (Hamby and Zimmer, 1988). Since the original publication of our results, we have refined the data set and the analyses. The genera represented are *Zea* (maize), *Tripsacum, Sorghum, Saccharum* (sugarcane), *Oryza* (rice), *Hordeum* (barley), *Avena* (oats), *Triticum* (wheat), and *Arundinaria* (bamboo); *Colocasia* (elephant's ear), another monocot in the order Arales, was used as an outgroup. Of the 1,714 positions aligned, 1,097 from the 18S gene, 604 from the 26S gene, and 13 gap sites, only 143 were variable (i.e., 1,571 sites were invariant). Of the 143 variable sites, 54 were specific to the outgroup, 26 were autapomorphies within the grasses, and six were variable but uninformative. The remaining 57 positions were variable and phylogenetically informative. Fifty-six of the informative sites changed via base substitution; only one of the gap positions was informative. At 33 of the variable and informative sites, the changes were restricted to transitions, whereas only transversions had occurred at 14 sites. At the other 9 sites, both transition and transversion events had occurred during the differentiation of these grasses.

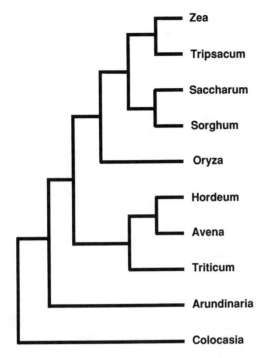

Figure 4.3. The most parsimonious arrangement of the nine grass genera based on rRNA sequence data. The length of the tree is 187 steps.

Although almost twice as many positions were sequenced within the 18S rRNA molecule, the 18S molecule had only slightly more variable sites (74 to 63) and fewer informative sites (22 to 34) in comparison to the 26S rRNA molecule. The most variable regions were those sequenced with 18E, 26D, and 26F. The region sequenced with 18G was the most conserved with only 8 variable sites among the 250 sequenced.

The aligned sequences were entered into PAUP and a heuristic search process found the tree shown in Fig. 4.3 to be the most parsimonious arrangement. Only one most parsimonious tree was found, with a length of 187 steps and an overall consistency index (excluding the contribution due to the autapomorphic sites) of 0.695. The branch separating the ingroups from the outgroup contributes 65 of the steps on the tree. Both a branch-and-bound (Hendy and Penny, 1982) search and an exhaustive search in which all possible topologies are tested, found the same shortest tree shown in Fig. 4.3. In our most parsimonious tree, *Arundinaria* branches first off the tree, leaving the other eight taxa to form a natural group. This group is split into two smaller monophyletic groups; one contains *Triticum*, *Avena*, and *Hordeum*; the other consists of *Oryza, Zea, Tripsacum, Sorghum*, and *Saccharum. Zea* and *Tripsacum* form a natural group as do *Sorghum* and *Saccharum*. These four genera also form another monophyletic group. There is one tree of 188 steps differing only in the placement of *Sorghum* and *Saccharum* relative to one another: instead of forming a monophyletic group, they form

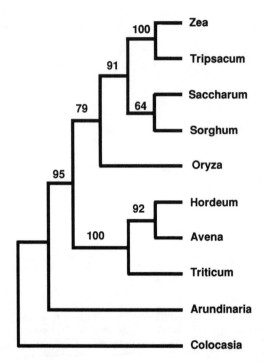

Figure 4.4. Results of 250 bootstrap replications of the grass rRNA sequence data. Nodes are labeled by the percentage of trees which support that node.

a grade with *Sorghum* between *Saccharum* and the node leading to *Zea* and *Tripsacum*.

Two hundred fifty bootstrap (Felsenstein, 1985) replications were performed on the grass data to see which monophyletic groups were best supported by the rRNA sequence data. The results of the bootstrap are shown in Fig. 4.4. Each node is labeled with the percentage of times out of 250 that the best tree(s) contained these nodes. In every bootstrap replication, *Zea* and *Tripsacum* were placed together as a monophyletic group, as were *Hordeum, Avena,* and *Triticum.* Ninety-five percent of the time, *Arundinaria* was placed as the most primitive of the grasses, with all the other grass taxa forming a monophyletic group. These are the only statistically significant groupings on the tree at the 95% confidence level.

We used the neighbor joining program of Saitou and Nei (1987) to compare a phenetic analysis to our cladistic one. The nucleotide sequence data were converted to pairwise distances by dividing the number of variable positions by the number of sites compared between each different pair of taxa, the overall dissimilarity, a method considered valid for species not separated by great evolutionary time (Nei, 1987). The distances were alternatively calculated by the Jukes-Cantor method (1969), which compensates for multiple mutations at the same locus (position), and by the Kimura (1980) two-parameter model, which gives more weight to less frequent transversions. Regardless of which distances were used, the topology of the resulting phenogram was the same as our cladogram.

In comparing our results to traditional classifications of the grass family, we find congruence between our most parsimonious tree and the classifications of Gould and Shaw (1985) who recognize six different subfamilies. Based on morphologic and nonmorphologic (e.g., biochemical and genetic) similarities, they propose the subfamilies Pooideae, Panicoideae, Chloridoideae, Bambusoideae, Arundinoideae, and Oryzoideae. They place *Zea, Tripsacum, Sorghum,* and *Saccharum* within Panicoideae; *Avena, Triticum,* and *Hordeum* within Pooideae; *Oryza* within Oryzoideae; and *Arundinaria* within Bambusoideae. Our tree is consistent with this scheme except that we place *Avena* and *Hordeum* as more closely related to one another than either is to *Triticum,* whereas Gould and Shaw place *Hordeum* and *Triticum* in the tribe Triticeae and *Avena* in Aveneae.

Our tree does not support the classification of Watson et al. (1985) who recognize only five subfamilies of grasses. They place the genera of the Oryzaneae as a tribe within Bambusoideae. Under their scheme, our results would show the Bambusoideae (represented by *Oryza* and *Arundinaria*) to be a paraphyletic group. The other groupings within our most parsimonious tree are consistent with their arrangement, except again for the relationship of *Triticum, Hordeum,* and *Avena.*

Wolfe et al. (1989) compared the sequences of three chloroplast genes of certain members of the grass family and found that the Panicoids grouped together and that the Pooids grouped together. They were unsure of the branching order of *Oryza* relative to the two other subfamilies.

The three classifications mentioned above are all based on phenetic analyses from which one cannot necessarily infer an evolutionary relationship. A cladistic analysis of the same characters used by Watson et al. (1985) showed that the Pooideae, Panicoideae, and Bambusoideae (including the tribe Oryzeae) were monophyletic assemblages (Kellogg and Campbell, 1987). Again, our results are not consistent with theirs with respect to *Oryza* and *Arundinaria*, but we do show the Pooideae and Panicoideae as represented by the genera in our analysis to be monophyletic groups.

Our results are congruent with another recent cladistic analysis of molecular sequence data within the grasses (Doebley et al., 1990). In this study, DNA sequences of the chloroplast-encoded *rbc*L gene, which codes for the large subunit of ribulose bisphosphate carboxylase, were compared among Panicoids, Pooids, and *Oryza* (Doebley et al. do not have any bamboo species in their analysis). Figure 4.5 shows that the monophyletic groups in their most parsimonious tree and their maximum-likelihood tree were the same as those in ours and showed the same relative branching order.

For the most part, we have been able to show that rRNA sequences can resolve relationships within a family, at least at the subfamily level, whose age is on the order of 50–70 myr (Wolfe et al., 1989). At the tribal level, our rRNA data did

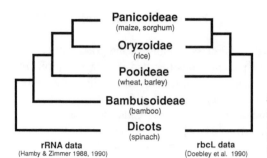

Panicoideae
(maize, sorghum)

Oryzoidae
(rice)

Pooideae
(wheat, barley)

Bambusoideae
(bamboo)

Dicots
(spinach)

rRNA data
(Hamby & Zimmer 1988, 1990)

rbcL data
(Doebley et al. 1990)

Figure 4.5. A comparison of the most parsimonious rRNA tree and the parsimony and maximum-likelihood tree of Doebley et al. (1990) based on sequences from the *rbc*L gene of various grass taxa.

not group the two members of the Triticeae together relative to *Avena*. The rRNA sequence data do not place *Oryza* with *Arundinaria*, although *Oryza* has moved throughout the tree as we added more data during the course of our project. It is possible that we simply do not have enough informative sites yet to place *Oryza* unequivocally and that we need to add more representatives of Oryzoideae and Bambusoideae. Preliminary results after the addition of *Secale, Brachyelytrum,* and *Diarrhena* show the positions of *Oryza* and *Arundinaria* to be unaffected (Issel et al., 1990).

Plant Ribosomal RNA Sequences and Seed Plant Evolution

Our main interests lie in determining the early evolutionary history of the flowering plants including the relationship between the gymnosperms and the angiosperms. Specifically we are asking these questions: (1) Are the gymnosperms a monophyletic group? (2) Are the angiosperms a monophyletic group? (3) What is the relationship of the Gnetales to the angiosperms? (4) Among the angiosperms, do the monocots and the dicots each form monophyletic groups? (5) Among angiosperms, which lineage(s) is/are basal? We have sequenced nuclear rRNAs of representatives of more than fifty extant seed plant genera. The gymnosperm representatives include *Ginkgo*, three conifers, three cycads, and all three genera of Gnetales (*Welwitschia, Gnetum,* and *Ephedra*). The flowering plants are represented by 29 dicot and 17 monocot genera. As outgroups we used *Equisetum* (horsetail) and *Psilotum*, two seedless vascular plants. The RNA sequences were determined for the same regions described above for the experiments within the grass family.

Comparisons of the sequence data for all 60 taxa are summarized in Table 4.2. Of the 1,701 nucleotide sites from the 18S and 26S rRNA molecules, 1,097 were constant and 604 were variable. Only 417 of the 604 variable sites were phylogenetically informative, the remainder were autapomorphies, the large majority of which occurred within the ingroups. All 13 gap sites were variable and informative. Thirty percent of the 18S sites were variable, and 20% of the 18S

Table 4.2. A summary of the rRNA sequence data for the 60 taxa in this study.

Primer	Region	Sites	Variable	Tn[a]	Tv[b]	MH[c]	Informative	Tn[a]	Tv[b]	MH[c]
18E	90–308	191	98	29	14	55	77	18	8	51
18G	300–554	250	52	19	14	19	37	11	8	18
18H	910–1134	215	41	18	8	15	25	9	3	13
18J	1210–1429	214	66	22	17	27	43	15	6	22
18L	1535–1766	227	75	35	20	20	42	21	7	14
18s subtotal		1097	332	123	73	136	224	74	32	118
26C	740–949	202	58	17	11	30	38	8	2	28
26D	1625–1836	202	108	41	25	42	80	30	14	36
26F	1960–2172	200	106	46	12	48	75	32	2	41
26s subtotal		604	272	104	48	120	193	70	18	105
18s+26s		1701	604	227	121	256	417	144	50	223
Gaps		13	13	—	—	—	13	—	—	—
TOTAL		1714	617	227	121	256	430	144	50	223

[a]Transition.
[b]Transversion.
[c]Multiply hit.

sites were informative. Forty-five percent of the 26S sites were variable, and 32% were informative. More than half of the variable sites and more than half of the informative sites were multiply hit (both transitions and transversions had occurred at these sites). The overall ratio of transitions-to-transversions was 1.9 to 1 in the variable sites, but within the informative sites there was a transitions-to-transversions ratio of approximately 3 to 1. The most variable regions were those sequenced with the 18E, 26D, and 26F primers. It is interesting to note that secondary structure calculations (Gerbi, 1985; Gutell and Fox, 1988) predict that the 18E region and the 26F region are within expansion segments (Clark et al., 1984); the other primer regions lie completely or mostly in regions of more conserved structure. Therefore, sequencing of additional regions in expansion segments offers the potential for higher resolution at lower taxonomic levels.

The number of taxa in the data set is so large that the only available tree inference option in PAUP is the heuristic search. Using the tree bisecting and reconnection swapping option and the simple sequence addition option, PAUP found the shortest tree to be 1,870 steps with an overall consistency index of 0.390. There were at least 20 different variations of the shortest tree. When the search was started again and an option was chosen in PAUP to save all trees that were one step longer than the shortest tree (i.e., to save the trees of length 1,870 and 1,871 steps), PAUP actually found seven trees that were 1,869 steps. Normally PAUP only performs branch swapping on trees of minimal length, and, in the case of the first search, these were trees of 1,870 steps. However, the second search showed that swapping on a nonminimal tree (1,871 steps) can lead ultimately to trees that are actually shorter. This second search, which was terminated after 5 days, found 2,358 trees that were 1,871 steps, 259 trees that were 1,870 steps, and 7 that were 1,869 steps.

The data were then converted by PAUP into a format for input into Hennig86. Surprisingly, Henning86 found two trees that were 1,867 steps, two steps shorter than the shortest trees found by PAUP. In past experiments with as many as 57 taxa, Hennig86 had found the same shortest trees as had been found by PAUP. Hennig86 also found two trees that were 1,868 steps long. One of the two trees of 1,867 steps (the most parsimonious) was then used as a starting topology for branch swapping in PAUP to see if PAUP could find other trees of 1,867 or if PAUP could rearrange the 1,867-step tree to a still shorter tree. PAUP could only find the other tree of 1,867 steps found by Hennig86. However, using the two trees of 1,868 steps as beginning topologies in separate PAUP searches ultimately identified 30 trees that were 1,868 steps long. Several of the 1,869-step trees were used as beginning swapping points in later PAUP searches, and all the resulting trees combined into one large file. The condense option of PAUP was used to make sure that all the trees were unique, and, after condensation, a total of 3,413 trees with overall lengths between 1,867 and 1,871 were found. Memory limits of the Macintosh computers (4.5 Mbyte) prevented searching any further out, there being too many trees five steps longer than the shortest trees. The 3,413

tech: limitations
di not sure I
understand

trees break down into 2,358 at 1,871 steps, 666 at 1,870 steps, 357 at 1,869 steps, 30 at 1,868 steps, and 2 at 1,867 steps. PAUP was unable to find any more trees of 1,868 or 1,867 steps, but there are more trees other than those already identified at lengths greater than 1,868 steps. This is certain because in the search for trees less than 1,871 steps, the program was terminated while swapping on tree #748 out of the more than 3,000 saved. Unfortunately, trees were being accumulated at the rate of about 500 a day, but PAUP was only able to swap on about 150 a day (the run was stopped on the fifth day) and PAUP became in danger of running out of memory, in which case all trees would have been lost.

One of the two shortest trees of 1,867 steps is shown in Fig. 4.6. The only difference between the two shortest trees is in the placement of *Sorghum* relative to *Saccharum*: in one tree they form a monophyletic group that is the sister group to the group that contains *Zea* and *Tripsacum*; in the other, they form a grade with *Sorghum* between *Saccharum* and the monophyletic grouping of *Zea* and *Tripsacum*. All other features of the two most parsimonious trees are identical, and, in the discussion that follows, all references to the most parsimonious tree are understood to include both variations.

In the most parsimonious tree, the gymnosperms are divided into two separate natural groups: one of these groups consists of the three genera of the order Gnetales, and the other is composed of the three conifers (*Pinus, Cryptomeria*, and *Juniperus*), *Ginkgo*, and the four cycads (*Cycas, Encephalartos*, and two *Zamias*). The three conifers form a natural group according to the rRNA sequences, as do the four cycads, and the conifers and cycads together form another monophyletic group. According to this arrangement, the gymnosperms are not a monophyletic assemblage of taxa, because the most recent common ancestor of all gymnosperms is also an ancestor of the angiosperms. This is not surprising, nor in conflict with traditional views of the origin of the angiosperms, which hold that the flowering plants are derived from within the gymnosperms (Cronquist, 1968; Takhtajan, 1969; Stebbins, 1974). The biological interpretation of the most parsimonious rRNA trees is in accordance with the view that the angiosperms arose from the gymnosperms.

Within the Gnetales, the rRNA data indicate that *Welwitschia* and *Gnetum* are more closely related to one another than either is to *Ephedra*, in agreement with the morphologic analyses of Crane (1985) and Donoghue and Doyle (1989). That the Gnetales themselves are a coherent natural group is unquestionably confirmed by the rRNA sequence data. In bootstrap tests with subsets of the data (20 taxa were eliminated in the interest of time), the Gnetales were grouped together in 99 out of 100 replications. Within the other gymnosperm clade, the rRNA data suggest that cycads and conifers are more closely related to one another than either is to *Ginkgo*. The morphologically based cladistic analyses do not agree with this placement, putting *Ginkgo* and the conifers into a monophyletic group, coniferopsids. An advantage enjoyed by these morphologic treatments is the inclusion of numerous fossil taxa, which has been shown to affect the placement

of extant taxa (Donoghue et al., 1989). A preliminary examination of the morphologic data for just the extant seed plant lineages concurs with the most parsimonious rRNA trees (Donoghue et al., unpublished data). Therefore, it is possible
that the relative placement of *Ginkgo* and cycads would change in the rRNA tree,
if fossil sequences were available.

In the shortest trees, the Gnetales are the earliest diverging seed plants and the
other gymnosperms (conifers, cycads, and *Ginkgo*) are the sister group of the
angiosperms. These results are not in concordance with cladistic analyses of
morphological data, in which Crane (1985) and Donoghue and Doyle (1989)
separately found that of the extant gymnosperms, the Gnetales were most closely
related to the flowering plants, united with them by such characteristics as reduced
gametophytes and vascular structure. In earlier analyses of rRNA sequences with
fewer taxa, we have almost always found the Gnetales to be the sister group of
the flowering plants, although this placement is normally favored by only one or
two steps over the placement of the other gymnosperm clade as sister to the
angiosperms. Our most parsimonious tree with 60 taxa can be rearranged to place
the Gnetales as the sister group to the angiosperms with a penalty of only one
step (Fig. 4.7). When the Gnetales are placed as the sister group of the angiosperms, the node which unites the Gnetales and angiosperms is supported by
more characters (17 versus 12) than the node in our most parsimonious tree,
which unites the conifer-cycad-*Ginkgo* clade with the angiosperms. In addition,
the characters that support the Gnetales and flowering plants in Fig. 4.7 are less
homoplasious than those that support the alliance of the other gymnosperms with
the flowering plants in Fig. 4.6 (0.410 versus 0.389).

Within the flowering plants, the cladistic analysis of the rRNA sequences
places members of the order Nymphaeales at the base of the angiosperm radiation
followed next by members of the order Piperales. In the shortest tree, the genera
of Nymphaeales, which represent the earliest divergence of the angiosperms,
include *Nymphaea, Nuphar, Cabomba*, and *Barclaya*, but not *Ceratophyllum* or
Nelumbo, which the rRNA data place in a different position. The former four
genera constitute a natural group without *Ceratophyllum* and *Nelumbo* in 3,413
trees up to four steps longer than the most parsimonious tree. In the bootstrap
with 40 taxa, *Barclaya* and *Nymphaea* were included, and they were placed
together in 100% of the replications; *Ceratophyllum* and *Nelumbo* were also
included in this bootstrap, and they were never grouped with *Barclaya* and
Nymphaea, nor were they placed with one another a significant number of times.
In another cladistic analysis (Donoghue and Doyle, 1989) the families containing
Nymphaea and *Cabomba* formed a natural group that did not include *Nelumbo*
(*Ceratophyllum* was not included in Donoghue and Doyle's study). A cladistic
treatment of morphologic characters of genera within the Nymphaeales (Ito,
1987) found *Nelumbo* to be distinct from the other members of the order and
found *Cabomba* to be more closely related to *Ceratophyllum* than to any of the
other genera of Nymphaeales. There are morphologic characteristics that support

the separation of *Nelumbo* from the Nymphaeales, notably the pollen of *Nelumbo* which is triaperturate, whereas the pollen of all other Nymphaeales is monosulcate, and Takhtajan (1969) does place *Nelumbo* in a separate order. The rRNA data then are consistent with cladistic morphologic treatments and some traditional classifications insofar as placing *Nelumbo* as separate from *Nymphaea, Cabomba, Nuphar*, and *Barclaya*, but not with respect to the placement of *Ceratophyllum*. It is possible that the addition of *Brasenia* will help to unite *Ceratophyllum* with the other Nymphaeales, because *Brasenia, Ceratophyllum*, and *Cabomba* constitute a natural group in Ito's (1987) analysis.

After Nymphaeales, the next branch to diverge from the rRNA tree leads to a natural grouping of the members of the order Piperales (in the sense of Takhtajan, 1969). In the rRNA tree, the genera *Piper, Peperomia*, and *Saururus* are united, and *Chloranthus*, which is considered by Cronquist (1968) to be a member of the Piperales, is placed elsewhere in the tree. The rRNA tree supports Takhtajan (1969) and Thorne (1976) who separate *Chloranthaceae* from the rest of the Piperales. The cladistic morphologic treatment of Donoghue and Doyle (1989) also separates Chloranthaceae from Piperaceae and Saururaceae.

The remaining 39 angiosperm taxa form two monophyletic sister groups. One of these groups contains all the monocot taxa plus *Nelumbo* and *Ceratophyllum*. Were it not for the presence of these two taxa, the monocots would constitute a natural group derived from within the dicots. With these water lilies present, the monocots cannot be considered a natural group. Within the monocots, the nine grasses are placed together in the same arrangement found in the analysis of the grasses alone. *Sabal* and *Hosta* form a natural group based on 18 shared characters, but according to traditional classifications, *Sabal* is more closely related to the two members of the family Araceae (*Colocasia* and *Pistia*), which form a natural group. The four aquatic monocots (*Sagittaria, Echinodorus, Najas*, and *Potamogeton*) also form a monophyletic group. *Nelumbo* and *Ceratophyllum* form a grade near the aquatic monocots. The rRNA data support the resemblance of these groups based on their aquatic habit and suggest that the first monocots were aquatic. Several key monocot lineages have not been sampled yet, so this remains a preliminary conjecture.

The other group of derived angiosperms (relative to Nymphaeales and Piperales) consists of the other members of the Magnoliidae subclass, as well as the other dicot subclasses. The two genera from Aristolochiales (*Aristolochia* and *Saruma*) are placed together in a natural group, as are three of the four members of the Magnoliales (*Magnolia, Liriodendron*, and *Asimina*). *Drimys*, the fourth Magnoliales, has never been placed close to any other member of its order in the rRNA trees until the recent addition of another species of *Drimys, D. aromatica* (Suh, personal communication). In phylogenetic analyses that include both *D. aromatica* and *D. winterii*, the two are allied and have moved closer to the rest of the order. The two legumes (*Glycine* and *Pisum*) form a natural group, but *Duchesnea* and *Petroselinum*, the other genera of the subclass Rosidae, do

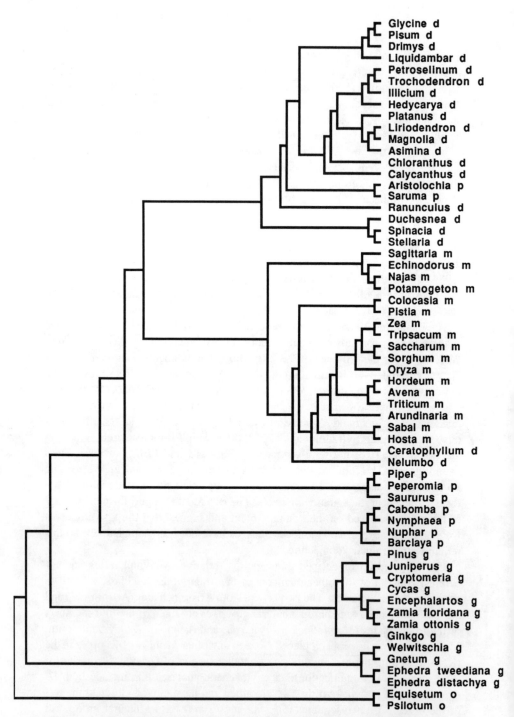

Figure 4.6. The most parsimonious tree found by Hennig86 for 60 taxa based on rRNA sequence data. The tree length is 1,867 steps with an overall consistency index of 0.39. After each taxon name is one letter indicating whether the genus is a dicot (d), monocot (m), paleoherb (p), gymnosperm (g), or outgroup (o). All taxa labeled as paleoherbs in this figure are dicots (see text for further discussion of the paleoherbs). It is also understood that all monocots are considered to be paleoherbs.

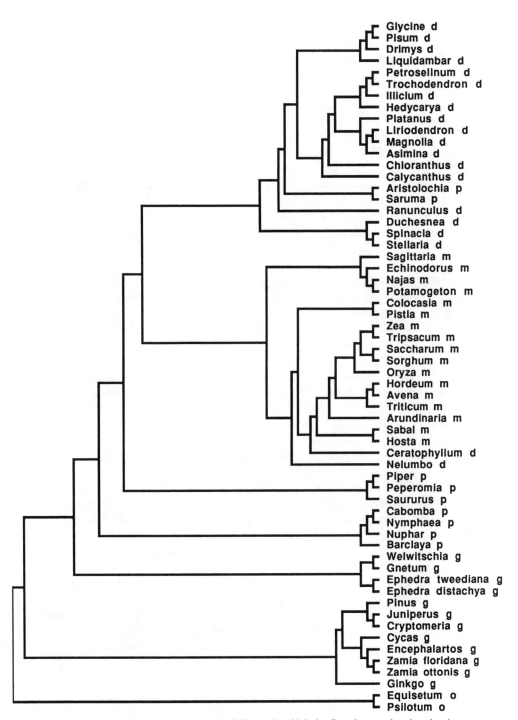

Figure 4.7. An alternative arrangement of 60 taxa in which the Gnetales are placed as the sister group of the flowering plants. The length of this tree is 1,868 steps, i.e., one step longer than the most parsimonious tree. The consistency index is 0.39.

not form a natural group with the legumes. The two genera of the subclass Caryophyllidae (*Stellaria* and *Spinacia*) form a monophyletic group. Much of the resolution within the remaining dicots is poor. Many of the branches are supported by few characters, and many of these characters are quite homoplasious. The various members of the subclass Hamamelidae (*Trochodendron, Platanus*, and *Liquidambar*) are paraphyletic according to the rRNA data, as are the members of the subclass Magnoliidae (Magnoliales, *Hedycharya*, and *Calycanthus*) and the subclass Ranunculidae (*Illicium, Ranunculus*, and *Chloranthus*). Donoghue and Doyle (1989) also found the Magnoliidae and Ranunculidae to be paraphyletic in their analysis, but they did find the Hamamelidae to group together. It is possible that the addition of other taxa closely related to those whose positions are inconsistent in the present analyses, that is, better sampling of the tree of higher dicots, will result in a more stable topology.

Above, we alluded to bootstrapping tests over all 60 taxa. Except for the consistent placement of the Gnetales and the flowering plants as well-supported monophyletic groups, no other conclusions could be drawn from the bootstrapping. The analysis suffered from the need to exclude some taxa in the interest of time (it required 21 days to bootstrap 100 replications on 40 of the 60 taxa), and the large amount of homoplasy in the data set. As an alternative to bootstrapping, we constructed majority-rule consensus trees by combining less parsimonious trees with the most parsimonious ones. We were able to collect trees up to four steps longer than the shortest tree, and to construct the consensus trees shown in Fig. 4.8. In the figure, key nodes are labeled with the percent of the contributing trees supporting that node. Figure 4.8A is the majority-rule consensus of the two trees that are 1,867 steps and the 30 that are 1,868 steps. Figure 4.8B is a majority-rule consensus of the 32 trees less than or equal to 1,868 steps and the 357 trees found at 1,869 steps. The majority-rule consensus trees shown in Fig. 4.8C and 8D were calculated from 1,055 and 3,413 trees, respectively. The first nodes to disintegrate with the addition of less parsimonious trees, as reflected by the dissolution of dichotomous branching into polychotomous branching, were among the higher dicots. This is consistent with the apparently poor resolution in this part of the tree. There does not appear to be sufficient information contained within the current rRNA data set to resolve completely the relationships within the higher dicots. On the other hand, within the angiosperms, the nodes that are best supported are the ones that place the Nymphaeales, Piperales, and the monocot group near the base of flowering plant evolution. The position of the Gnetales relative to the angiosperms shifts about depending on which set of trees is used for the consensus, again indicating the weakness of the placement of either gymnosperm group as sister to the flowering plants.

The neighbor-joining analysis (Saitou and Nei, 1987) yielded different results based on the manner in which the sequences were converted to distances. When the distance was simply equivalent to the dissimilarity (number different/number compared), the phenogram in Fig. 4.9 was inferred. When the data were corrected for possible multiple changes at individual loci (Jukes and Cantor, 1969) and adjusted to give more weight to transversions (Kimura, 1980), the resultant phenograms shared the topology shown in Fig. 4.10. In comparing the two topologies, there are slight differences among the more derived taxa, depending on how the distances were determined, but several features of the two phenograms are consistent. In both topologies, the Gnetales are placed as the sister group of the flowering plants, and the remaining gymnosperms form another monophyletic group. At the base of the flowering plants lie the Nymphaeales and Piperales, although the Piperales are split into two separate lineages, one consisting of *Saururus* alone, and the other comprising *Piper* and *Peperomia*. The phenetic analysis supports the placement of some members of the paleoherbs as the first flowering plants, specifically the Nymphaeales and Piperales. It also supports the placement of the Gnetales as the sister group of the angiosperms. There is no provision for calculating phenograms other than deriving the shortest one, so it is not possible to investigate alternative arrangements of taxa with distance data.

Conclusions

We have collected rRNA sequences from 60 different taxa and analyzed them cladistically with PAUP and Hennig86. The rRNA data do not support Beck's (1981) contention that the seed plants arose through two different events: one of which gave rise to the cycads, seed ferns, and angiosperms, whereas the other event gave rise to the other gymnosperms. Nor do the rRNA data support theories that the seed plants arose once but that the cycads and angiosperms are more closely related to one another than either is to any of the other gymnosperms. If the rRNA data supported either of these proposals, the cycads and angiosperms would form a monophyletic group to the exclusion of the other gymnosperms. This is not the case in either of the most parsimonious trees, nor is this topology found in any of the 3,413 trees within four steps of the shortest tree. All the trees found within four steps of the most parsimonious tree unite the conifers, cycads, and *Ginkgo*.

The rRNA sequence data strongly support the theories of a single origin for the flowering plants. In the most parsimonious trees and all 3,413 trees found within four steps of the shortest tree, the angiosperms constitute a monophyletic group. In a bootstrapping trial with 40 of the 60 taxa, the flowering plants were placed in a single clade 100 out of 100 replications. The branch that leads to the common ancestor of all the flowering plants is supported by more characters (42) than all but two other internal branches on the phylogenetic tree; one of these is the branch that separates the seedless plants from the seed plants. The characters that support this branch have a lower level of homoplasy than any other internal branch on the tree except again for the branch that separates the ingroups from the outgroups. The rRNA data are in strong support of a monophyletic origin of flowering plants and consequently a single origin for each of the features, such

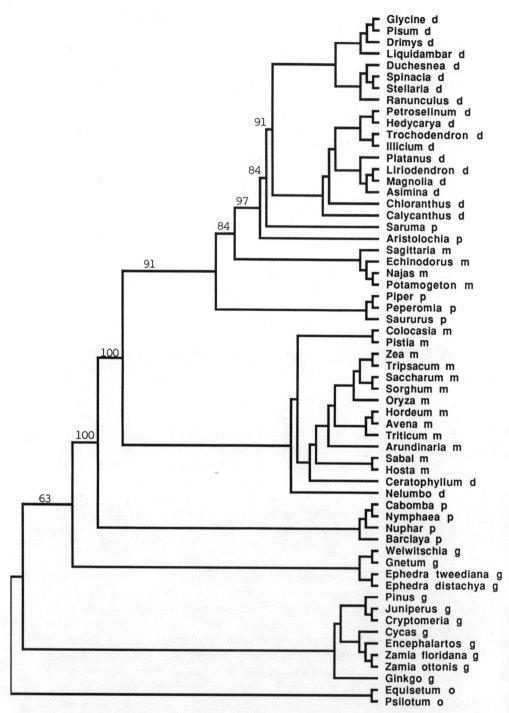

Figure 4.8A-D. Majority-rule consensus trees for all 60 taxa. Key nodes are labeled with the percent of trees which support the node.

A. Majority-rule consensus of 32 trees between 1,867–1,868 steps.

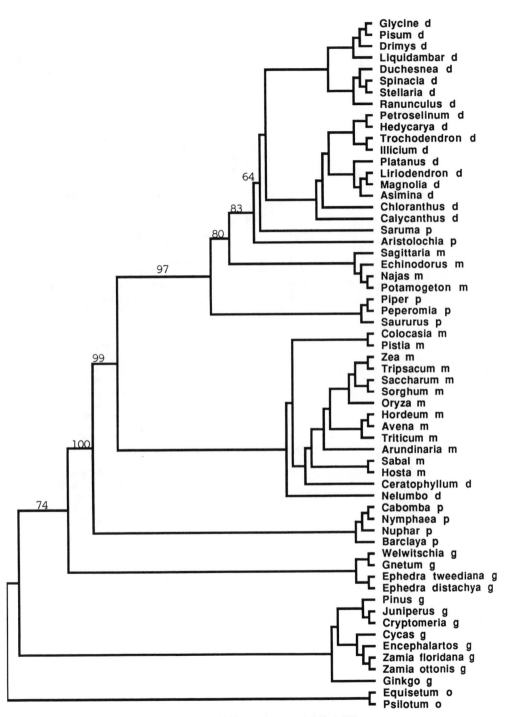

Figure 4.8B. Majority-rule consensus of 389 trees between 1,867–1,869 steps.

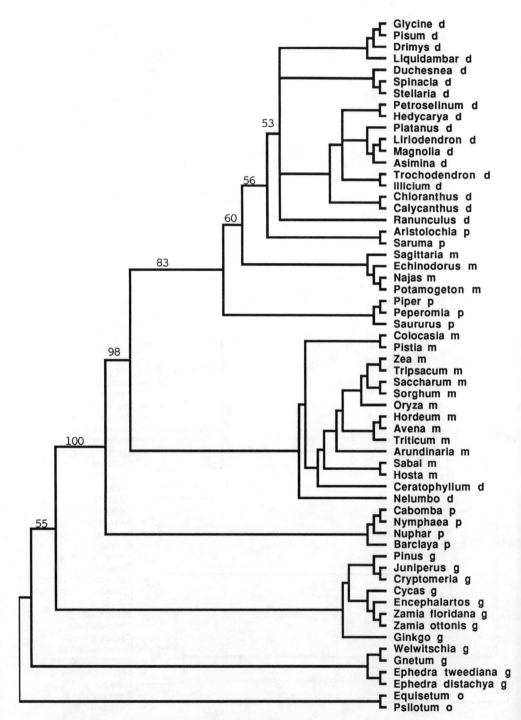

Figure 4.8C. Majority-rule consensus of 1,055 trees between 1,867–1,870 steps.

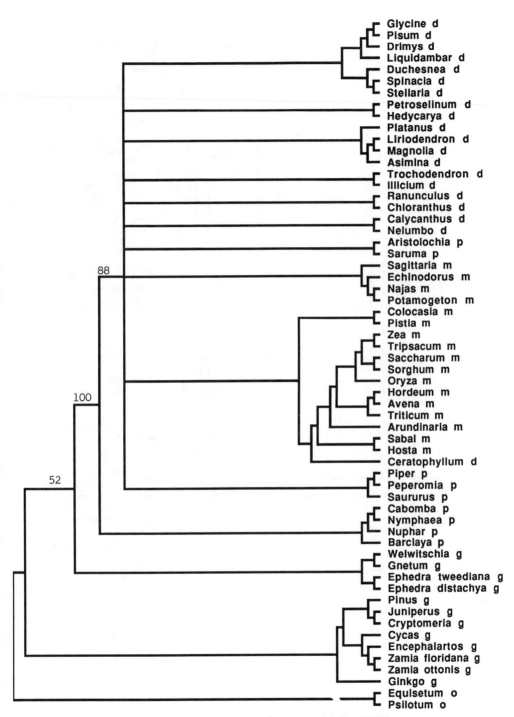

Figure 4.8D. Majority-rule consensus of 3,413 trees between 1,807 and 1,871 steps.

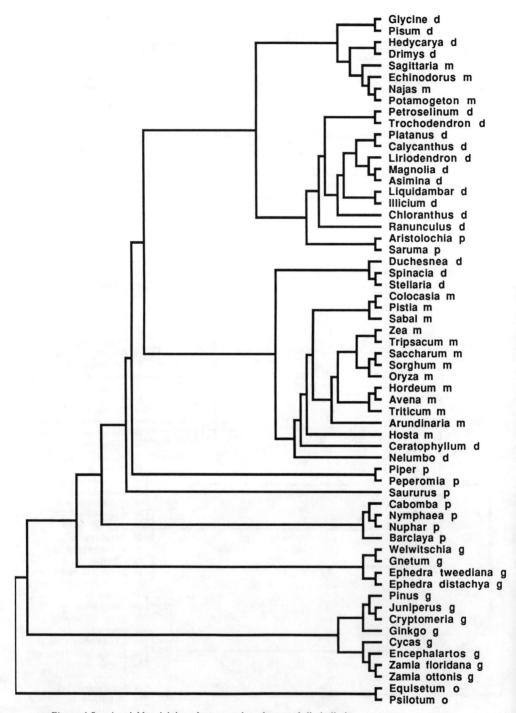

Figure 4.9. A neighbor-joining phenogram based on total dissimilarity.

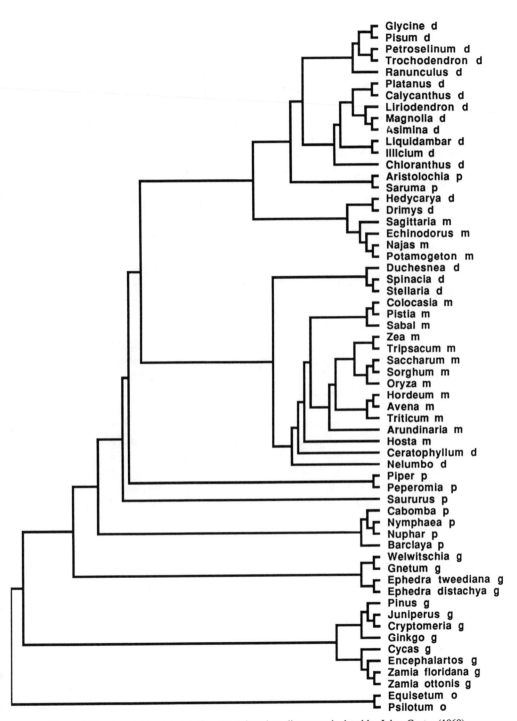

Figure 4.10. A neighbor-joining phenogram based on distance calculated by Jukes-Cantor (1969) formula or Kimura's (1980) two-parameter formula. The topologies of the two phenograms were identical irrespective of which formula was used to convert the nucleotide sequence data to distances.

as double fertilization, which are unique to the angiosperms, and the rRNA data refute theories of a multiple origin for the different groups of flowering plants (Meeuse, 1967). The rRNA data are consistent with a derivation of the flowering plants from one of the extinct seed fern lineages. However, if the flowering plants were derived from a seed fern group, it was not the same seed fern group that gave rise to cycads (unless all seed plants, or all seed plants except Gnetales, are descended from the same seed fern ancestor).

The rRNA data give strong support for the coherence of the Gnetales as a natural group. The most parsimonious rRNA trees do not place the Gnetales as the sister group of the angiosperms, although an insignificant penalty of one step is all that is required to reverse the position of the Gnetales and the remaining gymnosperms. In this alternative tree, the branch that unites the angiosperms with their most closely related gymnosperm (the Gnetales in this case) is supported by more characters and with less homoplasy than is the analogous branch in the most parsimonious tree. The Gnetales were often placed as the gymnosperms most closely related to angiosperms in preliminary analyses of rRNA sequences with fewer taxa, and they are also placed as sister to the flowering plants when there are 64 taxa (Suh, personal communication) and 72 taxa (Bult, personal communication). In addition, the distance analyses indicated that the Gnetales were the sister group of the angiosperms. Clearly, the placement of the Gnetales relative to the other gymnosperms and the angiosperms cannot unequivocally be resolved by the rRNA sequence data from eight primers alone.

The most interesting result of our rRNA sequence analysis is the support for certain members of the "paleoherb" group of Donoghue and Doyle (1989) at the base of the angiosperm radiation and the rejection of the traditional placement of the woody members of the Magnoliidae at the base. The paleoherbs include some herbaceous families of the subclass Magnoliidae: Piperaceae, Saururaceae, Nymphaeaceae, Cabombaceae, Aristolochiaceae, and Lactoridaceae, as well as the monocots. In their cladistic treatment of morphologic characters, Donoghue and Doyle found that the most parsimonious tree placed the woody members of the Magnoliidae at the base of the angiosperm radiation, but that with a penalty of only one step, the angiosperms could be rooted within Nymphaeales (Nymphaeaceae and Cabombaceae, but not Nelumboaceae), monocots, and Piperales (Piperaceae and Saururaceae, but not Chloranthaceae), all paleoherbs. This arrangement is very close to that of our most parsimonious tree and is not very different from the proposed arrangement of Burger (1977, 1981). The Nymphaeales and Piperales lie at the base of the angiosperm diversification according to rRNA sequence analysis. The rRNA sequences suggest that some members of the paleoherbs are the earliest diverging flowering plants, and that the rest of the angiosperms arose from within these paleoherbs. The basal arrangement of these paleoherb groups was supported by the most parsimonious tree and the majority of all trees up to four steps longer than the shortest tree. It was also supported by the distance analyses. Even in the tree in which the Magnoliales were forced to the base of the tree, the paleoherb groups were the next groups to evolve according to the rRNA data. The paleoherbs, according to the rRNA data, should no longer be referred to as a group, because they are not a natural assemblage. If they were a natural assemblage, and therefore, an appropriate group for plant classification systems, there would be one ancestor common to all paleoherbs that had no other descendents aside from the paleoherbs. This is not the case in the rRNA tree, because the common ancestor of the paleoherbs is also the common ancestor of the remaining flowering plants.

The rRNA data also suggest that neither the monocots nor the dicots are natural groups, and that the number of cotyledons on an emerging seedling axis should no longer be considered as a valid systematic character, at least not in the Hennigian sense.

Future Studies

Sequence information and new taxa need to be added to this study. Our long-term goal is to sequence the entire 18S and 26S coding regions for the key lineages of the angiosperm radiation. We also will collect parallel sequence information for the 16S and 23S chloroplast rDNAs. To accomplish these goals, we are using the polymerase chain reaction (PCR) (Saiki et al., 1985; Mullis and Faloona, 1987) to amplify nuclear or organellar rDNA from total DNA preps. Sequencing double-stranded PCR products has proven difficult. It is easier to sequence PCR product from an asymmetric amplification, but asymmetric amplifications do not always work in both directions. We plan to avoid these problems by cloning the PCR product into a standard vector and sequencing from a plasmid miniprep (Scharf, 1990).

We are adding more taxa to maximize the overlap between our data set and that of Doyle and Donoghue (1986) and Donoghue and Doyle (1989). We have already begun to combine our data sets for the seven major plant lineages (Donoghue et al., unpublished data). Eventually we will also combine our data sets for flowering plants. We have found that addition of new taxa in older lineages (>40 myr) stabilizes the tree topologies. We saw this with the monocot *Sagittaria*, which was consistently placed among the dicots in early analyses of the data (not shown). The addition of more monocot taxa to the data set finally settled the placement of *Sagittaria* among the other monocots. The necessity to "sample the tree" fully in producing a stable topology currently is a topic of intensive discussion (Donoghue et al., 1989; Penny et al., 1990).

References

Appels, R., and Dvorak, J. (1982) The wheat ribosomal DNA spacer region: its structure and variation in populations and among species. *Theor. Appl. Genet.* **63**, 337–348.

Appels, R., and Honeycutt, R.L. (1986) rDNA: evolution over a billion years. In: *DNA*

Systematics Vol. II: Plants (ed. S.K. Dutta), CRC Press, Boca Raton, FL., pp. 81–135.

Arnheim, N. (1983) Concerted evolution of multigene families. In: *Evolution of Genes and Proteins* (eds. M. Nei and R.K. Koehn), Sinauer Assoc., Sunderland, MA., pp. 38–61.

Birky, C.W., Jr., and Skavaril, R.V. (1976) Maintenance of genetic homogeneity in systems with multiple genomes. *Genet. Res.* **27**, 249–265.

Brown, D.D., Wensink, P.C., and Jordan, E. (1972) Comparison of the ribosomal DNA's of *Xenopus laevis* and *Xenopus mulleri*: the evolution of tandem genes. *J. Mol. Biol.* **63**, 57–73.

Buchheim, M.A., Turmel, M., Zimmer, E.A., and Chapman, R.L. (1990) Phylogeny of *Chlamydomonas* (Chlorophyta): An investigation based on cladistic analysis of nuclear 18S rRNA sequence data. *J. Phycol.* **26**, 689–699.

Burger, W.C. (1977) The piperales and the monocots. *Bot. Rev.* **43**, 345–393.

Burger, W.C. (1981) Heresy revived: the monocot theory of angiosperm origin. *Evol. Theo.* **5**, 189–225.

Cech, T.R. (1983) RNA splicing: three themes with variations. *Cell* **34**, 713–716.

Chapman, R.L., and Avery, D.W. (1989) Nuclear ribosomal RNA genes and the phylogeny of the Trentepohliales. *J. Phycol.* **25** (suppl.), 25.

Clark, C.G., Tague, B.W., Ware, V.C., and Gerbi, S.A. (1984) *Xenopus laevis* 28S ribosomal RNA: a secondary structure model and its evolutionary and functional implications. *Nucleic Acids Res.* **12**, 6197–6220.

Crane, P.R. (1985) Phylogenetic analysis of seed plants and the origin of angiosperms. *Ann. Missouri Bot. Gard.* **72**, 716–793.

Cronquist, A. (1968) *The Evolution and Classification of Flowering Plants*, Houghton Mifflin, Boston.

Dahlberg, A.E. (1989) The functional role of ribosomal RNA in protein synthesis. *Cell* **57**, 525–529.

Devereux, J., Haeverli, P., and Smithies, O. (1984) A comprehensive set of sequence analysis programs for the VAX. *Nucleic Acids Res.* **12**, 387–395.

DeWinter, R.F.J., and Moss, T. (1986) The ribosomal spacer in *Xenopus laevis* is transcribed as part of the primary ribosomal RNA. *Nucleic Acids Res.* **14**, 6041–6051.

Doebley, J., Durbin, M., Golenberg, E.M., Clegg, M.T., and Ma, D.P. (1990) Evolutionary analysis of the large subunit of carboxylase (*rbcL*) nucleotide sequence among the grasses (Gramineae). *Evolution* **44**, 1097–1108.

Donoghue, M.J., and Doyle, J.A. (1989) Phylogenetic studies of seed plants and angiosperms based on morphological characters. In: *The Hierarchy of Life* (eds. B. Fernholm, K. Bremer, and H. Jörnvall), Elsevier Science Publishers, Amsterdam, pp. 181–195.

Donoghue, M.J., Doyle, J.A., Gauthier, J., Kluge, A.G., and Rowe, T. (1989) The importance of fossils in phylogeny reconstruction. *Ann. Rev. Ecol. Syst.* **20**, 431–460.

Dover, G. (1982) Molecular drive: a cohesive mode of species evolution. *Nature* **299**, 111–117.

Dover, G.A., and Flavell, R.B. (1984) Molecular coevolution: DNA divergence and the maintenance of function. *Cell* **38**, 622–623.

Doyle, J.A., and Donoghue, M.J. (1986) Seed plant phylogeny and the origin of angio-sperms: an experimental cladistic approach. *Bot. Rev.* **52**, 321–431.

Doyle, J.J., and Beachy, R.N. (1985) Ribosomal gene variation in soybean (*Glycine max*) and its relatives. *Theor. Appl. Genet.* **70**, 369–376.

Eckenrode, V.K., Arnold, J., and Meagher, R.B. (1985) Comparison of the nucleotide sequence of soybean 18S rRNA with the sequences of other small-subunit rRNAs. *J. Mol. Evol.* **21**, 259–269.

Edman, J.C., Kovacs, J.A., Masur, H., Santi, D.V., Elwood, H.J., and Sogin, M.L. (1988) Ribosomal RNA sequence shows *Pneumocystis carinii* to be a member of the fungi. *Nature* **334**, 519–522.

Edward, K., and Koessel, H. (1981) The rRNA operon from *Zea mays* chloroplasts: nucleotide sequence of 23S rDNA and its homology with *E. coli* 23S rDNA. *Nucleic Acids Res.* **9**, 2853–2869.

Edwards, K., Bedbrook, K., Dyer, T., and Koessel, H. (1981) 4.5S rRNA from *Zea mays* shows a structural homology with the 3' end of prokaryotic 23S rRNA. *Biochem. Int.* **2**, 533–538.

Farris, J.S. (1986) *Hennig86 Manual*, Port Jefferson Station, New York.

Felsenstein, J. (1985) Confidence limits on phylogenies: an approach using the bootstrap. *Evolution* **39**, 783–791.

Field, K.G., Olsen, G.J., Lane, D.J., Giovannoni, S.J., Ghiselin, M.T., Raff, E.C., Pace, N.R., and Raff, R.A. (1988) Molecular phylogeny of the animal kingdom. *Science* **239**, 748–753.

Flavell, R.B. (1986) The structure and control of expression of ribosomal RNA genes. *Oxf. Surv. Pl. Mol. Cell Biol.* **3**, 251–274.

Flavell, R.B., O'Dell, M., and Thompson, W.F. (1983) Cytosine methylation of ribosomal RNA genes and nucleolus organizer activity in wheat. In: *Kew Chromosome Conference II* (eds. P.E. Brandham and M.D. Bennett), George Allen and Unwin, London, pp. 11–17.

Fogel, S., Mortimer, R., Lusnak, K., and Tavares, F. (1978) Meiotic gene conversion: a signal of the basic recombination event in yeast. *Cold Spring Harbor Symp. Quant. Biol.* **43**, 1325–1341.

Gerbi, S.A. (1985) Evolution of ribosomal RNA. In: *Molecular Evolutionary Genetics* (ed. R.J. MacIntyre), Plenum Press, New York, pp. 419–518.

Givens, J.F., and Phillips, R.L. (1976) The nucleolus organizer region of maize (*Zea mays* L). *Chromosoma* **57**, 103–117.

Gould, F.W., and Shaw, R.B. (1985) *Grass Systematics*, Texas A&M University Press, College Station, TX., pp. 111–130.

Gouy, M., and Li, W.-H. (1989a) Phylogenetic analysis based on rRNA sequences supports the archaebacterial rather than the eocyte tree. *Nature* **339**, 145–147.

Gouy, M., and Li, W.-H. (1989b) Molecular phylogeny of the kingdoms Animalia, Plantae, and Fungi. *Mol. Biol. Evol.* **6**, 109–122.

Grummt, I., Roth, E., and Paule, M.R. (1982) Ribosomal RNA transcription *in vitro* is species specific. *Nature* **296**, 173–174.

Gutell, R.R., and Fox, G.E. (1988) A compilation of large subunit RNA sequences presented in a structural format. *Nucleic Acids Res.* **16s**, r175–203.

Hamby, R.K., Issel, L.E., and Zimmer, E.A. (1987) Nuclear and organellar evolution in higher plants—ribosomal RNA as a marker molecule. *Genetics* **116**, s20.

Hamby, R.K., Sims, L.E., Issel, L.E., and Zimmer, E.A. (1988) Direct ribosomal RNA sequencing: optimization of extraction and sequencing methods for work with higher plants. *Plant Mol. Biol. Rep.* **6**, 175–192.

Hamby, R.K., and Zimmer, E.A. (1988) Ribosomal RNA sequences for inferring phylogeny within the grass family (Poaceae). *Plant Syst. Evol.* **160**, 29–37.

Hemleben, V., Ganal, M., Gerstner, J., Schiebel, K., and Torres, R.A. (1988) Organization and length heterogeneity of plant ribosomal RNA genes. In: *Architecture of Eukaryotic Genes* (ed. G. Kahl), VCH, Weinheim, Fed. Rep. Germany, pp. 371–383.

Hendy, M.D., and Penny, D. (1982) Branch and bound algorithms to determine minimal evolutionary tree. *Math. Biosci.* **59**, 277–290.

Hillis, D.M., and Dixon, M.T. (1989) Vertebrate phylogeny: evidence from 28S ribosomal DNA sequences. In: *The Hierarchy of Life* (eds. B. Fernholm, K. Bremer, and H. Jörnvall), Elsevier Science Publishers, Amsterdam, pp. 355–367.

Hood, L., Campbell, J.H., and Elgin, S.C.R. (1975) The organization, expression, and evolution of antibody genes and other multigene families. *Ann. Rev. Genet.* **9**, 305–353.

Hori, H., Lim, B.-L, and Osawa, S. (1985) Evolution of green plants as deduced from 5S rRNA sequences. *Proc. Natl. Acad. Sci. USA* **82**, 820–823.

Hori, H., and Osawa, S. (1987) Origin and evolution of organisms as deduced from 5S ribosomal RNA sequences. *Mol. Biol. Evol.* **4**, 445–472.

Hui, A.S., Eaton, D.H., and de Boer, H.A. (1988) Mutagenesis at the mRNA decoding site in the 16S ribosomal RNA using the specialized ribosome system in *Escherichia coli*. *EMBO J.* **7**, 4383–4388.

Ito, M. (1987) Phylogenetic systematics of the Nymphaeales. *Bot. Mag. Tokyo* **100**, 17–35.

Jacob, W.F., Santer, M., and Dahlberg, A.E. (1987) A single base change in the Shine-Dalgarno region of 16S rRNA of *Escherichia coli* affects translation of many proteins. *Proc. Natl. Acad. Sci. USA* **84**, 4757–4761.

Jorgensen, R.A., and Cluster, P.D. (1988) Modes and tempos in the evolution of nuclear ribosomal DNA: new characters for evolutionary studies and new markers for genetic and population studies. *Ann. Missouri Bot. Gard.* **75**, 1238–1247.

Jukes, T.H., and Cantor, C.R. (1969) Evolution of protein molecules. In: *Mammalian Protein Metabolism* (ed. H.N. Munro), Academic Press, New York, pp. 21–123.

Jupe, E.R., and Zimmer, E.A. (1990) Unmethylated regions in the intergenic spacer of maize and teosinte ribosomal RNA genes. *Plant Mol. Biol.* **14**, 333–347.

Kantz, T.S., Theriot, E.C., Zimmer, E.A., and Chapman, R.L. (1990) The Pleurastrophyceae and Micromonadophyceae: a cladistic analysis of nuclear rRNA sequence data. *J. Phycol.* **26**, 711–721.

Kellogg, E.A., and Campbell, C.S. (1987) Phylogenetic analysis of the Gramineae. In: *Grass Systematics and Evolution* (eds. T.R. Soderstrom, K.W. Hilu, C.S. Campbell, and M.E. Barkworth), Smithsonian Institution Press, Washington, DC, pp. 310–322.

Kimura, M. (1980) A simple method for estimating evolutionary rates of base substitutions through comparative studies of nucleotide sequences. *J. Mol. Evol.* **16**, 111–120.

Klein, H.L., and Petes, T.D. (1981) Intrachromosomal gene conversion in yeast. *Nature* **289**, 144–148.

Knaak, C., Hamby, R.K., Arnold, M.L., LeBlanc, M.D., Chapman, R.L., and Zimmer, E.A. (1990) Ribosomal DNA variation and its use in plant biosystematics. In: *Biological Approaches and Evolutionary Trends in Plants* (ed. S. Kawano), Academic Press, New York, pp. 135–158.

Kumazaki, T., Hori, H., and Osawa, S. (1983) Phylogeny of protozoa deduced from 5S rRNA sequences. *J. Mol. Evol.* **19**, 411–419.

Labhart, P., and Reeder, R.H. (1986) Characterization of three sites of RNA 3' end formation in the *Xenopus* ribosomal spacer. *Cell* **45**, 431–433.

Lake, J.A. (1988) Origin of the eukaryotic nucleus determined by rate-invariant analysis of rRNA sequences. *Nature* **331**, 184–186.

Lake, J.A. (1989) Origin of the eukaryotic nucleus determined by rate-invariant analyses of ribosomal RNA genes. In: *The Hierarchy of Life* (eds. B. Fernholm, K. Bremer, and H. Jörnvall), Elsevier Science Publishers, Amsterdam, pp. 87–101.

Larson, A., and Wilson, A.C. (1989) Patterns of ribosomal RNA evolution in salamanders. *Mol. Biol. Evol.* **6**, 131–154.

Li, W.-H., Luo, D.-C., and Wu, C.-I. (1985) Evolution of DNA sequences. In: *Molecular Evolutionary Genetics* (ed. R.J. MacIntyre), Plenum Press, New York, pp. 1–94.

Long, E.O., and Dawid, I.B. (1980) Repeated genes in eukaryotes. *Ann. Rev. Biochem.* **49**, 727–764.

Mascia, P.N., Rubenstein, I., Phillips, R.L., Wang, A.S., and Xiang, L.Z. (1981) Localization of the 5S rRNA genes and evidence for diversity in the 5S rDNA region of maize. *Gene* **14**, 205–215.

McCarroll, R., Olsen, G.J., Stahl, Y.D., Woese, C.R., and Sogin, M.L. (1983) Nucleotide sequence of the *Dictyostelium discoideum* small-subunit ribosomal ribonucleic acid inferred from the gene sequence: evolutionary implications. *Biochemistry* **22**, 5858–5868.

McClintock, B. (1934) The relationship of a particular chromosomal element to the development of the nucleoli in *Zea mays*. *Z. Zellforsch. Mikrosk. Anat.* **21**, 294 – 328.

Meeuse, A.D.J. (1967) Again: the growth habit of the early angiosperms. *Acta Bot. Neerl.* **16**, 33–41.

Messing, J., Carlson, J., Hagen, G., Rubenstein, I., and Oleson, A. (1984) Cloning and sequencing of the ribosomal RNA genes in maize: the 17S region. *DNA* **3**, 31–40.

Moazed, D., and Noller, H.F. (1987) Interaction of antibiotics with functional sites in 16S ribosomal RNA. *Nature* **327**, 389–394.

Moazed, D., and Noller, H.F. (1989) Interaction of tRNA with 23S rRNA in the ribosomal A, P, and E sites. *Cell* **57**, 585–597.

Mullis, K.B., and Faloona, F.A. (1987) Specific synthesis of DNA *in vitro* via a polymerase-catalyzed chain reaction. *Meth. Enzym.* **155**, 335–350.

Murgola, E.J., Hijazi, K.A., Goringer, H.U., and Dahlberg, A.E. (1988) Mutant 16S ribosomal RNA: a codon-specific translational suppressor. *Proc. Natl. Acad. Sci. USA* **85**, 4162–4165.

Nagylaki, T., and Petes, T.D. (1982) Intrachromosomal gene conversion and the maintenance of sequence homogeneity among repeated genes. *Genetics* **100**, 315–337.

Nairn, C.J., and Ferl, R.J. (1988) The complete nucleotide sequence of the small-subunit ribosomal RNA coding region for the cycad *Zamia pumila*: phylogenetic implications. *J. Mol. Evol.* **27**, 133–141.

Nei, M. (1987) *Molecular Evolutionary Genetics*, Columbia University Press, New York, p. 64.

Nickrent, D.L., and Franchina, C.R. (1989) Phylogenies of parasitic flowering plants (Santalales) using ribosomal RNA sequences. *Amer. J. Bot.* **76** (suppl.), 262.

Noller, H.F., Stern, S., Moazed, D., Powers, T., Svensson, P., and Changchien, L.-M. (1987) *Cold Spring Harbor Symp. Quant. Biol.* **52**, 695–708.

Ohta, T. (1983) On the evolution of multigene families. *Theor. Pop. Biol.* **23**, 216–240.

Ohta, T. (1984) Some models of gene conversion for treating the evolution of multigene families. *Genetics* **106**, 517–528.

Palmer, J.D. (1985) Evolution of chloroplast and mitochondrial DNA in plants and algae. In: *Molecular Evolutionary Genetics* (ed. R.J. MacIntyre), Plenum Press, New York, pp. 131–240.

Penny, D., Hendy, M.D., Zimmer, E.A., and Hamby, R.K. (1990) Trees from sequences: panacea or Pandora's box? *Austral. Syst. Bot.* **3**, 21–38.

Perasso, R., Baroin, A., Qu, L.H., Bachellerie, J.P., and Adoutte, A. (1989) Origin of the algae. *Nature* **339**, 142–144.

Petes, T.D. (1980) Unequal meiotic recombination within tandem arrays of yeast ribosomal DNA genes. *Cell* **19**, 765–774.

Phillips, R.L. (1978) Molecular cytogenetics of the nucleolus organizer region. In: *Maize Breeding and Genetics* (ed. D.B. Walden), John Wiley, New York, pp. 711–741.

Raff, R.A., Field, K.G., Olsen, G.J., Giovannoni, S.J., Lane, D.J., Ghiselin, M.T., Pace, N.R., and Raff, E.C. (1989) Metazoan phylogeny based on analysis of 18S ribosomal RNA. In: *The Hierarchy of Life* (eds. B. Fernholm, K. Bremer, and H. Jörnvall), Elsevier Science Publishers, Amsterdam, pp. 247–260.

Razin, A., and Riggs, A.D. (1980) DNA methylation and gene function. *Science* **210**, 604–610.

Reeder, R.H. (1984) Enhancers and ribosomal gene spacers. *Cell* **38**, 349–351.

Rivin, C.J., Cullis, C.A., and Walbot, V. (1986) Evaluating quantitative variation in the genome of *Zea mays. Genetics* **113**, 1009–1019.

Rogers, S.O., Honda, S., and Bendich, A.J. (1986) Variation in the ribosomal RNA genes among individuals of *Vicia faba. Plant Mol. Biol.* **6**, 339–345.

Rogers, S.O., and Bendich, A.J. (1987) Ribosomal RNA genes in plants: variability in copy number and in the intergenic spacer. *Plant Mol. Biol.* **9**, 509–520.

Saghai-Maroof, M.A., Soliman, K.M., Jorgensen, R.A., and Allard, R.W. (1984) Ribosomal DNA spacer-length polymorphisms in barley: Mendelian inheritance, chromosomal location, and population dynamics. *Proc. Natl. Acad. Sci. USA* **81**, 8014–8018.

Saiki, R.K., Scharf, S.J., Faloona, F., Mullis, K.B., Horn, G.T., Erlich, H.A., and Arnheim, N. (1985) Enzymatic amplification of β-globin genomic sequences and restriction site analysis for diagnosis of sickle cell anemia. *Science* **230**, 1350–1354.

Saitou, N., and Nei, M. (1987) The neighbor-joining method: a new method for reconstructing phylogenetic trees. *Mol. Biol. Evol.* **4**, 406–425.

Schaal, B.A., and Learn, G.H., Jr. (1988) Ribosomal DNA variation within and among plant populations. *Ann. Missouri Bot. Gard.* **75**, 1207–1216.

Scharf, S.J. (1990) Cloning with PCR. In: *PCR Protocols* (eds. M.A. Innis, D.H. Gelfand, J.J. Sninsky, and T.J. White), Academic Press, San Diego, pp. 84–91.

Schleifer, K.H., and Ludwig, W. (1989) Phylogenetic relationships among bacteria. In: *The Hierarchy of Life* (eds. B. Fernholm, K. Bremer, and H. Jörnvall), Elsevier Science Publishers, Amsterdam, pp. 103–117.

Schwartz, Z., and Koessel, H. (1980) The primary structure of 16S rDNA from *Zea mays* chloroplast is homologous to *Escherichia coli* 16S rRNA. *Nature* **283**, 739–742.

Shermoen, A.W., and Kiefer, B.I. (1975) Regulation in rDNA-deficient *Drosophila melanogaster. Cell* **4**, 275–280.

Smith, G.P. (1974) Unequal crossover and the evolution of multigene families. *Cold Spring Harbor Symp. Quant. Biol.* **38**, 507–513.

Smith, G.P. (1976) Evolution of repeated DNA sequences by unequal crossover. *Science* **191**, 528–535.

Sogin, M.L., Edman, U., and Elwood, H. (1989) A single kingdom of eukaryotes. In: *The Hierarchy of Life* (eds. B. Fernholm, K. Bremer, and H. Jörnvall), Elsevier Science Publishers, Amsterdam, pp. 133–143.

Stebbins, G.L. (1974) *Flowering Plants: Evolution Above the Species Level*, Belknap Press, Cambridge, MA.

Steffensen, D.M., and Patterson, E.B. (1979) Using translocations to map the 5S rRNA genes to chromosome 2L in maize. *Genetics* **9** (suppl.), s123.

Strauss, S.H., Palmer, J.D., Howe, G.T., and Doerksen, A.H. (1988) Chloroplast genomes of two conifers lack a large inverted repeat and are extensively rearranged. *Proc. Natl. Acad. Sci USA* **85**, 3898–3902.

Stringer, S.L., Hudson, K., Blase, M.A., Walzer, P.D., Cushion, M.T., and Stringer,

J.R. (1989) Sequence from ribosomal RNA of *Pneumocystis carinii* compared to those of four fungi suggests an ascomycetous affinity. *J. Protozool.* **36**, 14S–16S.

Swofford, D.L. (1989) PAUP *3.0*, Illinois Natural History Survey, Champaign, IL.

Sytsma, K.J., and Schaal, B.A. (1985) Phylogenetics of the *Lisianthius skinneri* (Gentianaceae) species complex in Panama utilizing DNA restriction fragment analysis. *Evolution* **39**, 594–608.

Takaiwa, F., and Sugiura, M. (1982) The complete nucleotide sequence of a 23S rRNA gene from tobacco chloroplasts. *Eur. J. Biochem.* **124**, 13–19.

Takaiwa, F., Oono, K., Iida, Y., and Sugiura, M. (1985a) The complete nucleotide sequence of a rice 25S ribosomal RNA gene. *Gene* **37**, 255–289.

Takaiwa, F., Oono, K., and Sugiura, M. (1985b) Nucleotide sequence of the 17S–25S spacer region from rice rDNA. *Plant Mol. Biol.* **4**, 355–364.

Takaiwa, F., Oono, K., and Sugiura, M. (1984) The complete nucleotide sequence of a rice 17S ribosomal RNA gene. *Nucleic Acids Res.* **12**, 5441–5448.

Takhtajan, A. (1969) *Flowering Plants—Origin and Dispersal*, Smithsonian Institution Press, Washington, DC.

Tanaka, Y., Dyer, T.A., and Brownlee, G.G. (1980) An improved direct RNA sequence method: its application to *Vicia faba* 5.8S ribosomal RNA. *Nucleic Acids Res.* **8**, 1259–1272.

Tartof, K.D. (1975) Redundant genes. *Ann. Rev. Genet.* **9**, 355–385.

Taylor, D., and Hickey, L.J. (1990) An Aptian plant with attached leaves and flowers: implications for angiosperm origin. *Science* **247**, 702–704.

Thorne, R.F. (1976) A phylogenetic classification of the Angiospermae. *Evol. Biol.* **9**, 35–106.

Tohdoh, N., and Sugiura, M. (1982) The complete nucleotide sequence of a 16S rRNA gene from tobacco chloroplasts. *Gene* **17**, 213–218.

Trifonov, E.N. (1987) Translation framing code and frame-monitoring mechanism as suggested by the analysis of mRNA and 16S rRNA nucleotide sequences. *J. Mol. Biol.* **194**, 643–652.

Turner, S., Burger-Wiersma, T., Giovannoni, S.J., Mur, L.R., and Pace, N.R. (1989) The relationship of a prochlorophyte *Prochlorothrix hollandica* to green chloroplasts. *Nature* **337**, 380–382.

Vincentz, M., and Flavell, R.B. (1989) Mapping of ribosomal RNA transcripts in wheat. *Plant Cell* **1**, 579–589.

Vossbrinck, C.R., Maddox, J.V., Friedman, S., Debrunner-Vossbrinck, B.A., and Woese, C.R. (1987) Ribosomal RNA sequence suggests microsporidia are extremely ancient eukaryotes. *Nature* **326**, 411–414.

Watanabe, J.-I., Hori, H., Tanabe, K., and Nakamura, Y. (1989) 5S ribosomal RNA sequence of *Pneumocystis carinii* and its phylogenetic association with "Rhizopoda/Myxomycota/Zygomycota group." *J. Protozool.* **36**, 16S–17S.

Watson, L., Clifford, H.T., and Dallwitz, M.J. (1985) The classification of Poaceae: subfamilies and supertribes. *Austral. J. Bot.* **33**, 433–484.

Weiss, R.B., Dunn, D.M., Atkins, J.F., and Gesteland, R.F. (1987) Slippery runs, shifty stops, backward steps, and forward hops: −2, −1, +1, +2, +5, and +6 ribosomal frameshifting. *Cold Spring Harbor Symp. Quant. Biol.* **52**, 687–693.

Weiss, R.B., Dunn, D.M., Dahlberg, A.E., Atkins, J.F., and Gesteland, R.F. (1988) Reading frame switch caused by base-pair formation between the 3′ end of 16S rRNA and the mRNA during elongation of protein synthesis in *Escherichia coli. EMBO J.* **7**, 1503–1507.

Wheeler, W.C., and Honeycutt, R.L. (1988) Paired sequence difference in ribosomal RNAs: evolutionary and phylogenetic implications. *Mol. Biol. Evol.* **5**, 90–96.

Whitfield, P.R., and Bottomley, W. (1983) Organization and structure of chloroplast genes. *Ann. Rev. Plant Physiol.* **34**, 279–310.

Woese, C.R. (1987) Bacterial evolution. *Microbiol. Rev.* **51**, 221–271.

Wolfe, K.H., Gouy, M., Yang, Y.-W., Sharp, P.M., and Li, W.-H. (1989) Date of the monocot-dicot divergence estimated from chloroplast DNA sequence data. *Proc. Natl. Acad. Sci. USA* **86**, 6201–6205.

Yakura, K., Kato, A., and Tanifuji, S. (1984) Length heterogeneity of the large spacer of *Vicia faba* is due to the differing number of a 325 bp repetitive sequence element. *Mol. Gen. Genet.* **193**, 400–405.

Zechman, F.W., Theriot, E.C., Zimmer, E.A., and Chapman, R.L. (1990) Phylogeny of the Ulvophyceae (Chlorophyta): Cladistic analysis of nuclear-encoded rRNA sequence data. *J. Phycol.* **26**, 700–710.

Zimmer, E.A., Hamby, R.K., Arnold, M.L., LeBlanc, D.A., and Theriot, E.C. (1989) Ribosomal RNA phylogenies and flowering plant evolution. In: *The Hierarchy of Life* (eds. B. Fernholm, K. Bremer, and H. Jörnvall), Elsevier Science Publishers, Amsterdam, pp. 205–214.

Zimmer, E.A., Martin, S.L., Beverley, S.M., Kan, Y.W., and Wilson, A.C. (1980) Rapid duplication and loss of genes coding for the α chains of hemoglobin. *Proc. Natl. Acad. Sci. USA* **77**, 2158–2162.

5

Evolution of the *Nor* and *5SDna* Loci in the Triticeae

R. Appels and *B. Baum*

The *Nor* and *5SDna* loci are composed of tandem arrays of units that carry the genes coding for RNA products, as well as associated spacers. Extensive sequence data bases are available for the RNA products of both loci in a wide range of organisms (reviewed in Appels and Honeycutt, 1986; Erdmann and Wolters, 1986; Huysmans and DeWachter, 1986; Woese, 1987). The available information at the nucleotide sequence level has been used in both phenetic and cladistic analyses to assess the relationships between organisms (Hendriks et al., 1986; Dams et al., 1987; Woese, 1987; Wolters and Erdmann, 1988; Van den Eyne et al., 1988; Hamby and Zimmer, 1988; Johnson and Baverstock, 1989; Gouy and Li, 1989). The spacer regions separating the genes at these loci have also been studied in considerable detail (reviewed in Appels and Honeycutt, 1986), at both the restriction fragment length polymorphisms (RFLP) banding pattern level and sequence level. They were found to evolve at a rate that appears to be much faster than the gene regions. The depth of analysis of the *Nor* and *5SDna* loci at the structure/function and evolutionary level is thus unique, and the available data suggest that the repetitive nature of the loci per se does not preclude them from being valuable taxonomic characters.

Studies on the *Nor* Loci in the Triticeae

The *Nor* loci of the Triticeae (Fig. 5.1) have been well studied over the past 15 years (reviewed in Appels and Honeycutt, 1986; see also Appels et al., 1986a,b; Lassner and Dvorak, 1986; Lassner et al., 1987; May and Appels, 1987; McIntyre et al., 1988; Barker et al., 1988; Appels et al., 1989; Vincentz and Flavell, 1989) and provide useful examples of the application of DNA analytic information to problems such as assessing the relationships between species. The

The authors are grateful to Drs. J. West and A.H.D. Brown for critical comments on an earlier version of this chapter.

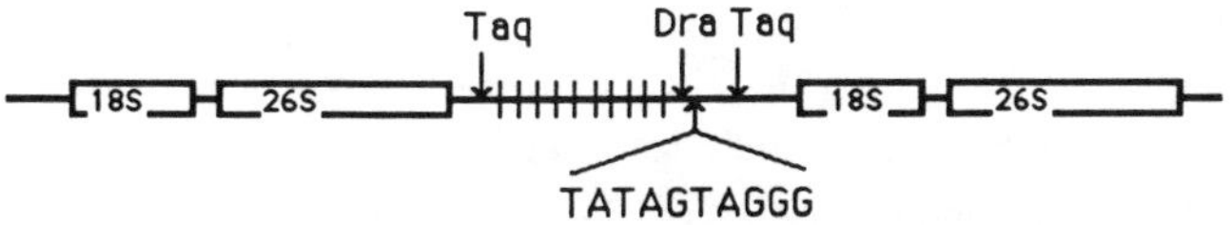

Figure 5.1. The rDNA repeating unit located at the *Nor* locus. The boxes represent the final, major 18S and 26S rRNA produced after processing of the precursor RNA (transcription start site is TATAGTAGGG). In the Triticeae, the repeat unit is 8–12 kb long. In most species of the Triticeae, the restriction endonuclease *Taq*1 sites indicated define the subrepeat spacer region plus a small part of the external transcribed spacer (ETS) and thus are extremely useful in assaying the polymorphism which occurs in this DNA segment. The ETS is defined by the start of transcription (TATAGTAGGG) at its 5' end and the start of the 18S gene at its 3' end. In some genera, such as *Hordeum,* a *Taq*1 site occurs in each of the spacer subrepeats and thus does not provide a useful assay of the subrepeat spacer region. In some genera, such as *Secale,* the restriction endonuclease *Dra* site provides a useful assay for the region near the transcription start site. Generally 600–2,000 units of rDNA are tandemly arranged at the *Nor* loci of species in the Triticeae.

studies fall into two broad levels of resolution, namely, that of variation in the DNA fragments generated by restriction enzymes, as assayed by specific probes (e.g., RFLPs), and variation at the level of DNA sequences. The RFLP analysis of *Nor* loci is very extensive, including many different species of lower and higher eukaryotes, and is dominated by variation in the spacer subrepeat region (see Fig. 5.1). Variation in the number of subrepeats and the presence or absence of restriction enzyme sites lead to different lengths of the DNA segment carrying homology to the subrepeat sequences. These studies have shown that the *Nor* region is a useful "fingerprint" region in distinguishing between individuals of a species (for example, see May and Appels, 1987). As discussed below, this level of analysis can also be extended to give some information about relationships at the DNA sequence level and thus becomes of greater interest in evolutionary problems. The studies of DNA segments from the *Nor* locus at the DNA sequence level are more limited, and a detailed analysis of the available data is presented below, following a discussion of RFLP studies as well as the types of computer analyses that are available.

Computer Analyses of Data Bases

Numerous approaches are now available for the computer analysis of restriction enzyme banding patterns (RFLPs) and sequence information. The basic premise underlying the RFLP type of analysis is that two fragments are regarded as equal, if their length, as determined by electrophoretic mobility, is identical. This assumption is generally valid only if the fragments are identified on the basis of stringent hybridization to well-characterized DNA molecules, such as small (200 bp) probes, which have been sequenced. The degrees of comparability in a phylogenetic study depend on the specificity of the probes being used in a way that is analogous to other taxonomic characters (Baum, 1973; Sneath, 1976). Examples of studies at this level include the study of relationships within *Drosophila* using single-copy sequences (Loukas et al., 1986), the use of rRNA probes for studying phylogenetic relationships between archaebacteria (Klenk et al., 1986), and the numerous studies utilizing chloroplast DNA as a phylogenetic marker (reviewed in Palmer, 1986; Palmer et al., 1988; see also Zurawski and Clegg, 1987; Govindaraju et al., 1988; Neale et al., 1988; Doyle et al., 1990; and several other chapters in this volume). It is important to note that ambiguities can arise due to the existence of regions that can evolve relatively quickly or exist in more than one location (for example chloroplast sequences in mitochodria and nuclei: Stern and Lonsdale, 1982; Cheung and Scott, 1989), and, in situations such as this, a thorough understanding of the structure of the DNA region under study is required.

The computer analysis of RFLP data can be carried out using phenetic procedures (for example, see the study of species of *Secale* by Reddy et al., 1990) employing a series of ordination and cluster analyses. Phenetic analyses also have been performed by Denny et al. (1988) on *Pseudomonas*. These workers assigned a unique number to each band and scored bands as characters either present or absent (0/1). A distance matrix was then computed based on Jaccard's coefficient; the matrix was analyzed via UPGMA clustering. Cladistic analyses (using PAUP) have also been carried out on RFLP-type data (e.g., Song et al., 1988a, b, on *Brassica*) and assume that common fragments reflect "shared derived restriction sites." Although the latter assumption is not always valid and requires a well-characterized situation at the molecular level to provide meaningful data, both phenetic and cladistic analyses of RFLP data have contributed to understanding the nature of change in the evolution of DNA sequences.

The RFLP data from a survey of variation at the *Nor* locus of species in the Triticeae (example shown in Fig. 5.2A) have been studied in our laboratory and scored so that every unique band was assigned a number, to generate a data matrix describing the variation (Clarke et al., unpublished data). The data matrix was subjected to the two-stage density clustering described by Sarle (1985) over a range of K (the number of neighbors for the Kth clustering) to find regions of relative stability and thus the possible number of clusters that exist in the data. A correlation matrix was also deduced from the above data matrix and subjected to principal component analysis (Kshirsagar, 1972) to check for trends of variation among the variables to detect taxonomic structure in the data. Pairwise similarities were then computed using the Gower (1971) similarity coefficient. Similarity coefficients were converted to distance coefficients and the resulting matrix used to detect taxonomic structure in a principal coordinate analysis. Finally, a canonical discriminant analysis (a method of ordination) was also carried out, using the preceding data matrices, on genera where more than one accession was available for study (the majority of genera examined). Figure 5.2B shows an

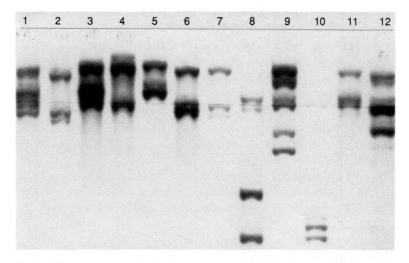

Figure 5.2A. Variation at the *Nor* loci from species of the Triticeae based on measuring the length of DNA fragments showing sequence similarity to the spacer subrepeat region from an rDNA unit isolated from the *Nor* locus of *Australopyrum retrofractum*. DNA samples (10 μg) were digested with the restriction endonuclease *Taq*1, electrophoresed, transferred to nitrocellulose and hybridized with a ^{32}P probe originating from *Australopyrum retrofractum* rDNA spacer. Lanes 1 to 12 contain DNA from *Pseudoroegeneria libanotica, Elymus ciliare, P. spicata, Thinopyrum elongatum, Agropyron cristatum, Elymus trachycaulus, Triticum monococcum, Critesion chilense, Hordeum spontaneum, Critesion californicum, P. libanotica,* and *H. vulgare,* respectively.

example of how some of the information from such a data set can be summarized in the form of a two-dimensional representation. The amount of variation in the data is clear but useful groupings such as the "H" samples are evident. A further summary of the data is possible, where the information from all representatives of a genus is pooled into a three-dimensional representation (discussed in Reddy et al., 1990), and this is illustrated in Fig. 5.2C. Higher dimensions are definitely required to portray the distribution of populations in space, but these are difficult to express visually.

In the analysis of DNA sequence data bases, the programs in the PHYLIP package (Felsenstein, 1989) have been used extensively and were employed in the comparisons reported in this chapter. In contrast to RFLP analyses, the comparison of DNA sequences is at a very high level of resolution and, as a result, is more complex. A basic prerequisite for a comparison is an appropriate alignment of the sequences based on either the primary sequence (for example, Smith, 1987) or secondary structure (reviewed in Waterman, 1988). Following alignment, numerical methods are used to infer phylogenetic relationships between the sequences under study (reviewed in Olsen, 1988). It is important to note that an accepted procedure in the alignment of sequences is the introduction

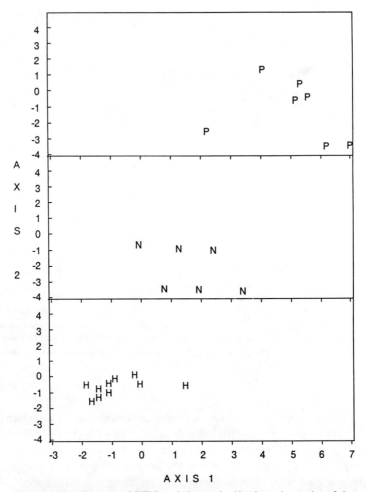

Figure 5.2B. Analysis of RFLP variation at the *Nor* locus in species of the Triticeae. The data were analyzed as discussed in the text, and sample data for three genera, *Agropyron* (P), *Psathyrostachys* (N), and *Hordeum* (H), are illustrated. Other genera, such *Pseudoroegeneria*, show variation over a wider range than illustrated; the reason for this has not, to date, been investigated. The two axes provide an estimate of the relationship among the samples analyzed, in arbitrary units.

of gaps to maximize similarity. In computer programs such as PHYLIP, a gap is not treated as a variable that contributes to the final tree of relationships between the sequences being compared. Generally, a gap of 10 or more bases in one sequence relative to another is clearly a single-step event and may represent a derived state. Alternatively, where transposable elements are suspected to be involved, the sequence with an apparent gap may represent the primitive state.

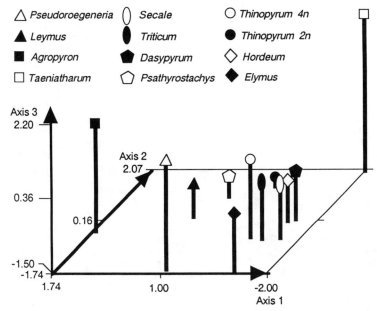

Figure 5.2C. Summary of the size variation at the *Nor* loci of the Triticeae. The analysis was carried out as discussed in the text with relationships among the genera being presented in arbitrary units, in three axes, and taking into account 86% of the variation present in the data base.

In alignment programs, however, smaller gaps are often introduced, which are not as readily interpreted in terms of clear deletion events. Thus, the concept of homology in an analyses of DNA sequences may not always correspond to the concept of homology in classical taxonomy (for an extensive discussion comparing the different concepts of homology see Patterson, 1988).

RFLP Analyses

An extensive survey (Clark et al., unpublished data) of the Triticeae has demonstrated that the major size class of rDNA has a large spacer region, approximately 3.4 kb long. Polymorphisms exist within species, but those with similar genomes (as defined by chromosome pairing, Löve, 1984) tend to be more similar to each other than to species of a different genome. In a study where RFLPs from uncloned DNA are being analyzed, an additional level of resolution is readily available through monitoring the behavior of the probe used to assay the RFLPs under more stringent conditions of DNA–DNA hybridization. The molecular basis for the enhanced resolution is explained in Fig. 5.3A (taken from Appels et al., 1986b), using three different probes hybridized either to heterologous DNA (from wheat) or homologous DNA. The available data show

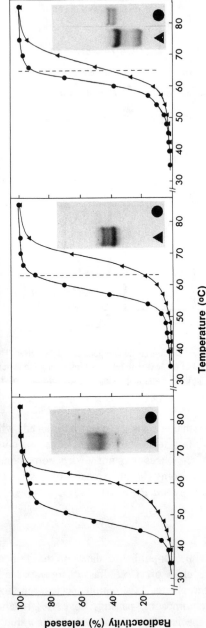

Differential loss of radioactivity from heterologous hybrids (●——●) relative to homologous hybrids (▲——▲)

Temperature (°C)

Radioactivity (%) released

Figure 5.3A. Molecular basis for the increase in resolution at the *Nor* loci using stringent hybridization washes. Under standard conditions of hybridization, a DNA-DNA heteroduplex can form with 10% to 15% of the base pairs mismatched but the thermal stability of this duplex is lowered by 1–2°C relative to the homoduplex for every mismatched base. The differences in thermal stability can be readily demonstrated by stepwise increases in the temperature at which a duplex between a ^{32}P-DNA probe and genomic DNA bound to nitrocellulose are exposed, and monitoring the release of the radioactivity. It is evident from the examples shown in the three panels that the heterologous hybrid in all cases breaks down at a lower temperature than the homologous hybrid. A critical feature of the observations demonstrated in the panels is that, at the midpoints of the radioactivity release curves for the homologous hybrids (T_m), most of the radioactivity has been released from the heterologous hybrid. The temperature at which the difference between heterologous and homologous hybrids is at a maximum is defined as a stringent hybridization wash, and, in the case of an unknown situation (with respect to the heterologous hybrid), the behavior of the ^{32}P-DNA probe provides an indication of the sequence similarity between the homologous and heterologous sequences. The inserts in the panels show the effect of stringent hybridization washes: prior to the wash the heterologous and homologous hybrids were very difficult to distinguish, whereas, as illustrated, the differences were very marked after the washes.

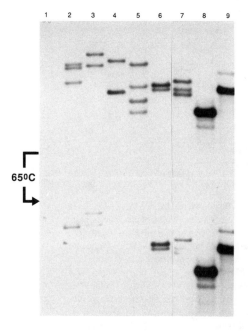

65°C

Figure 5.3B. An example of the use of the technique described in Fig. 5.3A. DNA samples (10 μg) were digested with the restriction endonuclease *Taq* 1, electrophoresed, transferred to nitrocellulose, and hybridized with a ^{32}P probe originating from *Australopyrum retrofractum* rDNA spacer. Lanes 1 to 9 contain DNA from *Brachypodium sylvaticum, Elymus scabrus, E. semicostatus, E. longearistatus, E. melantherus, E. tsuksiensis, Australopyrum pectinatum,* and *A. retrofractum,* respectively. After a standard hybridization in 3 × SSC, 50% formamide the filters were treated at 65°C, as indicated, and reexposed to x-ray film. Photograph kindly supplied by B.C. Clarke.

that for various rDNA spacer probes from the Triticeae, conditions of stringency for the DNA–DNA hybrid can be defined so that at a particular temperature 80% to 90% of the probe hybridized to heterologous DNA can be removed while only 40% to 50% of the homologous hybrid is lost (for further examples see Appels and Dvorak, 1982; Gill and Appels, 1988). A recent example of such a study is the utilization of probes from the *Australopyrum retrofractum Nor* locus to define *Nor* loci in other Triticeae species that were closely related to it (Clarke et al., unpublished data). The information obtained suggested that certain *Nor* loci from some *Elymus* species are more closely related to the *A. retrofractum* probe, at the DNA sequence level, than are others. In a broader context, this procedure, using DNA probes from a number of different species, allows the building up of a set of interrelated *Nor* relationships for species in the Triticeae to interpret evolutionary trends in the tribe; the S genome in the tetraploid *Elymus trachycaulus,* for example, has clearly originated from *Pseudoroegeneria spicata* as judged from the behavior of the rDNA probe from *P. spicata* in a survey of members of the Triticeae (Gill et al., 1988). The recent sequencing work on rDNA clones from *Hordeum bulbosum* (Procuniere and Kasha, 1990) and *Critesion bogdanii* (Knutsson and Appels, unpublished data) plus the probes already available (Appels et al., 1986a; Gill and Appels, 1988) indicate that this type of analysis can be applied extensively in the Triticeae to survey large numbers of samples. This type of survey work may be particularly useful in identifying *Nor* loci that may be of further interest regarding cloning and sequencing.

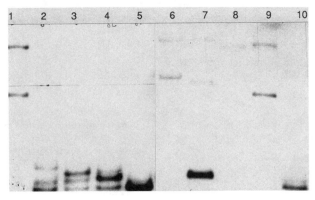

Figure 5.4. Characteristic RFLP patterns for the *Nor* loci of *Hordeum sensu lato*. DNA (10 μg), isolated from the leaves of the species indicated, were digested with the restriction endonuclease *Bam*HI and electrophoresed in a standard 1% agarose gel to separate DNA fragments according to their length. Following transfer to nitrocellulose membrane, a ^{32}P-DNA probe (pArWR4.T5, Clarke and Appels, unpublished data) was used to detect DNA segments with sequence similarity to the ETS region of the rDNA cistron. Lanes 1 to 10 contain DNA from *Haynaldia villosa, Critesion brevisubulatum, C. brevisubulatum, C. brevisubulatum, C. brevisubulatum, C. secalinum, C. bogdanii, C. marinum, C. marinum,* and *C. roshevitzii,* respectively. See also Fig. 5.2A for additional examples of *Hordeum sensu lato* species.

In certain situations, a spacer RFLP can group a subset of species. For example, in Fig. 5.4, *Hordeum* species are differentiated according to the presence or absence of an extra *Bam*HI site (Knutsson and Appels, unpublished data). Preliminary sequence information on rDNA clones from *H. bogdanii* (syn. *Critesion bogdanii*) has suggested that the short *Bam*HI fragment characteristic of the *H. bogdanii* group of species is due to an extra *Bam*HI site rather than to an exceptionally short spacer region (Knutsson and Appels, unpublished data) as previously suggested (McIntyre et al., 1988). Appels et al. (1980) and Molnar et al. (1989) also found that additional *Bam*HI sites were characteristic of many of the *Hordeum* species. Molnar et al (1989) suggested that the *Hordeum* species can be subdivided on the basis of the distribution of *Bam*HI sites in their rDNA units; it seems likely, however, that these maps will require confirmation by sequencing and by using probes that differentiate between *Nor*s before the subdivisions can be used to interpret the evolution of the *Nor* locus in *Hordeum*.

Sequence Analyses

Landmarks in the Analysis of rDNA Sequences

Certain sequences show high levels of evolutionary conservation and thus provide valuable starting points for the alignments required for sequence comparisons and subsequent phenetic or cladistic studies. Within the rRNA gene regions, short stretches of sequences have remained invariant in eukaryote evolution and also are recognizable in prokaryotes (Lane et al., 1985; Hamby and Zimmer, 1988). The core regions of conserved primary structure, as well as other regions of conserved secondary structure, provide binding sites for ribosomal proteins and possibly active sites for protein synthesis. Experimentally, the highly conserved sequences provide the basis of protocols for the rapid sequencing of rRNA using reverse transcriptase (Youvan and Hearst, 1981; Qu et al., 1983; Hamby and Zimmer, 1988) to establish sequence data bases for use in taxonomy.

In the Triticeae, the subrepeats of the spacer region have a conserved region, which provides a marker for the alignment of sequences in carrying out comparative studies—the consensus sequence of this conserved region is GCGCCATG-GAAAACTGGGCAAACCAC (McIntyre et al., 1988). Because this sequence is also recognizable in maize rDNA (Toloczyki and Feix, 1986), the region may have functional significance and requires further investigation.

In the so-called external transcribed spacer (ETS) region, the start of transcription can be clearly recognized in a wide range of organisms (Toloczyki and Feix, 1986). The sequence is TATAGTAGGGG in the plants studied and provides a clear start to the ETS; the conserved 5′ end of the 18S rRNA gene, TACCTGGTT-GATC, defines the end of this region so that sequence comparisons among a wide range of species can be carried out. In addition to the primary structure, comparisons at the secondary structure level can also be performed. A computer folding (Zucker and Stiegler, 1981) of a region near the middle of the ETS recognized in the Triticeae species and maize is shown in Fig. 5.5. The predicted secondary structures are strikingly similar. The primary sequences (TTGTACTG in the Triticeae and CTGCCGTG in maize), which appear as a looped-out structure at the top of the hairpin structure, do not have obvious similarities, but their appearance in similar secondary structure conformations may indicate a common biological property. Although the biological significance of computer-generated secondary structures is not clear (discussed in Zucker and Stiegler, 1981; Gerbi et al., 1989), comparative data at this level have the potential for revealing conservation of an aspect of the sequence that may not be immediately obvious from the primary sequence data (Gutell et al., 1985; Olsen, 1988). Procuniere and Kasha (1990) have noted that in the ETS of *Hordeum bulbosum* a characteristic hairpin structure downstream from the transcription start is bounded by sequences that are recognizable in wheat (GCTTTTG, GTTTGCAGGGTTGCCTC in *H. bulbosum*, and CCTTCCT, CTTCC-CAACGTTGCCTC in wheat). The conservation of the GTTGCCTC sequence in particular, in a position adjoining a hairpin structure of variable length, could indicate another region of further biological interest.

Relationships Based on Comparisons of the
Spacer Subrepeat

Early work (reviewed in Appels and Honeycutt, 1986) has shown that different parts of the rDNA region evolve at different rates. The gene regions are relatively

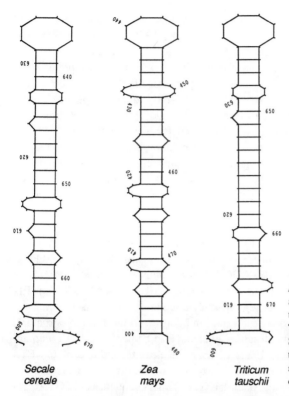

Secale
cereale

Zea
mays

Triticum
tauschii

Figure 5.5. Computer folding of a DNA segment in the external transcribed spacer (ETS) region. The FOLD algorithm (Zucker and Stiegler, 1981), as implemented in the GCG (University of Wisconsin) suite of programs, was used to obtain the structures shown.

conserved whereas the subrepeat spacer regions (see Fig. 5.1) are sufficiently fast in their rates of evolution to provide resolution between the *Nor* loci of species within a genus such as *Triticum*. The work in the Triticeae has, in recent years, provided a modest data base, comprising sequence data from 12 *Nor* loci, for more detailed study. To carry out an analysis, a consensus sequence for the available spacer subrepeats from each of the *Nor* loci was determined. Alignment of the sequences used the landmark sequence GCGCCATGGAAAACTGGG-CAAACCAC discussed in the preceding section. To carry out the alignment two approaches were used:

Smith's iterative multiway   The alignment program described in Smith (1987) is based on an iterative multiway approach, with realignment of the sequences following the branches of the UPGMA tree. Leading or trailing gaps were not weighted. The existence of a gap in the alignment was initially given a penalty of 1, whereas a single sequence mismatch was given a penalty of 0.1; final penalties were assigned by trial and error to optimize the alignment of the sequences using the shortest attainable length of the aligned sequences as a guide.

Higgins and Sharp's CLUSTAL   This alignment program differs from Smith (1987) in that sequences were aligned using the Wilbur and Lipman (1984)

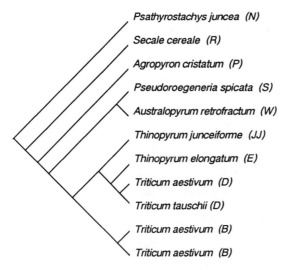

Psathyrostachys juncea (N)

Secale cereale (R)

Agropyron cristatum (P)

Pseudoroegeneria spicata (S)

Australopyrum retrofractum (W)

Thinopyrum junceiforme (JJ)

Thinopyrum elongatum (E)

Triticum aestivum (D)

Triticum tauschii (D)

Triticum aestivum (B)

Triticum aestivum (B)

Figure 5.6. Phylogenetic relationships based on the rDNA spacer subrepeat sequences. Available sequences were aligned and compared using a cladistic approach, to obtain the tree shown (as discussed in text). The letters, in brackets, following the species names represent the genome designations according to Löve (1984).

procedures based on the order of clustering in the dendrogram. This method allowed large numbers of sequences to be handled and has been shown empirically (Higgins and Sharp, 1988, 1989) to align regions of similar secondary structure. Although the output format of this alignment program necessitated laborious editing for input into cladistic programs, the algorithm allowed alignments to be more readily optimized (using the shortest attainable length of the aligned sequences as a guide).

The aligned matrix was input into the branch-and-bound algorithm (Hendy and Penny, 1982) as implemented by the program DNAPENNY in PHYLIP version 3.2 (Felsenstein, 1989). *Psathyrostachys* was used to root the tree, following Baum (1983), and the single most parsimonious tree is shown in Fig. 5.6. The tree shows features which are consistent with evolution within the Triticeae as assessed from independent characters (West et al., 1988). Notable features of the tree include the separation of *Secale* from the *Triticum* species and the fact that the sequences from *T. tauschii* (the D genome donor of wheat, *T. aestivum*) and the D genome *Nor* locus of wheat are grouped together. The close relationship between *Pseudoroegeneria* and *Australopyrum* is discussed further in a later section.

The available sequences of subrepeats from non-Triticeae species (*Vicia faba*, Kato et al., 1985; *Daucus carota*, Taira et al., 1988; *Cucurbita maxima*, Kelly and Siegel, 1989; *Vigna radiata*, Gerstner et al., 1988; *Raphanus sativus*, Delcasso-Tremousaygue et al., 1988; *Zea mays*, McMullen et al., 1986, Toloczyki and Feix, 1986) were also analyzed together with a consensus sequence from the Triticeae spacer subrepeats. No single most parsimonious tree was obtained, suggesting that the sequences are too divergent and that any similarity that exists is difficult to differentiate from random.

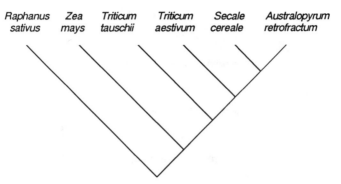

Figure 5.7. Phylogenetic relationships based on the analysis of the external transcribed spacer (ETS) region. Available sequences were aligned and compared using a cladistic approach, to obtain the tree shown (as discussed in text).

Relationships Based on Comparisons of the External Transcribed Spacer Region

The data base for sequences from the ETS region is not large; data are limited to *Vigna radiata* (Gerstner et al., 1988), *Raphanus sativus* (Delcasso-Tremousaygue et al., 1988), *Zea mays* (McMullen et al., 1986; Toloczyki and Feix, 1986), *Triticum tauschii* (Vinizky et al., unpublished data), *T. aestivum* (Lassner and Dvorak, 1986; Barker et al., 1988); *Secale cereale* (Appels et al., 1986b), and *Australopyrum retrofractum* (Clarke and Appels, unpublished data). The alignment and cladistic analyses on this small data base were carried out as described in the previous section and provided the tree shown in Fig. 5.7. The sequence for the ETS from *V. radiata* was not used due to ambiguity in defining the start of the ETS. Two ETS sequences, from *R. sativus* and *Z. mays*, contained large gaps after alignment and because a gap carries no information, the data matrix was also run in a reduced form of 780 bp, instead of 1,215 bp, to remove the regions that generated the gap. Both runs generated the same tree topology.

A comparison of the trees in Fig. 5.6 and 5.7 suggests that in Fig. 5.6 the *Nor* from *S. cereale* may be displaced further from the *Triticum* species than the *A. retrofractum* in Fig. 5.7. Although the trees in Fig. 5.6 and 5.7 are difficult to compare due to the different numbers of species in the database, we believe, on the basis of survey work carried out with probes from adjoining regions, such as the ETS and spacer subrepeats (Clarke et al., unpublished data), that different tree topologies with different sequences (even though they are assaying the same locus) will prove to be a common occurrence. The observation could result from different rates of evolution of the sequences (Appels and Dvorak, 1982). In addition, the regions are most likely to have different functional constraints acting on them and thus lead, fortuitously, to different phylogenetic relationships;

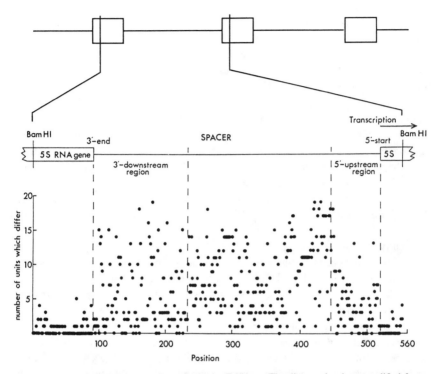

Figure 5.8. The 5S DNA repeating units in the Triticeae. The diagram has been modified from Scoles et al. (1987) and summarizes the variation in nucleic acid sequence in the spacer region from 5S DNA units studied from a wide range of species. Each position in the sequence was scored for the number of 5S DNA units that differed from the consensus (determined for the entire data base). The results were then plotted as a graph with the *y* axis showing the number of units which differed and the *x* axis the position (1–552) in the consensus sequence.

secondary structure may be more important in the ETS product, for example, than for any short-lived product from the spacer subrepeat.

The *5SDna* Loci of the Triticeae

Published studies on the *5SDna* loci in species of the Triticeae (Kota and Dvorak, 1986; Gill et al., 1988; Scoles et al., 1988; Dvorak et al., 1989; Lagudah et al., 1989; Reddy and Appels, 1989) have shown that, within a species, two separate loci for the 5SDNA can coexist and that these loci are characterized by certain size classes of unit. At each locus, the respective size class of the subunit is tandemly arranged, and different parts of the unit accumulate changes during the course of evolution at different rates (Fig. 5.8). In *Secale* (Reddy and Appels, 1989), *Triticum tauschii* (Lagudah et al., 1989), and *T. monococcum* (Dvorak et al., 1989) the long 5SDNA units (469–496 bp long) are located on group 5 chromosomes, whereas the short units (327–469 bp long) are located on group 1 chromosomes. In the analysis of 5SDNA RFLPs, certain restriction endonucleases have been found to distinguish particular chromosomal loci. Dvorak et al. (1989), for example, showed that the restriction enzyme *Sst* characterized the 5A locus of *T. monococcum* and hexaploid wheat (*T. aestivum*). Sequencing of 5SDNA units from *T. monococcum* (Scoles et al., 1988) localized this restriction enzyme site in the spacer region of the 482-bp unit which was analyzed. The ability to relate RFLPs of a *5SDna* locus from a diploid species to that of the analogous locus in a hexaploid indicates the relative stability of the DNA at these loci. Similar findings have been made by Gill et al. (1988) for a *5SDna* locus from *Pseudoroegeneria spicata* (diploid genome SS), which occurs in *Elymus trachycaulus* (tetraploid genome SSHH) and by Lagudah et al. (1989) for a locus from *T. tauschii* (diploid genome DD) which is found in *T. aestivum* (hexaploid genome AABBDD).

Ideally the short and long size classes of 5SDNA in all species studied should be assigned to a chromosome and treated as separate characters in a comparative study if they are located on different chromosomes. However, the assignment of 5SDNA units to a certain size class, and thus to either a group 1 or group 5 chromosome locus, requires a detailed analysis using chromosome substitution lines (or *in situ* hybridization). At present this information about the two size classes (repeat lengths ranging from 327 to 496 bp, Fig. 5.9) in all the Triticeae species is limited, and thus we have *tentatively* assigned 5SDNA units 469 bp and larger to the group 5 chromosome locus and units smaller than 468 bp to the chromosome 1 locus.

RFLP Analyses

The Triticeae species surveyed with the *Nor* loci probes were also studied with 5SDNA probes and a detailed study of RFLP variation at the *5SDna* loci carried out using the computer programs discussed earlier. The RFLP variation appeared to be less at this locus, although the basic repeat unit ranged in length from 327 to 496 bp. The study indicated that genera tend to have certain characteristic "fingerprints" for the 5SDNA present in their genomes, even though it is not possible to deduce phylogenetic relationships from the RFLPs per se. The work that characterized RFLPs under stringent conditions of hybridization (Gill et al., 1988; Lagudah et al., 1989), as described for *Nor* loci, predicted that at a sequence level the *5SDna* loci of species should be readily distinguishable from each other. As discussed in the following section, this has been confirmed by direct sequencing. The ability to demonstrate this at the level of RFLPs using total genomic DNA and specific DNA probes is important because it indicates that the sequences of the units used in a number of studies represent the large proportion of the population of units at a particular locus.

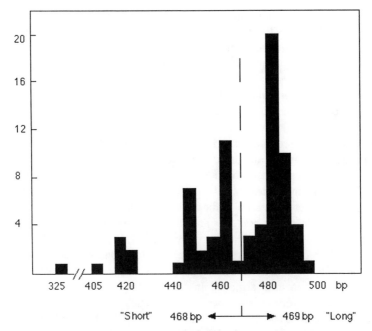

Figure 5.9. Size distribution of 5S DNA units in the Triticeae. The sizes for the long units indicate a symmetrical distribution about a modal size of 485 bp. Units from *Secale cereale, Triticum monococcum,* and *T. tauschii* in the long size range, as defined in the diagram, have been formally shown to be located on group 5 chromosomes (as discussed in the text). The short units (468 bp and less) are a more heterogeneous group with respect to length, but within this group units of 327 bp (*T. monococcum*), 419 bp (*T. tauschii*), and 465 bp (*S. cereale*) in length have been formally located to group 1 chromosomes (as discussed in the text).

Sequence Analyses

Finding the Consensus Sequence

The DNA sequence data base available for the entire 5SDNA unit from many species in the Triticeae (Scoles et al., 1988; Lagudah et al., 1989; Reddy and Appels, 1989; Clarke et al., unpublished data) includes three to five units from each species. To determine a consensus sequence for a group of 5SDNA sequences, an alignment was carried out since it has been found that single or groups of bases are often simply deleted from a unit relative to the other units. The alignment programs discussed earlier were run without weighting leading, or trailing, gaps to maximize the positional similarities. We found that the alignments were improved (as judged from the minimal increase in the length of the aligned sequences) by arranging the highly conserved sections of the 5S RNA genes in a block at one end rather than leaving them in the form in which they were cloned, with parts of the gene at the leading and trailing regions. The *Bam*HI site used for cloning is located 32 bases from the 5′ start of the 5S RNA gene. The consensus sequence for a group of units from a species was determined by assessing the most prominent base at a given position.

Relationships between members of the Triticeae based on the 5SDna locus

The DNAPARS program in PHYLIP was used to compare 75 5SDNA sequences; in this program all sites are inspected, and any site that is invariant in all the sequences entered is excluded from further computations, whereas all other sites (including apparent deletions) are used to construct a relationship. Many trees (45) were generated in the analysis, but these differed from each other mainly at the ends of the branches and in general the overall topology was the same for the entire set. A bootstrap tree using DNABOOT in PHYLIP was computed. The tree obtained is also the majority-rule consensus tree with reliability of the branches indicated by the number of times the respective branches appeared in the bootstrap samples. In our tree, represented as a network, the basic branches are very reliable, because the branch leading to the short units is present in 100% of the bootstrap samples, the branch for the *Hordeum–Critesion* group in 95% of the samples, the *Haynaldia* branch in 100% of the samples, and the branch for the long units in 85% of the samples. The genera within the groups are also very reliable as are the individual sequences within a species. It is only the branches among the species within genera that are less reliable. Since numerous sequences of the 5SDNA from *Triticum* species are currently becoming available (Clarke and Appels, unpublished data), we have chosen to summarize the data as an unrooted network (Fig. 5.10). It is evident from the network that the separation between short and long units is confirmed and that the units from *Hordeum* and *Dasypyrum* occupy an intermediate position. The units in *H. vulgare* and *D. villosa* are located on group 5 chromosomes (Scoles et al., 1988), and thus they fit the definition of being long units; it will be of interest to see whether further sequence work confirms their intermediate position. The relatively close position of *Australopyrum* to *Elymus* and *Pseudoroegeneria* is consistent with observations made on the *Nor* locus comparisons (Fig. 5.6) and is discussed further below.

Discussion

The positions of genera in the various phylogenetic trees shown in this chapter (Figs. 5.6, 5.7, and 5.10), and elsewhere (West et al., 1988; Kellogg, 1989) differ from each other depending on the parameter used to assess the relationships. In view of the fact that two *adjoining* DNA sequences, such as the *Nor* spacer subrepeat and ETS, assaying the same locus, can provide a different relationship

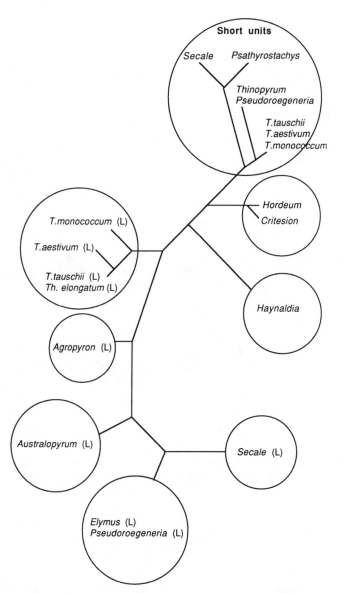

Figure 5.10. Relationships among members of the Triticeae based on the 5S DNA loci. As discussed in the text the relationships are presented as unrooted networks with the letter L following those units which were classified into the long category. A number of different analyses have indicated that the long and short 5S DNA units should be treated as separate characters and, consistent with this, the network shown appears to be two networks superimposed on each other.

between the *Nor* loci of *Triticum, Secale*, and *Australopyrum*, it seems unlikely that molecular parameters will provide new stability in establishing relationships between whole organisms. (We note that this is not the case when relationships at a much broader level, comparing plants, animals, and lower eukaryotes, are studied using different parts of the rRNA gene regions (e.g., see McCarroll et al., 1983)). In certain situations, however, the behavior of a single character can throw new light on a specific evolutionary question. In particular we can, for example, ask if the properties of the *Nor* locus in *Australopyrum* are consistent with the possibility that *Australopyrum* is a Gondwanan relic or a relatively recent introduction (discussed in Scoles et al., 1988, and West et al., 1988). *Australopyrum* is largely limited to southeastern Australia, although as discussed below there is a possibility that it may also occur in New Guinea. The relationships at the *Nor* locus, as assessed by the spacer subrepeat, demonstrate that *A. retrofractum* groups with *Pseudoroegeneria spicata* (Fig. 5.6). The genera *Secale, Agropyron*, and *Triticum* fall outside this group. The relationships based on a *5SDna* locus (the long category and tentatively assigned to the group 5 chromosome) are not exactly the same but are consistent in suggesting a relatively close relationship between *Australopyrum* and *Elymus/Pseudoroegeneria* (Fig. 5.10). Because *Pseudoroegeneria, Secale, Triticum*, and *Agropyron* are all genera from the Northern Hemisphere, the available information on relationships based on the *Nor* locus studies would suggest that *Australopyrum* is not a Gondwanan relic. However, if *Australopyrum* were a Gondwanan relic, the genera of the Triticeae also would have to be extremely ancient. The above argument is not a strong one; however, it does raise the possibility that *Australopyrum* found its way into Australia in relatively recent times, possibly via an Asian route. In this regard, it is of interest that Veldkamp and Scheindelen (1989) have recently described *Australopyrum* among species of the Triticeae occurring in New Guinea. Confirmation of this observation will open up the intriguing possibility of a significant relationship between the Australian and Asian Triticeae species.

Few data are currently available to calibrate the rate of change in a DNA segment, such as the spacer subrepeat region found in the rDNA units of the *Nor* locus (for a recent discussion, see Qu et al., 1988). A particular point of interest is the length of time that has separated *Australopyrum* from the other Triticeae species. Vinizky et al. (unpublished data) have compared the sequences of the spacer subrepeat regions from the *Nor* locus on chromosome 5D of *Triticum tauschii* (the D genome donor to the hexaploid *T. aestivum*, genomic formula AABBDD) and from the *Nor* locus on chromosome 5D of *T. aestivum*. The ends of the 26S rDNA gene regions differed by six mutations (1.51%), the start of the 18S gene region by one mutation (0.25%), and the spacer subrepeat region by ten mutations (1.09%). The absence of a clear difference between the gene and spacer regions in terms of the percentage of mutational change contrasts to the known differences in the rates of accumulation of change in these regions in the Triticeae (Appels and Dvorak, 1982), as discussed at the outset of this chapter.

It thus seems possible that the mutational differences between the rDNA units from the two sources of chromosome 5D *Nor* loci reflect a certain level of "noise" in the analysis due to the fact that a single unit from a locus composed of tandemly repeated units, from a single accession of each of the species studied, was examined. It *is* known, however, that the rDNA units from the two chromosome 5D loci have been separated for at least 8,000 to 10,000 years, because the hexaploid *T. aestivum* appeared in primitive agriculture systems; it appears that no significant differences have accumulated in this time. Comparing the *Nor* locus sequence information for *Australopyrum* to that from *T. tauschii* shows no significant differences in the gene regions, whereas the spacer subrepeat region shows differences at 18.1% of the positions. The *Australopyrum Nor* locus has therefore been separated from that of *T. tauschii* for a period of time that greatly exceeds the separation of the *T. tauschii Nor* locus (on chromosome 5) and the *Nor* locus on chromosome 5D in *T. aestivum*, although an exact estimate of this time is not yet possible. As noted by Scoles et al. (1988), it will be of interest to investigate further the Triticeae species in Australia and Asia at a number of loci to see if sequence changes in DNA can be further calibrated in time.

In concluding this review, it is evident that in attempts to understand the evolution of species in a tribe such as the Triticeae, it is more important to consider the evolution of specific loci. In this way the evolution of linkage groups in both diploids and polyploids can be compared. Within a linkage group, a number of different DNA sequence families will coexist, ranging from single-copy genes coding for specific protein markers, groups of genes controlling a morphologic character, families of genes such as found at the *Nor* and *5SDna* loci, families of tandemly repeated sequences such as found at the ends of chromosomes (*Ter* loci), and families of dispersed repetitive sequences, including transposable elements. The DNA sequences interact via protein and other biological molecules to modulate the final morphology of the individual in which they are located. The available evidence indicates that each individual DNA sequence, or family of sequences, is under different functional constraints, and thus no single character can provide the basis for the relationships between whole organisms. In using loci such as the *Nor* and *5SDna* for studying the relationships between diploid and polyploid species, it is worth noting that, although instability with respect to numbers of repetitive units at these loci has been observed (e.g., Brettell et al., 1985), the loci are generally stable at the level of the DNA sequence of individual repetitive units; a similar conclusion was reached in the analysis of the *5SDna* locus of *Drosophila* (Samson and Wegnez, 1989).

References

Appels, R., and Dvorak, J. (1982) The wheat ribosomal DNA spacer region: its structure and variation in populations and among species. *Theor. Appl. Genet.* **63**, 337–348.

Appels, R., and Honeycutt, R.L. (1986) rDNA: evolution over a billion years. In: *DNA Systematics. Vol. 2,* (ed. S.K. Dutta), CRC Press, Boca Raton, FL., pp. 81–135.

Appels, R., Gerlach, W., Dennis, E., Swift, H., and Peacock, W. (1980) Molecular and chromosomal organization of DNA sequences coding for the ribosomal RNAs in cereals. *Chromosoma* **78**, 293–311.

Appels, R., McIntyre, C.L., and Clarke, B.C. (1986a) Alien chromatin in wheat: ribosomal DNA spacer probes for detecting specific nucleolar organiser region loci introduced into wheat. *Can. J. Genet. Cytol.* **28**, 665–672.

Appels, R., Moran, L.B., and Gustafson, J.P. (1986b) Structure of DNA from the rye (*Secale cereale*) Nor-R1 locus and its behavior in wheat backgrounds. *Can. J. Genet. Cytol.* **28**, 673–685.

Appels, R., Reddy, P., McIntyre, C.L., Moran, L.B., Frankel, O.H., and Clarke, B.C. (1989) The molecular-cytogenetic analysis of grasses and its application to studying relationships among species of the Triticeae. *Genome* **31**, 122–133.

Barker, R.F., Harberd, N.P., Jarvis, M.G., and Flavell, R.B. (1988) Structure and evolution of the intergenic spacer in ribosomal DNA repeat units of wheat. *J. Mol. Biol.* **21**, 1–17.

Baum, B.R. (1973) The concept of relevance in taxonomy with special emphasis on automatic classification. *Taxon* **22**, 329–332.

Baum, B.R. (1983) A phylogenetic analysis of the tribe Triticeae (Poaceae) based on morphological characters of the genera. *Can. J. Bot.* **61**, 518–535.

Brettell, R.I.S., Pallotta, M.A., Gustafson, J.P., and Appels, R. (1985) Variation in the Nor loci in triticale derived from tissue culture. *Theor. Appl. Genet.* **71**, 637–643.

Cheung, W.Y., and Scott, N.S. (1989) A contiguous sequence in spinach nuclear DNA is homologous to three separated sequences in chloroplast DNA. *Theor. Appl. Genet.* **77**, 625–633.

Dams, E., Yamada, T., De Baere, R., Huysmans, E., Vandenberghe, A., and DeWachter, R. (1987) Structure of 5S rRNA in Actinomycetes and relatives and evolution of eubacteria. *J. Mol. Evol.* **25**, 255–260.

Delcasso-Tremousaygue, D., Grellet, F., Panabieres, F., Ananiev, E., and Delseny, M. (1988) Structural and transcriptional characterization of the external spacer of a ribosomal RNA nuclear gene from a higher plant. *Eur. J. Biochem.* **466**, 1–10.

Denny, T.P., Gilmour, M.N., and Selander, R.K. (1988) Genetic diversity and relationships of two pathovars of *Pseudomonas syringae*. *J. Gen. Microbiol.* **134**, 1949–1960.

Doyle, J.J., Doyle, J.L., and Brown, A.H.D. (1990) A chloroplast DNA phylogeny of the wild perennial relatives of soybean (*Glycine* subgenus *Glycine*): congruence with morphological and crossing groups. *Evolution* **44**, 371–389.

Dvorak, J., Zhang, H.-B., Kota, R.S., and Lassner, M. (1989) Organization and evolution of the 5S ribosomal RNA gene family in wheat and related species. *Genome* **32**, 1003–1016.

Erdmann, V.A., and Wolters, J. (1986) Collection of published 5S, 5.8S, and 4.5S ribosomal RNA sequences. *Nucleic Acids Res.* **14**, r1–r59.

Felsenstein, J. (1989) *PHYLIP: Phylogeny Inference Package*, Version 3.2, Dept of Genetics, SK50, University of Washington, Seattle, WA.

Gerbi, S.A., Savino, R., Stebbins-Boaz, B., Jeppesen, C., and Rivera-Leon, R. (1989) A role for U3 in the nucleolus? In: *Ribosomes* (eds. W. Hill, P. Moore, D. Schlessinger, A. Dahlberg, J. Warner, and R. Garrett), American Society for Microbiology, Washington, DC.

Gerstner, J., Schiebel, K., von Waldburg, G., and Hemleben, V. (1988) Complex organization of the length heterogeneous 5' external spacer of mung bean (*Vigna radiata*) ribosomal DNA. *Genome* **30**, 723–733.

Gill, B.S., and Appels, R. (1988) Relationships between Nor-loci from different Triticeae species. *Plant Syst. Evol.* **160**, 77–90.

Gill, B.S., Morris, K.L., and Appels, R. (1988) Assignment of the genomic affinities of chromosomes from polyploid *Elymus* species added to wheat. *Genome* **30**, 70–82.

Gouy, M., and Li, W.-H. (1989) Molecular phylogeny of the kingdoms Animalia, Plantae and Fungi. *Mol. Biol. Evol.* **6**, 109–122.

Govindaraju, D.R., Wagner, D.B., Smith, G.P., and Dancik, B.P. (1988) Chloroplast DNA variation within individual trees of a *Pinus banksiana–Pinus contorta* sympatric region. *Can. J. For. Res.* **18**, 1347–1350.

Gower, J.C. (1971) A general coefficient of similarity and some of its properties. *Biometrics* **27**, 857–871.

Gutell, R.R., Weiser, B., Woesse, C.R., and Noller, H.F. (1985) Comparative anatomy of 16S-like ribosomal RNA. *Prog. Nucleic Acid Res. Mol. Biol.* **32**, 155–216.

Hamby, R.K., and Zimmer, E.A. (1988) Ribosomal RNA sequences for inferring phylogeny within the grass family (Poaceae). *Plant Syst. Evol.* **160**, 29–37.

Hendriks, L., Huysmans, E., Vandenberghe, A., and DeWachter, R. (1986) Primary structures of the 5S ribosomal RNAs of 11 arthropods and applicability of 5S RNA to the study of metozoan evolution. *J. Mol. Evol.* **24**, 103–109.

Hendy, M.D., and Penny, D. (1982). Branch and bound algorithms to determine minimal evolutionary trees. *Math. Biosci.* **59**, 277–290.

Higgins, D.G., and Sharp, P.M. (1988) CLUSTAL: a package for performing multiple sequence alignment on a microcomputer. *Gene* **73**, 237–244.

Higgins, D.G., and Sharp, P.M. (1989) Fast and sensitive multiple sequence alignments on a microcomputer. *CABIOS* **5**, 151–153.

Huysmans, E., and DeWachter, R. (1986) Compilation of small ribosomal subunits of RNA sequences. *Nucleic Acid Res.* **14**, r73–r118.

Johnson, A.M., and Baverstock, P.R. (1989) Rapid ribosomal RNA sequencing and the phylogenetic analysis of Protists. *Parasitol. Today* **5**, 102–105.

Kato, A., Yakura, K., and Tanifuji, S. (1985) Repeated DNA sequences found in the large spacer of *Vicia faba* rDNA. *Biochim. Biophys. Acta* **825**, 411–415.

Kellogg, E.A. (1989) Comments on genomic genera in the Triticeae (Poaceae). *Amer. J. Bot.* **76**, 796–805.

Kelly, R.J., and Siegel, A. (1989) The *Cucurbita maxima* ribosomal DNA intergenic spacer has a complex structure. *Gene* **80**, 239–248.

Klenk, H.-P., Haas, B., Schwass, V., and Zillig, W. (1986) Hybridization homology; a new parameter for the analysis of phylogenetic relations, demonstrated with the Urkingdom of the archaebacteria. *J. Mol. Evol.* **24**, 167–173.

Kota, R.S., and Dvorak, J. (1986) Mapping of a chromosomal pairing gene and 5S rRNA genes in *Triticum aestivum* L. by a spontaneous deletion in chromosome arm 5. *Can. J. Genet. Cytol.* **28**, 266–271.

Kshirsagar, A.M. (1972) *Multivariate Analysis*. New York.

Lagudah, E.S., Clarke, B.C., and Appels, R. (1989) Phylogenetic relationships of *Triticum tauschii*, the D genome donor of hexaploid wheat. 4. Variation and chromosomal location of 5SDNA. *Genome* **32**, 1017–1025.

Lane, D.J., Pace, B., Olsen, G.J., Stahl, D.A., Sogin, M.L., and Pace, N.R. (1985) Rapid determination of 16S ribosomal RNA sequences for phylogenetic analyses. *Proc. Natl. Acad. Sci. USA* **82**, 6955–6959.

Lassner, M., and Dvorak, J. (1986) Preferential homogenization between adjacent and alternate subrepeats in wheat rDNA. *Nucleic Acids Res.* **14**, 5499–5512.

Lassner, M., Anderson, O., and Dvorak, J. (1987) Hypervariation associated with a 12-nucleotide direct repeat and inferences on intergenomic homogenization of ribosomal RNA gene spacers based on the DNA sequence of a clone from the wheat Nor-D3 locus. *Genome* **29**, 770–781.

Loukas, M., Delidakis, C., and Kafatos, F.C. (1986) Genomic blot hybridization as a tool of phylogenetic analysis: evolutionary divergence in the genus *Drosophila*. *J. Mol. Evol.* **24**, 174–188.

Löve, A. (1984) Conspectus of the Triticeae. *Feddes Repert.* **95**, 425–521.

May, C.E., and Appels, R. (1987) Variability and genetics of spacer DNA sequences between the ribosomal RNA genes of hexaploid wheat (*Triticum aestivum*). *Theor. Appl. Genet.* **74**, 617–624.

McCarroll, R., Olsen, G.J., Stahl, Y.D., Woesse, C.R., and Sogin, M.L. (1983) Nucleotide sequence of the *Dictyostelium discoideum* small-subunit ribosomal ribonucleic acid inferred from the gene sequence: evolutionary implications. *Biochemistry* **22**, 5858–5868.

McIntyre, C.L., Clarke, B.C., and Appels, R. (1988) DNA sequence analyses of the ribosomal spacer regions in the Triticeae. *Plant Syst. Evol.* **160**, 91–104.

McMullen, M.D., Hunter, B., Phillips, R.L., and Rubenstein, I. (1986) The structure of the maize ribosomal DNA spacer region. *Nucleic Acids Res.* **14**, 4953–5968.

Molnar, S.J., Gupta, P.K., Fedak, G., and Wheatcroft, R. (1989) Ribosomal DNA repeat unit polymorphism in 25 *Hordeum* species. *Theor. Appl. Genet.* **78**, 387–392.

Neale, D.B., Saghai-Maroof, M.A., Allard, R.W., Zhang, Q., and Jorgensen, R.A. (1988) Chloroplast DNA diversity in populations of wild and cultivated barley. *Genetics* **120**, 1105–1110.

Olsen, G.J. (1988) Phylogenetic analysis using ribosomal RNA. In: *Ribosomes* (eds. H.F. Noller, Jr. and K. Moldave), *Meth. Enzymol.* **164**, 793–812.

Olsen, G.J., and Woese, C.R. (1989) A brief note concerning archaebacterial phylogeny. *Can. J. Microbiol.* **35**, 119–123.

Palmer, J.D. (1986) Chloroplast DNA and phylogenetic relationships. In: *DNA Systematics, Vol. 2* (ed. S.K. Dutta), CRC Press, Boca Raton, FL. pp. 63–80.

Palmer, J.D., Jansen, R.K., Michaels, H.J., Chase, M.W., and Manhart, J.R. (1988) Chloroplast DNA variation and plant phylogeny. *Ann. Missouri Bot. Gard.* **75**, 1180–1206.

Patterson, C. (1988) Homology in classical and molecular biology. *Mol. Biol. Evol.* **5**, 603–625.

Procunier, J.D. and Kasha, K.J. (1991) Organization of the intergenic spacer region of rRNA genes of *Hordeum bulbosum* L., submitted.

Qu, L.-H., Michot, B., and Bachellerie, J.-P. (1983) Improved methods for structure probing in large RNAs: a rapid "heterologous" sequencing approach is coupled to the direct mapping of nuclease accessible sites. *Nucleic Acids Res.* **11**, 5903–5920.

Qu, L.-H., Nicolson, M., and Bachellerie, J.-P. (1988) Phylogenetic calibration of the 5' terminal domain of large rRNA achieved by determining twenty eucaryotic sequences. *J. Mol. Evol.* **28**, 113–124.

Reddy, P., and Appels, R. (1989) A second locus for the 5S multigene family in *Secale* L.: sequence divergence in two lineages of the family. *Genome* **32**, 456–467.

Reddy, P., Appels, R., and Baum, B.R. (1990) Ribosomal DNA spacer-length variation in *Secale* (Poaceae). *Plant Syst. Evol.*, in press.

Samson, M.-L., and Wegnez, M. (1989) An approach to study the evolution of the *Drosophila* 5S ribosomal genes using P-element transformation. *J. Mol. Evol.* **28**, 517–523.

Sarle, W.S. (1985) Two-stage density linkage. In: *SAS-User's Guide: Statistics*, SAS Institute Inc., Cary, NC., p. 267.

Scoles, G.J., Gill, B.S., Xin, Z.-Y., Clarke, B.C., McIntyre, C.L., Chapman, C., and Appels, R. (1988) Frequent duplication and deletion events in the 5S RNA genes and the associated spacer region of the Triticeae. *Plant Syst. Evol.* **160**, 105–122.

Smith, D.K. (1987) *Concept and Use of Software for Pair-wise and Multiple Genetic Sequence Alignment*, Technical Rept. No. 11, Canberra College of Advanced Education, ACT, Australia.

Sneath, P.H.A. (1976) Phenetic taxonomy at the species level and above. *Taxon* **25**, 437–450.

Song, K.M., Osborn, T.C., and Williams, P.H. (1988a) Brassica taxonomy based on nuclear restriction fragment length polymorphisms (RFLPs). 1. Genome evolution of diploid and amphidiploid species. *Theor. Appl. Genet.* **75**, 784–794.

Song, K.M., Osborn, T.C., and Williams, P.H. (1988b) Brassica taxonomy based on nuclear restriction fragment length polymorphisms (RFLPs). 2. Preliminary analysis of subspecies within *B. rapa* (syn. *campestris*) and *B. oleracea. Theor. Appl. Genet.* **76**, 593–600.

Stern, D.B., and Lonsdale, D.M. (1982) Mitochondrial and chloroplast genomes of maize have a 12-kilobase DNA sequence in common. *Nature* **299**, 698–702.

Taira, T., Kato, A., and Tanifuji, S. (1988) Difference between two major size classes of carrot rDNA repeating units is due to reiteration of sequences of about 460bp in the large spacer. *Mol. Gen. Genet.* **213**, 170–174.

Toloczyki, C., and Feix, G. (1986) Occurrence of 9 homologous repeat units in the external spacer region of a nuclear maize rDNA gene unit. *Nucleic Acids Res.* **14**, 4969–4986.

Van den Eyne, H., De Baere, R., De Roeck, E., Van de Peer, Y., Vandenberghe, A., Willekens, P., and de Wachter, R. (1988) The 5S ribosomal RNA sequences of a red algal rhodoplast and gymnosperm chloroplast. Implications for the evolution of plastids and cyanobacteria. *J. Mol. Evol.* **27**, 126–132.

Veldkamp, J.F., and Scheindelen, van H.J. (1989) *Australopyrum, Brachypodium,* and *Elymus* (Gramineae) in Malesia. *Blumea* **34**, 61–76.

Vincentz, M., and Flavell, R.B. (1989) Mapping of ribosomal RNA transcripts in wheat. *The Plant Cell* **1**, 579–589.

Waterman, M.S. (1988) Computer analysis of nucleic acid sequences. In: *Ribosomes* (eds. H.F. Noller, Jr. and K. Moldave), *Meth. Enzymol.* **164**, 765–793.

West, J.G., McIntyre, C.L., and Appels, R. (1988) Evolution and systematic relationships in the Triticeae (Poaceae). *Plant Syst. Evol.* **160**, 1–28.

Wilbur, W.J., and Lipman, D.J. (1984) The context dependent comparison of biological sequences. *SIAM J. Appl. Math.* **44**, 557–567.

Woese, C.R. (1987) Bacterial evolution. *Microbiol. Rev.* **51**, 221–271.

Wolters, J., and Erdmann, V.A. (1988) Cladistic analysis of ribosomal RNAs–the phylogeny of eukaryotes with respect to the endosymbiotic theory. *Biosystems* **21**, 209–214.

Youvan, D., and Hearst, J. (1981) A sequence from *Drosophila melanogaster* 18S rRNA bearing the conserved hyper-modified nucleoside amU: analysis by reverse transcription and high-performance liquid chromatography. *Nucleic Acids Res.* **9**, 1723–1741.

Zucker, M., and Stiegler, P. (1981) Optimal computer folding of large RNA sequences using thermodynamics and auxillary information. *Nucleic Acids Res.* **9**, 133–148.

Zurawski, G., and Clegg, M.T. (1987) Evolution of higher-plant chloroplast DNA-encoded genes: implications for structure-function and phylogenetic studies. *Ann. Rev. Plant Physiol.* **38**, 391–418.

6

Intraspecific Chloroplast DNA Variation: Systematic and Phylogenetic Implications

Douglas E. Soltis, Pamela S. Soltis, and
Brook G. Milligan

Systematists have long sought methodologies that would facilitate phylogenetic reconstruction based on at most a few representative collections per species. As a result of their perceived invariance within species, molecular techniques have become popular tools. Further study, however, has often revealed additional intraspecific variation that must be considered in systematic studies. For example, during the 1960s and 1970s, flavonoid chemistry emerged as the tool of choice. Later, detailed investigations of single species proved that, for many plant groups, flavonoids were much more variable than originally thought and were therefore less useful (reviewed in Bohm, 1987). In contrast, the presence of intraspecific protein variation detected by electrophoresis has long been recognized and so has never been used extensively by plant systematists. An insightful discussion of this is given by Crawford (1989).

Restriction site analysis of chloroplast DNA (cpDNA) has emerged as the most popular technique in plant systematics for phylogenetic reconstruction below the family level. Foremost of the advantages offered by this molecular marker is its conservative rate of evolution, both structurally and in DNA sequence (Curtis and Clegg, 1984; Gillham et al., 1985; Palmer, 1985a, b, 1987; Clegg et al., 1986, 1990; Wolfe et al., 1987; Zurawski and Clegg, 1987). This molecule is regarded by many systematists as the ideal tool, because it offers the possibility of inferring accurate phylogenies based on one or a few samples of each species. However, this view is only tenable for taxa that exhibit little or no intraspecific variation.

This chapter reviews examples of intraspecific variation in the chloroplast genome from a wide array of plant groups. In fact, intraspecific variability has been extensive in some instances and has been detected in a large proportion of those species for which several populations have been analyzed. Furthermore, in many studies these observations were made using sampling schemes that were grossly inadequate for detecting intraspecific variation. Indeed, that was not the goal in most cases. This suggests that intraspecific variation in plastid DNA may be more prevalent than currently appreciated.

This work was supported in part by NSF grants BSR-8620444 to PSS and BSR-8516721 and BSR-8717471 to DES. We greatly appreciate the help of those investigators who shared unpublished data.

The presence of intraspecific variability in cpDNA raises a number of important questions for the plant systematist and evolutionary biologist. We deal with the most fundamental of these questions, including (1) how often and in which taxa has variation been observed, (2) what is the genetic nature of the variation, (3) what is the magnitude of intraspecific sequence divergence, (4) is the variation sufficient to be useful for population-level studies, and (5) is cpDNA variation concordant with variation in other traits such as morphology or nuclear-encoded markers? Of particular importance to the subject of this volume is the implication of the documented intraspecific cpDNA variation for plant systematists interested primarily in phylogenetic relationships among congeneric species.

Distribution of Intraspecific
Chloroplast DNA Variation

Although the chloroplast genome is conservative in its evolution, reports of intraspecific variation are frequent. Table 6.1 summarizes nearly 60 examples of such variation from over 15 plant families ranging from ferns to conifers to a diversity of angiosperms. This is a considerable number given the short history of cpDNA as a tool in systematics (Atchison et al., 1976, and Vedel et al., 1976, represent the beginnings of this field). It is even more remarkable given the sampling designs typically used, which are biased against detecting intraspecific variation. Most of the studies cited in Table 6.1 did not attempt to survey large numbers of either conspecific individuals or populations, and thus were unlikely to detect intraspecific variation. Nevertheless, variation was detected. Additionally, several of the early studies revealed intraspecific variation, even though they employed only one or a few restriction endonucleases (e.g., Scowcroft, 1979; Metzlaff et al., 1981; Bowman et al., 1983).

The first attempt to analyze a large number of conspecific individuals and populations was by Banks and Birky (1985). Among approximately 100 individual plants of *Lupinus texensis,* representing 21 populations from throughout central Texas, the authors detected three variant forms. Two of the variants, represented by single plants, differed from the wild-type form by a single restriction site gain or loss. The third variant, represented by all 11 plants from a single population, differed from the wild-type form both by the gain of one restriction site also found in another variant and by a 100-bp deletion. In retrospect, the title of this now classic paper, "Chloroplast DNA diversity is low in a wild plant, *L. texensis,"* was in some ways unfortunate. As indicated by Banks and Birky, the amount of variation detected is low compared to that found either in animal mitochondrial DNA or in nuclear DNA from plants and animals (reviewed by Wolfe et al., 1987; Zurawaski and Clegg, 1987). However, the number of restriction endonucleases employed in their analysis (seven) is low compared to the number commonly used today (15–30). Thus, even more extensive cpDNA variation may be detectable in this species with additional enzymes. Unfortu-

Table 6.1. Examples of intraspecific cpDNA variation.

Taxon	Comments	Source
Aegilops speltoides	Variation detected with 4 endonucleases	Bowman et al. (1983)
Aegilops triuncialis	Analysis of 26 accessions of this allotetraploid revealed three different cytoplasms. Two of these are due to multiple origins involving different maternal parents	Murai and Tsunewaki (1986) Ogihara and Tsunewaki (1982, 1988)
Astragalus molybdenus	High levels of variation; geographic paritioning	Lavin et al. (unpublished)
Beta macrocapra, B. maritima	Variation detected with 7 endonucleases	Kishima et al. (1987)
Brassica campestris	Different accessions differ by as many as 4 restriction site mutations	Palmer et al. (1983), Kemble (1987)
B. napus	Variation in this tetraploid may be due in part to introgression	Palmer et al. (1983), Kemble (1987)
B. nigra	Different accessions differ by as many as seven restriction site mutations	Palmer et al. (1983)
Calocedrus decurrens	Polymorphism detected with only 3 endonucleases	Ali et al. (1991)
Daucus carota	Difference in restriction profile between ssp. *sativus* and ssp. *gummifer* for *BamH*I; few endonucleases employed	Matthews et al. (1984)
Dioscorea bulbifera	Evidence for polyphyletic origin of Asian cultivated yam	Terauchi (1989)
Gliricidia sepium	3 distinct cytoplasms were identified	Lavin et al. (unpublished)
Glycine max, G. soja	Numerous cultivars and introductions were screened; 6 different cpDNAs were identified	Close et al. (1989)
G. tabacina	Considerable variation; evidence for multiple origins of the polyploid cytotype	Doyle et al. (1990a, b)
Helianthus annuus	4 restriction site mutations detected	Rieseberg et al. (1988)
H. bolanderi	"Weedy" and "serpentine" races were well-differentiated; variation also present within the "serpentine" race	Rieseberg et al. (1988)
H. debilis ssp. *cucumerifolius*	Evidence for introgression	Rieseberg et al. (1990b)
H. petiolaris	Variation within *H. petiolaris* suggested that *H. neglectus* is a recent derivative of *H. petiolaris* ssp. *fallax*	Rieseberg et al. (1990a)
Heuchera micrantha	14 restriction site mutations, 3 length mutations; evidence for multiple origins of tetraploid cytotype	Soltis et al. (1989a)
H. grossulariifolia	High levels of variation; multiple origins of tetraploid cytotype	Wolf et al. (1990)
Hordeum vulgare ssp. *spontaneum, H. vulgare* ssp. *vulgare*	Data suggest a cytoplasmic bottleneck during domestication; intra- as well as interpop-ulational variation observed in ssp. *spontaneum*	Clegg et al. (1984a), Holwerda et al. (1986), Neale et al. (1988)
Lisianthius skinneri	1 restriction site mutation, 1 length mutation	Sytsma and Schaal (1985)
Lupinus texensis	First detailed study of numerous individuals within a species	Banks and Birky (1985)

(*continued*)

Table 6.1. (Continued)

Taxon	Comments	Source
Lycopersicon peruvianum	Implications for relationship to *L. chilense* and *L. chmielewskii*, evidence for introgression	Palmer and Zamir (1982)
Medicago sativa	Intraplant cpDNA heterogeneity	Johnson and Palmer (1989)
M. scutellata	Intraplant cpDNA heterogeneity	Johnson and Palmer (1989)
Nicotiana debueyi	First report of intraspecific cpDNA variation; 1 endonuclease used	Scowcroft (1979)
Orzya latifolia	2 strains differed; 1 endonuclease used	Ichikawa et al. (1986)
O. sativa	cpDNA differences between ecospecies; intraplant cpDNA heterogeneity	Ichikawa et al. (1986), Ishii et al. (1986)
Pelargonium zonale hort.	1 endonuclease employed; 3 plastome types	Metzlaff et al. (1981)
Persea americana	Mutations distinguished the 3 presently recognized varieties; the data also suggest that *P. americana* is not monophyletic	Furnier et al. (1990)
Pinus contorta, P. banksiana	Numerous individuals examined; 6 cpDNA variants observed in each; novel cpDNA polymorphism and variation within individual trees in a zone of sympatry	Wagner et al. (1987), Govindaraju et al. (1988, 1989)
Pisum humile	Distribution of variation suggests that *P. sativum* was domesticated from northern populations of this species	Palmer et al. (1985)
P. sativum	Evidence for introgression	Palmer et al. (1985)
P. elatius	Minor variation	Palmer et al. (1985)
Polystichum munitum	Minor variation	Soltis and Soltis (1987)
Pseudotsuga menziesii	Polymorphism detected with only 3 restriction endonucleases	Ali et al. (1991)
Pteridium aquilinum	17 restriction site mutations detected; support for the 2 subspecies recognized by Tryon (1941)	Tan and Thomson (1990)
Salix exigua	4 distinct cpDNA lineages that differ by as many as 8 restriction site mutations	Brunsfield et al. (unpublished)
S. interior	6 distinct cpDNA genomes that differ by as many as 18 restriction site mutations	Brunsfeld et al. (unpublished)
S. melanopsis	5 distinct cpDNA genomes that differ by as many as 13 restriction site mutations	Brunsfeld et al. (unpublished)
S. taxifolia	Evidence for introgression	Brunsfeld et al. (unpublished)
Sequoia sempervirens	Polymorphism detected with only 3 restriction endonucleases	Ali et al. (1991)
Solanum tuberosum ssp. *tuberosum, S. tuberosum* ssp. *andigena*	Variation within and between subspecies; a geographic cline of cpDNA was detected from the Andean region to coastal Chile: cpDNA data support the Andean origin of Chilean ssp. *tuberosum*	Hosaka et al. (1984), Hosaka (1986), Hosaka and Hanneman (1988a, b)

(*continued*)

Table 6.1. (Continued)

Taxon	Comments	Source
Solanum stenotomum, *S. goniocalyx,* *S. phureja*	The large amount of cpDNA variation within these diploids suggests that this was the original source of cpDNA variation in the tetraploid *S. tuberosum* ssp. *andigena*	Hosaka and Hanneman (1988a, b)
Sorghum bicolor	6 cpDNA types among 9 accessions, defined by 9 restriction site mutations and 1 length mutation; evidence for introgression with *S. halpense*	Duvall and Doebley (1990)
Tellima grandiflora	Evidence for introgression, well differentiated northern and southern lineages	Soltis et al. (1991a, b)
Tragopogon dubius	2 restriction site mutations observed in a small geographic area	Soltis and Soltis (1989a)
T. miscellus	Variation in this allotetraploid is due to multiple origins involving different maternal parents	Soltis and Soltis (1989a)
Trifolium pratense	Multiple genotypes within populations; permitted analysis of geographic differentiation and estimation of gene flow	Milligan (1991)
T. subterraneum	Several different cpDNA genotypes present	Milligan et al. (1989, unpublished)
Triticum monococcum	Length mutations observed; 4 endonucleases employed	Bowman et al. (1983)
Tolmiea menziesii	High levels of variation; no cytoplasmic gene flow between diploids and tetraploids; cytoplasmically distinct groups of diploids occupy distinct geographic areas; evidence for long-distance seed dispersal	Soltis et al. (1989b)
Zea mays ssp. *mays,* *Z. mays* ssp. *mexicana, Z. mays* ssp. *parviglumis*	5 cpDNA genotypes; evidence for introgression and/or phylogenetic sorting	Doebley et al. (1987), Doebley (1990), Timothy et al. (1979)
Z. perennis	Evidence for introgression involving an undiscovered *Zea* species	Doebley (1989)

nately, however, the findings of Banks and Birky were interpreted by some to indicate that intraspecific plastid DNA variation is so low as to be inconsequential. As a result, their results discouraged some investigators from undertaking detailed surveys of single species. For those interested in phylogenetic relationships among species, the use of small samples seemed justified.

Surprisingly few single species have been the subject of extensive cpDNA analyses. In fewer than 25 taxa, 100 or more individuals or 10 or more populations have been analyzed: *Glycine latifolia, G. microphylla, G. tabacina* (Doyle et al., 1990b), *Helianthus annuus* (Rieseberg et al., 1988), *Heuchera micrantha* (Soltis et al., 1989a), *H. grossulariifolia* (Wolf et al., 1990), *Hordeum vulgare* subsp. *vulgare* and spp. *spontaneum* (Clegg et al., 1984a; Holwerda et al., 1986; Neale et al., 1988), *L. texensis* (Banks and Birky, 1985), *Pennisetum glaucum* (Clegg et al., 1984b; Gepts and Clegg, 1988), *Pinus contorta* and *P. banksiana* (Wagner et al., 1987), *Salix exigua, S. interior,* and *S. melanopsis* (Brunsfeld et al., unpublished data), *Tellima grandiflora* (Soltis et al., 1991a), *Tolmiea menziesii* (Soltis et al., 1989b), *Tragopogon dubius* (Soltis and Soltis, 1989a), *Trifolium pratense* (Milligan, 1991), *T. subterraneum* (Milligan et al., unpublished data), and *Zea mays* ssp. *mays,* ssp. *mexicana,* and ssp. *parviglumis* (Doebley et al., 1987; Doebley, 1989, 1990, unpublished data). In part, the paucity of these studies may be due to the time-consuming nature of large-scale population surveys.

In all but one detailed study of conspecific individuals and/or populations, intraspecific variation in the plastid genome has been reported. The only species in which no variation was detected was *Pennisetum glaucum* (Clegg et al., 1984b; Gepts and Clegg, 1989). Several of the examples in which detailed surveys revealed variation, for example, *Helianthus annuus* (Rieseberg et al., 1988) and *Tragopogon dubius* (Soltis and Soltis, 1989a), involve members of Compositae, a family well known for an apparently low rate of cpDNA sequence divergence, presumably due to its relatively recent origin and radiation (Jansen and Palmer, 1988; see also Chapter 11, this volume). This suggests that intraspecific variation in the chloroplast genome may be more prevalent than currently believed. As reviewed in a later section of this chapter, this variation has provided important evolutionary insights.

Most examples of intraspecific cpDNA variation involve differences among, rather than within, populations. This may be primarily a result of the sampling schemes commonly employed, which tend to minimize the number of individuals studied within each population. As a result, quantitative estimates of the frequency of intrapopulational variation are not possible. Nevertheless, several studies have detected intrapopulational variation in cpDNA, despite commonly sampling only a few individuals per population. For example, the first in-depth analysis of intraspecific plastid DNA variation (*Lupinus texensis*) revealed variability in two populations, even though only two and six individuals were sampled from each (Banks and Birky, 1985). Intrapopulational variation has also been observed in one population of *Zea perennis* (Doebley, 1989), one population of *Pisum humile* (Palmer et al., 1985), several populations of *Medicago sativa* (Johnson and Palmer, 1989), and a number of populations of *Hordeum vulgare* ssp. *spontaneum* (Neale et al., 1988). In addition, Wagner et al. (1987) detected intrapopulational variation in *Pinus contorta* and *P. banksiana,* even though their sampling scheme was not specifically designed to detect such variation. Polymorphism was found in 28 of the 43 allopatric populations from which more than one individual was sampled (see also Govindaraju et al., 1989). The most extensive case of intrapopulational variation has been observed in *Trifolium pratense,* in which each population examined was polymorphic and some contained at least nine different genotypes (Milligan, 1991). The presence of intrapopulational variation in these taxa is significant and suggests that cpDNA variation within populations may be more common than traditionally maintained.

Several recent studies have revealed variation within single plants having paternal or biparental inheritance of the plastid genome. Govindaraju et al. (1988) analyzed cpDNA within each of six trees occurring in a region of sympatry between *P. banksiana* and *P. contorta*. Four trees exhibited variation among samples from different parts of the crown. It is not clear, however, whether this variation represents somatic mutation or occasional biparental inheritance of the chloroplast genome. Two types of intraplant heterogeneity were observed in *Medicago* (Johnson and Palmer, 1989). In one type, two DNAs differed by a single *Xba*I restriction site, and plants possessing both forms were detected in *M. sativa* ssp. *varia,* ssp. *caerulea,* and ssp. *sativa*. The second type, involving frequent insertions or deletions, was detected within plants of *M. scutellata*. Intraplant heterogeneity has also been reported for rice (Moon et al., 1987), although the evidence for this is not conclusive (Nishizawa and Harai, 1987).

Together these studies indicate that intraspecific variation in the chloroplast genome has been detected in a wide taxonomic range of plant species, including representatives of groups thought to exhibit relatively little variation. This variation has been observed at all levels: among populations, within populations, and within individual plants. Importantly, much of the observed variation was revealed by studies that were designed to detect neither intraspecific nor intrapopulational differences. One can only conclude that current estimates of the prevalence of intraspecific variability in the chloroplast genome are underestimates.

Genetic Nature of Intraspecific Chloroplast DNA Variation

The nature of intraspecific cpDNA polymorphism is typically limited to restriction site changes and insertion/deletion mutations. For example, based on the data available from four species with samples greater than 10 genomes (*Pennisetum glaucum, Lupinus texensis, Hordeum vulgare,* and *Zea mays*), Clegg (1989) found that insertions and/or deletions accounted for one-third to one-half of all population-level variation with restriction site changes accounting for the remainder. In several cases, however, intraspecific variation has been ascribed to an inversion. For example, one population of *Pisum humile* is distinguished from other conspecific populations by an inversion of between 2.2 kb and 5.2 kb (Palmer et al., 1985). Similarly, four of 12 analyzed populations of *Salix melanopsis* are characterized by an inversion of between 1.5 kb and 6.5 kb (Brunsfeld et al., unpublished data). Such intraspecific rearrangements of the chloroplast genome appear to be rare, however.

In Table 6.2, we summarize the nature of intraspecific cpDNA variation for a greater number of species than analyzed by Clegg (1989). The results of this larger analysis support his earlier finding that there is a high incidence of length

Table 6.2. *Intraspecific cpDNA variation: site changes versus length mutations.*

Species	Site changes	Length Mutations	Source
Dioscorea bulbifera	1	10	Terauchi (1989)
Helianthus annuus	3	0	Rieseberg et al. (1988)
Helianthus bolanderi (serpentine)	1	0	Rieseberg et al. (1988)
Helianthus petiolaris	1	0	Rieseberg et al. (1990)
Heuchera grossulariifolia	7	3	Wolf et al. (1990)
Heuchera micrantha	14	3	Soltis et al. (1989a)
Hordeum vulgare	3	1	Neale et al. (1988)
Lupinus texensis	3	1	Banks and Birky (1985)
*Lycopersicon peruvianum**	4	0	Palmer and Zamir (1982)
Pennisetum glaucum	0	0	Gepts and Clegg (1989)
Pisum elatius	1	1	Palmer et al. (1985)
Pisum humile†	5	6	Palmer et al. (1985)
Pisum sativum	4	6	Palmer et al. (1985)
Salix exigua	8	2	Brunsfeld et al. (unpublished)
*Salix interior**	20	3	Brunsfeld et al. (unpublished)
*Salix melanopsis**	17	2	Brunsfeld et al. (unpublished)
Sorghum bicolor	9	1	Duvall and Doebley (1990)
Tolmiea menziesii	7	5	Soltis et al. (1989b)
Tragopogon dubius	2	0	Soltis and Soltis (1989)
Zea mays	4	2	Doebley (1990; unpublished)
*Zea perennis**	15	3	Doebley (1989)

*Intraspecific cpDNA variation is hypothesized to be due, entirely or in part, to introgression

†An inversion was detected in one population of this species.

mutation within single species. However, species differ considerably in the occurrence of length mutations versus restriction site mutations. For example, in *L. texensis, H. vulgare,* and *Z. mays* the incidence of length mutations reported was lower than that of restriction site mutations. In contrast, within species such as *Dioscorea bulbifera, Pisum sativum,* and *P. humile,* more length mutations than restriction site changes were reported. Several comparisons among species have also revealed significantly higher numbers of length mutations than restriction site mutations (Gorden et al., 1982; Bowman et al., 1983). For example, 14 length mutations and only 1 restriction site mutation were found among 14 species in *Triticum* and *Aegilops* (Bowman et al., 1983).

These data suggest that in most cases the incidence of intraspecific length variation in cpDNA is relatively high. The available estimates may, however, underestimate the actual incidence (Takaiwa and Sugiura, 1982; Palmer et al., 1985). Whereas the majority of restriction site mutations are easily detected, many length mutations are probably less than 50 bp and thus would easily go undetected in many of the gel systems currently used.

If variation in length is as prevalent as suggested by these data, a question of considerable importance arises with regard to the utility of length mutations in intraspecific phylogenetic analyses: how reliably can one recognize homologous length mutations? Some investigators have emphasized the difficulties involved and have cautioned against the use of length mutations in phylogenetic reconstruction (Palmer, 1985a, b; Palmer et al., 1985; Sytsma and Gottlieb, 1986). Several analyses have demonstrated, however, that length mutations provide assessments of phylogenetic relationship that are concordant with those obtained via restriction site analysis (e.g., Doebley et al., 1987; Soltis et al., 1989a, b; Wolf et al., 1990). Although the investigator should proceed cautiously, available data suggest that length mutations can be safely used in phylogenetic analyses within species. In part, the utility of length mutations will depend on the amount of length variation present. If there are few length mutations and they occur in different parts of the genome, then interpretation is straightforward. If, however, there are many length mutations in a single region, then they are best not used. Given the seeming prevalence of length variation, it may provide a useful tool for intraspecific studies of evolution and population biology.

Magnitude of Intraspecific cpDNA Variation

Although intraspecific cpDNA variation is frequent in occurrence (Table 6.1), the magnitude is seldom very high and varies among species. The low mean value of plastid DNA diversity is not surprising given that interspecific comparisons have demonstrated that the chloroplast genome diverges very slowly in primary sequence compared with nuclear genes and animal mitochondrial genes (Curtis and Clegg, 1984; Gillham et al., 1985; Palmer, 1985a, b; Wolfe et al., 1987; Zurawski and Clegg, 1987; Birky, 1988; Clegg, 1989; Clegg et al., 1990). The considerable difference among species in levels of intraspecific cpDNA variation is clearly evident from Table 6.2, and ranges from no variation revealed by a detailed survey of pearl millet, *Pennisetum glaucum* (Clegg et al., 1984b; Gepts and Clegg, 1989), to relatively high levels of variation reported for such taxa as *Heuchera micrantha* (Soltis et al., 1989a), *H. grossulariifolia* (Wolf et al., 1990), *Salix interior* and *S. melanopsis* (Brunsfeld et al., unpublished data), *Sorghum bicolor* (Duvall and Doebley, 1990), *Trifolium pratense* (Milligan, 1991), several species of *Glycine* subgenus *Glycine* (Doyle et al., 1990b), and *Zea perennis* (Doebley, 1990a).

Several investigators have reported intraspecific restriction site variation in cpDNA in terms of sequence divergence (Nei and Li, 1979; Nei, 1987). These values also reveal considerable variation among species. Obviously, sequence divergence in the monomorphic *P. glaucum* is 0.0%. Low levels of sequence divergence were reported for *Z. mays* (0.00%–0.06%) (Doebley et al., 1987), *Helianthus annuus* (0.00%–0.08%) (Rieseberg et al., 1988), and *Lupinus texensis* (mean = 0.026%) (Banks and Birky, 1985). The highest values reported to date are for three species of Saxifragaceae, which all have surprisingly similar mean values: *Heuchera micrantha* (0.000%–0.206%; mean of 0.074%) (Soltis et al., 1989a), *H. grossulariifolia* (0.000%–0.299%; mean of 0.071%) (Wolf et al., 1990), and *Tolmiea menziesii* (0.000%–0.144%; mean of 0.076%) (Soltis et al., 1989b). Comparable values have also been determined for several species of *Salix* (Brunsfeld et al., unpublished data): *S. exigua* (0.00%–0.08%; mean of 0.04%), *S. melanopsis* (0.00%–0.16%; mean of 0.07%); *S. interior* (0.00%–0.19%; mean of 0.09%). The values of sequence divergence for populations of these taxa begin to approach those reported for congeneric species in other groups, such as closely related species of *Lycopersicon* (0.0%–0.7%) (Palmer and Zamir, 1982), *Lisianthius* (0.0%–0.3%) (Sytsma and Schaal, 1985), and *Helianthus* (0.0%–0.4%) (Rieseberg et al., 1988). Nonetheless, even the highest levels of cpDNA sequence divergence for single plant species are still lower than those reported for the mitochondrial genome in animals (reviewed in Birky, 1988).

The level of intraspecific cpDNA variation for some taxa (e.g., *Brassica napus, Pisum sativum, Salix melanopsis,* and most notably *S. taxifolia* and *Zea perennis*) may result from introgression. Inflated levels of intraspecific variation due to introgression will probably be common based on the fact that introgression has already been invoked frequently to explain discrepancies between cpDNA-based phylogenies and those inferred from other types of data. This topic will be discussed later in this chapter and also in greater detail by Rieseberg and Brunsfeld in Chapter 7.

Why sequence divergence values are so variable among species is still unclear. It may be that sequence divergence in some groups is higher due to the greater antiquity of their lineages. As with other estimates of intraspecific plastid DNA variation, however, so few species have been examined intensively that few generalities can be made. Obviously, if only one or two populations of a species are sampled, as is frequently the case, there is little or no chance of discovering high levels of sequence divergence. Undoubtedly as more species are examined in greater detail, high levels of sequence divergence comparable to that seen in the Saxifragaceae and Salicaceae will be revealed in other plant groups.

Few patterns of intraspecific cpDNA variation can be discerned, except that the level of variation is itself variable. This is evident even with the single family Poaceae. In one grass species (*Pennisetum glaucum*), no variation was detected (Clegg et al., 1984b; Gepts and Clegg, 1989), whereas in several others (*Hordeum vulgare, Sorghum bicolor, Zea mays,* and *Z. perennis*) relatively high levels of variation have been observed (see Table 6.1), including polymorphism within populations (Neale et al., 1988). Variation is also evident among plant families, with members of the Salicaceae and Saxifragaceae typically exhibiting much greater variation than members of the Compositae. While it may be safe to assume that high levels of intraspecific variation will not be common in the Compositae, phylogenetic analyses of families such as Saxifragaceae, Salicaceae, and Poaceae should take into account the possibility of high intraspecific cpDNA variation. However, the recent documentation of extremely high levels of cpDNA variation within some species of Compositae (Jansen et al., Chapter 11, this volume) indicates the need for caution even in a family well known for its low rate of cpDNA sequence divergence. These results reemphasize the need for assessing levels of intraspecific variation prior to choosing accessions for phylogenetic analysis at the species level or higher (Neale et al., 1988; Clegg, 1989; Soltis et al., 1989a; Doyle et al., 1990b).

Evolutionary Processes and Phylogenetic Relationships

The rate of change of cpDNA in plants is intrinsically lower than that of mitochondrial DNA in animals (Palmer, 1985a, b, 1987; Wolfe et al., 1987; Birky, 1988). The low levels of intraspecific cpDNA variation reported for several plant species led Clegg and others (Clegg et al., 1986; Clegg, 1987) to suggest that it was too low to facilitate the analysis of microevolutionary processes in plants to a degree comparable to that achieved for animals with mitochondrial DNA (Avise, 1986, 1989; Avise et al., 1987). Although this conclusion still appears to be warranted, sufficient variation nonetheless exists in the plastid genome of many species to permit the elucidation of intraspecific and interspecific evolutionary processes.

Intraspecific cpDNA variation has, for example, provided information regarding the domestication of several crops, the origin of both polyploids and diploids, introgression, and genetic differentiation both among and within populations. In this section, we demonstrate that although levels of cpDNA variation are often low, sufficient polymorphism frequently exists within plant species to permit the resolution of microevolutionary processes on a scale that in some instances begins to approach that achieved for animals via analysis of mitochondrial DNA.

Domestication of Crop Species

Intraspecific variation in the chloroplast genome has provided important insights into the evolution and domestication of several crop plants (see Chapter 9, this volume). For example, cultivated barley (*Hordeum vulgare* ssp. *vulgare*) exhibits much lower levels of plastid variation than its wild progenitor (*H. vulgare* ssp. *spontaneum*), a situation suggesting that barley experienced a cytoplasmic bottleneck during domestication (Clegg et al., 1984a; Neale et al., 1988). Cultivated barley may well trace its origin to a small number of cytoplasmic types.

A second example concerns the origins of several allotetraploid Brassicas: *B. carinata* (Abyssinian mustard), *B. juncea* (leaf mustard), and *B. napus* (oil rape, rutabaga) (Erickson et al., 1983; Palmer et al., 1983). Chloroplast DNA variation within two of the diploid parental species (*B. nigra* and *B. campestris*) implicated specific populations of these two diploids that must have contributed the cytoplasm to each of the three allotetraploids.

Using only three restriction enzymes, high cpDNA diversity was detected in the Andean cultivated tetraploid potato, *Solanum tuberosum* ssp. *andigena* (Hosaka and Hanneman, 1988a, b). Furthermore, a geographic cline in cpDNA types was found for the tetraploid potato from the Andes to coastal Chile. Several different cpDNA types were also detected in ssp. *tuberosum*, as well as in the cultivated diploids *S. stenotomum*, *S. goniocalyx*, *S. phureja*, and the wild, weedy *S. chacoense*. The cpDNA variation detected suggested that (1) the Andean cultivated tetraploid potato, ssp. *andigena*, could have arisen independently many times from cultivated diploid populations and (2) ssp. *tuberosum* arose in the Andes.

Another example of the potential phylogenetic utility of intraspecific cpDNA variation concerns an analysis of the avocado, *Persea americana*, and related species (Furnier et al., 1990). The cultivars of avocado are members of three varieties (*P. americana* var. *americana*, var. *drymifolia*, and var. *guatemalensis*) whose origins and relationships are unclear. Some have suggested that *P. americana* var. *guatemalensis* is a variety of the related species *P. nubigena*, whereas others have suggested that *P. nubigena* actually represents a variety of *P. americana* (reviewed in Furnier et al., 1990). Chloroplast DNA data clarified some of these relationships. For example, all three varieties of *P. americana* possess distinct plastid genotypes. Significantly, these data also suggest that *P. americana* is not monophyletic; *P. floccosa*, *P. nubigena*, and *P. steyermarkii* are contained as clades within *P. americana*. Finally, the cpDNA data in conjunction with nuclear ribosomal DNA data suggest that var. *guatemalensis* arose as a hybrid of *P. steyermarkii* and *P. nubigena*.

Intraspecific variation has also elucidated the origins of several other crops. In a study of the garden pea (*Pisum sativum*) (Palmer et al., 1985), a wild population of *P. humile* from northern Israel and 12 of 13 cultivated lines of *P. sativum* form a single lineage based on cpDNA data. Southern populations of *P. humile*, in contrast, show a closer relationship to *P. elatius*. These observations support an earlier hypothesis that the cultivated pea was domesticated primarily from northern populations of *P. humile*. In *Zea mays*, no cpDNA differences were detected between maize (*Z. mays* ssp. *mays*) and some teosintes (*Z. mays* ssp. *mexicana* and ssp. *parviglumis*), a result consistent with the hypothesis that maize is a domesticated form of teosinte (Doebley et al., 1987; Doebley, 1990). Finally, Terauchi (1990) detected intraspecific variation in the plastid genome within *Dioscorea bulbifera*. Not only do African and Asian populations of this wild species have different genomes, but also an Asian cultivated yam, *D. bulbifera* var. *sativa*, is polyphyletic in origin.

Derivation of Polyploids

One area in which intraspecific cpDNA polymorphisms offer tremendous potential is the analysis of both auto- and allopolyploids (see also Chapter 8, this volume). In the autotetraploid *Tolmiea menziesii*, for example, cpDNA data provided several new insights into autopolyploid evolution (reviewed in Soltis et al., 1989b). In other cases cpDNA variation among diploid populations has facilitated the unambiguous documentation of multiple origins of both autopolyploids (e.g., *Heuchera micrantha*, Soltis et al., 1989a; *H. grossulariifolia*, Wolf et al., 1990) and allopolyploids (e.g., the *Glycine tabacina* complex, Doyle et al., 1990a). Documentation of multiple origins of autoploids is particularly significant because successful autopolyploid events have traditionally been considered to be extremely rare in nature (reviewed in Soltis and Soltis, 1989b, 1990; see also Chapter 8, this volume).

Intraspecific cpDNA polymorphisms have also been used to implicate specific diploid populations in the origin of naturally occurring allotetraploids, including the cultivated Brassicas discussed earlier (Palmer et al., 1983). A second example concerns the origin of the allotetraploid *Tragopogon miscellus*. Not only were multiple origins detected involving different species as the maternal parent—*T. dubius* in one case and *T. pratensis* in the other—but two distinct plastid genotypes within the diploid *T. dubius* identified certain populations as probable contributors to the allotetraploid *T. miscellus* (Soltis and Soltis, 1989a, 1990).

Origin of Species and Intraspecific Phylogenetic Analysis

Intraspecific cpDNA variation has also revealed the origins of diploid species and phylogenetic relationships among populations. For example, one study provided important evidence regarding the origin of *Helianthus neglectus* from specific populations of *H. petiolaris* (Rieseberg et al., 1990a). A cpDNA polymorphism revealed the presence of two lineages of *H. petiolaris*. Populations of *H. neglectus* shared identical genotypes with one of these lineages, suggesting its derivation from *H. petiolaris* ssp. *falax*. This hypothesis is also supported by other lines of evidence. In this instance, cpDNA data elucidated a progenitor-derivative relationship comparable to what has been achieved with allozymes in other groups (see review by Crawford, 1989).

Helianthus bolanderi was considered to comprise two ecologically distant races, a "serpentine" race and a "weedy" race. Rieseberg et al. (1988) found that eight or nine cpDNA restriction sites differed between the two races depending on the specific populations compared. The mean sequence divergence between races, 0.30%, was almost as high as between either race and *H. annuus* (0.35% for the weedy and 0.40% for the serpentine race), prompting the suggestion that the two "races" of *H. bolanderi* be treated as distinct species (Rieseberg, 1990). This conclusion is also supported by evidence from morphology, crossability, geographic distribution, and fatty acid composition of achenes.

A similar situation with *Tolmiea menziesii* illustrates the utility of both chloroplast and nuclear markers for distinguishing what may be cryptic species. This species comprises diploid and tetraploid cytotypes that, although geographically distinct, are nearly indistinguishable both morphologically and by numerous nuclear-encoded characters such as allozymes (reviewed in Soltis et al., 1989b; Soltis and Soltis, 1990). Yet the two cytotypes differ by a minimum of three cpDNA restriction site mutations and three length mutations. Mean sequence divergence between the two cytotypes is 0.08%, a value similar to that reported for some congeneric species of angiosperms. This high degree of divergence, together with crossability data and the apparent absence of cytoplasmic gene flow (reviewed in Soltis et al., 1989b; Soltis and Soltis, 1990), suggest that the cytotypes of *T. menziesii* are probably best considered distinct species. It is important to note that this conclusion is based on the concordance of several different types of data, not solely on the differentiation of the chloroplast genomes.

Pteridium aquilinum provides another example of the potential phylogenetic information revealed by cpDNA variation within species. Seventeen restriction site mutations were found within this widespread, morphologically variable species (Tan and Thomson, 1990). This is considerable intraspecific variation, especially given that fewer than 10 endonucleases were used. Although this study is still in progress, results to date indicate two distinct lineages within *P. aquilinum*, one from the southern hemisphere and one from the northern hemisphere. These lineages agree with Tryon's (1941) recognition of two subspecies, ssp. *aquilinum* and ssp. *caudatum* (reviewed in Page, 1976).

The intraspecific cpDNA variation detected in *Lycopersicon peruvianum* (Palmer and Zamir, 1982) presents a situation similar to that observed in *Persea americana*, discussed in the previous section. Analysis of six accessions of *L. peruvianum* revealed three different genotypes within this single species. This intraspecific variation completely encompasses all of the variation observed in *L. chilense* and *L. chmielewskii*, suggesting that the latter two be relegated to positions within the *L. peruvianum* complex (Palmer and Zamir, 1982). A similar analysis of cpDNA variation in *Ipomopsis tenuituba* indicated that this species is actually polyphyletic. Some populations were apparently derived from *I. arizonica*, whereas other populations were derived from *I. aggregata* (Wolf et al., unpublished data). Again, these conclusions must be viewed with caution given the process of differentiation of organellar genomes during speciation.

The utility of cpDNA analyses for revealing phylogenetic relationships among conspecific populations is clear. Although much remains to be discovered and the interpretation of these data must be refined, progenitor-derivative relationships can be recognized from patterns of plastid genotypes. This should prove to be a particularly fertile field for evolutionary biologists interested in understanding the process of speciation.

Introgressive Hybridization

Perhaps most intriguing from a systematic viewpoint is the relatively frequent documentation of introgressive hybridization via intraspecific variation in plastid

DNA (see Chapter 7, this volume). Although the field of cpDNA systematics is relatively young and many investigators have relied on only one or a few samples per species, introgression has been implicated in the origin of some populations of the following taxa: *Brassica napus* (Palmer et al., 1983), *Helianthus neglectus* (Rieseberg et al., 1990a), *H. annuus* ssp. *texanus* (Rieseberg et al., 1990b), *Lycopersicon chmielewskii* (Palmer and Zamir, 1982), *Pisum sativum* (Palmer et al., 1985), *Salix melanopsis* and *S. taxifolia* (Brunsfeld et al., unpublished data), *Tellima grandiflora* (Soltis et al., 1991a, b), *Zea perennis* (Doebley, 1990), *Populus nigra* (Smith and Sytsma, 1990), and several species of *Quercus* (Whittemore, personal communication). In most of these cases, evidence for introgression was obtained by chance simply because several representative samples of each species were included in the analyses. For many of these taxa, the investigators were not attempting to test for introgression, nor was it even speculated in advance to be a possible concern. In the case of *Helianthus bolanderi*, in contrast, analysis of cpDNA variation provided no compelling support for introgression, even though it was long suspected to have occurred (Rieseberg et al., 1988). The frequent documentation of introgression, particularly in taxa where it was not suspected, has tremendous implications for sampling strategies and the investigation of phylogenetic relationships at the species level. This topic will be covered in greater detail in the last portion of this chapter.

Genetic Differentiation among Populations

Extensive variation in the plastid genome has also facilitated studies of differentiation among populations. This has been useful for studying hybrid zones and recognizing long-distance gene flow. Wagner et al. (1987) analyzed cpDNA variation for two markers from throughout the ranges of lodgepole pine (*Pinus contorta*) and jack pine (*Pinus banksiana*), two species that share an extensive zone of sympatry in which they hybridize. Although gene flow between the two species has been reported for nuclear genes based on analyses of morphologic, terpenoid, and allozymic markers, there was no evidence of interspecific organellar gene flow in the allopatric populations of either species. The absence of cytoplasmic gene flow in this instance is particularly intriguing because the chloroplast genome is paternally inherited in pines. Thus, pollen movement could potentially transfer the plastid genome between species. Several possible explanations for this apparent contradiction are provided by Wagner et al. (1987), one of which is that nuclear genes may sometimes be able to cross species borders that organellar DNA cannot. Subsequent analyses revealed novel cpDNA variants as well as apparent recombinant types in the range of sympatry between *P. contorta* and *P. banksiana* (Govindaraju et al., 1989). The unusual chloroplast types detected in the zone of sympatry suggested that the genetic complexity of hybrid zones may not be limited to the nuclear genome.

In *Tolmiea menziesii,* diploid and tetraploid cytotypes occur in close geographic

proximity in central Oregon (Fig. 6.1). All tetraploid populations are characterized by the same chloroplast genome, distinct from any found in diploid populations. Despite the fact that populations of the two cytotypes occur within several
hundred meters of each other, neither cytotype possesses the chloroplast genome characteristic of the other (Soltis et al., 1989b). Because the chloroplast
genome is maternally inherited in the Saxifragaceae (Soltis et al., 1990), movement of pollen between diploid and tetraploid plants will not transfer the chloroplast genome. Nevertheless, anomalous cpDNAs due to introgression have been
reported in a number of species characterized by maternal inheritance (noted
above and reviewed in Chapter 7 by Rieseberg and Brunsfeld. The absence of
cytoplasmic gene flow between diploid and tetraploid *Tolmiea,* together with
crossability data (Soltis et al., 1989b), suggests that these cytotypes are reproductively isolated in nature.

The variation in chloroplast genomes among populations of *T. menziesii* has
also revealed an example of long-distance gene flow. Three different groups of
diploid populations can be identified based on their plastid genotypes; two of
these groups occupy distinct geographic areas. One cytoplasmic type is centered
in the Siskiyou and Klamath Mountains, whereas a second type is found primarily
along the coast of southern Oregon and northern California. One population
having the coastal cytoplasm occurs as a disjunct in the Siskiyou Mountains of
southern Oregon (Fig. 6.1), a situation indicative of occasional long-distance
seed dispersal. These data also exemplify the potential of cpDNA investigations
to provide novel information at the populational level. *Tolmiea* had been exhaus-
tively examined using allozymes, flavonoids, and other nuclear markers, none of
which differentiated the cytotypes or revealed geographically distinct groups of
diploid populations.

In *Tellima grandiflora* the pattern of intraspecific cpDNA variation has afforded
the opportunity to integrate the phylogenetic analysis of populations with biogeography (Soltis et al., 1991a). This type of analysis has been possible in animals
via analysis of the mitochondrial genome and has led to a new discipline referred
to by Avise and coworkers as "intraspecific phylogeography" (Avise et al., 1987;
Avise, 1989). A clear intraspecific phylogeographic pattern is present in *T.
grandiflora.* Two major cpDNA clades exist that are extremely well differentiated, differing by a minimum of 18 restriction site mutations and several length
mutations. One clade occupies the northern portion of the range of *Tellima,*
whereas the second major clade occurs largely in the southern portion of the
species' range (Fig. 6.2). Several different lineages of the southern clade exist
that differ by two or three restriction site mutations. Although these lineages have
well-differentiated chloroplast genomes, populations of *T. grandiflora* are nearly
uniform for nuclear-encoded allozymes (Rieseberg and Soltis, 1987). A cpDNA
investigation at the generic level suggests that the southern clade in *T. grandiflora*
obtained its chloroplast genome via ancient hybridization with a species of *Mitella*
(Soltis et al., 1991b).

The presence of a continuous geographic distribution in *T. grandiflora,* with a major genetic (cpDNA) discontinuity, suggests that past biogeographic processes have initiated the intraspecific population structure. Significantly, several populations of *T. grandiflora* that possess the typical "southern" cpDNA pattern also occur disjunctly to the north on Prince of Wales Island, Alaska, and the Olympic Peninsula, Washington (Fig. 6.2). Both of these geographic areas are proposed glacial refugia. The cpDNA results for *T. grandiflora* suggest that past glaciation may have created discontinuities in the geographic distribution of this species, eventually resulting in the cpDNA discontinuity that exists today. These results for *T. grandiflora* indicate that the discipline of "intraspecific phylogeography," which has provided tremendous insights into microevolutionary processes in animals (via the analysis of mtDNA), may similarly be applicable to plants through investigation of the chloroplast genome.

Genetic Diversity within Populations

Variation in the chloroplast genome has also been useful for quantifying genetic diversity within populations of some species. One example is an analysis of wild barley (*Hordeum vulgare* ssp. *spontaneum*) from Israel and Iran (Neale et al., 1988). Among 245 accessions from 30 populations, three plastid genotypes were detected. Nineteen of the populations were monomorphic, six were dimorphic, and the remaining five contained all three genotypes (Fig. 6.3). Almost half of the samples consisted of five or fewer accessions; of these, most were monomorphic, although one was dimorphic, and one contained all three genotypes. This suggests that the detection of within-population diversity was constrained by the sampling scheme and that additional variation is likely to exist in wild barley. Nevertheless, distinct differences were observed among populations in plastid genotype frequencies (Fig. 6.3). Although populations in the same geographic region were similar to each other, regional differences existed. One genotype predominated north of the Sea of Galilee, a second genotype predominated in the Lower Jordan Valley, whereas the third genotype was scattered in distribution. Thus, cpDNA diversity does not appear to be randomly distributed in this species.

The presence of plastid DNA variation within and among populations offers new research possibilities. Recently, Golenberg and Clegg (unpublished data) investigated associations between the organellar variants discussed above and nuclear variants in *H. vulgare* ssp. *spontaneum,* both to ascertain whether adaptive patterns of variation may exist and to determine whether patterns of organellar and nuclear variation are correlated. Both allozymes and nuclear ribosomal DNA variants were used as nuclear markers. In some instances, the level of intrapopulational variability in the two cpDNA sites examined is almost as high as that observed for the two most variable nuclear markers (a ribosomal DNA locus and the isozyme locus *Esterase-3*), and much higher than that observed for the remaining nuclear markers. In two of five populations examined in detail, the

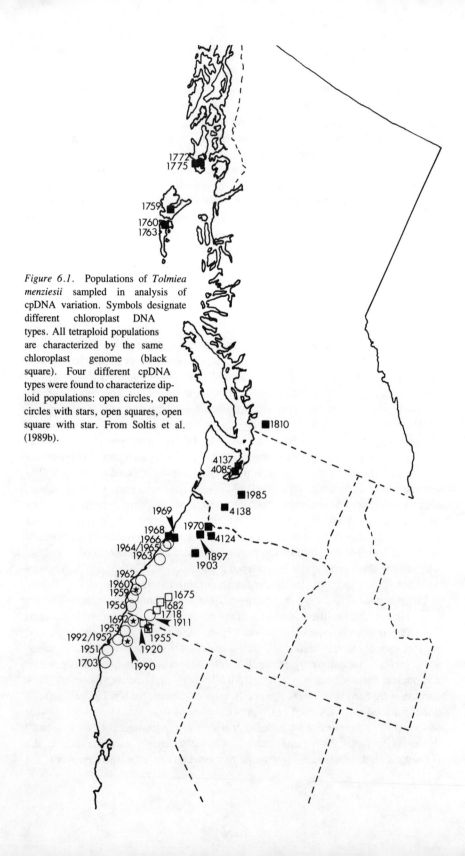

Figure 6.1. Populations of *Tolmiea menziesii* sampled in analysis of cpDNA variation. Symbols designate different chloroplast DNA types. All tetraploid populations are characterized by the same chloroplast genome (black square). Four different cpDNA types were found to characterize diploid populations: open circles, open circles with stars, open squares, open square with star. From Soltis et al. (1989b).

Figure 6.2. Populations of *Tellima grandiflora* sampled in analysis of cpDNA variation. Symbols designate different chloroplast DNA types. Populations of the northern clade are designated by an open circle; black squares, circles, and stars represent the three different branches of the southern clade. The northern and southern clades differ by a minimum of 18 restriction site mutations. From Soltis et al. (1991a).

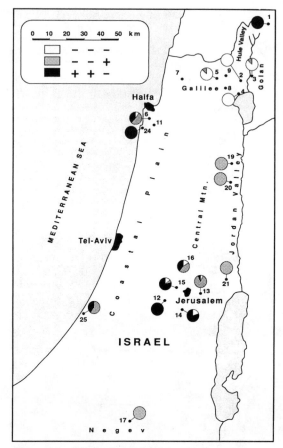

Figure 6.3. Distribution of *Hordeum vulgare* ssp. *spontaneum* samples in Israel and frequency pie diagrams for the three variable restriction sites. Numbers correspond to population numbers (given in Neale et al., 1988). Frequency pie diagrams are not shown for populations 7–11, 18, 22, and 23 because sample sizes were less than 5. The location of population 16 was not known with certainty, but is somewhere within the Central Mountain Region. The symbols + and − refer to presence or absence of restriction sites in the order *Hind*III, *Eco*RI, and *Bcl*I. From Neale et al. (1988).

two chloroplast sites are not significantly associated with each other in either population, yet the different chloroplast sites are associated with nuclear markers. In these populations, the two chloroplast sites appear to recombine relative to each other, a possibility that must be excluded given the uniparental inheritance of the plastid genome in this species. Golenberg and Clegg suggest instead a high rate of localized mutation for at least one cpDNA locus. If correct and of general occurrence, this possibility will substantially alter the interpretation of cpDNA data as a useful marker for phylogenetic reconstruction.

The greatest opportunities for new research possibilities due to high levels of intrapopulational cpDNA variation have been realized in the genus *Trifolium* (Fabaceae). To date, two species have been studied in detail. A map of the chloroplast genome from *T. subterraneum* "Tallarook" indicates that the genome has been largely rearranged relative to other angiosperms, including *Medicago sativa* (Milligan et al., 1989). Since the genome from *M. sativa* is structurally identical to the presumed ancestor of both *Medicago* and *Trifolium,* the rearrangements can be ascribed to the clover lineage. This suggests a greatly increased rate of cpDNA evolution within *Trifolium* and provides an indication that intraspecific variation may be present and useful for studies of population biology.

Initial worldwide surveys of *T. subterraneum* have identified several different cpDNA genotypes based on their restriction fragment profiles. It is important to distinguish nucleotide sequence change from structural rearrangement as a cause of these observations. Maps of each genotype are being constructed and, although incomplete, suggest that different genotypes are identical in much of the chloroplast genome. Variation is restricted to small regions, a pattern indicative of either site-specific nucleotide mutation or site-specific rearrangements. These potential mechanisms are being explored to investigate the dynamics of plastid DNA evolution. To the extent that the origin of new genotypes influences the patterns observed in comparisons among species, this research will be important for studies seeking to reconstruct phylogenetic relationships. Regardless of the underlying mechanism, however, readily identifiable intraspecific variability is evident within *T. subterraneum.*

An even more extreme, and potentially more useful, example of intraspecific variation is found within *T. pratense.* In what may be the most extensive within-population sampling of cpDNA, five to nine different genotypes have been identified from single populations (Milligan, 1991). Two genotypes are found in all populations; one is quite common in each population ($p > 0.7$), and the second is much less common in each ($p < 0.1$). The remainder of the genotypes are rare and usually found in a single population. This extreme level of cpDNA diversity may be a manifestation at the population level of the accelerated rate of rearrangement found in *T. subterraneum* (Milligan et al., 1989), or may be due to a greatly increased rate of nucleotide substitution. Genomic maps are being constructed to distinguish between these alternatives.

This high level of intraspecific genetic diversity in the plastid genome of *T. pratense* provides a unique opportunity for advancing studies of population biology. The variation is currently being used to quantify the geographic differentiation of red clover. For example, the application of several measures of gene flow (Slatkin and Barton, 1989) predicts moderate migration rates among populations ($Nm > 7$). As the first study of cpDNA diversity to provide this type of information, it is unclear how representative these results might be. Additional population studies are being undertaken to determine what fitness effects, if any, might be associated with variation among cytoplasms with specific plastid genotypes. Although assumed to be neutral, the consequences of cpDNA variation have been impossible to study in other taxa. If, however, cpDNA does influence fitness, the accuracy of plastid DNA phylogenies may require reevaluation.

These examples of intraspecific plastid genetic diversity within *Hordeum vulgare* ssp. *spontaneum* and species of *Trifolium* illustrate that in at least some cases the genome will prove useful for studies at the population level. Due to its uniparental transmission genetics, the genome will provide information unavailable from the nuclear genome. Furthermore, studies of cpDNA diversity within species may provide critical information for determining appropriate parameters involved in the evolutionary models used for phylogenetic reconstruction. Without such a basis, inferences based on cpDNA are subject to unknown errors. Although in many cases these parameters will remain unmeasurable, those taxa that exhibit extensive intraspecific variation provide an invaluable opportunity both to place cpDNA systematics on a solid mechanistic base and to obtain critical information about population biology.

Discrepancies between cpDNA Data and Other Lines of Evidence

The frequent occurrence of intraspecific cpDNA variation raises the obvious question: are the observed levels and patterns of plastid variation concordant with other lines of evidence, such as data from morphology and other nuclear-encoded characters? This question has, in part, been addressed briefly in the preceding section. In some cases, data from the chloroplast genome are correlated with other lines of evidence at the intraspecific level, but in many instances the data sets do not agree. Discordance of plastid and nuclear markers should not, however, necessarily be viewed with alarm, for it can provide new evolutionary information at the intra- or interspecific level not revealed by either type of marker alone. An important example of this point is the role of such discordances in revealing several possible instances of introgression (discussed briefly above and in detail in Chapter 7 by Rieseberg and Brunsfeld).

One might assume *a priori* that morphologically variable taxa are most likely to exhibit high levels of intraspecific cpDNA variation, and several examples support this contention. *Heuchera micrantha, H. grossulariifolia, Lycopersicon peruvianum, Trifolium pratense,* and *Zea mays* are all morphologically variable, and intraspecific cpDNA variation was detected in all taxa (Palmer and Zamir, 1982; Doebley et al., 1987; Soltis et al., 1989a; Wolf et al., 1990; Milligan, 1991). In contrast, however, cpDNA variation was insufficient to resolve relationships among several diploid species of *Antennaria* from the eastern United States, even though these species are quite distinct morphologically (Michaels, personal communication).

Conversely, diploid *Tolmiea menziesii* seems morphologically uniform, or at least no more morphologically variable than the tetraploid cytotype. Yet, plastid DNA variation was detected within the diploid but not the tetraploid. It also seems unlikely that the extensive cpDNA variation detected within and among populations of wild barley (Neale et al., 1988) could have been predicted, when collections taken from around the world of another grass species, *Pennisetum glaucum,* exhibited no such variation (Clegg et al., 1984b; Gepts and Clegg, 1989). Even within a single section of the genus *Machaeranthera,* a member of a family (Compositae) believed to have a low rate of cpDNA divergence, variation in the chloroplast genome was not predictable from morphologic variation. *Machaeranthera pinnatifida* and *M. gracilis* are virtually indistinguishable morphologically, yet cpDNA variation exists both within *M. pinnatifida* and between the two species. In contrast, much less cpDNA variation is evident between the two morphologically divergent species *M. gracilis* and *M. stenoloba* (Morgan, personal communication). In an analysis of the three diploid species composing the B genome of *Glycine* subgenus *Glycine,* patterns of cpDNA polymorphism showed little agreement with the morphologically based taxonomic classification for any of the three species or for most infraspecific groupings (Doyle et al., 1990b; this topic is also discussed in Chapter 10, this volume).

Furthermore, introgression can introduce considerable cpDNA variability into what appears to be a morphologically uniform assemblage (see Chapter 7 by Rieseberg and Brunsfeld). This point is aptly demonstrated by investigations of *Zea perennis* and *Helianthus neglectus.* Doebley (1989) detected individuals of *Z. perennis* that possessed a foreign chloroplast genome, although morphologically, cytologically, and allozymically the plants appeared to be typical of *Z. perennis* (Fig. 6.4). Similarly, Rieseberg et al. (1990a) detected a single individual of *H. neglectus* that had the chloroplast genome of *H. annuus,* although it was morphologically similar to *H. neglectus* and possessed the nuclear ribosomal DNA repeat of *H. neglectus.*

As with morphologic traits, no necessary correlation exists between variation in nuclear markers and variation in plastid markers. Diverse lines of cultivated barley (*Hordeum vulgare* ssp. *vulgare*) and its wild progenitor (*H. vulgare* ssp. *spontaneum*) that had been selected for comparable levels of nuclear diversity as measured by isozyme loci had very different levels of cpDNA diversity, with the cultivar having significantly lower diversity than the wild progenitor (Clegg et al., 1984a; see also Neale et al., 1988). *Heuchera micrantha* exhibited relatively low levels of nuclear variation compared to other angiosperms based on isozyme markers (Ness et al., 1989), yet this species possesses higher cpDNA diversity than most species analyzed to date (Soltis et al., 1989a). It is difficult to predict, therefore, based on either morphologic or isozyme variation, which taxa will exhibit high levels of intraspecific cpDNA variation.

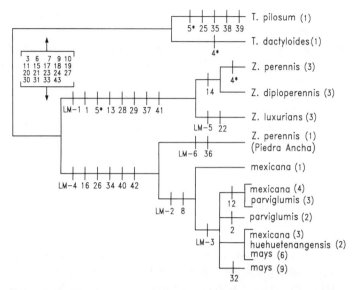

Figure 6.4. Wagner parsimony tree for *Zea* chloroplast genome types based on 41 restriction site loss/gain and six insertion/deletion mutations. The identification numbers of restriction site and length (insertion/deletion) mutations appear along the branch segments (Doebley et al., 1987; Doebley, unpublished data). Only two of the 41 restriction site mutations (numbers 4 and 5) are homoplastic, each involving the parallel loss of a *Dra*I site. The number of collections possessing a particular type appears parenthetically after the specific or subspecific epithets. From Doebley (1990).

In several instances restriction site analyses of cpDNA have revealed patterns at the intraspecific level that are inconsistent with other lines of evidence. For example, discordance of cpDNA and nuclear-encoded characters was observed for diploid and autotetraploid *Tolmiea menziesii* (Soltis et al., 1989b). If autotetraploid *Tolmiea* were of relatively recent origin, it would be expected that diploids and tetraploids would be identical or nearly so for numerous characteristics. This is, in fact, the case for morphologic and chromosomal characters, flavonoids, anthocyanins, allozymes, and restriction fragment patterns of 5S and 18S–25S nuclear ribosomal DNA (reviewed in Soltis et al., 1989b; see also Chapter 8, this volume). However, diploid and tetraploid *Tolmiea* possess divergent chloroplast genomes. Furthermore, as noted earlier, several plastid genotypes are present in the diploid cytotype, whereas the tetraploid is cytoplasmically uniform (Fig. 6.1). The discordance of organellar and nuclear-encoded characters may reflect the effects of a cytoplasmic bottleneck associated with polyploidization and/or glaciation. By chance, a plastid genotype present only rarely in the diploid may have been fixed in the tetraploid. Tetraploid plants characterized by this genotype subsequently spread throughout the Pacific Northwest of North America following the bottleneck. Loss of nuclear genetic variation was not incurred because the tetraploid is an autoploid with tetrasomic inheritance. Thus, nuclear genetic variability would not be lost as rapidly following a bottleneck.

Another possible source of discordance between organellar and nuclear data is the progressive extinction of organellar lineages following speciation in a polymorphic species, a process termed "phylogenetic sorting" by Avise (1986). An excellent example among animals was revealed by a mitochondrial DNA analysis of two sibling species of deer mice, *Peromyscus maniculatus* and *P. polionotus* (Avise et al., 1983). Some populations of *P. maniculatus* are genetically more similar to *P. polionotus* than to geographically separated populations of *P. maniculatus*. Because the ranges of the two species do not overlap, there is no compelling reason to suspect introgression. Instead, phylogenetic sorting of mitochondrial DNA lineages during speciation could explain the results. As Avise et al. (1983; reviewed in Avise, 1986) found, *P. maniculatus* is paraphyletic with regard to *P. polionotus* in terms of maternal ancestry; *P. polionotus* is monophyletic with respect to *P. maniculatus,* but forms a subclade within the larger *P. maniculatus–polionotus* assemblage. Avise (1986) also discusses another possible maternal genealogy in which two hypothetical sibling species are actually polyphyletic.

Theoretical models reinforce these conclusions. For example, the phylogenetic distributions of matriarchal lineages can lack concordance with species boundaries, particularly when speciation events have been very recent (Neigel and Avise, 1985). Tajima (1983) similarly concluded that closely related species can exhibit a polyphyletic or paraphyletic relationship in terms of mitochondrial DNA due to phylogenetic sorting, even without the occurrence of introgression (reviewed in Avise, 1986).

An example of discordance that may be due in part to phylogenetic sorting in plants is provided by Doebley (1990). Chloroplast DNA data agree with other lines of evidence in suggesting a close relationship among the subspecies of *Zea mays* (ssp. *mays, mexicana, parviglumis,* and *hueuhuetenangensis*) (Fig. 6.4). However, cpDNA data did not differentiate between *Z. mays* ssp. *mexicana* and *parviglumis,* which are distinct in morphology, allozymes, and ecology. Rather, of the five cpDNA genotypes present in these subspecies, two are found in both ssp. *mexicana* and *parviglumis.* Two possible factors could be involved in the discordance of organellar and nuclear data: introgression and phylogenetic sorting. Following the phylogenetic sorting scenario (see Doebley, 1990), the ancestor of subspecies *mexicana* and *parviglumis* may have been polymorphic for the cpDNA mutations; as the subspecies diverged, the original polymorphisms were retained in both subspecies. Doyle et al. (1990b) have similarly invoked phylogenetic sorting as one possible explanation for the discrepancies between cpDNA and morphologic markers in the B genome species group of *Glycine* subgenus *Glycine.*

These few examples indicate that discrepancies have been recognized between variation detected within the plastid genome and variation detected using other genetic markers. It is encouraging, however, that this discordance has often been recognized as useful for providing additional information about the processes of speciation and evolution within species. As additional information regarding intraspecific variation in the chloroplast genome accumulates, its utility for these studies should increase. However, the complexity of the process of phylogenetic sorting, together with our general ignorance of cpDNA differentiation during speciation, suggests that interpretation of discordances should be made with caution.

Implications of Intraspecific cpDNA Variation for Phylogenetic Reconstruction at the Species Level

The previous sections have illustrated that although the chloroplast genome is conservative in its rate of evolution, intraspecific cpDNA variation is nonetheless frequent in occurrence and often sufficiently extensive to permit the elucidation of evolutionary processes within species. An obvious implication of these findings is the importance of assessing the level of variation within species prior to initiating phylogenetic analyses using cpDNA at the species level. Available data argue against the use of only a single accession per species for phylogenetic reconstruction at the species level or higher until intraspecific variation has been adequately estimated. Morphologic uniformity or low levels of nuclear genetic variability within species should not be used as arguments for sampling only one individual per species. As shown in the previous section, there is no necessary correlation among these factors and the level of cpDNA variation.

One might argue that because phylogenetic reconstruction is based only on shared, derived characters, intraspecific cpDNA variation will not be problematic to the systematist interested in phylogenetic relationships among species. In many instances, this will probably be true, but two processes suggest caution: introgression and phylogenetic sorting. Either can lead to discordances between organellar-based phylogenies and nuclear-based phylogenies. This should be of major concern to all investigators using organellar DNA at the species level or above.

Analysis of an actual data set should underscore the potential impact of discordance on phylogenetic reconstruction at the species level. An excellent genetic data base, including both plastid and nuclear markers, is available for *Zea* (reviewed in Doebley, 1990). Doebley (1989) detected eight individuals from one population of *Z. perennis* that possessed a markedly different chloroplast genome from other collections of this species, differing by 15 restriction site mutations and two length mutations (Fig. 6.4). The same eight individuals were morphologically typical of *Z. perennis;* they were also tetraploid, as is *Z. perennis,* rather than diploid, as are other taxa of *Zea*. The cpDNA data clearly argue for the incorporation of a foreign chloroplast genome into the nuclear background of *Z. perennis*.

The plants having the atypical plastid genome are positioned with Z. *mays,* whereas the remaining collections show a close relationship to Z. *diploperennis.* Considerable biosystematic data, including morphology, allozymes, and cytology, support a close relationship of Z. *perennis* and Z. *diploperennis.* A close relationship of Z. *perennis* and Z. *mays,* in contrast, does not agree with other lines of evidence. This represents a clear discordance between organellar- and nuclear-based phylogenies.

Because of similar discordances, introgression has already been invoked frequently, even though cpDNA systematics is a young field and many studies precluded the possibility of detecting it by employing only one or a few samples per species. Furthermore, in most instances the finding of introgression was unexpected (e.g., Palmer et al., 1983, 1985; Doebley, 1990; Rieseberg et al., 1990a; Brunsfeld et al., unpublished data). Most of these putative examples of introgression were detected because (1) several populations per species were sampled, (2) considerable biosystematic data were already available, and (3) the discordance of data sets was striking. In contrast, application of cpDNA analysis to a poorly understood group for which nuclear-encoded characters, such as allozymes, have not been examined would not permit the detection of introgression. If only one sample per species were included in the DNA analysis and if by chance it had a foreign cytoplasm due to introgression, a fallacious phylogeny would result. As systematists apply cpDNA analyses to poorly understood and/ or taxonomically complex groups for which nuclear-encoded data bases are not available, discordances due to introgression or phylogenetic sorting may be a frequent problem. It is in these cases in particular that adequate intraspecific sampling of populations is required. The very fact that cpDNA possesses a uniparental mode of inheritance, thereby raising the possibility of these major discordances, emphasizes the need for adequate sampling rather than minimizes it.

Conclusions

Despite the conservative nature of cpDNA evolution, restriction site and length variation has been detected within many species. In fact, intraspecific cpDNA polymorphism has been reported in nearly 60 species, a considerable number given the brief history of cpDNA systematics and the fact that many studies have employed only one or a few samples per species.

Although the levels of intraspecific cpDNA variation typically are considerably lower than the levels reported for animal mitochondrial DNA variation, nonetheless sufficient plastid DNA variation exists within many species to permit the elucidation of evolutionary processes. Intraspecific variation has, for example, demonstrated multiple origins of autopolyploids, suggested possible cases of introgression, pinpointed certain diploid populations that most likely contributed the cytoplasm to allopolyploids, permitted the analysis of "intraspecific phylogeography," identified specific populations of wild crop-relatives as likely progenitors of domesticated species, and indicated that some species experienced a cytoplasmic bottleneck during domestication.

Intraspecific cpDNA variation has also facilitated studies of population-level phenomena, such as gene flow, long-distance seed dispersal, and genetic diversity within and among populations. Extensive variation within, as well as among, populations of *Hordeum vulgare* ssp. *spontaneum* and certain members of Fabaceae offers particularly exciting research possibilities. For example, multiple plastid genotypes are present within single populations of *H. vulgare* ssp. *spontaneum* and *Trifolium pratense*. Additionally, different populations of *T. pratense* contain unique genotypes. This provides an opportunity to study the process of geographic differentiation and to estimate the rate of gene flow among populations. It may also provide an opportunity to determine whether cpDNA variants exhibit fitness differences. These processes bear directly on the interpretation of cpDNA evolution with regard to phylogenetic studies.

The pattern of cpDNA variation within species is not always concordant with other data sets. There is no necessary correlation, for example, between levels of morphologic and/or isozyme variation and cpDNA variation. Discordance between plastid and nuclear markers has been detected in several species and, while of concern to the systematist, often provides novel evolutionary information that would not have been obtained via the analysis of either organellar or nuclear markers alone.

One obvious implication of intraspecific cpDNA variation for the plant systematist is the need to assess adequately the level of intraspecific cpDNA variation prior to initiating phylogenetic analysis at the species level or higher. In many groups of plants, it will be essential to include several representative populations per species and perhaps even many conspecific individuals. Sampling of only one population per species may lead to erroneous phylogenetic conclusions; it will certainly lead to fewer insights at the species level and below.

When intraspecific cpDNA variation is present, the populations chosen for analysis can occasionally have profound evolutionary implications for the resulting phylogenies. Two processes, introgression and phylogenetic sorting, suggest the need for adequate sampling at the intraspecific level, as well as the importance of using nuclear-encoded markers in analyzing species relationships. Cytoplasmic introgression, in particular, which appears to be a frequent phenomenon based on cpDNA studies conducted so far, can lead to fallacious phylogenetic conclusions when only one or a few populations are employed. In addition, the presence of polymorphism within an ancestral taxon, followed by phylogenetic sorting, may explain the discordance of nuclear and plastid data sets for closely related species.

Available data therefore suggest caution to the systematist interested primarily in phylogenetic relationships at the species level or higher. Clearly, restriction site analysis of cpDNA is a powerful tool for analyzing phylogenetic relationships.

However, adequate sampling of populations for cpDNA variation is a necessity, and the importance of the concomitant analysis of nuclear-encoded characters, such as allozymes and/or nuclear ribosomal DNA, should not be underestimated.

References

Ali, I.F., Neale, D.B., and Marshall, K.A. (1991) Chloroplast DNA restriction fragment length polymorphism in *Sequoia sempervirens* D. Don Endl., *Pseudotsuga menziesii* (Mirb.) Franco, *Calocedrus decurrens* Torr., and *Pinus taeda* L. *Theor. Appl. Gen.* **81**, 83–89.

Atchinson, B.A., Whitfeld, P.R. and Bottomley, W. (1976) Comparison of chloroplast DNAs by specific fragmentation with EcoRI endonuclease. *Mol. Gen. Genet.* **148**, 263–269.

Avise, J.C. (1986) Mitochondrial DNA and the evolutionary genetics of higher animals. *Phil. Trans. Roy. Soc. Lond. [Biol.]* **312**, 325–342.

Avise, J.C. (1989) Gene trees and organismal histories: a phylogenetic approach to population biology. *Evolution* **43**, 1192–1208.

Avise, J.C., Shapira, J.F., Daniel, S.W., Aquadro, C.F., and Lansman, R.A. (1983) Mitochondrial DNA differentiation during the speciation process in *Peromyscus*. *Mol. Biol. Evol.* **1**, 30–56.

Avise, J.C., Arnold, J., Ball, R.M., Bermingham, E., Lamb, T., Neigel, J.E., Reeb, C.A., and Saunders, N.C. (1987) Intraspecific phylogeography: the mitochondrial bridge between population genetics and systematics. *Ann. Rev. Ecol. Syst.* **18**, 489–522.

Banks, J.A., and Birky, C.W., Jr. (1985) Chloroplast DNA diversity is low in a wild plant, *Lupinus texensis*. *Proc. Natl. Acad. Sci. USA* **82**, 6950–6954.

Birky, C.W. (1988) Evolution and variation in plant chloroplast and mitochondrial genomes. In: *Plant Evolutionary Biology* (eds. L.D. Gottlieb and S.K. Jain), Chapman and Hall, London, pp. 23–53.

Bohm, B.A. (1987) Intraspecific flavonoid variation. *Bot. Rev.* **53**, 197–279.

Bowman, C.M., Bonnard, G., and Dyer, T.A. (1983) Chloroplast DNA variation between species of *Triticum* and *Aegilops*. Location of the variation on the chloroplast genome and its relevance to the inheritance and classification of the cytoplasm. *Theor. Appl. Genet.* **65**, 247–262.

Clegg, M.T. (1987) Preface, plant molecular evolution. *Amer. Natur.* **130**, S1–S5.

Clegg, M.T. (1989) Molecular diversity in plant populations. In: *Plant Population Genetics, Breeding, and Genetic Resources* (eds. A.H.D. Brown, M.T. Clegg, A.L. Kahler, and B.S. Weir), Sinauer Associates, Sunderland, MA, pp. 98–115.

Clegg, M.T., Brown, A.H.D., and Whitfeld, P.R. (1984a) Chloroplast DNA diversity in wild and cultivated barley: implications for genetic conservation. *Genet. Res.* **43**, 339–343.

Clegg, M.T., Rawson, J.R., and Thomas, K. (1984b) Chloroplast DNA variation in pearl millet and related species. *Genetics* **106**, 449–461.

Clegg, M.T., Ritland, K., and Zurawski, G. (1986) Processes of chloroplast DNA evolution. In: *Evolutionary Processes and Theory* (eds. S. Karlin and E. Nevo), Academic Press, New York, pp. 275–294.

Clegg, M.T., Learn, G.H., and Golenberg, E.M. (1990) Molecular evolution of the chloroplast DNA. In: *Evolution at the Molecular Level* (eds. R.K. Selander, A.G. Clark, and T.S. Whittam), Sinauer Associates, Sunderland, MA.

Close, P.S., Shoemaker, R.C., and Keim, P. (1989) Distribution of restriction site polymorphism within the chloroplast genome of the genus *Glycine,* subgenus *Soja. Theor. Appl. Genet.* **77,** 768–776.

Crawford, D.J. (1989) Isozymes and plant systematics. In: *Isozymes in Plant Biology* (eds. D.E. Soltis and P.S. Soltis), Dioscorides, Portland, OR, pp. 146–164.

Curtis, S.E., and Clegg, M.T. (1984) Molecular evolution of chloroplast DNA sequences. *Mol. Biol. Evol.* **1,** 291–301.

Doebley, J. (1989) Molecular evidence for a missing wild relative of maize and the introgression of its chloroplast genome into *Zea perennis. Evolution* **43,** 1555–1559.

Doebley, J. (1990) Molecular systematics of *Zea* (Gramineae). *Maydica* **35,** 143–150.

Doebley, J., Renfroe, W., and Blanton, A. (1987) Restriction site variation in the *Zea* chloroplast genome. *Genetics* **117,** 139–147.

Doyle, J.J., Doyle, J.L., Grace, J., and Brown, A.H.D. (1990a) Multiple origins of polyploids in the *Glycine tabacina* complex inferred from chloroplast DNA polymorphism. *Proc. Natl. Acad. Sci. USA* **87,** 714–717.

Doyle, J.J., Doyle, J.L., and Brown, A.H.D. (1990b) Chloroplast DNA polymorphism and phylogeny in the B genome of *Glycine* subgenus *Glycine* (Leguminosae). *Amer. J. Bot.* **77,** 772–782.

Duvall, M.R., and Doebley, J. (1990) Restriction site variation in the chloroplast genome of *Sorghum* (Poaceae). *Syst. Bot.* **15,** 472–480.

Erickson, L.R., Straus, N.A., and Beversdorf, W.D. (1983) Restriction patterns reveal origins of chloroplast genomes in *Brassica* amphiploids. *Theor. Appl. Genet.* **65,** 201–206.

Furnier, G.R., Cummings, M.P., and Clegg, M.T. (1990) Evolution of avocados as revealed by DNA restriction site variation. *J. Hered.* **81,** 183–188.

Gepts, P., and Clegg, M.T. (1989) Genetic diversity in pearl millet (*Pennisetum glaucum* (L.) R.Br.) at the DNA sequence level: consequences for genetic conservation. *J. Hered.* **80,** 203–208.

Gillham, N.W., Boynton, J.E., and Harris, E.H. (1985) Evolution of plastid DNA. In: *DNA and Evolution: Natural Selection and Genome Size* (ed. T. Cavalier-Smith), Wiley, New York, pp. 220–351.

Gordon, K.H.J., Crouse, E.J., Bonhert, H.J., and Herrmann, R.G. (1982) Physical mapping of differences in chloroplast DNA of the five wild type plastomes in *Oenothera* subsection *Euoenothera. Theor. Appl. Genet.* **61,** 373–384.

Govindaraju, D.R., Wagner, D.B., Smith, G.P., and Dancik, B.P. (1988) Chloroplast

DNA variation within individual trees of a *Pinus banksiana–Pinus contorta* sympatric region. *Can. J. For. Res.* **18**, 1347–1350.

Govindaraju, D.R., Dancik, B.P., and Wagner, D.B. (1989) Novel chloroplast DNA polymorphism in a sympatric region of two pines. *J. Evol. Biol.* **2**, 49–59.

Holwerda, B.C., Jana, S., and Crosby, W.L. (1986) Chloroplast and mitochondrial DNA variation in *Hordeum vulgare* and *Hordeum spontaneum*. *Genetics* **114**, 1271–1291.

Hosaka, K. (1986) Who is the mother of the potato?—restriction endonuclease analysis of chloroplast DNA of cultivated potatoes. *Theor. Appl. Genet.* **72**, 606–618.

Hosaka, K., and Hanneman, R.E., Jr. (1988a) The origin of the cultivated tetraploid potato based on chloroplast DNA. *Theor. Appl. Genet.* **76**, 172–176.

Hosaka, K., and Hanneman, R.E., Jr. (1988b) Origin of chloroplast DNA diversity in the Andean potatoes. *Theor. Appl. Genet.* **76**, 333–340.

Hosaka, K., Ogihara, Y., Matsubayashi, M., and Tsunewaki, K. (1984) Phylogenetic relationship between the tuberous *Solanum* species as revealed by restriction endonuclease analysis of chloroplast DNA. *Jap. J. Genet.* **59**, 349–369.

Ichikawa, H., Hirai, A., and Katayama, T. (1986) Genetic analyses of *Oryza* species by molecular markers for chloroplast genomes. *Theor. Appl. Genet.* **72**, 353–358.

Ishii, T., Terachi, T., and Tsunewaki, K. (1986) Restriction endonuclease analysis of chloroplast DNA from cultivated rice species, *Oryza sativa* and *O. glaberrima*. *Jap. J. Genet.* **61**, 537–541.

Jansen, R.K., and Palmer, J.D. (1988) Phylogenetic implications of chloroplast DNA restriction site variation in the Mutisieae (Asteraceae). *Amer. J. Bot.* **75**, 751–764.

Johnson, L.B., and Palmer, J.D. (1989) Heteroplasmy of chloroplast DNA in *Medicago*. *Plant Mol. Biol.* **12**, 3–11.

Kemble, R.J. (1987) A rapid single leaf, nucleic acid assay for determining the cytoplasmic oranelle complement of rapeseed and related *Brassica* species. *Theor. Appl. Genet.* **73**, 364–370.

Kishima, Y., Mikami, T., Hirai, A., Sugiura, M., and Kinoshita, T. (1987) *Beta* chloroplast genomes: analysis of Fraction I protein and chloroplast DNA variation. *Theor. Appl. Genet.* **73**, 330–336.

Matthews, B.F., Wilson, K.G., and DeBonte, L.R. (1984) Variation in culture, isoenzyme patterns and plastid DNA in the genus *Daucus*. *In Vitro* **20**, 38–44.

Metzlaff, M., Börner, T., and Hagemann, R. (1981) Variation of chloroplast DNAs in the genus *Pelargonium* and their biparental inheritance. *Theor. Appl. Genet.* **60**, 37–41.

Milligan, B.G. (1991) Chloroplast DNA diversity within and among populations of *Trifolium pratense*. *Curr. Genet.*, in press.

Milligan, B.G., Hampton, J.N., and Palmer, J.D. (1989) Dispersed repeats and structural reorganization in subclover chloroplast DNA. *Mol. Biol. Evol.* **6**, 355–368.

Moon, E., Kao, T.-H, and Wu, R. (1987) Rice chloroplast DNA molecules are heterogeneous as revealed by DNA sequences of a cluster of genes. *Nucleic Acids Res.* **15**, 611–630.

Murai, K., and Tsunewaki, K. (1986) Molecular basis of genetic diversity among cytoplasms of *Triticum* and *Aegilops* species. IV. ctDNA variation in *Ae. triuncialis*. *Heredity* **57**, 335–339.

Neale, D.B., Wheeler, N.C., and Allard, R.W. (1986) Paternal inheritance of chloroplast DNA in Douglas fir. *Can. J. For. Res.* **16**, 1152–1154.

Neale, D.B., Saghai-Maroof, M.A., Allard, R.W., Zhang, Q., and Jorgensen, R.A. 1988. Chloroplast DNA diversity in populations of wild and cultivated barley. *Genetics* **120**, 1105–1110.

Nei, M. (1987) *Molecular Evolutionary Genetics,* Columbia University Press, New York.

Nei, M., and Li, W. (1979) Mathematical model for studying genetic variation in terms of restriction endonucleases. *Proc. Natl. Acad. Sci. USA* **76**, 5269–5273.

Neigel, J., and Avise, J.C. (1985) Phylogenetic relationships of mitochondrial DNA under various demographic models of speciation. In: *Evolutionary Processes and Theory* (eds. E. Nevo and S. Karlin), Academic Press, New York, pp. 515–534.

Ness, B.D., Soltis, D.E., and Soltis, P.S. 1989. Autopolyploidy in *Heuchera micrantha* Dougl. (Saxifragaceae). *Amer. J. Bot.* **76**, 614–626.

Nishizawa, J., and Hirai, A. (1987) Nucleotide sequence and expression of the gene for the large subunit of rice ribulose-1,5-bisphosphate carboxylase. *Jap. J. Genet.* **62**, 389–395.

Ogihara, Y., and Tsunewaki, K. (1982) Molecular basis of the genetic diversity of the cytoplasm in *Triticum* and *Aegilops*. I. Diversity of the chloroplast genome and its lineage revealed by the restriction pattern of ct-DNAs. *Jap. J. Genet.* **57**, 371–396.

Ogihara, Y., and Tsunewaki, K. (1988) Diversity and evolution of chloroplast DNA in *Triticum* and *Aegilops* as revealed by restriction fragment analysis. *Theor. Appl. Genet.* **76**, 321–332.

Page, C.N. (1976) The taxonomy and phytogeography of bracken—a review. *Bot. J. Linn. Soc.* **73**, 1–34.

Palmer, J.D. (1985a) Evolution of chloroplast and mitochondrial DNA in plants and algae. In: *Monographs in Evolutionary Biology: Molecular Evolutionary Genetics* (ed. R.J. MacIntyre), Plenum, New York, pp. 131–140.

Palmer, J.D. (1985b) Comparative organization of chloroplast genomes. *Ann. Rev. Genet.* **19**, 325–354.

Palmer, J.D. (1987) Chloroplast DNA evolution and biosystematic uses of chloroplast DNA variation. *Amer. Natur.* **130**, S6–S29.

Palmer, J.D., and Zamir, D. (1982) Chloroplast DNA evolution and phylogenetic relationships in *Lycopersicon*. *Proc. Natl. Acad. Sci. USA* **79**, 5006–5010.

Palmer, J.D., Shields, C.R., Cohen, D.B., and Orton, T.J. (1983) Chloroplast DNA evolution and the origin of amphidiploid *Brassica* species. *Theor. Appl. Genet.* **65**, 181–189.

Palmer, J.D., Jorgensen, R.A., and Thompson, W.F. (1985) Chloroplast DNA variation and evolution in *Pisum:* patterns of change and phylogenetic analysis. *Genetics* **109**, 195–213.

Palmer, J.D., Jansen, R.K., Michaels, H.J., Chase, M.W., and Manhart, J.R. (1988) Chloroplast DNA variation and plant phylogeny. *Ann. Missouri Bot. Gard.* **75,** 1180–1206.

Rieseberg, L.H. (1990) Hybridization in rare plants: insights from case studies in *Helianthus* and *Cercocarpus*. In: *Conservation of Rare Plants: Biology and Genetics* (eds. D.A. Falh and K.E. Holsinger), Oxford University Press, Oxford.

Rieseberg, L.H., and Soltis, D.E. (1987) Allozymic differentiation between *Tolmiea menziesii* and *Tellima grandiflora* (Saxifragaceae). *Syst. Bot.* **12,** 154–161.

Rieseberg, L.H., Soltis, D.E., and Palmer, J.D. (1988) A molecular reexamination of introgression between *Helianthus annuus* and *H. bolanderi* (Compositae). *Evolution* **42,** 227–238.

Rieseberg, L.H., Carter, R., and Zona, S. (1990a) Molecular tests of the hypothesized hybrid origin of two diploid *Helianthus* species (Asteraceae). *Evolution,* **44,** 1498–1511.

Rieseberg, L.H., Beckstrom-Sternberg, S., and Doan, K. (1990b) *Helianthus annuus* ssp. *texanus* has chloroplast DNA and nuclear ribosomal RNA genes of *Helianthus debilis* ssp. *cucumerifolius*. *Proc. Natl. Acad. Sci. USA* **87,** 593–597.

Scowcraft, W.R. (1979) Nucleotide polymorphism in chloroplast DNA of *Nicotiana debneyi*. *Theor. Appl. Genet.* **55,** 133–137.

Smith, R.L., and Sytsma, K.J. (1990) Evolution of *Populus nigra* L. (sect. *Aigeiros*): introgressive hybridization and the chloroplast contribution of *Populus alba* L. (sect. *Populus*). *Amer. J. Bot.* **77,** 1176–1187.

Slatkin, M., and Barton, N.H. (1989) A comparison of three indirect methods for estimating average levels of gene flow. *Evolution* **43,** 1349–1368.

Soltis, D.E., and Soltis, P.S. (1989a) Allopolyploid speciation in *Tragopogon:* insights from chloroplast DNA. *Amer. J. Bot.* **76,** 1119–1124.

Soltis, D.E., and Soltis, P.S. (1989b) Genetic consequences of autopolyploidy in *Tolmiea* (Saxifragaceae). *Evolution* **43,** 586–594.

Soltis, D.E., and Soltis, P.S. (1990) Chloroplast DNA and nuclear rDNA variation: insights into autopolyploidy and allopolyploidy. In: *Biological Approaches and Evolutionary Trends in Plants* (ed. S. Kawano), Academic Press, San Diego, pp. 97–117.

Soltis, D.E., Soltis, P.S., Collier, T.G., and Edgerton, M. (1991b) Chloroplast DNA variation within and among genera of the *Heuchera* group Saxifragaceae: evidence for chloroplast transfer and paraphyly. *Amer. J. Bot.,* in press.

Soltis, D.E., Soltis, P.S., and Ness, B.D. (1989a) Chloroplast DNA variation and multiple origins of autopolyploidy in *Heuchera micrantha* (Saxifragaceae). *Evolution* **43,** 650–656.

Soltis, D.E., Soltis, P.S., Ranker, T.A., and Ness, B.D. (1989b) Chloroplast DNA variation in a wild plant, *Tolmiea menziesii*. *Genetics* **121,** 819–826.

Soltis, D.E., Soltis, P.S., and Ness, B.D. (1990) Maternal inheritance of the chloroplast genome in *Heuchera* and *Tolmiea* (Saxifragaceae). *J. Hered.* **81,** 168–170.

Soltis, D.E., Mayer, M., Soltis, P.S., and Edgerton, M. (1991a) Chloroplast DNA variation and intraspecific phylogeography in *Tellima grandiflora* (Saxifragaceae). *Amer. J. Bot.*, in press.

Soltis, P.S., and Soltis, D.E. (1987) Genetic differentiation between two closely related species of *Polystichum* (Dryopteridaceae). *Amer. J. Bot.* **74,** 755.

Sytsma, K.J., and Gottlieb, L.D. (1986) Chloroplast DNA evolution and phylogenetic relationships in *Clarkia* sect. *Peripetasma* (Onagraceae). *Evolution* **40,** 1248–1261.

Sytsma, K.J., and Schaal, B.A. (1985) Phylogenetics of the *Lisianthius skinneri* (Gentianaceae) species complex in Panama utilizing DNA restriction fragment analysis. *Evolution* **39,** 594–608.

Tajima, F. (1983) Evolutionary relationships of DNA sequences in finite populations. *Genetics* **105,** 437–460.

Takaiwa, F., and Sugiura, M. (1982) Nucleotide sequence of the 16S–23S spacer region in an rRNA gene cluster from tobacco chloroplast DNA. *Nucleic Acids Res.* **10,** 2665–2676.

Tan, M.K., and Thomson, J.A. (1990) Evolutionary studies of the chloroplast genome in *Pteridium. Bracken Biology and Management* (eds. J.A. Thomson and R.T. Smith), Australian Institute of Agricultural Science, Sydney, pp. 95–103.

Terauchi, R. (1989) Chloroplast DNA variation in *Dioscorea bulbifera* L. *4th International Symposium of Plant Biosystematics,* p. 37, abstract.

Timothy, D.H., Levings, C.S., III, Pring, D.R., Conde, M.F., and Kermicle, J.L. (1979) Organelle DNA variation and systematic relationships in the genus *Zea:* Teosinte. *Proc. Natl. Acad. Sci. USA* **76,** 4220–4224.

Tryon, R.M. (1941) A revision of the genus *Pteridium. Rhodora* **43,** 1–31.

Vedel, F., Quetier, F., and Bayen, M. (1976) Specific cleavage of chloroplast DNA from higher plants by EcoRI restriction nuclease. *Nature* **263,** 440–442.

Wagner, D.G., Furnier, G.R., Saghai-Maroof, M.A., Williams, S.M., Dancik, D.P., and Allard, R.W. (1987) Chloroplast DNA polymorphisms in lodgepole and jack pines and their hybrids. *Proc. Natl. Acad. Sci. USA* **84,** 2097–2100.

Wolf, P.G., Soltis, D.E., and Soltis, P.S. (1990) Allozymic and chloroplast DNA variation in *Heuchera micrantha. Amer. J. Bot.* **77,** 232–244.

Wolfe, K.H., Li, W.-H., and Sharp, P.M. (1987) Rates of nucleotide substitution vary greatly among plant mitochondrial, chloroplast, and nuclear DNAs. *Proc. Natl. Acad. Sci. USA* **84,** 9054–9058.

Zurawski, G., and Clegg, M.T. (1987) Evolution of higher plant chloroplast DNA-encoded genes: implications for structure-function and phylogenetic studies. *Ann. Rev. Plant Physiol.* **38,** 391–418.

7

Molecular Evidence and
Plant Introgression

Loren H. Rieseberg and *Steven J. Brunsfeld*

The role of introgression (interspecific gene exchange) in plant evolution has been the source of much discussion and debate (Anderson, 1949; Heiser, 1949, 1965, 1973; Stebbins, 1950, 1959, 1969; Grant, 1971; Rieseberg et al., 1988). Numerous authors have hypothesized that introgression could provide a significant avenue for the interspecific transfer of genetic adaptations (Anderson, 1949; Heiser, 1949, 1965; Baker, 1965, 1974; Stebbins, 1965; Levin, 1975). As such, introgression could contribute to an increase (or decrease) in the relative fitness of an introgressed taxon (Arnold et al., 1990; Lewontin and Birch, 1966; Levin and Bulinska-Radomska, 1988), allow the colonization of new habitats (Harlan and de Wet, 1963; Heiser, 1965), and/or result in the origin and establishment of new types (Heiser, 1949, 1973; Stebbins, 1950). Introgression could also result in the breakdown of isolating barriers between two partially isolated taxa and their subsequent merger (Grant, 1971).

To determine whether introgression actually plays an important role in plant evolution, two types of information are needed. First, estimates are needed of the frequency of occurrence and extent of introgression in plants. Second, where introgression occurs, the adaptive significance of the process must be determined. Although both types of information are essential, documentation of the occurrence and extent of introgression in a group of organisms is necessary before studies of its adaptive significance in that group are meaningful.

Surprisingly, the frequency of occurrence, extent, and therefore, evolutionary significance of introgression in plants are unclear (Heiser, 1973). This is not due to a lack of proposed examples in the plant literature (some of the best examples are discussed in Heiser, 1973). Rather, we often lack the necessary information to confirm or deny the existence of introgression in these cases or to estimate

We thank A. Liston, T. Philbrick, J. Wendel, and S. Zona for many helpful suggestions. The research on introgression in *Helianthus* and *Salix* has been aided by NSF grants BSR-8722643, BSR-8601450, and BSR-8612789.

precisely the extent of introgression when detected. Reasons for this are (1) alternative hypotheses to introgression, such as convergence or the joint retention of the ancestral condition (symplesiomorphy), are difficult to refute and (2) low levels of introgression are theoretically difficult to detect. For example, the mean number of nuclear markers per backcross individual is halved in each successive generation, and, for a given backcross generation, the mean number of cytoplasmic markers will be one-half that of nuclear genes (Avise and Saunders, 1984).

Molecular markers currently provide the best means for analyzing ambiguous cases of introgression (Rieseberg et al., 1988; Doebley, 1989). Molecular markers tend to be neutral (Kimura, 1982), whereas morphologic characters often converge when exposed to similar selective pressures. Specific mutations for molecular characters can easily be polarized by comparison to related taxa, thus allowing one to distinguish between hypotheses of introgression and symplesiomorphy. Theoretically, an almost unlimited number of independent molecular markers that differentiate related forms could be obtained, allowing the detection of even extremely low levels of introgression. Furthermore, molecular markers provide the opportunity to monitor both nuclear and cytoplasmic gene flow.

In this review, we discuss the utility of various types of molecular markers for the study of introgression and examine evidence for introgression in plants as revealed by nucleic acid studies.

Types of Molecular Markers

Cytoplasmic Markers

Organellar genomes (chloroplast and mitochondrial) are useful for the study of introgression because they often contain multiple, nonrecombinant, molecular markers (Doebley, 1989; Rieseberg et al., 1990a). This characteristic of cytoplasmic genomes greatly increases the potential for the *simultaneous* appearance of multiple markers in an introgressed individual (Avise and Saunders, 1984). Clearly, if a putative introgressant possessed multiple, linked markers of a potential hybridizer, the probability that this situation could be attributed to symplesiomorphy or convergence would be minimized.

In some instances, cytoplasmic gene flow occurs in the absence of nuclear gene flow, as observed for mitochondrial DNA (mtDNA) in mouse (Ferris et al., 1983). Obviously, introgression would go undetected in such cases if cytoplasmic markers were not analyzed. Conversely, in *Pinus banksiana* and *P. contorta,* it is suggested that introgression of nuclear markers occurred without a concomitant exchange of cytoplasmic markers (Wagner et al., 1987).

Various characteristics of the chloroplast genome simplify the analysis of plant chloroplast (cpDNA) variation relative to plant mtDNA variation (Palmer, 1985). For this reason, most studies of cytoplasmic introgression in plants have relied on cpDNA rather than mtDNA markers. Nevertheless, due to the frequency of structural rearrangements observed between mtDNAs of closely related forms and the concomitant difficulty in finding structural or sequence differences between cpDNAs of closely related plant species, it seems likely that mtDNA markers could be valuable for future studies of introgression in plants.

Nuclear Markers

Nuclear markers are of more direct importance to the study of introgression than cytoplasmic markers, because most nuclear genes are inherited in a simple Mendelian fashion and will be additively combined in hybrids or introgressants. Furthermore, by using multiple, tightly linked nuclear markers, information comparable to that described above for nonrecombinant cytoplasmic markers can be obtained (Doebley, personal communication; Rieseberg et al., 1990a). Finally, a virtually unlimited supply of nuclear markers is potentially available.

Isozymes

Isozymes are useful for detecting hybridization and introgression because allozymes, like other nuclear markers, will be additively combined in hybrids or introgressants (Gallez and Gottlieb, 1982; Crawford and Ornduff, 1989). Nevertheless, the utility of isozyme markers relative to restriction fragment length polymorphisms (RFLP) is limited by the generally low number of diagnostic alleles distinguishing closely related forms (e.g., Rieseberg et al., 1988, 1990b) and the difficulty of defining (e.g., is the locus or the allele the character?) and polarizing isozymic character states (Buth, 1984; Avise, 1989). Although isozymes have been used effectively in a number of studies of introgression in plants (e.g., Doebley et al., 1984; Rick et al., 1974; Rieseberg et al., 1988), it seems likely that they will be supplemented and possibly replaced by nuclear RFLP markers in future studies. Thus, isozyme evidence pertaining to introgression will not be discussed herein.

Ribosomal RNA Genes

Higher plants possess two families of ribosomal RNA genes (rRNA): the 18S, 5.8S, and 25S gene family, and the 5S gene family (reviewed in Doyle et al., 1984; Appels and Honeycutt, 1986; Schaal and Learn, 1988; see Chapters 4 and 5, this volume). Herein, we will follow standard terminology (Appels and Honeycutt, 1986) and refer to those DNA sequences encoding the former gene family as rDNA and those encoding the latter gene family as 5S DNA.

Both families of rRNA genes have been useful for studies of hybridization and introgression for several reasons. First, rRNA genes are organized as arrays of tandemly repeated genes (reviewed in Doyle et al., 1984; Appels and Honeycutt,

1986; Schaal and Learn, 1988). The high copy number of rRNA genes obviates the technical problems sometimes associated with detection of single-copy sequences. This feature also allows both the determination of the percentages of parental contributions and the detection of extremely low levels of foreign rRNA genes. For example, 2% Johnsongrass contamination in a Piper Sudangrass DNA was detected using a maize rDNA probe (Springer et al., 1989). Second, the highly conserved rRNA coding sequences are separated by a highly variable (both in length and sequence) intergenic spacer (reviewed in Doyle et al., 1984; Appels and Honeycutt, 1986; Schaal and Learn, 1988). Thus, both restriction site and length polymorphisms can generally be found that distinguish closely related forms. Finally, several studies (e.g., Sytsma and Schaal, 1985; Rieseberg et al., 1988, 1990a, b) have demonstrated that restriction fragment changes in DNA sequences encoding rRNA can generally be polarized by comparison to related species.

Although complete additivity of rDNA repeat types is usually observed in hybrids (Doyle et al., 1985; Doyle and Doyle, 1988; Rieseberg et al., 1990a, b; Liston et al., 1990), some anomalies have been observed. For example, some synthetic F_1 hybrids between maize and teosinte did not inherit all the rDNA length variants of their two parents, whereas other progeny did (Zimmer et al., 1988). Similarly, in synthetic F_1 hybrids between *Zea mays* and *Z. diploperennis*, codominant inheritance of 5S DNA patterns was observed in only two of the six hybrids examined (Zimmer et al., 1988). Three of the four remaining individuals had the repeat type of both parents, but the *Z. diploperennis* type was underrepresented. The fourth individual lacked the *Z. diploperennis* 5S repeat type. It is noteworthy that all individuals were hybrids for rDNA.

Another consideration with the use of rDNA markers for the study of introgression is the possibility of gene conversion. This is important because the frequency of introgressed rDNA genes in an individual, population, or taxon could be increased or decreased because of this process. Gene conversion has been experimentally demonstrated in eukaryotes (Klein, 1984). Furthermore, biased gene conversion has been used to account for patterns of rDNA introgression in grasshopper (Arnold et al., 1988; Marchant et al., 1988).

RFLP Markers

The recent development of RFLP markers provides an almost unlimited number of qualitative genetic markers that are easily characterized and completely penetrant (Botstein et al., 1980; Beckmann and Soller, 1983; Keim et al., 1989). The availability of such markers representing all major linkage groups for a specific nuclear genome should allow detailed and quantitative estimates of the actual frequency of occurrence and extent of introgression in a specific group of plants. The possible use of linked RFLP markers to track the introgression of specific chromosomes or chromosome segments is particularly intriguing. For example, genetic models for the transgression of genes across a chromosomal sterility barrier could be tested using this method. At least one detailed study of introgression in plants using nuclear RFLP markers (Keim et al., 1989) demonstrates the potential of this approach.

Hypervariable Sequences

Hypervariable loci are regions of the nuclear genome that contain tandem repeats of short DNA segments (minisatellites). Variability arises from differences in the number and length of these repeats (Vassart et al., 1987). Hypervariable sequence analysis (DNA fingerprinting) facilitates the analysis of a large number of highly variable DNA loci in a single test (Burke and Bruford, 1987). DNA fingerprinting could greatly reduce the labor and expense of detailed nuclear RFLP studies of introgression. Although most studies of hypervariable sequence variation have concentrated on animals, recent studies have demonstrated the potential of DNA fingerprinting for evolutionary studies in plants (Rogstad et al., 1988).

Molecular Evidence and Introgression

The process of introgression was first extensively studied and named by Anderson (Anderson and Hubricht, 1938; Anderson, 1949). Since then, numerous examples of introgression have been reported, many reviews on the subject have appeared, and three types or effects of introgression have been proposed: localized introgression (Stebbins, 1950; Grant, 1971; Heiser, 1973), dispersed introgression (Heiser, 1973), and race or species formation through introgression (Stebbins, 1950; Grant, 1971; Heiser, 1973). These terms are relative, however, and difficult to define unambiguously. Ancient hybridization or introgression events are particularly difficult to classify because the actual type of introgression cannot be determined. One of the most effective methods for detecting ancient events of introgression is by the reconstruction of molecular phylogenies using more than one molecule and multiple population samples. Thus, some of the most convincing cases of introgression in plants have been inferred from molecular phylogenetic analysis (e.g., Palmer et al., 1983; Doebley, 1989).

Localized Introgression

Most reported cases of introgression involved species with large regions of sympatry (Heiser, 1973). Hybrid swarms occur frequently, thus giving the appearance of extensive introgression. The actual extent of interspecific gene flow, however, appears to be localized, the hybrid swarms ephemeral, and the long-term evolutionary significance has been considered slight (Heiser, 1973). Nevertheless, this type of introgression may be of greater significance, if the interspecific gene flow extends over a greater distance than is superficially apparent, if the repeated occurrence of hybridization extends over large geographic areas or over a long time frame, or if a stabilized introgressant is derived from one of these localized pockets of hybridization and backcrossing.

Several studies have used nucleic acid markers to document the occurrence of hybridization in natural plant populations. Doyle et al. (1985) examined repeat length differences in rDNA and 5S DNA in *Tolmiea menziesii, Tellima grand-iflora,* and a putative intergeneric hybrid. The putative hybrid combined the repeat lengths of *Tolmiea* and *Tellima* for both rDNA and 5S DNA, substantiating the occurrence of hybridization between the two genera. Similarly, rDNA markers were used to demonstrate the occurrence of hybridization in a zone of overlap between two species of *Claytonia* (Doyle and Doyle, 1988). Surprisingly, fewer hybrids were detected by rDNA analysis than were predicted based on evidence from leaf morphology. Doyle and Doyle (1988) suggested that the discrepancy between the leaf morphology and rDNA data was probably due to differences in the inheritance of these traits. Tandem repeats of rRNA genes occur in a few clusters throughout the genome and are inherited in a Mendelian fashion. In contrast, leaf morphology may be controlled by numerous genes. Thus, one or a few rDNA markers might be rapidly lost by backrossing to one of the parental species, but genes affecting leaf morphology might persist over longer periods of time. A similar discrepancy between morphologic predictions and rDNA evidence was observed in *Lotus* (Liston et al., 1990). Only four rDNA-based hybrids were detected out of a total of 38 individuals examined from adjacent populations of the rare island endemic *Lotus scoparius* ssp. *traskiae* and the more widespread *Lotus argophyllous* ssp. *ornithopus.* At least seven hybrids were predicted based on morphologic criteria.

A detailed study of a hybrid swarm or localized introgression has been carried out in *Populus* (Keim et al., 1989). Random nuclear RFLP markers were used to examine the occurrence, extent, and direction of introgression in a hybrid swarm of *Populus fremontii* and *P. angustifolia*. By examining approximately 11 markers per genet, the genetic contribution of the *fremontii* and *angustifolia* parents to genomes of genets in the hybrid area could be measured precisely. Of the 30 genets examined in the zone of overlap, 24 appeared to be F_1 hybrids or the result of backcrosses between F_1 hybrids and *P. angustifolia*. Five individuals were homozygous *P. fremontii,* and one individual was homozygous *P. angusti-folia.* No individuals could be classified as either backcrosses to *P. fremontii* or as progeny of further crossing between hybrids or between hybrids and backcrossed individuals. The authors concluded that gene flow between the parental popula-tions is asymmetric and provided evidence that crosses between hybrid trees, or between *P. fremontii* and hybrids, are not productive due to genetic incompatibil-ity. It is difficult to determine whether the directional introgression observed locally between *P. fremontii* and *P. angustifolia* will have a significant evolution-ary effect, although several potential consequences are discussed (Keim et al.,

1989). Nevertheless, it has been demonstrated that hybrid cottonwood trees are much more susceptible to aphids than are pure populations of either parent, and that the hybrid zone described above appears to be acting as a pest sink (Whitham, 1989).

Evidence from a range of characters (morphology, allozymes, flavonoids, cpDNA, rDNA, chromosome numbers, and artificial hybridization and backcrossing experiments) was used to reexamine an often-cited example of hybridization and introgression in plants involving two introduced thistle species, *Carduus nutans* ($2n = 16$) and *C. acanthoides* ($2n = 22$) (Warwick et al., 1989). All data sets were consistent with the hypothesis of bidirectional introgression in populations within a stable hybrid zone. However, no evidence of introgression was observed outside of the hybrid zone, suggesting that interspecific gene flow is only possible through extension of the hybrid zone rather than via dispersed introgression.

Dispersed introgression of morphologic, terpene, and allozyme characters has been reported (Critchfield, 1985) for allopatric populations of lodgepole pine (*Pinus contorta*) and jack pine (*Pinus banksiana*). In contrast, no evidence of dispersed cpDNA introgression was observed in these same species (Wagner et al., 1987) despite extensive sampling of both sympatric and allopatric populations (including some of the same populations on which previous reports of introgression were based). A few trees found in the zone of sympatry had the morphology of one species and the chloroplast genome of the other species, suggesting the occurrence of local introgression. Several factors may contribute to the apparent contradiction between the cpDNA data and evidence from previous morphologic, allozyme, and terpene studies: (1) differences in sample size, (2) studies of morphologic, terpene, and allozyme characters may have detected intraspecific polymorphism rather than introgression, and (3) nuclear genes may sometimes be able to cross species barriers that cpDNA is unable to cross. The reverse situation, that organelle DNA may sometimes cross species borders that nuclear genes do not appear to cross, has been reported for mitochondrial DNA in animals (Ferris et al., 1983; Powell, 1983).

Perhaps the most intriguing aspect of the *Pinus* study was the presence of a number of unusual cpDNA phenotypes in and adjacent to the zone of sympatry (potential mechanisms for this phenomenon are discussed by Wagner et al., 1987). These data are of interest because it is now clear that there is an elevated frequency of novel allozymes in hybrid swarms or hybrid zones (reviewed in Barton and Hewitt, 1985). A contrasting situation was observed in a detailed study of chloroplast DNA variation in a 45-year-old hybrid swarm containing F_1 and later generation plants of *H. annuus* and *H. bolanderi* (Rieseberg et al., 1988; Rieseberg, unpublished data). Although 30 individuals were analyzed using 36 restriction enzymes, 29 of the 30 cpDNA types observed had previously been identified in allopatric populations of the two parents. The only "novel" cpDNA was identical to that of *H. exilis,* a closely related species—indicating that the hybrid swarm was actually composed of three species rather than two. It may be significant that the chloroplast genome of *Pinus* is paternally inherited, whereas maternal inheritance of cpDNA has been observed in *Helianthus* (Rieseberg et al., 1991).

Dispersed Introgression

Dispersed introgression is gene flow that extends some distance from the area of hybridization (Heiser, 1973) and is usually considered to occur between species that are largely allopatric. Hybridization followed by repeated backcrossing could eventually result in clinal variation throughout the combined ranges of the taxa. Because hybridization between such species only occurs in a limited area of sympatry, the resulting introgression appears, in most cases, to be highly localized. Nevertheless, dispersed introgression could have effects throughout the range of a plant species. However, there are few unequivocal cases of dispersed introgression in plants (Heiser, 1973).

As discussed in the previous section, dispersed cpDNA introgression was not observed in allopatric populations of lodgepole and jack pine—in contrast to positive evidence for dispersed introgression from allozyme, morphologic, and terpene data sets. Likewise, no evidence for dispersed introgression was observed in populations of *Helianthus niveus* ssp. *canescens* and *H. petiolaris* ssp. *fallax* using cpDNA and rDNA data (Beckstrom-Sternberg et al., 1991). *Helianthus niveus* ssp. *canescens* occurs in sandy soils ranging from western Colorado and Utah south to New Mexico and Arizona. As noted by Heiser et al. (1969), "*H. niveus* ssp. *canescens* presents a somewhat anomalous situation with plants in the eastern part of the range appearing more similar to *H. petiolaris* than to the other two subspecies of *H. niveus*." Heiser et al. (1969) suggested that hybridization might provide a possible explanation for the apparent intergradation in New Mexico, although he did not find any natural hybrids. Extensive field work in New Mexico also failed to document the presence of hybrids in this area (Rieseberg, unpublished data). Alternatively, Heiser et al. (1969) suggested that the "intergrading material represents types similar to those that might be the common ancestor of both *H. petiolaris* and *H. niveus* ssp. *canescens*."

Molecular evidence supports the latter hypothesis. No evidence of hybridization or introgression was observed between *H. petiolaris* ssp. *fallax* and *H. niveus* ssp. *canescens*. In fact, the two subspecies could not be consistently differentiated from each other in terms of cpDNA, rDNA, or allozymes. In contrast, diagnostic rDNA and cpDNA markers do distinguish the other two subspecies of *H. niveus* from ssp. *fallax* and ssp. *canescens*. The fact that ssp. *fallax* and ssp. *canescens* are intersterile due to a chromosomal sterility barrier would argue against a change in their present classification. Thus, the most parsimonious explanation for the genetic homogeneity between populations of ssp. *canescens* and ssp. *fallax* is that the zone of contact between these taxa is primary rather than secondary in origin. Thus, some form of parapatric speciation (Murray, 1972; White, 1978) may be most consistent with the molecular and biological data observed.

Positive evidence for dispersed introgression has been reported in *Iris* (Arnold et al., 1990). *Iris fulva* and *I. hexagona* have been viewed as a classic example of introgressive hybridization (Anderson, 1949). Yet, a subsequent study, based upon morphologic and cytologic data, concluded that there was no evidence for a transfer of genetic material between the two parental species (Randolph et al., 1967). More recently, Arnold et al. (1990) examined both allopatric and parapatric populations of the two species using diagnostic rDNA markers. They concluded that localized introgression of ribosomal sequences had occurred into both *I. fulva* and *I. hexagona*. In addition, they detected rDNA markers from each species in allopatric populations of the alternate species, thus providing strong positive evidence for bidirectional dispersed introgression.

Race or Species Formation through Introgression

Although introgression can produce convergence between previously distinct species, there is some evidence that the stabilization of hybrid derivatives also can occur, resulting in the creation of new types (Stebbins, 1950). This is particularly likely when a new or intermediate habitat is available for the stabilized introgressant (Anderson, 1949; Stebbins, 1950). Both new genes (Golding and Strobeck, 1983) and new genotypes (Hunt and Selander, 1973; Sage and Selander, 1979) may be produced through recombination, thus increasing the genetic diversity of the species and possibly altering their adaptive potential (Lewontin and Birch, 1966).

What has long been thought to represent a "classic" example of the origin of a new type through introgression involves the three annual sunflowers of California, *H. bolanderi, H. exilis,* and *H. annuus*. Based on morphologic and cytologic evidence, Heiser (1949) hypothesized that *H. bolanderi* originated recently by the introgression of genes from the recently introduced *H. annuus* into *H. exilis*. As a test of this hypothesis, populations of all three species were investigated with reference to allozyme, cpDNA, and rDNA variation (Rieseberg et al., 1988). Although numerous diagnostic molecular markers were used, no evidence of interspecific gene transfer away from hybrid swarms was observed. Furthermore, *H. bolanderi* had a unique and divergent chloroplast genome when compared to the cpDNAs of *H. exilis* and *H. annuus* (Fig. 7.1), which is evidence suggesting that *H. bolanderi* was not derived through introgression during the past several hundred years as hypothesized but is relatively ancient in origin.

Likewise, cpDNA and rDNA evidence was used to examine the hypothesized origin of *Solanum raphanifolium* by diploid hybridization between *S. canasense* and *S. megistacrolobum* in the overlap zone between the two putative parental species in southern Peru (Spooner et al., 1991). Phylogenetic analysis of both cpDNA and rDNA indicates that the putative parental species form a monophy-

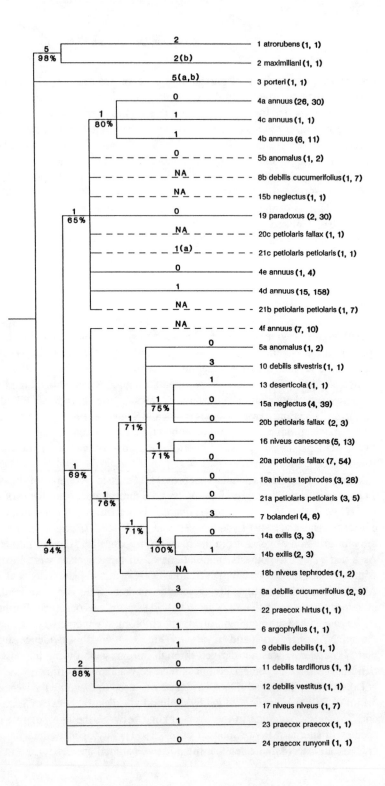

letic clade with *S. raphanifolium* as a sister group. These data indicate that *S. raphanifolium* is not of recent hybrid origin between *S. canasense* and *S. megistacrolobum* and possibly not even closely related to them.

In a contrasting study, Heiser (1951) hypothesized that *Helianthus annuus* ssp. *texanus* was derived by the introduction of *H. annuus* into Texas and subsequent introgression of genes from *H. debilis* ssp. *cucumerifolius* into *H. annuus*. Although often considered to be one of the best cases of introgression in plants, alternative hypotheses to introgression, such as convergence and symplesiomorphy, could not be ruled out in the original study. Diagnostic cpDNA and rDNA markers were used to examine allopatric populations of *H. annuus,* typical populations of *H. debilis* ssp. *cucumerifolius,* as well as 14 populations of *H. annuus* ssp. *texanus* (Rieseberg et al., 1990a). Thirteen of the 14 populations of *H. annuus* ssp. *texanus* had cpDNA (Fig. 7.1) and/or rDNA markers of *H. debilis* ssp. *cucumerifolius,* providing strong support for the hypothesized introgressive origin of *H. annuus* ssp. *texanus.* Furthermore, cpDNA and rDNA introgression from *H. annuus* into *H. debilis* ssp. *cucumerifolius* was also observed (Fig. 7.1).

Another well-documented example (Brunsfeld et al., 1990) involves a race of *Salix taxifolia* (section *Longifoliae*) found near Fort Davis, Texas. This species is primarily distributed in central and southern Mexico, but scattered populations occur in isolated, moist habitats as far north as southern Arizona and Texas. The Fort Davis site lies within the range of *S. interior,* a wide-ranging, eastern North American member of section *Longifoliae.* Common-garden experiments revealed that the Fort Davis material differed morphologically from Mexican *S. taxifolia* and appeared to approach *S. interior* in several characteristics. Molecular analysis revealed that the population had the cpDNA of *S. interior* (Fig. 7.2) and a mixture of allozymes characteristic of both *S. taxifolia* and *S. interior.* Surprisingly, the Fort Davis population also had allozymes characteristic of *S. exigua* and *S. melanopsis,* two other members of section *Longifoliae.* Lastly, the Fort Davis population currently appears to be hybridizing with *S. goodingii,* a member of section *Humboldtianae,* which is also present at the Fort Davis site. Thus, the Fort Davis race appears to be at least the result of gene flow into a relict population of *S. taxifolia* via wind-dispersed seed from a neighboring population of *S. interior,* and possibly the result of introgression of as many as four or five willow species.

Figure 7.1. Chloroplast DNA-based phylogenetic tree of *Helianthus* section *Helianthus.* Population designations and taxon are given at the ends of the branches, and the numbers of mutations are given above the branches. Percentages indicate the number of times that a monophyletic group occurred in 100 bootstrap samples. Lower-case letters indicate parallel site changes. Dashed lines indicate discrepancies between morphologic classification and cpDNA type that are thought to result from cytoplasmic introgression. NA = number of autapomorphies unknown because only "diagnostic" mutations were surveyed. Numbers in parentheses following taxon designations are the numbers of populations and individuals sampled, respectively.

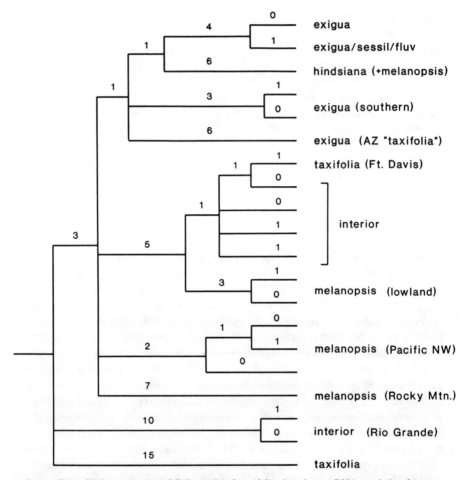

Figure 7.2. Phylogenetic tree of *Salix* section *Longifoliae* based on cpDNA restriction fragment differences. Names at ends of branches include the specific epithets and cytoplasmic races discussed in text. The number of mutations appears above each branch.

A less clear-cut example of the introgressive origin of a new type involves the origin of emmer (AABB) wheat (*Triticum turgidum*). Although there is ample evidence that *T. monococcum* (AA) is the source of the A genome in emmer wheat (Gill and Kimber, 1974; Fernandez de Caleya et al., 1976; Jones et al., 1982; Kerby and Kuspira, 1987), there is little agreement regarding the origin of the B genome. What appears certain based on evidence from nucleo-cytoplasmic interactions and fraction I proteins (reviewed in Graur et al., 1989) is that the donor of the B genome is also the donor of the cytoplasm of emmer wheat. Unfortunately, cpDNA and mtDNA evidence are not clear as to the cytoplasmic donor of the B genome. Original reports that *Aegilops longissima* was the donor of the B genome (Ogihara and Tsunewaki, 1982; Tsunewaki and Ogihara, 1983) have since been retracted in favor of a form of *Ae. speltoides* (Ogihara and Tsunewaki, 1988). It should be noted, however, that a form of *Ae. speltoides* possessing a cpDNA identical to that of emmer wheat has yet to be discovered. Mitochondrial DNA evidence eliminates both of these species as cytoplasmic donors, but does not suggest an alternative candidate (Graur et al., 1989). A possible solution to this problem has been proposed by Gill and Chen (1987). They suggest, based on evidence from chromosome banding studies, that *T. timopheevii* (AAGG) originated as an allotetraploid of *Ae. speltoides* (GG) and *T. monococcum* (AA), and that *T. turgidum* was derived from *T. timopheevii* by introgressive hybridization with an unknown diploid species, which contributed its distinctive cytoplasm, chromosome 4B or a substantial portion of it, and additional chromosome fragments.

A special case of hybrid and/or introgressive speciation, termed recombinational speciation, has been proposed by Grant (1971). Grant proposed that new homozygous chromosome arrangements can be obtained by the recombination of two or more separable arrangements that distinguish the parental species. The new recombinant types will be fertile *inter se* but at least partially sterile with both parents.

Rieseberg et al. (1990b) used allozyme, cpDNA, and rDNA evidence to test the hypothesis that both *Helianthus neglectus* and *H. paradoxus* were derived by recombinational speciation from *H. annuus* and *H. petiolaris*. All four species are annual, diploid outcrossers, and are almost completely reproductively isolated by chromosomal sterility barriers (Heiser et al., 1969). The alleles observed in *H. neglectus* were not a combination of those observed in *H. annuus* and *H. petiolaris* but were simply a subset of those found in *H. petiolaris*. Furthermore, *H. neglectus* had the same rDNA type as *H. petiolaris* (Fig. 7.3), and 39 of the 40 *H. neglectus* individuals assayed had the same cpDNA as that found in two populations of *H. petiolaris* ssp. *fallax*. The remaining individual had a chloroplast genome typical of *H. annuus* (Fig. 7.1). Rieseberg et al. (1990b) suggested that the anomalous cpDNA may be due to recent maternal introgression from wild or domesticated *H. annuus* into *H. neglectus* because *H. neglectus* grows along a relatively well-traveled highway where contamination might occur. Nevertheless, these data suggest that *H. neglectus* may be a recent derivative of *H. petiolaris* ssp. *fallax*, rather than a stabilized hybrid derivative as originally proposed.

In contrast, *H. paradoxus* combined the allozymes of *H. annuus* and *H. petiolaris* and had no unique allozymes. Furthermore, *H. paradoxus* combined the rDNA repeat types of both proposed parents (Fig. 7.3) and had the chloroplast genome of *H. annuus* (Fig. 7.1). These data provide compelling evidence that *H. paradoxus,* in contrast to *H. neglectus,* was derived via recombinational speciation. It is noteworthy that the rDNA repeat types of both parents were found in all individuals of *H. paradoxus* in apparently equivalent copy numbers

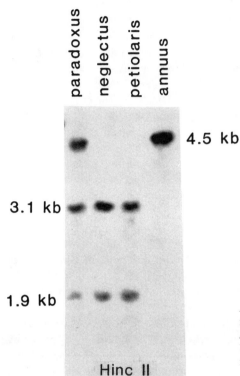

Figure 7.3. Photograph of *Hinc*II rDNA restriction fragments in *Helianthus*. The *Hinc*II fragments were visualized by hybridizing with 3.55-kb and 2.45-kb *Sac*I rDNA fragments from *H. argophyllus*.

(Fig. 7.3); no evidence of segregation was observed. To account for the "fixed" rDNA heterozygosity observed, Rieseberg et al. (1990b) suggested that either (1) the rDNA locus in *H. paradoxus* had been duplicated or (2) more than one locus encodes ribosomal RNA genes in the annual *Helianthus* species and that the *annuus* repeat type had become fixed at one locus and the *petiolaris* type at the second locus in *H. paradoxus*. The presence of both parental rDNA types in the hybrid taxon has important implications to the study of introgression, because it has not been clear whether more than one type of rDNA will be maintained in a hybrid taxon over time due to the homogenizing effects of concerted evolution (Marchant et al., 1988).

Inferences from Molecular Phylogenetic Studies

One of the most exciting aspects of the reconstruction of molecular phylogenies has been the discovery of unexpected and sometimes ancient events of introgression, as a result of nonconcordance between cytoplasmic-based (cpDNA/mtDNA) and nuclear-based (rDNA, RFLP, karyotypes, morphology, allozymes, etc.) phylogenies. It should be noted that, given further study, many of these introgression events probably could be classified as one of the three types of introgression discussed previously.

Discrepancies between cpDNA- and nuclear-based phylogenies obtained for two accessions of *Brassica napus* led to the hypothesis that these accessions of *B. napus* obtained their cytoplasm by introgression from an as-yet unidentified *Brassica* species (Palmer et al., 1983). A similar result was observed for the mtDNAs of these accessions (Palmer, 1988). This result is important because these two accessions appear to represent wild-type *B. napus*. In other words, introgression had not been suspected based on morphologic or other available evidence. In addition, phylogenetic analysis of nuclear RFLP data suggests that introgression also may have occurred between *B. campestris* and *B. oleracea*, the diploid parents of *B. napus* (Song et al., 1988).

A similar situation has been reported in pea (Palmer et al., 1985). Twelve of 13 cultivated lines of pea (*Pisum sativum*) had the chloroplast genome of northern populations of *P. humile*, the putative wild progenitor of the domesticated pea. The remaining cultivated line had a cpDNA more similar to that of *P. elatius* and southern populations of *P. humile*, possibly indicating that introgression occurred during or after the domestication of the pea.

A number of discrepancies observed between morphologic and cpDNA-based phylogenies for *Gossypium* may be the result of introgression or hybrid speciation (Wendel et al., 1991; Wendel and Albert, 1991). The first example involves the Australian (C-genome or G-genome) species, *Gossypium bickii*. This species, along with *G. nelsoni* and *G. australe*, are morphologically similar, arid-zone species included in section *Hibiscoidea* of subgenus *Sturtia*. These three species are quite distinct morphologically from the rest of the genus. Yet, the cpDNA of *G. bickii* is most similar to that of *G. sturtianum* of section *Sturtia*. A second example involves the arborescent D-genome diploid, *G. aridum*. This species is morphologically similar to other arborescent species such as *G. schwendemanii*, *G. lobatum*, and *G. laxum*, but belongs to the same cpDNA lineage as two morphologically very different species (*G. klotschianum* and *G. davidsonii*). Likewise, the northwestern Australian species *G. cunninghamii* is morphologically similar to members of section *Grandicalyx*, but cpDNA data suggest that its cytoplasm came from some other taxon in the clade leading to *G. sturtianum* (section *Sturtia*).

In the Hawaiian silverswords, nonconcordance between karyologic and cpDNA data suggests introgression (Baldwin et al., 1990). Chromosome data suggest that *Dubautia scabra* ($n = 14$) is basal to all $n = 13$ species of *Dubautia* and serves as a bridge to other *Dubautia* with $n = 14$. Chloroplast DNA analysis of pooled population samples of *D. scabra* from two localities (Kilauea and Pu'u Huluhulu) revealed that *D. scabra* from both populations had a plastome identical to that of *D. ciliolata* ($n = 13$) from Kilauea, placing *D. scabra* at the end of the cpDNA clade of species with $n = 13$. *Dubautia scabra* is sympatric with *D. ciliolata* at both localities, and hybridization between the two taxa has been documented using flavonoid evidence (Crins et al., 1988). These data suggest that the observed discrepancy between chromosome rearrangements may be due to introgression.

The situation is not this simple, however. *Dubautia ciliolata* from Pu'u Huluhulu had a cpDNA that differed slightly from that found in *D. scabra* and the other population of *D. ciliolata*. Perhaps the most plausible explanation for this pattern of cpDNA variation is that *D. scabra* "captured" the *D. ciliolata* cytoplasm through introgression at Kilauea. The population of *D. scabra* at Pu'u Huluhulu was then derived from a previously introgressed population of *D. scabra* similar to that found at Kilauea. Additional possible explanations for the observed cpDNA variation are discussed in Baldwin et al. (1990).

A second example of putative introgression in Hawaiian Madiinae involves the silverswords and greenswords of the genus *Argyroxiphium* (Baldwin et al., 1990). Bog greenswords (*A. grayanum*) from West Maui had a cpDNA type identical to that of a sympatric species of bog silversword (*A. caliginis*) and different from that of a bog greensword population from East Maui. Hybridization between bog greenswords and silverswords has been documented previously and appears to be the most parsimonious explanation for the patterns of cpDNA variation observed. A polyphyletic origin of greenswords from East and West Maui cannot be ruled out, however.

Phylogenetic analysis of cpDNA variation in *Fuchsia* also produced a surprising result, possibly due to introgression (Sytsma et al., 1991). The cpDNAs of *F. perscendens, F. excorticata,* and their natural hybrid were almost identical, although the two parental species are very different with respect to habit (liana and large tree, respectively). Sytsma et al. (1991) suggested two possible explanations for this result: (1) divergence between the species occurred very recently but was accompanied by extreme and rapid change in habit, or (2) introgression.

Doebley (1989) found an atypical chloroplast genome in 8 of 10 plants examined of *Zea perennis* from Piedra Ancha. Chromosomally and morphologically, however, these plants were typical *Z. perennis*. Comparison of the atypical cpDNA to other cpDNA types previously described for *Zea* (Doebley et al., 1987b) revealed that this genome type was not present in any *Zea* examined to date. Although this cpDNA type is most similar to those found in the subspecies of *Z. mays,* it is distinguished from *Z. mays* by five or six mutations. These data suggest that the foreign cytoplasm must have come from a missing taxon, potentially located in the regions surrounding the population from Piedra Ancha.

Introgression also appears to have played a role in the origin of the cytoplasmic male sterile-S (cms-S) cytoplasm of *Z. mays* (Doebley and Sisco, 1989). The cpDNAs or cms-C and -T were identical to those found in the majority of maize accessions and in ssp. *parviglumis,* the probable progenitor of maize (Doebley et al., 1987b). In contrast, the cpDNAs of cms-S cytoplasms were the same as those found in three collections of *Z. mays* ssp. *mexicana* (Central Plateau Teosinte) from Copandiro, Michoacan, Mexico. Doebley and Sisco (1989) suggested two possible explanations for the presence of the same cytoplasm in cms-S and Copandiro teosinte: (1) multiple domestications of *Z. mays* or (2) introgression. Introgression seems more plausible, because it is consistent with other data that suggest that maize was domesticated once from ssp. *parviglumis* and not from ssp. *mexicana* (Doebley et al., 1987b). In this context, it is interesting to note that discrepancies between cytoplasmic groupings and presumed racial affinities have been noted previously in both teosinte (Timothy et al., 1979) and maize (Weissinger et al., 1983).

Although several well-documented cases of introgression (or the absence of introgression) have already been described for the annual species of *Helianthus,* nonconcordance between morphologic classification and cpDNA type was observed for populations or individuals of three other taxa (Fig. 7.1): *H. petiolaris* ssp. *fallax,* *H. petiolaris* ssp. *petiolaris,* and *H. anomalus* (Rieseberg et al., 1991). Two individuals from a population of *H. petiolaris* ssp. *fallax* had the chloroplast genome of *H. annuus* but the morphology of *H. petiolaris* ssp. *fallax* (Fig. 7.1). Likewise, a single individual of *H. petiolaris* ssp. *petiolaris* from southern California was detected that possessed the cpDNA of *H. annuus,* although morphologically the plant appeared to be typical of *H. petiolaris* ssp. *petiolaris.*

One of the most interesting cases of introgression in *Helianthus* involves the source population of *H. petiolaris* ssp. *petiolaris* from which hybrid sunflower CMS is derived. Rieseberg and Seiler (1990) demonstrated that the hybrid sunflower had the chloroplast genome of *H. annuus* rather than that of *H. petiolaris.* This was a surprising finding given that sunflower CMS is thought to be derived from *H. petiolaris* (Leclercq, 1969). Because cpDNA is inherited maternally in *Helianthus* (Rieseberg et al., 1991), the most likely explanation for the lack of *H. petiolaris* cpDNAs in the domesticated hybrid sunflower was that the source population for sunflower CMS was a hybrid or introgressive population of *H. annuus* and *H. petiolaris* (Rieseberg and Seiler, 1990). This hypothesis was confirmed by analysis of seven individuals from the source population of *H. petiolaris,* from which sunflower CMS was derived (Heiser, personal communication). All seven individuals had the chloroplast genome of *H. annuus.*

The last case involves *H. anomalus,* a sunflower restricted in range to northern Arizona and southern Utah. Four individuals from a single population were analyzed with regard to maternal ancestry. Two very different cpDNAs were detected. One cpDNA was like that of *H. annuus,* whereas the second cpDNA was similar to that of other southwestern sunflowers, such as *H. petiolaris, H. niveus,* and *H. deserticola* (Fig. 7.1). Because only one population was analyzed, it was difficult to deduce whether the *annuus* cpDNA type or the *petiolaris* cpDNA type was alien to *H. anomalus.* Based on morphology and geographic distribution, we might suggest that the *petiolaris* type cpDNA is the native cpDNA for *H. anomalus* and the *annuus* type cpDNA is alien to *H. anomalus.* Evidence from nuclear rDNA (Rieseberg, unpublished data), however, suggests that *H. anomalus* may be a stabilized hybrid derivative of *H. annuus* and *H. petiolaris*.

It is notable that *H. annuus* is involved in all introgression events observed in section *Helianthus*. This is consistent with the fact that *H. annuus* is geographically widespread and sympatric with most of the taxa in section *Helianthus*. These data may also provide some support for the concept of *H. annuus* as a compilospecies (Heiser, 1965), a term introduced by Harlan and de Wet (1963) to denote those species that are "genetically aggressive, plundering related species of their heredities. . . ." Heiser (1965) further suggested that *H. annuus* has "contributed genes to those species with which it has hybridized, often allowing them to expand their ranges also." Chloroplast DNA evidence is consistent with these ideas, but it is not clear whether nuclear introgression actually has accompanied cytoplasmic introgression in most of these cases.

It is possible that some of the discrepancies between morphologic classification and cpDNA type attributed to introgression in *Helianthus* (Fig. 7.1) may actually result from phylogenetic sorting (see Chapter 6, this volume). This explanation seems less parsimonious than introgression, however, given that all the cases of introgression discussed involved sympatric taxa—even though members of section *Helianthus* (with the exception of *H. annuus*) are largely allopatric. The phylogenetic sorting scenario is also difficult because it would require an extremely polymorphic ancestor (in terms of cpDNA) from which almost all members of section *Helianthus* must have been derived (Fig. 7.1). Nevertheless, it is clear that phylogenetic sorting has played an important role in the evolution of *Helianthus* (Rieseberg et al., 1991) and may partially account for the fact that all polytypic species in section *Helianthus* are polyphyletic in terms of maternal ancestry (Fig. 7.1).

An excellent example of ancient hybrization (Smith and Sytsma, 1990; Sytsma, 1990) involves the European poplars (*P. nigra* and *P. alba*) and the American cottonwoods (*P. deltoides* and *P. fremontii*). The cpDNAs of *P. nigra* were quite similar to those of *P. alba* but very different from those of the American cottonwoods. In contrast, the rDNA of *P. nigra* was distinct from both the cottonwoods and *P. alba,* suggesting that the origin of *P. nigra* included ancient hybridization with *P. alba* as the maternal parent. These data are consistent with evidence from artificial hybridization experiments, in which successful crosses between *P. nigra* and *P. alba* are possible, but only with the latter as the maternal parent.

A second apparent example of ancient hybridization or introgression was revealed in a cpDNA-based phylogenetic tree of *Salix* (Fig. 7.2; Brunsfeld, 1990). The lowland race of *S. melanopsis,* widespread in western North America, was found to share five mutations with *S. interior* of eastern North America. Two other cytoplasmic races of *S. melanopsis* lacked the mutations, but were similar to the lowland race in morphology and allozymes. An early hybridization event may have allowed the lowland race to "capture" the cytoplasm of the other more distantly related *S. interior*. Hybridization must have occurred early, because both lineages possess additional autapomorphic mutations (Fig. 7.2).

Another example in *Salix* illustrates that cytoplasmic capture has also occurred more recently (Fig. 7.2). In the upper Sacramento River drainage of California, a population of *S. melanopsis* was found to contain the highly diverged cpDNA of *S. hindsiana*. Although several individuals of *S. hindsiana* were present at the site, there was no indication from morphologic or allozyme data that introgression had occurred in the *S. melanopsis* population (Brunsfeld, 1990)—evidence suggesting that repeated backcrossing had occurred to *S. melanopsis*.

Another possibly ancient case of hybridization or introgression has been examined in *Clarkia* (Sytsma et al., 1990). Lewis and Lewis (1955) suggested that *Clarkia* section *Fibula* had a hybrid origin between diploid species of sections *Phaeostoma* and *Sympherica*. In support of this hypothesis, Sytsma et al. (1990) found that section *Fibula* has a chloroplast genome similar to that of one species of section *Phaeostoma*. Alternatively, the cpDNA evidence may simply indicate that section *Phaeostoma,* as presently circumscribed, is paraphyletic.

Discussion

The use of molecular markers has greatly increased our ability to detect and quantify interspecific gene exchange. In contrast to two decades ago (Grant, 1971; Heiser, 1973), there is now unequivocal evidence in plants for the occurrence of localized introgression, dispersed introgression, and the origin of new types through hybridization and introgression. Furthermore, relatively ancient events of hybridization or introgression have been detected through molecular phylogenetic analysis, suggesting that homoploid reticulate evolution could have profound effects both on phylogenetic reconstruction and phyletic evolution in plants.

Although rigorous molecular analyses are still few, there appear to be fewer good cases of dispersed introgression than localized introgression (if the establishment and spread of stabilized introgressants is omitted and examples from phylogenetic studies are classified separately). A similar situation was observed by Heiser (1973). It seems, however, that many of the apparent examples of introgression from phylogenetic studies, if studied in more detail, could be classified as examples of dispersed introgression. This certainly would be the case if dispersed introgression were defined as widespread introgression in both space and time.

The data presented here also indicate the need for adequate sampling strategies in molecular phylogenetic studies. The potential danger of inadequate sampling not only may result from inherently high levels of cpDNA (see Chapter 6 by Soltis et al., this volume) and rDNA variability within some species, but also from introgression. This danger is particularly acute for the clonally inherited chloroplast genome, where evidence of introgression would be most easily missed. Unfortunately, however, analysis of many population samples in phylogenetic studies is often deterred because it is expensive and time consuming. Although analysis of the entire chloroplast genome of several populations per species may be the most appropriate method of analyzing intraspecific molecular variation, this process can be supplemented by surveying additional population samples with regard to key diagnostic mutations, rather than the entire battery of restriction enzymes surveyed in the original phylogenetic study. Although it might be argued that introgression with unknown taxa might be missed using this approach, this seems unlikely if autapomorphic mutations were included in the populational survey. The probability of detecting events of introgression could also be increased by analyzing both nuclear and cytoplasmic genes.

These data further indicate that morphologic evidence may be useful for studies of localized introgression but cannot be considered reliable for the detection and analysis of dispersed introgression, stabilized introgressants, or ancient hybridization events, such as those detected by molecular phylogenetic analysis.

Conclusions

Heiser (1973) suggested that with the availability of new tools from biochemistry and genetics, contributions toward an understanding of the evolutionary significance of introgression could be expected. During the past 10 years, integration of nuclear markers, such as allozymes and rDNA, with analysis of cytoplasmic markers, such as cpDNA and mtDNA, have borne out Heiser's predictions. Nonetheless, more study is needed before Heiser's (1973) conclusions that introgession "may play a very significant role; but it must be admitted, there is as yet no strong evidence to support such a claim," can be modified significantly. The following sorts of investigations are needed:

1. Phylogenetic studies based on both cytoplasmic and nuclear genes that include samples of multiple populations and individuals. This may be the most effective way to detect numerous new cases of introgression and to provide an actual estimate of the frequency of occurrence of introgression in plants. This is also the only way to document the extent and impact of introgression in each individual case.

2. Molecular genetic studies of previously proposed cases of introgression based on other data sets (e.g., morphology, cytology, secondary chemistry, etc.). Many of these cases are difficult to evaluate in the absence of detailed molecular genetic data.

3. Detailed molecular analyses of known examples of introgression (e.g., from molecular phylogenetic studies) to determine the actual extent of introgression in these cases. Obviously, these studies should use multiple nuclear and cytoplasmic markers and examine numerous populations and individuals from throughout the ranges of the introgressed species.

4. Studies of the relative fitness of hybrids and introgressants relative to "pure" populations of the parental species in groups where the occurrence and extent of introgression is well documented.

5. Studies of the effect of introgression on reproductive isolating barriers. These studies would be most useful for groups where introgression had been documented previously.

6. Studies examining microevolutionary forces that may be of importance in determining the direction and magnitude of gene flow within and among populations. Such analyses are necessary before any rigorous claims can be made as to the evolutionary significance of introgression.

References

Anderson, E. (1949) *Introgressive Hybridization,* Wiley, New York.

Anderson, E., and Hubricht, L. (1938) Hybridization in *Tradescantia* III. The evidence for introgressive hybridization. *Amer. J. Bot.* **25**, 396–402.

Appels, R., and Honeycutt, R.L. (1986) rDNA evolution over a billion years. In: *DNA Systematics* (ed. S.K. Dutta), CRC Press, Boca Raton, FL, pp. 81–135.

Arnold, M.L., Contreras, N., and Shaw, D.D. (1988) Biased gene conversion and asymmetrical introgression between subspecies. *Chromosoma* **96**, 368–371.

Arnold, M.L., Bennett, B.D., and Zimmer, E.A. (1990) Natural hybridization between *Iris fulva* and *I. hexagona:* pattern of ribosomal DNA variation. *Evolution,* **44**, 1512–1521.

Avise, J.C. (1989) Gene trees and organismal histories: a phylogenetic approach to population biology. *Evolution* **43**, 1192–1208.

Avise, J.C., and Saunders, N.C. (1984) Hybridization and introgression among species of sunfish (*Lepomis*): analysis of mitochondrial DNA and allozyme markers. *Genetics* **108**, 237–255.

Baker, H.G. (1965) Characteristics and modes of origin of weeds. In: *The Genetics of Colonizing Species* (eds. H.G. Baker and G.L. Stebbins), Columbia University Press, New York, pp. 147–172.

Baker, H.G. (1974) The evolution of weeds. *Ann. Rev. Ecol. Syst.* **5**, 1–24.

Baldwin, B.G., Kyhos, D.W., and Dvorak, J. (1990) Chloroplast DNA evolution and adaptive radiation in the Hawaiian silversword alliance (Madiinae, Asteraceae). *Ann. Missouri Bot. Gard.* **77**, 96–109.

Barton, N.H., and Hewitt, G.M. (1985) Analysis of hybrid zones. *Ann. Rev. Ecol. Syst.* **16**, 113–148.

Beckmann, J.S., and Soller, M. (1983) Restriction fragment length polymorphisms in genetic improvement: methodologies, mapping and costs. *Theor. Appl. Genet.* **67**, 34–43.

Beckstrom-Sternberg, S., Rieseberg, L.H., and Doan, K. (1991). Gene lineage analysis of populations of *Helianthus niveus* and *H. petiolaris. Plant Syst. Evol.,* in press.

Botstein, D., White, R.L., Skolnick, M., and Davis, W. (1980) Construction of a genetic map in man using restriction fragment length polymorphisms. *Amer. J. Hum. Genet.* **32**, 314–331.

Brunsfeld, S.J. (1990) Systematics and evolution in *Salix* section *Longifoliae.* Ph.D. Thesis, Washington State University, Pullman.

Burke, T., and Bruford, M.W. (1987) DNA fingerprinting in birds. *Nature* **327**, 150–151.

Buth, D.G. (1984) The application of electrophoretic data in systematic studies. *Ann. Rev. Ecol. Syst.* **15**, 501–522.

Crawford, D.J., and Ornduff, R. (1989) Enzyme electrophoresis and evolutionary relationships among three species of *Lasthenia* (Asteraceae: Heliantheae). *Amer. J. Bot.* **76**, 289–296.

Crins, W.J., Bohm, B.A., and Carr, G.D. (1988) Flavonoids as indicators of hybridization in a mixed population of lava-colonizing Hawaiian tarweeds (Asteraceae: Heliantheae: Madiinae). *Syst. Bot.* **13**, 567–571.

Critchfield, W.B. (1985) The late quaternary history of lodgepole and jack pine. *Can. J. For. Res.* **15**, 749–772.

Doebley, J. (1989) Molecular evidence for a missing wild relative of maize and the introgression of its chloroplast genome into *Zea perennis. Evolution* **43**, 1555–1559.

Doebley, J., and Sisco, P.H. (1989) On the origin of the maize male sterile cytoplasms: it's completely unimportant, that's why it's so interesting. *Maize Genet. Newsl.* **63**, 108–109.

Doebley, J., Goodman, M.M., and Stuber, C.W. (1984) Isoenzyme variation in *Zea* (Gramineae). *Syst. Bot.* **9**, 203–218.

Doebley, J., Goodman, M.M., and Stuber, C.W. (1987a) Patterns of isozyme variation between maize and Mexican teosinte. *Econ. Bot.* **41**, 234–246.

Doebley, J., Renfroe, W., and Blanton, A. (1987b) Restriction site variation in the *Zea* chloroplast genomes. *Genetics* **117**, 139–147.

Doyle, J.J., and Doyle, J.L. (1988) Natural interspecific hybridization in eastern North American *Claytonia. Amer. J. Bot.* **75**, 1238–1246.

Doyle, J.J., Beachy, R.N., and Lewis, W.H. (1984) Evolution of rDNA in *Claytonia* polyploid complexes. In: *Plant Biosystematics: 40 Years Later* (ed. W.F. Grant), Academic Press, Toronto, pp. 321–341.

Doyle, J.J., Soltis, D.E., and Soltis, P.S. (1985) An intergeneric hybrid in the Saxifragaceae: evidence from ribosomal RNA genes. *Amer. J. Bot.* **72**, 1388–1391.

Fernandez de Caleya, R., Hernandez-Lucas, C., Carbonero, P., and Garcia-Olmedo, F. (1976) Gene expression in allopolyploids: genetic control of lipopurothionins in wheat. *Genetics* **83**, 687–699.

Ferris, S.D., Sage, R.D., Huang, C.-M., Nielson, J.T., Ritte, U., and Wilson, A.C. (1983) Flow of mitochondrial DNA across a species boundary. *Proc. Natl. Acad. Sci. USA* **80**, 2290–2294.

Gallez, G.P., and Gottlieb, L.D. (1982) Genetic evidence for the hybrid origin of the diploid plant *Stephanomeria diegenesis. Evolution* **36**, 1158–1167.

Gill, B.S., and Chen, P.D. (1987) Role of cytoplasm-specific introgression in the evolution of the polyploid wheats. *Proc. Natl. Acad. Sci. USA* **84,** 6800–6804.

Gill, B.S., and Kimber, G. (1974) Giemsa C-banding and the evolution of wheat. *Proc. Natl. Acad. Sci. USA* **71,** 4086–4090.

Golding, G.B., and Strobeck, C. (1983) Increased number of alleles found in hybrid populations due to intragenic recombination. *Evolution* **37,** 17–29.

Grant, V. (1971) *Plant Speciation,* Columbia University Press, New York.

Graur, D., Bogher, M., and Breiman, A. (1989) Restriction endonuclease profiles of mitochondrial DNA and the origin of the B genome of bread wheat, *Triticum aestivum. Heredity* **62,** 335–342.

Harlan, J., and de Wet, J.M.J. (1963) The compilospecies concept. *Evolution* **17,** 497–501.

Heiser, C.B. (1949) Study in the evolution of the sunflower species *Helianthus annuus* and *H. bolanderi. Univ. Calif. Publ. Bot.* **23,** 157–196.

Heiser, C.B. (1951) Hybridization in the annual sunflowers: *Helianthus annuus* × *debilis* var. *cucumerifolius. Evolution* **5,** 42–51.

Heiser, C.B. (1965) Sunflowers, weeds and cultivated plants. In: *The Genetics of Colonizing Species* (eds. H.G. Baker and G.L. Stebbins), Academic Press, New York, pp. 391–401.

Heiser, C.B. (1973) Introgression re-examined. *Bot. Rev.* **39,** 347–366.

Heiser, C.B., Smith, D.M., Clevenger, S., and Martin, W.C. (1969) The North American Sunflowers (*Helianthus*). *Mem. Torr. Bot. Club* **22,** 1–218.

Hunt, W.G., and Selander, R.K. (1973) Biochemical genetics of hybridization in European house mice (*Mus musculus*). *Heredity* **31,** 11–33.

Jones, B.L., Lookhart, G.L., Mak, A., and Cooper, D.B. (1982) Sequences of purothionins and their inheritance in diploid, tetraploid and hexaploid wheats. *J. Hered.* **73,** 143–144.

Keim, P., Paige, K.N., Whitham, T.G., and Lark, K.G. (1989) Genetic analysis of an interspecific hybrid swarm of *Populus:* occurrence of unidirectional introgression. *Genetics* **123,** 557–565.

Kerby, K., and Kuspira, J. (1987) The phylogeny of the polyploid wheats *Triticum aestivum* (bread wheat) and *Triticum turgidum* (macaroni wheat). *Genome* **29,** 722–737.

Kimura, M. (1982) *Molecular Evolution, Protein Polymorphism and Evolutionary Theory,* Japan Scientific Societies Press, Tokyo.

Klein, H.L. (1984) Lack of association between intrachromosomal gene conversion and reciprocal exchange. *Nature* **310,** 748–753.

Leclercq, P. (1969) Une sterilité male cytoplasmique chez le tournesol. *Ann. Amelior. Pl.* **19,** 99–106.

Levin, D.A. (1975) Interspecific hybridization, heterozygosity and gene exchange in *Phlox. Evolution* **29,** 37–51.

Levin, D.A., and Bulinska-Radomska, Z. (1988) Effects of hybridization and inbreeding on fitness in *Phlox. Amer. J. Bot.* **75,** 1632–1639.

Lewis, H., and Lewis, M.W. (1955) The genus *Clarkia. Univ. Calif. Publ. Bot.* **20,** 241–392.

Lewontin, R.C., and Birch, L.C. (1966) Hybridization as a source of variation for adaptation to new environments. *Evolution* **20,** 315–336.

Liston, A., Rieseberg, L.H., and Mistretta, O. (1990) Ribosomal DNA evidence for hybridization between island endemic species of *Lotus. Biochem. Syst. Ecol.* **18,** 239–244.

Marchant, A.D., Arnold, M.L., and Wilkinson, P. (1988) Gene flow across a chromosomal tension zone. I. Relicts of ancient hybridization. *Heredity* **61,** 321–328.

Murray, J. (1972) *Genetic Diversity and Natural Selection,* Oliver and Boyd, Edinburgh.

Ogihara, Y., and Tsunewaki, K. (1982) Molecular basis of the genetic diversity in the cytoplasm in *Triticum* and *Aegilops.* I. Diversity of the chloroplast genome and its lineage revealed by the restriction pattern of ctDNA. *Jap. J. Genet.* **57,** 371–396.

Ogihara, Y., and Tsunewaki, K. (1988) Diversity and evolution of chloroplast DNA in *Triticum* and *Aegilops* as revealed by restriction fragment analysis. *Theor. Appl. Genet.* **16,** 321–332.

Palmer, J.D. (1985) Chloroplast DNA and molecular phylogeny. *BioEssays* **2,** 263–267.

Palmer, J.D. (1988) Intraspecific variation and multicircularity in *Brassica* mitochondrial DNAs. *Genetics* **118,** 341–351.

Palmer, J.D., Shields, C.R., Cohen, D.B., and Orton, T.J. (1983) Chloroplast DNA evolution and the origin of amphidiploid *Brassica* species. *Theor. Appl. Genet.* **65,** 181–189.

Palmer, J.D., Jorgenson, R.A., and Thompson, W.F. (1985) Chloroplast DNA variation and evolution in *Pisum:* patterns of change and phylogenetic analysis. *Genetics* **109,** 195–213.

Powell, J.R. (1983) Interspecific cytoplasmic gene flow: evidence from *Drosophila. Proc. Natl. Acad. Sci. USA* **80,** 492–495.

Randolph, L.G., Nelson, I.S., and Plaisted, R.L. (1967) Negative evidence of introgression affecting the stability of Louisiana *Iris* species. *Cornell Univ. Agr. Expt. Stat. Mem.* **398,** 1–56.

Rick, C., Zobel, R.W., and Fobes, J.F. (1974) Four peroxidase loci in red-fruited tomato species: genetics and geographic distribution. *Proc. Natl. Acad. Sci. USA* **71,** 835–839.

Rieseberg, L.H., and Seiler, G. (1990) Molecular evidence and the origin and development of the domesticated sunflower (*Helianthus annuus* L.). *Econ. Bot.* **445,** 79–91.

Rieseberg, L.H., Soltis, D.E., and Palmer, J.D. (1988) A molecular re-examination of introgression between *Helianthus annuus* and *H. bolanderi. Evolution* **42,** 227–238.

Rieseberg, L.H., Beckstrom-Sternberg, S., and Doan, K. (1990a) *Helianthus annuus* ssp. *texanus* has chloroplast DNA and nuclear ribosomal RNA genes of *Helianthus debilis* ssp. *cucumerifolius. Proc. Natl. Acad. Sci. USA* **87,** 593–597.

Rieseberg, L.H., Carter, R., and Zona, S. (1990b) Molecular tests of the hypothesized hybrid origin of two diploid *Helianthus* species. *Evolution* **44**, 1498–1511.

Rieseberg, L.H., Beckstrom-Sternberg, S., Liston, A., and Arias, D. (1991) Phylogenetic and systematic inferences from chloroplast DNA and isozyme variation in *Helianthus* section *Helianthus* (Asteraceae). *Syst. Bot.* **16**, 50–76.

Rogstad, S.H., Patton II, J.C., and Schaal, B.A. (1988) M13 repeat probe detects DNA minisatellite-like sequences in gymnosperms and angiosperms. *Proc. Natl. Acad. Sci. USA* **85**, 9176–9178.

Sage, R.D., and Selander, R.K. (1979) Hybridization between species of the *Rana pipiens* complex in central Texas. *Evolution* **33**, 1069–1088.

Schaal, B.A., and Learn, G.H. (1988) Ribosomal DNA variation within and among plant populations. *Ann. Missouri Bot. Gard.* **75**, 1207–1216.

Song, K.M., Osborn, T.C., and Williams, P.H. (1988) *Brassica* taxonomy based on nuclear restriction fragment length polymorphisms (RFLPs). *Theor. Appl. Genet.* **75**, 784–794.

Smith, R.L., and Sytsma, K.J. (1990) Evolution of *Populus nigra* L. (sect. *Aigeiros*): introgressive hybridization and the chloroplast contribution of *Populus alba* L. (sect. *Populus*). *Amer. J. Bot.* **77**, 1176–1187.

Spooner, D.M., Sytsma, K.J., and Smith, J.F. (1991) A molecular reexamination of diploid hybrid speciation of *Solanum raphanifolium*. *Evolution* **45**, 757–764.

Springer, P.S., Zimmer, E.A., and Bennetzen, J.L. (1989) Genomic organization of the ribosomal DNA of sorghum and its close relatives. *Theor. Appl. Genet.* **77**, 844–850.

Stebbins, G.L. (1950) *Variation and Evolution in Plants,* Columbia University Press, New York.

Stebbins, G.L. (1959) The role of hybridization in evolution. *Proc. Amer. Phil. Soc.* **103**, 231–251.

Stebbins, G.L. (1965) Colonizing species of the native California flora. In: *The Genetics of Colonizing Species* (eds. H.G. Baker and G.L. Stebbins), Columbia University Press, New York, pp. 173–191.

Stebbins, G.L. (1969) The significance of hybridization for plant taxonomy and evolution. *Taxon* **18**, 26–35.

Sytsma, K.J. (1990) DNA and morphology: inference from plant phylogeny. *Trends Ecol. Evol.* **5**, 104–110.

Sytsma, K.J., and Schaal, B.A. (1985) Phylogenetics of the *Lisianthius skinneri* (Gentianaceae) species complex in Panama utilizing DNA restriction fragment analysis. *Evolution* **39**, 594–608.

Sytsma, K.J., Smith, J.F., and Gottlieb, L.D. (1990) Phylogenetics in *Clarkia* (Onagraceae): restriction site mapping of chloroplast DNA. *Syst. Bot.* **15**, 280–295.

Sytsma, K.J., Smith, J.F., and Berry, P.E. (1991) Biogeography and evolution of morphology, breeding systems, and flavonoids in the Old World *Fuchsia* sect. *Skinnera* (Onagraceae): evidence from chloroplast DNA. *Syst. Bot.* **16**, 257–269.

Timothy, D.H., Levings III, C.S., Pring, D.R., Conde, M.G., and Kermicle, J.L. (1979)

Organelle DNA variation and systematic relationships in the genus *Zea:* Teosinte. *Proc. Natl. Acad. Sci. USA* **76,** 4220–4224.

Tsunewaki, K., and Ogihara, T. (1983) The molecular basis of genetic diversity among cytoplasms of *Triticum* and *Aegilops* species. II. On the origin of polyploid wheat cytoplasms as suggested by chloroplast DNA restriction fragment patterns. *Genetics* **109,** 155–171.

Vassart, G., Georges, M., Monsieur, R., Brocas, H., Lequarre, A.S., and Christophe, D. (1987) A sequence in M13 phage detects hypervariable minisatellites in human and animal DNA. *Science* **235,** 683–684.

Wagner, D.B., Furnier, G.R., Saghai-Maroof, M.A., Williams, S.M., Dancik, B.P., and Allard, R.W. (1987) Chloroplast DNA polymorphisms in lodgepole and jack pines and their hybrids. *Proc. Natl. Acad. Sci. USA* **84,** 2097–2100.

Warwick, S.I., Bain, J.F., Wheatcroft, R., and Thompson, B.K. (1989) Hybridization and introgression in *Carduus nutans* and *C. acanthoides* reexamined. *Syst. Bot.* **14,** 476–494.

Weissinger, A.K., Timothy, D.H., Levings III, C.S., and Goodman, M.M. (1983) Patterns of mitochondrial DNA variation in indigenous maize races of Latin America. *Genetics* **104,** 365–379.

Wendel, J.F., and Albert, V.A. (1991) Phylogenetics of the cotton genus (*Gossypium* L.): character-state weighted parsimony analysis of chloroplast-DNA restriction site data and its systematic and biogeographic implications. *Syst. Bot.,* submitted.

Wendel, J.F., Stewart, J. McD., and Rettig, J.H. (1991) Molecular evidence for homoploid reticulate evolution among Australian species of *Gossypium. Evolution* **45,** 694–711.

White, J.D. (1978) *Modes of Speciation,* W.H. Freeman and Company, San Francisco.

Whitham, T. (1989) Plant hybrid zones as sinks for pests. *Science* **244,** 1490–1493.

Zimmer, E.A., Jupe, E.R., and Walbot, V. (1988) Ribosomal gene structure, variation and inheritance in maize and its ancestors. *Genetics* **120,** 1125–1136.

8

Molecular Data and
Polyploid Evolution in Plants

Pamela S. Soltis, Jeff J. Doyle, and *Douglas E. Soltis*

Polyploidy is a significant force in plant evolution. Approximately 47% to 52% of all angiosperm species are polyploid (V. Grant, 1981). Estimates of the frequency of polyploidy in pteridophytes range from 43.5% for the ferns alone (Vida, 1976) to 95% for pteridophytes as a whole (V. Grant, 1981), suggesting ancient polyploidy in several lineages of pteridophytes. In byrophytes polyploidy is common in mosses, but rare in liverworts (V. Grant, 1981). In contrast to angiosperms, pteridophytes, and bryophytes, polyploidy in gymnosperms is very rare and sporadic. Polyploidy has not been detected in cycads or ginkgo, and only 1.5% of the species of Coniferales are polyploid (Khoshoo, 1959). In the Gnetales, tetraploidy is common in *Ephedra* and rare or nonexistent in *Gnetum* and *Welwitschia* (Delevoryas, 1980).

Because of its high frequency in the plant kingdom, particularly in angiosperms and pteridophytes, polyploidy has been the subject of several investigations into the mode(s) of origin of polyploids (e.g., Harlan and de Wet, 1975; Mendiburu and Peloquin, 1976; de Wet, 1980), the nature of polyploidy (auto- versus allopolyploidy) (Kihara and Ono, 1926; Stebbins, 1947), the genetic attributes of polyploids (Muller, 1914; Stebbins, 1947), the genetic, biochemical, and evolutionary consequences of polyploidy (Roose and Gottlieb, 1976; Levin, 1983; D. Soltis and Soltis, 1989a), and polyploidization as a mode of speciation (Levin, 1983; D. Soltis and Rieseberg, 1986). Furthermore, systematists have attempted to unravel the patterns of polyploidy and reticulate evolution in taxonomically troublesome species complexes. These diverse endeavors have relied upon data from a variety of sources, including morphology, cytogenetics, and enzyme electrophoresis. The advent of molecular techniques provides a direct

We thank Dan Crawford for helpful comments on the manuscript. JJD acknowledges the contributions of Jane L. Doyle and A.H.D. Brown. This work was supported by NSF grants BSR-8620444 to PSS, BSR-8516721 and BSR-8717471 to DES, and BSR-8516630 and BSR-8805630 to JJD.

assessment of the genomic constitutions of polyploids and permits evaluation of the origins, evolutionary processes, and molecular consequences of polyploidy.

Two primary classes of DNA data have been used to address questions of polyploid evolution in plants: chloroplast DNA (cpDNA) and the nuclear-encoded ribosomal RNA genes (rDNA). These DNAs differ in their modes of inheritance, the information they provide, and the assumptions their analyses carry. Chloroplast DNA is typically maternally inherited in angiosperms, although biparental inheritance has been observed in a few taxa (Kirk and Tilney-Basset, 1978; Sears, 1980, 1983; Corriveau and Coleman, 1988). In conifers cpDNA is paternally inherited (Neale et al., 1986; Szmidt et al., 1987). The mode of cpDNA inheritance in pteridophytes has not been analyzed genetically, although ultrastructural studies suggest either maternal or biparental inheritance of the plastid in ferns (Sears, 1980, 1983). Analysis of cpDNA variation in allopolyploid ferns suggests uniparental inheritance of the chloroplast genome (Ranker et al., 1989; P. Soltis et al., 1991). Chloroplast DNA data therefore typically provide uniparental, usually maternal, phylogenetic markers and permit the designation of a maternal parent in analyses of hybridization and polyploidy. In contrast, rDNA is biparentally inherited. Ribosomal DNA comprises two major classes, the 5S rDNA and the 18S-25S rDNA, both of which occur as tandemly repeated, multigene families that are present in thousands of copies (Long and Dawid, 1980; Goldsborough et al., 1981). Both classes of rDNA have been employed in analyses of polyploidy (e.g., J. Doyle et al., 1984; D. Soltis and Doyle, 1987; Gill and Appels, 1988; J. Doyle and Brown, 1989). The biparental inheritance of both classes of rDNA permits the assessment of all parental contributions to polyploid species. Furthermore, when used in conjunction with the uniparentally inherited chloroplast genome, rDNA markers can provide novel insights into polyploid evolution (D. Soltis and Soltis, 1990a).

Analyses of both cpDNA and rDNA rely on several assumptions. As in studies of polyploidy using other characters (e.g., morphology, isozymes, etc.), we assume that little or no divergence has occurred in the cpDNA or rDNA between progenitor and derivative species since the polyploidization event. Although generally justified, this assumption is occasionally violated using any source of data. However, as long as cladistic relationships between diploids and polyploid derivatives are not obscured by homoplasious mutations, the diploid progenitors should be identifiable regardless of the number of autapomorphies that appear on the tree.

Several additional assumptions hold for analysis of both classes of rDNA. First, for rDNA markers to be most useful in analyses of polyploidy, we must assume that all of the tandemly repeated genes of each rDNA class within an individual are identical. Although this uniformity has been observed within individuals of many plant species, intraindividual variation has also been detected (e.g., Rafalski et al., 1982; J. Doyle et al., 1984; P. Soltis and Soltis, 1991). Second, because rDNA is biparentally inherited, additivity of parental rDNA patterns is expected in hybrids and allopolyploids, and this has been observed in several cases (e.g., J. Doyle et al., 1985; J. Doyle and Brown, 1989; P. Soltis and Soltis, 1991). However, in some hybrids (e.g., *Zea mays* × *Z. luxurians*; Zimmer et al., 1988), the rDNA of only a single parent is present (Arnheim et al., 1980; Appels and Honeycutt, 1986; Flavell, 1986). Although intriguing from the perspective of molecular evolution, this occasional lack of additivity in hybrids and/or polyploids is a disadvantage for molecular analyses of the origins of polyploid species.

In this chapter, we review the available literature on DNA analyses of polyploidy and address the following questions:

1. What do DNA data reveal regarding cases of proposed "ancient polyploidy"?
2. How have DNA data contributed to a better understanding of species relationships in polyploid complexes?
3. Have DNA data allowed an assessment of multiple versus unique origins of polyploid species?
4. What implications do the DNA data have for the process of polyploidization?
5. What are the molecular consequences of polyploidy?
6. What are the future prospects of molecular systematics and studies of polyploid evolution?

Ancient Polyploidy

Estimates of the frequency of polyploidy are often based solely on a criterion of chromosome number: plants with approximately $n = 13$ or greater are typically considered polyploid (e.g., Stebbins, 1971; Goldblatt, 1980; V. Grant, 1981). Using this criterion, several families of angiosperms and 95% of all pteridophytes would be considered polyploid. Furthermore, because these lineages lack members with low (i.e., less than approximately $n = 13$) chromosome numbers, these groups have been considered "ancient polyploids," regardless of their actual or even relative ages, because the diploid progenitors are now extinct. Ancient polyploidy has been implicated in several families of primitive (Takhtajan, 1969, 1980; Cronquist, 1981; Dahlgren, 1983; Thorne, 1983) angiosperms, including Calycanthaceae, Cercidiphyllaceae, Lauraceae, Magnoliaceae, Myristicaceae, and Trochondendraceae, with chromosome numbers of $n = 11$ to 19. Salicaceae and Hippocastanaceae, more evolutionarily advanced (Cronquist, 1981), woody families, with $n = 19$ and 20, respectively, have also been considered ancient polyploids (Stebbins, 1971). The mean haploid chromosome number for all major lineages of pteridophytes, including Psilotophyta, class Aglossopsida of Microphyllophyta, Arthrophyta, and most Pteridophyta (classification *sensu* Bold et al., 1986), is $n = 55$ (Klekowski and Baker, 1966); all lineages have been considered ancient polyploids (Klekowski and Baker, 1966; V. Grant, 1981).

Electrophoretic analyses of isozyme number have demonstrated polyploid gene expression in families of woody angiosperms with high chromosome numbers (D. Soltis and Soltis, 1990b), supporting the designation of these families as ancient polyploids. In contrast, all lineages of pteridophytes have diploid isozyme expression despite their high chromosome numbers (Haufler and Soltis, 1986; D. Soltis, 1986; D. Soltis and Soltis, 1988a; P. Soltis and Soltis, 1988), suggesting that pteridophytes (1) are not ancient polyploids or (2) have undergone extensive gene silencing that has restored diploid gene expression (see D. Soltis and Soltis, 1989b, for discussion).

To discriminate between hypotheses (1) and (2) it is necessary to examine the genome itself, not merely an expressed portion of the genome. Of particular value are single-copy or nearly single-copy genes. If pteridophytes are of ancient polyploid origin, they should possess multiple, and perhaps several nonfunctional, copies of a gene regarded as "single-copy" in diploid angiosperms. If pteridophytes are not of ancient polyploid origin, but are true diploids, they should exhibit approximately the same number of copies of the gene as observed in diploid angiosperms. Multiple copies of defective genes for the chlorophyll a/b-binding (CAB) proteins have been detected in the genome of the fern *Polystichum munitum* ($n = 41$) (Pichersky et al., 1990). In angiosperms, CAB genes occur as a small gene family (Pichersky et al., 1989) of three to 16 copies (Dunsmuir, 1985; Leutwiler et al., 1986). Only a single defective CAB gene has been detected in scores of CAB genes analyzed in angiosperms (Pichersky et al., 1985). Several hypotheses could explain the multiple defective CAB genes in *P. munitum* (Pichersky et al., 1990). First, several copies of a diploid complement of CAB genes have mutated to a nonfunctional state. Second, some defective CAB genes may have been specifically amplified in the genome of *P. munitum* during evolution. Finally, the defective CAB genes may reflect the diploidization of a polyploid genome. Although the numerous CAB genes in *P. munitum* are consistent with a hypothesis of ancient polyploidy and subsequent gene silencing in ferns, additional data for other genes and other species must be obtained to evaluate whether or not ferns and other pteridophytes are of ancient polyploid origin. Nonetheless, the CAB gene data provide a novel view of the genome of pteridophytes.

Although certainly more recent in origin than the lineages of ancient polyploids discussed above, *Glycine* ($n = 20$; Leguminosae) may also be of ancient polyploid origin, based on a diploid ancestor with $n = 10$ or $n = 11$ (Hymowitz, 1970; Hadley and Hymowitz, 1973; Bingham et al., 1976). Molecular data support the ancient tetraploidy of *Glycine*. Leghemoglobin (Lb) genes in the kidney bean, *Phaseolus vulgaris* ($n = 11$), exist as a tandem cluster of four genes (Lee and Verma, 1984). In both soybean (*Glycine max*) and its closest wild relative *G. soja*, Lb genes occur in two loci, each consisting of four tandemly repeated genes. This organization of Lb genes has been attributed to an ancient event of tetraploidization (Lee and Verma, 1984) and provides an opportunity for studying molecular evolution following gene duplication. In fact, Lb pseudogenes have been described in both *G. max* and *G. soja* (Lee and Verma, 1984). The structure and organization of the actin gene family also support ancient tetraploidy in *Glycine* (Hightower and Meagher, 1985). The six actin genes in soybean fall into three classes with two genes each. Each pair of genes may represent homoeologous genes contributed by ancestral diploid species.

Origins of Polyploids

Autopolyploidy and allopolyploidy, as originally defined (Kihara and Ono, 1926), represent ends of a continuum between duplicate sets of genetically homologous and nonhomologous chromosomes. Although the definitions and classifications of polyploids are controversial (Darlington, 1932; Muntzing, 1936; Clausen et al., 1945; Stebbins, 1947; Lewis, 1980), it is generally agreed that allopolyploidy is more widespread in plants than is autopolyploidy (Stebbins, 1947; Manton, 1950; Grant, 1981). The examples below illustrate the value of DNA data for elucidating the origins of both auto- and allopolyploids (see also D. Soltis and Soltis, 1990a).

Autopolyploidy

DNA restriction site variation has been analyzed in both diploid ($2n = 14$) and tetraploid ($2n = 28$) cytotypes of *Heuchera micrantha* (D. Soltis et al., 1989a), *H. grossulariifolia* (Wolf et al., 1990), and *Tolmiea menziesii* (D. Soltis and Doyle, 1987; D. Soltis et al., 1989b). All three species satisfy systematic, genetic, and cytogenetic definitions of autopolyploids (D. Soltis, 1984; D. Soltis and Bohm, 1986; D. Soltis and Rieseberg, 1986; P. Soltis and Soltis, 1986; D. Soltis and Soltis, 1988b, 1989a, c; Ness et al., 1989; Wolf et al., 1989, 1990; D. Soltis and Hauber, unpublished data).

High levels of cpDNA restriction site variation among populations of *H. micrantha* permitted a phylogenetic analysis of interpopulational relationships (D. Soltis et al., 1989a). Tetraploid populations possess chloroplast genomes identified among populations of diploid *H. micrantha,* providing additional support for the autoploid origin of the tetraploid cytotype.

The related *H. grossulariifolia* also exhibits high levels of interpopulational cpDNA restriction site variation (Wolf et al., 1990). As in *H. micrantha,* tetraploid populations of *H. grossulariifolia* possess cpDNAs exhibited by diploid populations, further supporting an autoploid origin of the tetraploid.

Autotetraploid evolution in *Tolmiea menziesii* has produced a different pattern of DNA diversity. Diploid and tetraploid populations possess 5S and 18S–25S ribosomal RNA genes of identical repeat length and restriction profile, providing further evidence for autoploidy (D. Soltis and Doyle, 1987). High levels of cpDNA restriction site variation were observed in *T. menziesii* (D. Soltis et al., 1989b). However, in contrast to the species of *Heuchera,* the cytotypes of *T.*

menziesii did not possess identical chloroplast genomes. In fact, all tetraploid populations are identical and differ from all diploid populations by at least three restriction site mutations and three length mutations. The mean sequence divergence between diploid and tetraploid populations is 0.08%, a value comparable to that obtained for many congeneric species of angiosperms. Furthermore, phylogenetic analysis of interpopulational relationships indicates that the tetraploid populations possess the primitive chloroplast genome. Several hypotheses may explain the cytoplasmic divergence between diploid and tetraploid *T. menziesii*; the most likely were provided by D. Soltis et al. (1989b) as follows:

1. The chloroplast genomes of the two cytotypes have diverged substantially since the origin of the autotetraploid.
2. The diploid source of the tetraploid's cytoplasm is now extinct.

Chloroplast DNA variation was also used to assess cytoplasmic gene flow in *T. menziesii* (D. Soltis et al., 1989b). Several distinct cytoplasmic genotypes occupy discrete geographic areas, suggesting limited seed flow among regions. Furthermore, despite the close proximity (a few meters) of diploid and tetraploid individuals in their region of parapatry in central Oregon, no cytoplasmic gene flow has been observed between diploid and tetraploid cytotypes. The cpDNA analysis of *T. menziesii* provided novel insights into autopolyploid evolution, revealing historic and evolutionary phenomena not obtained using a suite of nuclear-encoded markers.

Allopolyploidy

Chloroplast DNA and rDNA markers have been used successfully to identify the cytoplasmic donor (e.g., Erickson et al., 1983; Palmer et al., 1983; Hilu, 1988; D. Soltis and Soltis, 1989d; Wendel, 1989; J. Doyle et al., 1990d; P. Soltis et al., 1991) and parental nuclear genomes (e.g., J. Doyle et al., 1984; J. Doyle and Brown, 1989; P. Soltis and D. Soltis, 1991) of allopolyploids. The complementary use of cpDNA and rDNA not only can identify the ancestors of allopolyploids but also can illustrate modes of polyploid evolution (e.g., D. Soltis and Soltis, 1989d, 1990a; P. Soltis and Soltis, 1991).

The majority of allopolyploid plant species examined molecularly are cultivated plants or possible progenitors of cultivated plants (Table 8.1; see also chapter by Doebley) and include polyploid species of finger millet (Hilu, 1988), coffee (Berthou et al., 1983), *Brassica* (Erickson et al., 1983; Palmer et al., 1983), cotton (Wendel, 1989), potato (Hosaka, 1986), and wheat (Ogihara and Tsunewaki, 1988).

Fewer wild allopolyploid species have been investigated (Table 8.1). The cytoplasmic donors of several allotetraploid species of ferns have been identified. Although ultrastructural studies and genetic analyses of putative cpDNA-encoded mutations suggest maternal or biparental inheritance of the chloroplast genome in ferns (Sears, 1980, 1983), inheritance studies of the chloroplast genome itself have not been conducted. Thus, the cytoplasmic donors of allopolyploid ferns should not be considered "maternal parents" as they are in analyses of allopolyploid angiosperms for which maternal inheritance of cpDNA has been demonstrated. Cytoplasmic donors have been identified for the allotetraploids *Phanerophlebia juglandifolia* and *P. auriculata* (Yatskievych et al., 1988) and *Hemionitis pinnatifida* (Ranker et al., 1989), although the second diploid parent has not been identified for any of these tetraploids. Allozymic data have confirmed that the diploid progenitors of the allotetraploids *Polystichum californicum* and *P. scopulinum* are *P. dudleyi* and *P. imbricans* and *P. lemmonii* and *P. imbricans,* respectively (P. Soltis et al., 1991). Interpopulational cytoplasmic uniformity was observed within both tetraploid species; the cpDNA donor to *P. californicum* was *P. imbricans* and that to *P. scopulinum* was *P. lemmonii* (P. Soltis et al., 1991).

Allopolyploid species of *Draba* from Scandinavia have been analyzed for rDNA variation (Brochmann et al., 1989b; D. Soltis and Soltis, 1990a). Conservation of rDNA repeat lengths and restriction profiles among morphologically and isozymically distinct species precluded the identification of parental genomes for most allopolyploids. However, the allohexaploid *D. lactea* exhibited the rDNA profile of the diploid *D. fladnizensis,* a presumed progenitor of *D. lactea* that possessed a distinctive rDNA profile. Although isozymic fixed heterozygosity was observed in *D. lactea,* the allohexaploid did not show additivity of rDNA profiles.

The complementary use of cpDNA and rDNA markers confirmed the origins of the allotetraploids *Tragopogon mirus* and *T. miscellus* (D. Soltis and Soltis, 1989d; P. Soltis and Soltis, 1991). *Tragopogon mirus* combines the rDNA profiles of the diploids *T. dubius* and *T. porrifolius* (P. Soltis and Soltis, 1991) but is cytoplasmically uniform, all populations possessing the chloroplast genome of *T. porrifolius* (D. Soltis and Soltis, 1989d). *Tragopogon miscellus* possesses the rDNAs of *T. dubius* and *T. pratensis* (P. Soltis and Soltis, unpublished data). Most populations exhibit the chloroplast genome of *T. pratensis*; however, the maternal parent of two populations from Pullman, Washington, is *T. dubius* (D. Soltis and Soltis, 1989d).

The use of both rDNA and cpDNA has contributed to a better picture of the complex evolutionary origins of polyploid species in *Glycine* subgenus *Glycine* (Leguminosae). Subgenus *Glycine* is an almost wholly Australian group of perennial species that includes diploids ($2n = 38, 40$) and three polyploid taxa. Two of these polyploids, *G. tabacina* and *G. tomentella,* are broadly distributed, not only in Australia, but, unlike their diploid relatives, also throughout the southern and westcentral Pacific.

On the basis of chromosome pairing in artificially produced intraspecific hybrids (Singh and Hymowitz, 1985a, b; Singh et al., 1987), *Glycine tabacina* includes two races. Genome designations of AAB_2B_2 and BBB_2B_2 were assigned

Table 8.1. *dna data and the origins of polyploids.*

Taxon	cpDNA Donor Identified	rDNA Evidence of Ancestry	DNA Evidence of Multiple Origins	References
Autopolyploids				
Heuchera micrantha (4x)	diploid *H. micrantha*	—	yes	D. Soltis et al. (1989a)
H. grossulariifolia (4x)	diploid *H. grossulariifolia*	—	yes	Wolf et al. (1990)
Tolmiea menziesii (4x)	uncertain	—	no	D. Soltis et al. (1989b)
Zea perennis (4x)	*Z. diploperennis*	—	no	Doebley et al. (1987)
Allopolyploids				
Brassica carinata (4x)	*B. nigra*	—	no	Palmer et al. (1983)
B. juncea (4x)	*B. campestris*	—	no	Palmer et al. (1983)
B. napus (4x)	*B. campestris*, unidentified donor	—	yes	Palmer et al. (1983), Erickson et al. (1983)
Coffea arabica (4x)	*C. eugenioides*	—	no	Berthou et al. (1983)
Draba lactea (6x)	—	*D. fladnizensis* only	yes	Brochmann et al. (1989b), D. Soltis and Soltis (1990a)
Eleusine coracana (4x)	*E. indica*	—	no	Hilu (1988)
Gossypium barbadense (4x)	A genome diploid (*G. arboreum* or *G. herbaceum*)	—	no	Wendel (1989)
G. hirsutum (4x)	A genome diploid (*G. arboreum* or *G. herbaceum*)	—	no	Wendel (1989)
Glycine tabacina (4x) race AAB$_2$B$_2$	A genome diploids	D4 diploid of *G. tomentella*, B genome diploid	no	J. Doyle et al. (1990a, 1990c, 1990d)
race BBB$_2$B$_2$	B genome diploids	2 B genome diploids	yes	J. Doyle et al. (1990a, 1990c, 1990d)

Glycine tomentella (4x)			
T_1 race	D_3 diploid *G. tomentella*	D_1/D_2 diploid *G. tomentella*, D_3 diploid *G. tomentella*	yes	J. Doyle and Brown (1989), J. Doyle et al. (1990a)
T_2 race	D_3/D_4 diploid *G. tomentella*	D_3/D_4 diploid *G. tomentella*	no	J. Doyle and Brown (1989), J. Doyle et al. (unpublished)
T_3 race	D_3 diploid *G. tomentella*	D_3 diploid *G. tomentella*	possible	J. Doyle and Brown (1989), J. Doyle et al. (unpublished)
T_4 race	D_3/D_5 diploid *G. tomentella*	D_3/D_5 diploid *G. tomentella*	possible	J. Doyle and Brown (1989), J. Doyle et al. (unpublished)
T_5 race	—	D_3 diploid *G. tomentella*, unidentified diploid,	no	J. Doyle and Brown (1989), J. Doyle et al. (unpublished)
T_6 race	—	D_1/D_2 diploid *G. tomentella*, D_5 diploid *G. tomentella*	—	J. Doyle and Brown (1989), J. Doyle et al. (1990a)
G. hirticaulis (4x)	A genome diploids	—	possible	J. Doyle et al. (1990c)
Hemionitis pinnatifida (4x)	diploid *H. pinnatifida*	—	—	Ranker et al. (1989)
Phanerophlebia juglandifolia (4x)	diploid *P. juglandifolia*	—	—	Yatskievych et al. (1988)
P. auriculata (4x)	*P. remotispora/P. nobilis*	—	—	Yatskievych et al. (1988)
Polystichum californicum (4x)	*P. imbricans*	—	no	P. Soltis et al. (1991)
P. scopulinum (4x)	*P. lemmonii*	—	no	P. Soltis et al. (1991)
Solanum tuberosum ssp. *andigena* (4x)	A type cp-genome diploid, S type cp-genome diploid	—	yes	Hosaka (1986)
Tragopogon mirus (4x)	*T. porrifolius*	*T. porrifolius, T. dubius*	yes	D. Soltis and Soltis (1989d), P. Soltis and Soltis (1991)
T. miscellus (4x)	*T. pratensis, T. dubius*	*T. pratensis, T. dubius*	yes	D. Soltis and Soltis (1989d), P. Soltis and Soltis (unpublished)

(*continued*)

Table 8.1. (Continued)

Taxon	cpDNA Donor Identified	rDNA Evidence of Ancestry	DNA Evidence of Multiple Origins	References
Aegilops kotschyi (4x)	Ae. searsii	—	—	Ogihara and Tsunewaki (1988)
Ae. variabilis (4x)	Ae. searsii	—	—	Ogihara and Tsunewaki (1988)
Ae. ovata (4x)	Ae. umbellulata/mutica group	—	—	Ogihara and Tsunewaki (1988)
Ae. ventricosa (4x)	Ae. squarrosa	—	—	Ogihara and Tsunewaki (1988)
Ae. cylindrica (4x)	Ae. squarrosa	—	—	Ogihara and Tsunewaki (1988)
Ae. triuncialis (4x)	Ae. caudata, Ae. umbellulata	—	yes	Ogihara and Tsunewaki (1988)
Ae. biuncialis (4x),	Ae. umbellulata	—	—	Ogihara and Tsunewaki (1988)
Ae columnaris (4x),				
Ae. triaristata (4x),				
Ae. juvenilis (6x)	4x Ae. crassa	—	—	Ogihara and Tsunewaki (1988)
Ae. crassa (6x),				
Ae. vavilovii (6x)				
Ae. triaristata (6x)	4x Ae. triaristata	—	—	Ogihara and Tsunewaki (1988)
Triticum araraticum (4x),	Aegilops aucheri	—	—	Ogihara and Tsunewaki (1988)
T. timopheevi (4x),				
T. dicoccoides* (4x)				
T. dicoccoides* var. spontaneum,	Ae. aucheri/speltoides group	—	—	Ogihara and Tsunewaki (1988)
T. dicoccum				
T. zhukovskyi (6x)	T. araraticum/timopheevi group	—	—	Ogihara and Tsunewaki (1988)
T. aestivum (6x),	T. dicoccoides var.	—	—	Ogihara and Tsunewaki (1988)
T. spelta (6x)	spontaneum/dicoccum group			

*Note multiple origins of T. dicoccoides

186

to denote their allopolyploid origins and a mutual progenitor genome. The two polyploids differ in both morphology and geographic range; the AAB_2B_2 race occupies a more northerly (tropical) distribution. Analysis of nuclear 18S–25S rDNA profiles from these two polyploid types showed both to be fixed hybrids (J. Doyle and Beachy, 1985; J. Doyle et al., 1990c). Both types have one major and one minor repeat class, and, within each polyploid, the same repeat class consistently predominates. Whether this phenomenon is due to concerted evolution, selective elimination of a repeat class, or simply to consistent differences in the number of repeats present in the diploid progenitors is at present unknown.

One rDNA repeat class is shared between the two polyploids, supporting the hypothesis of a shared diploid progenitor. A survey of rDNA profiles of over 400 accessions representing all diploid *Glycine* species has revealed strong candidates for the diploid genome donors of both polyploids. The shared B_2B_2 genome appears to have been donated by a divergent member of the B diploid genome group, a recently discovered taxon allied with diploid *G. tabacina*. The B_2B_2 rDNA repeat is always the major repeat in the BBB_2B_2 polyploid and always the minor repeat in the AAB_2B_2 group. The BB genome of the BBB_2B_2 polyploid could have been donated by any of the B genome group of three currently recognized species, *G. latifolia*, *G. microphylla*, or diploid *G. tabacina*; accessions of all three species possess rDNA maps similar to those of the minor repeat in the BBB_2B_2 polyploid. The AA genome donor appears to have been a member of the D_4 isozyme group of *G. tomentella*, which has an rDNA map unique among the sample of A genome diploids.

A survey of cpDNA variation in the subgenus revealed considerable variation and delimited three major plastome groups that corresponded well to the A, B, and C diploid genome groups identified by artificial hybridization studies (J. Doyle et al., 1990a). All of nearly 100 accessions of polyploid *G. tabacina* had either A- or B-type plastomes (J. Doyle et al., 1990d). Furthermore, plastome groups, rDNA classification, and results of artificial intraspecific hybridization studies are completely concordant. The AAB_2B_2 polyploid had an A plastome donor, whereas the BBB_2B_2 polyploid had a B plastome donor.

The *G. tomentella* complex, for which extensive systematic data are available, includes not only eudiploid and eutetraploid cytotypes, but also aneuploids at both the diploid and tetraploid levels, leading to chromosome numbers of $2n = 38, 40, 78$, and 80. Isozyme data have been used to divide the complex into four diploid and six tetraploid races that correlate well with cytogenetic and crossing results (J. Grant et al., 1984; M. Doyle and Brown, 1985; M. Doyle et al., 1986).

Considerable spacer-length variation in 5S DNA (J. Doyle and Brown, 1989) was correlated with diploid isozyme groups. Several of the polyploid races were fixed hybrids, combining spacer-length variants found in diploids. These data, along with analyses of meiotic chromosome pairing in artificial interracial hybrids, suggested origins of most of the polyploid cytotypes from within the pool of known *G. tomentella* diploids. Interestingly, aneuploidy appears to have occurred only once in the complex, as the two aneudiploid races, D_1 and D_2 ($2n = 38$), which are very similar isozymically, have identical 5S spacer-length classes. The D_1/D_2 cytotype, in turn, is implicated in the origins of both aneutetraploid ($2n = 78$) isozyme races, most likely hybridizing with the D_3 group ($2n = 40$) to form the T_1 race and with the D_5 group ($2n = 40$) to form the T_6 race. A note of caution in these interpretations is necessary, however, as one diploid with $2n = 40$ was found with an apparent "fixed hybrid" pattern identical to that of the T_1 tetraploid.

A lack of differentiation among the diploids in the 18S–25S rDNA cistrons, combined with occasional intraracial polymorphism, has limited the use of 18S–25S rDNA for assessing the origins of the polyploid races of *G. tomentella*. However, in at least one polyploid, additivity expected on the basis of crossing studies, isozyme variation, and 5S fixed hybridity is not found for the 18S–25S rDNA, despite the divergence of both rDNA restriction site map and spacer length in the putative diploid progenitors. These data clearly suggest caution in interpreting *lack* of apparent fixed hybridity for rDNA as evidence of autopolyploidy, just as it is clearly important to recognize that internal genomic heterogeneity may not always be due to hybridity.

Multiple Origins of Polyploid Species

Polyploid species have traditionally been considered genetically depauperate, perhaps representing evolutionary dead-ends. This perspective is based on, among other assumptions, the premise that polyploidization is a very rare event and that each polyploid species had a unique origin. However, allozymic data for several polyploid species (e.g., Werth et al., 1985a, b; Ranker et al., 1989; P. Soltis et al., 1991) indicate instead surprisingly high levels of segregating genetic variation. Genetic variability of polyploid species may have arisen via mutation or through the introduction of allelic variation in unreduced, heterozygous gametes (e.g., Gastony, 1986) and/or multiple origins of the polyploid involving genetically distinct progenitors (e.g., Werth et al., 1985b; Ranker et al., 1989).

The detection of multiple origins of a polyploid species depends on the genetic variability of the progenitor(s). Although allozymic and morphologic data have been useful in documenting multiple origins of some polyploid species (Werth et al., 1985b; Ranker et al., 1989), these markers have provided at best equivocal evidence for multiple origins of others (Ownbey and McCollum, 1953; Roose and Gottlieb, 1976). The complementary use of rDNA and cpDNA markers has proven to be a valuable approach not only for unraveling the parentage of polyploid species but also for detecting multiple origins of polyploids.

Multiple origins of several polyploid species have been documented using DNA markers. Among these are (1) autopolyploids (species of *Heuchera*), (2) allopolyploids that had been studied intensively without unequivocal evidence for multiple origins (species of *Tragopogon*), (3) allopolyploids of multiple origin on a local scale (species of *Draba* and *Tragopogon*), and (4) allopolyploids for which other markers suggested multiple origins but underestimated the number of polyploidization events (species of *Glycine*).

Heuchera micrantha and *H. grossulariifolia* are well-documented examples of autopolyploids (Ness et al., 1989; D. Soltis and Soltis, 1989c; Wolf et al., 1989, 1990). High levels of cpDNA variation in both species permitted phylogenetic analyses of relationships among diploid and tetraploid populations. In *H. micrantha* var. *diversifolia* three branches of the phylogenetic tree held both diploid and tetraploid populations (D. Soltis et al., 1989a). *Heuchera micrantha* var. *pacifica* also contains both diploid and tetraploid populations, suggesting perhaps an additional origin. However, populations of var. *pacifica* are cytoplasmically identical to the ancestral *H. micrantha* and undifferentiated from one lineage of var. *diversifolia*. Autopolyploidy has occurred at least three, and perhaps four, times in *H. micrantha*. In *H. grossulariifolia*, both diploid and tetraploid populations appear in three lineages (Wolf et al., 1990), indicating at least three origins of autopolyploids in this species as well. Although multiple origins had been demonstrated previously for allopolyploids, these are the first reports of multiple origins of autoploidy. Multiple origins of autopolyploids are particularly significant because, unlike allopolyploidy, autopolyploidy has, with few exceptions (see Levin, 1983), traditionally been considered a rare, and perhaps unimportant, evolutionary force in angiosperms (Stebbins, 1971; V. Grant, 1981; but see Lewis, 1980).

Tragopogon mirus and *T. miscellus* (Compositae) are allotetraploids of recent origin, having arisen within the current century in the Palouse region of eastern Washington and western Idaho of the United States (Ownbey, 1950). Allozymic (Roose and Gottlieb, 1976) and morphologic (Ownbey and McCollum, 1953) markers suggested multiple origins of *T. mirus* and *T. miscellus*, respectively, although the data were equivocal.

Analysis of restriction site variation in the 18S–25S rDNA provided conclusive evidence of at least two origins of *T. mirus* within the geographic confines of the Palouse (P. Soltis and Soltis, 1991). All populations of *T. mirus* combined the rDNAs of its diploid progenitors, *T. dubius* and *T. porrifolius*. Two rDNA restriction profiles were observed within *T. porrifolius*, and each rDNA pattern was found in at least one population of *T. mirus*, indicating a minimum of two origins of the allotetraploid. A companion study of cpDNA variation in *T. mirus* and its progenitors was uninformative regarding multiple origins of the tetraploid (D. Soltis and Soltis, 1989d). All populations of *T. mirus* had the chloroplast genome of *T. porrifolius*; the cytoplasmic uniformity of *T. porrifolius* precluded an interpretation of single versus multiple origins of *T. mirus*.

Tragopogon miscellus combined the rDNA profiles of its diploid progenitors, *T. dubius* and *T. pratensis* (P. Soltis and Soltis, unpublished data). However, because no intraspecific rDNA variation was found in either diploid, no markers were available for detecting multiple origins of the tetraploid. In contrast, cpDNA markers provided conclusive evidence for at least two origins of *T. miscellus* (D. Soltis and Soltis, 1989d). The majority of populations of *T. miscellus* possesses the chloroplast genome of *T. pratensis,* whereas two populations of *T. miscellus,* both from Pullman, Washington, have the chloroplast genome of *T. dubius.* Furthermore, interpopulational cpDNA variation in *T. dubius* permitted the identification of the parental populations of the Pullman *T. miscellus.* Despite intensive previous study of both *T. mirus* and *T. miscellus* (Ownbey, 1950; Ownbey and McCollum, 1953; Roose and Gottlieb, 1976), multiple origins of these tetraploids had not been demonstrated conclusively. The complementary use of rDNA and cpDNA markers provides definitive evidence for multiple origins of both *T. mirus* and *T. miscellus* within the narrow geographic limits of the Palouse.

Many polyploid species have circumboreal or circumpolar distributions (Ehrendorfer, 1980). Because the diploid relatives of these species often have large geographic distributions as well (Ehrendorfer, 1980), multiple origins of the polyploids on a large geographic scale might be anticipated. In an investigation of the origins of several circumpolar species of *Draba* (Brassicaceae) from the Nordic flora, multiple origins of *D. lactea, D. cacuminum, D. norvegica,* and *D. corymbosa* were detected on a relatively small geographic scale using both allozymic and DNA markers (Brochmann et al., 1989b; D. Soltis and Soltis, 1990a). *Draba lactea* is an allohexaploid (Brochmann et al., 1989b) with a circumpolar distribution. Allozymic similarity among putative diploid progenitors prevented the identification of the parents of *D. lactea,* although *D. fladnizensis* is a probable progenitor on the basis of both allozymes and morphology. Restriction site analysis of rDNA supports the role of *D. fladnizensis* and also demonstrates two independent origins of *D. lactea.* Populations of both *D. fladnizensis* and *D. lactea* from southern Norway possessed rDNA restriction profiles distinct from those of both species from northern Norway. The presence of two rDNA lineages within the diploid *D. fladnizensis* agrees with crossability data for this species (Brochmann et al., 1989a), which suggests two intersterile races within *D. fladnizensis.* Each race of *D. fladnizensis* has apparently contributed rDNA to the geographically separated populations of the hexaploid *D. lactea.* Thus, multiple origins of polyploid species of *Draba* occur on a very local level, suggesting that many circumpolar polyploid species may be polyphyletic in origin.

Previous studies of *Glycine* subgenus *Glycine* recognized several races within each of the polyploids *G. tabacina* and *G. tomentella.* These data alone suggested possible multiple origins of each tetraploid species. However, the extent of independent polyploidization events within each tetraploid "species" was not discovered until DNA restriction site variation was examined in large samples of *G. tabacina, G. tomentella,* and their possible diploid progenitors.

A survey of 65 BBB_2B_2 polyploid accessions of *G. tabacina* revealed nine different plastome types, of which six were identical to cpDNA types found among diploid accessions (J. Doyle et al., 1990c). Thus, at least six independent origins appear likely for the BBB_2B_2 polyploid race; given the extreme diversity within the B diploid group, it seems reasonable that diploids exist with plastomes like those of the remaining polyploid cpDNAs, so even more origins may be possible. The diploid progenitors of the BBB_2B_2 polyploid appear to have included representatives of all three of the currently recognized species in the B genome group. Multiple origins of the BBB_2B_2 polyploid were detected based on cpDNA variation, despite no obvious nuclear rDNA variation, a situation contrasting with common expectations.

All crosses in both the AAB_2B_2 and BBB_2B_2 allopolyploids appear to have taken place with the B_2B_2 diploid as pollen donor. There is currently no explanation for this finding, although the phylogenetic position of that diploid, which appears basal to the main radiation of BB diploids on the basis of cpDNA (J. Doyle et al., 1990a, b), may be noteworthy. Finally, whereas cpDNA and 18S–25S rDNA have proven useful tools for the elucidation of the evolutionary history of this group, 5S rDNA variation has not been successfully used because of within-individual complexity. Although the diploid taxa implicated in the origins of these polyploids generally have only a single major 5S repeat, polyploid individuals generally have a minimum of four types, making analysis difficult.

As with the two major polyploid races of *G. tabacina,* the multiplicity of tetraploid races combining different diploid entities of *G. tomentella* may be considered to represent "multiple origins." However, in both cases, the diploids involved are probably genetically divergent enough to represent "good" biological species. Thus, the two tetraploid races of *G. tabacina* and the six tetraploid races of *G. tomentella* are themselves perhaps best considered different species, merely sharing some common progenitor genomes. However, some races of *G. tomentella* appear to have had several independent origins. As in *G. tabacina,* cpDNA has proven to be sufficiently polymorphic to elucidate these origins. The T_1 tetraploid race appears to be a uniform group on the basis of isozyme data, similar morphology, the formation of hybrids in artificial crossing studies, an identifiable geographic and ecological range in Australia, and the uniformity of 5S rDNA repeat length and 18S–25S rDNA repeat map. Based on the 5S rDNA results, the origin of this group probably involved hybridization between the D_1/D_2 group ($2n = 38$) and the D_3 group ($2n = 40$). The latter diploid race possesses several cpDNA variants, each of which appears in accessions of the T_1 polyploid, suggesting a minimum of three origins of this otherwise homogeneous tetraploid. Chloroplast DNA polymorphism, potentially due to multiple origins, occurs in at least two other polyploid races of *G. tomentella.*

A third polyploid in the subgenus is the newly described *G. hirticaulis* (Tindale and Craven, 1988). Currently, little is known about the affinities or origins of this species. However, a phylogenetic analysis of its cpDNA indicates that it is closely related to A genome species such as *G. tomentella* (J. Doyle et al., 1990b). Furthermore, although only three accessions have been surveyed for cpDNA variation, all three represent different plastome types. Given the evolutionary history of the other polyploids in the subgenus, it will not be surprising if each of these variants is found among different A genome diploid accessions, giving further evidence that multiple polyploid origins are the rule, and not the exception, in *Glycine* subgenus *Glycine*.

Although multiple origins of polyploid species seem the rule, a unique origin of a polyploid is difficult to demonstrate unequivocally. However, several polyploids appear to have had unique origins. The cytoplasmic uniformity among the 16 populations of the autotetraploid *Tolmiea menziesii* suggests that this cytotype had a unique origin (D. Soltis et al., 1989b). The allozymic diversity of the tetraploid could have originated via heterozygous, unreduced gametes.

Several allopolyploid species of ferns have been shown to have had multiple origins, yet the allotetraploids *Polystichum californicum* and *P. scopulinum* may each have had a single origin. Both species exhibit multiple electrophoretic phenotypes, suggesting as many as seven distinct origins of each species (P. Soltis et al., 1991). However, the diversity of electrophoretic phenotypes may have arisen following the production of a highly heterozygous allotetraploid and subsequent allelic segregation. Both *P. californicum* and *P. scopulinum* are cytoplasmically uniform, despite sampling from throughout their ranges, suggesting a unique origin for each species (P. Soltis et al., 1991). Alternatively, the hybridization event leading to each allopolyploidization may only occur successfully in a single direction, producing cytoplasmically uniform allotetraploids despite multiple origins.

Polyploid Evolution

The Polyploidization Process

Based on restriction site data, and to a lesser extent allozymic data, polyploid species are typically polyphyletic with two or more independent origins. This implies that polyploidization is quite common, at least in some groups, and suggests that perhaps plant systematists should begin to focus on the mechanisms and processes of polyploidization (e.g., Gastony, 1986; D. Soltis and Soltis, 1989d; P. Soltis and Soltis, 1991). Although several pathways have been proposed for the origin of polyploid species (e.g., Harlan and deWet, 1975; Mendiburu and Peloquin, 1976; deWet, 1980), few schemes have been tested with empirical data. However, the high frequency of polyploidization, even on a small geographic scale, suggests that polyploidy recurs in some groups. In *Tragopogon mirus* and *T. miscellus,* allopolyploidization occurs when the rare diploid progenitor (*T. porrifolius* or *T. pratensis,* respectively) is the maternal parent and the common progenitor (*T. dubius*) is the pollen parent (D. Soltis and Soltis, 1989d). However, the point at which chromosome doubling occurs is unknown. The production of unreduced pollen has been detected in low frequency in the diploid cytotypes of *Heuchera grossulariifolia* (Wolf et al., 1990) and *Tolmiea menziesii*

(D. Soltis and Soltis, 1988b); however, the pathway to polyploidy has not been elucidated for these or most other polyploids.

Genetic Variability of Polyploids

The polyphyletic nature of many polyploid species also implies that most polyploids are not genetically uniform. Multiple origins of polyploid species from genetically distinct diploid progenitors incorporate genetic variability into the polyploids without reliance on mutation to introduce allelic variation. Thus, although genetic variability is typically lower in polyploids than in their diploid progenitors, polyploids are generally not as depauperate genetically as some have suggested. The genetic variability of polyploid species may be partially responsible for their evolutionary success.

Molecular Consequences of Polyploidy

Because of their relative ease of analysis, ribosomal RNA genes provide a model for evaluating the molecular consequences of polyploidization. Although intraplant variation in rDNA repeat length has been observed in some species (Rafalski et al., 1982; J. Doyle et al., 1984; Zimmer et al., 1988; King and Schaal, 1989; P. Soltis and Soltis, 1991), considerable homogeneity exists among tandemly repeated genes in most diploid species (Appels and Dvorak, 1982a, b; Ellis et al., 1984; Snape et al., 1984; Saghai-Maroof et al., 1984; Flavell, 1986). This homogeneity may reflect unequal crossing over (Smith, 1976; Dover, 1982; Ohta and Dover, 1983) and gene conversion (Dover, 1982; Ohta and Dover, 1983).

Plants of hybrid or polyploid origin typically exhibit additivity of the rDNAs of their parents (e.g., J. Doyle et al., 1985; J. Doyle and Brown, 1989; Zimmer et al., 1988; P. Soltis and Soltis, 1991). However, some diploid artificial hybrids, such as *Zea mays* × *Z. luxurians* and *Z. mays* × *Z. diploperennis,* possess the rDNA of a single parent (Zimmer et al., 1988), suggesting that homogenization of rDNA or the elimination of a locus may occur quite rapidly. Deletion of an rDNA locus may explain many nonadditive rDNA profiles in hybrids and allopolyploids, particularly those of recent origin. Nuclear organizer region loci are typically terminally located on chromosomes and are surrounded by heterochromatin, making these loci susceptible to gene loss. In several interspecific hybrids and some polyploids, both parental rDNAs are present but the genes are differentially expressed (Navashin, 1934; McClintock, 1934; Keep, 1962; Kasha and Sadasivaiah, 1971; Wallace and Langridge, 1971; Darvey and Driscoll, 1972; Jessop and Sabrahmanyam, 1984). In hexaploid *Triticum* a hierarchy of relative rDNA gene expression occurs among loci (Longwell and Svihla, 1960; Martini et al., 1982; Martini and Flavell, 1985), resulting in competition among loci. Although most genes are not present in the large numbers or tandem organization of rRNA genes, rDNA may provide a valuable model for assessing the molecular consequences of polyploidization. Differential expression and competition for polymerases and other substances may also occur among loci encoding other products in allopolyploids. Through analysis of rDNA and other gene systems it may be possible to ascertain a molecular basis for the success of polyploids in nature. Comparisons of sequences of single-copy genes in polyploids and their diploid progenitors may also elucidate rates of molecular evolution. Furthermore, sequence analysis, such as those of the CAB genes in the fern *Polystichum munitum* (Pichersky et al., 1990), may serve as a model for studies of gene silencing.

Conclusions

Chloroplast DNA and rDNA markers are valuable tools for the analysis of polyploid species. Molecular markers have been used successfully to discern the parentage of both auto- and allopolyploids and to document multiple origins of polyploid species. The complementary use of cpDNA and rDNA markers provides the greatest information on polyploid evolution. In addition to rDNA, nuclear RFLPs may also be valuable for studying the origins of polyploids; to date, this approach has been underutilized.

In most cases examined to date, the assumptions expressed in the opening paragraphs of this chapter regarding the evolution of cpDNA and rDNA markers seem justified. Exceptions include the cpDNA divergence between the diploid and tetraploid cytotypes of *Tolmiea menziesii* (D. Soltis et al., 1989b), the presence of a single parental rDNA profile in hybrids of *Zea* (Zimmer et al., 1988), and lack of additivity of parental 18S–25S rDNAs in tetraploid species of *Glycine* (J. Doyle and Brown, unpublished data). Thus, analyses of polyploidy using molecular markers should be performed with caution. However, because the assumptions are frequently justified, discrepancies such as those observed for *Tolmiea, Zea,* and *Glycine* point to potentially exciting events of molecular and/ or organismal evolution.

Because of their reliability, DNA data should be used routinely to investigate the parentage of polyploid species. Once molecular data have documented multiple origins for an allo- or autopolyploid complex, comparative studies of ecophysiology, population ecology, and related topics could be addressed in the populations of different origin.

Molecular data are also keys to discovering the intricacies of the polyploidization process and the molecular consequences of polyploidization. The frequency with which polyploidization occurs in at least some groups of plants suggests that polyploid species are not the static, genetically uniform entities they were once considered to be. Genetic variability may be introduced into polyploid species recurrently from their diploid progenitors, perhaps providing the metabolic flexibility to account for the success of polyploid species in nature.

References

Appels, R., and Dvorak, J. (1982a) The wheat ribosomal DNA spacer region: its structure and variation in populations and among species. *Theor. Appl. Genet.* **63,** 337–348.

Appels, R., and Dvorak, J. (1982b) Relative rates of divergence of spacer and gene sequences within the rDNA region of species in the Triticeae: implications for the maintenance of homogeneity of a repeated gene family. *Theor. Appl. Genet.* **63,** 361–365.

Appels, R., and Honeycutt, R.L. (1986) rDNA: evolution over a billion years. In: *DNA Systematics,* Vol. II, *Plants* (ed. S.K. Dutta), CRC Press, Boca Raton, FL, pp. 81–135.

Arnheim, N., Krystal, M., Schmickel, R., Wilson, G., Ryder, O., and Zimmer, E. (1980) Molecular evidence for genetic exchanges among ribosomal genes on non-homologous chromosomes in man and apes. *Proc. Natl. Acad. Sci. USA* **77,** 7323–7327.

Bingham, E.T., Cutter, G.L., and Beversdorf, W.D. (1976) Creating genetic variability: tissue culture and chromosome manipulation. In: *World Soybean Research* (ed. L.D. Hill), Interstate Printers, Danville, IL, pp. 246–261.

Bold, H.C., Alexopoulos, C.J., and Delevoryas, T. (1986) *Morphology of Plants and Fungi,* Harper and Row, New York.

Brochmann, C., Borgen, L., and Stedje, B. (1989a) Chromosome numbers and crossing experiments in Nordic populations of *Draba* (Brassicaceae). In: *Biological Approaches and Evolutionary Trends in Plants* (abstract), 4th International Symposium of Plant Biosystematics, p. 39.

Brochmann, C., Soltis, P.S., and Soltis, D.E. (1989b) Evolutionary trends in Nordic populations of *Draba* (Brassicaceae). In: *Biological Approaches and Evolutionary Trends in Plants* (abstract), 4th International Symposium of Plant Biosystematics, p. 39.

Clausen, J.D., Keck, D., and Hiesey, W.H. (1945) Experimental studies on the nature of plant species. II. Plant evolution through amphidiploidy and autopolyploidy with examples from the Madiinae. *Carnegie Inst. Wash. Publ.* **546,** 1–174.

Berthou, F., Mathieu, C., and Vedel, F. (1983) Chloroplast and mitochondrial DNA variation as an indicator of phylogenetic relationships in the genus *Coffea* L. *Theor. Appl. Genet.* **65,** 77–84.

Corriveau, J.L., and Coleman, A.W. (1988) Rapid screening method to detect potential biparental inheritance of plastid DNA and results for over 200 angiosperm species. *Amer. J. Bot.* **75,** 1443–1458.

Cronquist, A. (1981) *An Integrated System of Classification of Flowering Plants,* Columbia University Press, New York.

Dahlgren, R. (1983) General aspects of angiosperm evolution and macrosystematics. *Nordic J. Bot.* **3,** 119–149.

Darlington, C.D. (1932) *Recent Advances in Cytology,* Churchill, London.

Darvey, N.L., and Driscoll, C.J. (1972) Nucleolar behaviour in *Triticum. Chromosoma* (Berl.) **36,** 131–139.

Delevoryas, T. (1980) Polyploidy in gymnosperms. In: *Polyploidy: Biological Relevance* (ed. W.H. Lewis), Plenum Press, New York, pp. 215–218.

de Wet, J.M.J. (1980) Origins of polyploids. In: *Polyploidy: Biological Relevance* (ed. W.H. Lewis), Plenum Press, New York, pp. 3–15.

Doebley, J., Renfroe, W., and Blanton, A. (1987) Restriction site variation in the *Zea* chloroplast genome. *Genetics* **117**, 139–147.

Dover, G.A. (1982) Molecular drive: a cohesive mode of species evolution. *Nature* **229**, 111–117.

Doyle, J.J., and Beachy, R.N. (1985) Ribosomal gene variation in soybean (*Glycine*) and its relatives. *Theor. Appl. Genet.* **70**, 369–376.

Doyle, J.J., Beachy, R.N., and Lewis, W.H. (1984) Evolution of rDNA in *Claytonia* polyploid complexes. In: *Plant Biosystematics* (ed. W.F. Grant), Academic Press, Ottawa, pp. 321–341.

Doyle, J.J., and Brown, A.H.D. (1989) 5S Nuclear ribosomal gene variation in the *Glycine tomentella* polyploid complex (Leguminosae). *Syst. Bot.* **14**, 398–407.

Doyle, J.J., Doyle, J.L., and Brown, A.H.D. (1990a) A chloroplast DNA phylogeny of the wild perennial relatives of the soybean (*Glycine* subgenus *Glycine*): congruence with morphological and crossing groups. *Evolution* **44**, 371–389.

Doyle, J.J., Doyle, J.L., and Brown, A.H.D. (1990b) Chloroplast DNA polymorphism and phylogeny in the B genome of *Glycine* subgenus *Glycine* (Leguminosae). *Amer. J. Bot.* **77**, 772–782.

Doyle, J.J., Doyle, J.L., Brown, A.H.D., and Grace, J.P. (1990c) Multiple origins of polyploids in the *Glycine tabacina* complex inferred from chloroplast DNA polymorphism. *Proc. Natl. Acad. Sci. USA* **87**, 714–717.

Doyle, J.J., Doyle, J.L., Grace, J.P., and Brown, A.H.D. (1990d) Reproductively isolated polyploid races of *Glycine tabacina* (Leguminosae) had different chloroplast genome donors. *Syst. Bot.* **15**, 173–181.

Doyle, J.J., Soltis, D.E., and Soltis, P.S. (1985) Ribosomal gene variation in *Tolmiea, Tellima,* and their intergeneric hybrid. *Amer. J. Bot.* **72**, 1388–1391.

Doyle, M.J., and Brown, A.H.D. (1985) Numerical analysis of isozyme variation in *Glycine tomentella. Biochem. Syst. Ecol.* **13**, 413–419.

Doyle, M.J., Grant, J., and Brown, A.H.D. (1986) Reproductive isolation between isozyme groups of *Glycine tomentella* (Leguminosae), and spontaneous doubling in their hybrids. *Austral. J. Bot.* **34**, 523–535.

Dunsmuir, P. (1985) The petunia chlorophyll *a/b* binding protein genes: a comparison of *Cab* genes from different gene families. *Nucleic Acids Res.* **13**, 2503–2518.

Ellis, T.H.N., Davies, D.R., Castleton, J.A., and Bedford, I.D. (1984) The organisation and genetics of rDNA length variants in peas. *Chromosoma* (Berl.) **91**, 74–81.

Ehrendorfer, F. (1980) Polyploidy and distribution. In: *Polyploidy: Biological Relevance* (ed. W.H. Lewis), Plenum Press, New York, pp. 45–60.

Erickson, L.R., Strauss, N.A., and Beversdorf, W.B. (1983) Restriction patterns reveal origins of chloroplast genomes in *Brassica* amphidiploids. *Theor. Appl. Genet.* **65**, 201–206.

Flavell, R.B. (1986) Ribosomal RNA genes and control of their expression. In: *Oxford Surveys of Plant Molecular and Cell Biology*, Vol. 3 (ed. B.J. Miflin), Oxford University Press, Oxford, pp. 251–275.

Gastony, G.J. (1986) Electrophoretic evidence for the origin of fern species by unreduced spores. *Amer. J. Bot.* **73**, 1563–1569.

Gill, B.S., and Appels, R. (1988) Relationships between *Nor*-loci from different *Triticeae* species. *Plant Syst. Evol.* **160**, 77–89.

Goldblatt, P. (1980) Polyploidy in angiosperms: monocotyledons. In: *Polyploidy: Biological Relevance* (ed. W.H. Lewis), Plenum Press, New York, pp. 219–239.

Goldsborough, P.B., Ellis, T.H.N., and Cullis, C.A. (1981) Organization of the 5S RNA genes in flax. *Nucleic Acids Res.* **9**, 5895–5904.

Grant, J.E., Brown, A.H.D., and Grace, J.P. (1984) Cytological and isozyme diversity in *Glycine tomentella* Hayata (Leguminosae). *Austral. J. Bot.* **32**, 665–677.

Grant, V. (1981) *Plant Speciation*, Columbia University Press, New York.

Hadley, H.H., and Hymowitz, T. (1973) Speciation and cytogenetics. In: *Soybeans: Improvement, Production, and Uses* (ed. B.E. Caldwell), American Society of Agronomists, Madison, WI, pp. 97–116.

Harlan, J.R., and de Wet, J.M.J. (1975) On Ö. Winge and a prayer: the origins of polyploidy. *Bot. Rev.* **41**, 361–390.

Haufler, C.H., and Soltis, D.E. (1986) Genetic evidence suggests that homosporous ferns with high chromosome numbers are diploid. *Proc. Natl. Acad. Sci. USA* **83**, 4389–4393.

Hightower, R.C., and Meagher, R.B. (1985) Divergence and differential expression of soybean actin genes. *EMBO J.* **4**, 1–8.

Hilu, K.W. (1988) Identification of the "A" genome of finger millet using chloroplast DNA. *Genetics* **118**, 163–167.

Hosaka, K. (1986) Who is the mother of the potato?—restriction endonuclease analysis of chloroplast DNA of cultivated potatoes. *Theor. Appl. Genet.* **72**, 606–618.

Hymowitz, T. (1970) On the domestication of the soybean. *Econ. Bot.* **24**, 408–421.

Jessop, C.M., and Sabrahmanyam, N.C. (1984) Nucleolar number variation in *Hordeum* species; their haploids and interspecific hybrids. *Genetica* **64**, 93–100.

Kasha, K.J., and Sadasivaiah, R.S. (1971) Genome relationships between *Hordeum vulgare* L. and *H. bulbosum* L. *Chromosoma* (Berl.) **35**, 264–287.

Keep, E. (1962) Satellite and nucleolar numbers in hybrids between *Ribes nigrum* and *R. grossularia* and in their backcrosses. *Canad. J. Genet. Cytol.* **4**, 206–218.

Khoshoo, T.N. (1959) Polyploidy in gymnosperms. *Evolution* **13**, 24–39.

Kihara, H., and Ono, T. (1926) Chromosomenzahlen und systematische Gruppierung der *Rumex*—Arten. *Zeit. Zellfrosch. Mikr. Anat.* **4**, 475–481.

King, L.M., and Schaal, B.A. (1989) Ribosomal-DNA variation and distribution in *Rudbeckia missouriensis*. *Evolution* **43**, 1117–1119.

Kirk, J.T.O., and Tilney-Bassett, R.A.E. (1978) *The Plastids: Their Chemistry, Structure, Growth,* and *Inheritance,* Elsevier, Amsterdam.

Klekowski, Jr., E.J., and Baker, H.G. (1966) Evolutionary significance of polyploidy in the Pteridophyta. *Science* **135,** 305–307.

Lee, J.S., and Verma, D.P.S. (1984) Structure and chromosomal arrangement of leghemoglobin genes in kidney bean suggest divergence in soybean leghemoglobin gene loci following tetraploidization. *EMBO J.* **3,** 2745–2752.

Leutwiler, L.S., Meyerowitz, E.M., and Tobin, E.M. (1986) Structure and expression of three light-harvesting chlorophyll a/b-binding protein genes in *Arabidopsis thaliana. Nucleic Acids Res.* **14,** 4051–4076.

Levin, D.A. (1983) Polyploidy and novelty in flowering plants. *Amer. Natur.* **122,** 1–25.

Lewis, W.H. (1980) Polyploidy in species populations. In: *Polyploidy: Biological Relevance* (ed. W.H. Lewis), Plenum Press, New York, pp. 103–144.

Long, E.O., and Dawid, I.B. (1980) Repeated genes in eukaryotes. *Ann. Rev. Biochem.* **49,** 727–764.

Longwell, A.C., and Svihla, G. (1960) Specific chromosomal control of the nucleolus and of the cytoplasm in wheat. *Exp'l. Cell Res.* **20,** 294–312.

Manton, I. (1950) *Problems of Cytology and Evolution in the Pteridophyta,* Cambridge University Press, London.

Martini, G., and Flavell, R.B. (1985) The control of nucleolus volume in wheat; a genetic study at three developmental stages. *Heredity* **54,** 111–120.

Martini, G., O'Dell, M., and Flavell, R.B. (1982) Partial inactivation of wheat nucleolus organisers by the nucleolus organiser chromosomes from *Aegilops umbellulatus. Chromosoma* (Berl.) **84,** 687–700.

McClintock, B. (1934) The relationship of a particular chromosomal element to the development of the nucleoli in *Zea mays. Zeit. Zellforsch. mik Anat.* **21,** 294–328.

Mendiburu, A.O., and Peloquin, S.J. (1976) Sexual polyploidization and depolyploidization: some terminology and definitions. *Theor. Appl. Genet.* **48,** 137–143.

Muller, H.J. (1914) A new mode of segregation in Gregory's tetraploid primulas. *Amer. Natur.* **48,** 508–512.

Muntzing, A. (1936) The evolutionary significance of autopolyploidy. *Hereditas* **21,** 263–378.

Navashin, M. (1934) Chromosomal alterations caused by hybridisation and their bearing upon certain genetic problems. *Cytologia* **5,** 169–203.

Neale, D.B., Wheeler, N.C., and Allard, R.W. (1986) Paternal inheritance of chloroplast DNA in Douglas fir. *Can. J. For. Res.* **16,** 1152–1154.

Ness, B.D., Soltis, D.E., and Soltis, P.S. (1989) Autopolyploidy in *Heuchera micrantha* Dougl. (Saxifragaceae). *Amer. J. Bot.* **76,** 614–626.

Ogihara, Y., and Tsunewaki, K. (1988) Diversity and evolution of chloroplast DNA in *Triticum* and *Aegilops* as revealed by restriction fragment analysis. *Theor. Appl. Genet.* **76,** 321–332.

Ohta, T., and Dover, G.A. (1983) Population genetics of multigene families that are dispersed in two or more chromosomes. *Proc. Natl. Acad. Sci. USA* **80**, 4079–4083.

Ownbey, M. (1950) Natural hybridization and amphiploidy in the genus *Tragopogon*. *Amer. J. Bot.* **37**, 487–499.

Ownbey, M., and McCollum, G. (1953) Cytoplasmic inheritance and reciprocal amphiploidy in *Tragopogon*. *Amer. J. Bot.* **70**, 788–796.

Palmer, J.D., Shields, C.R., Cohen, D.B., and Orton, T.J. (1983) Chloroplast DNA evolution and the origin of amphidiploid *Brassica*. *Theor. Appl. Genet.* **65**, 181–189.

Pichersky, E., Bernatzky, R., Tanksley, S.D., Breidenbach, R.B., Krausch, A.P., and Cashmore, A.R. (1985) Molecular characterization and genetic mapping of two clusters of genes encoding chlorophyll a/b-binding proteins in *Lycopersicon esculentum* (tomato). *Gene* **40**, 247–258.

Pichersky, E., Brock, T.G., Nguyen, D., Hoffman, N.E., Piechulla, B., Tanksley, S.D., and Green, B.R. (1989) A new member of the CAB gene family: structure, expression and chromosomal location of *Cab-8*, the tomato gene encoding the Type III chlorophyll a/b-binding polypeptide of photosystem I. *Plant Mol. Biol.* **12**, 257–270.

Pichersky, E., Soltis, D.E., and Soltis, P.S. (1990) Defective CAB genes in the genome of a homosporous fern. *Proc. Natl. Acad. Sci. USA* **87**, 195–199.

Rafalski, J.A., Wiewiorowski, M., and Soll, D. (1982) Organization and nucleotide sequence of nuclear 5S rRNA genes in yellow lupine (*Lupinus luteus*). *Nucleic Acids Res.* **10**, 7635–7642.

Ranker, T.A., Haufler, C.H., Soltis, P.S., and Soltis, D.E. (1989) Genetic evidence for allopolyploidy in the neotropical fern *Hemionitis pinnatifida* (Adiantaceae) and the reconstruction of an ancestral genome. *Syst. Bot.* **14**, 439–447.

Roose, M.L., and Gottlieb, L.D. (1976) Genetic and biochemical consequences of polyploidy in *Tragopogon*. *Evolution* **30**, 818–830.

Saghai-Maroof, M.A., Soliman, K., Jorgensen, R.A., and Allard, R.W. (1984) Ribosomal DNA spacer length polymorphisms in barley: Mendelian inheritance, chromosomal location, and population dynamics. *Proc. Natl. Acad. Sci. USA* **81**, 8014–8018.

Sears, B.B. (1980) Elimination of plastids during spermatogenesis and fertilization in the plant kingdom. *Plasmid* **4**, 233–255.

Sears, B.B. (1983) Genetics and evolution of the chloroplast. *Stadler Symp.* **15**, 119–139.

Singh, R.J., and Hymowitz, T. (1985a) Intra- and interspecific hybridization in the genus *Glycine* subgenus *Glycine* Willd.: chromosome pairing and genome relationships. *Z. Pflanzenzucht.* **95**, 289–310.

Singh, R.J., and Hymowitz, T. (1985b) The genomic relationships among six wild perennial species of the genus *Glycine* subgenus *Glycine* Willd. *Theor. Appl. Genet.* **71**, 221–230.

Singh, R.J., Kollipara, K.P., and Hymowitz, T. (1987) Polyploid complexes of *Glycine tabacina* (Labill.) Benth. and *G. tomentella* Hayata revealed by cytogenetic analysis. *Genome* **29**, 490–497.

Smith, G.P. (1976) Evolution of repeated sequences of unequal crossover. *Science* **191,** 528–535.

Snape, J.W., Flavell, R.B., O'Dell, M., Hughes, W.G., and Payne, P.I. (1984) Intrachromosomal mapping of the nucleolar organiser region relative to three marker loci on chromosome 1B of wheat (*Triticum aestivum*). *Theor. Appl. Genet.* **69,** 263–270.

Soltis, D.E. (1984) Autopolyploidy in *Tolmiea menziesii* (Saxifragaceae). *Amer. J. Bot.* **71,** 1171–1174.

Soltis, D.E. (1986) Genetic evidence for diploidy in *Equisetum. Amer. J. Bot.* **73,** 908–913.

Soltis, D.E., and Bohm, B.A. (1986) Flavonoid chemistry of diploid and tetraploid cytotypes of *Tolmiea menziesii* (Saxifragaceae). *Syst. Bot.* **11,** 293–297.

Soltis, D.E., and Doyle, J.J. (1987) Ribosomal RNA gene variation in diploid and tetraploid *Tolmiea menziesii* (Saxifragaceae). *Biochem. Syst. Ecol.* **15,** 75–78.

Soltis, D.E., and Rieseberg, L.H. (1986) Autopolyploidy in *Tolmiea menziesii* (Saxifragaceae): genetic insights from enzyme electrophoresis. *Amer. J. Bot.* **73,** 310–318.

Soltis, D.E., and Soltis, P.S. (1988a) Are lycopods with high chromosome numbers ancient polyploids? *Amer. J. Bot.* **75,** 238–247.

Soltis, D.E., and Soltis, P.S. (1988b) Electrophoretic evidence for tetrasomic inheritance in *Tolmiea menziesii* (Saxifragaceae). *Heredity* **60,** 375–382.

Soltis, D.E., and Soltis, P.S. (1989a) Genetic consequences of autopolyploidy in *Tolmiea* (Saxifragaceae). *Evolution* **43,** 586–594.

Soltis, D.E., and Soltis, P.S. (1989b) Polyploidy, breeding systems, and genetic differentiation in homosporous pteridophytes. In: *Isozymes in Plant Biology* (eds. D.E. Soltis and P.S. Soltis), Dioscorides Press, Portland, OR, pp. 241–258.

Soltis, D.E., and Soltis, P.S. (1989c) Tetrasomic inheritance in *Heuchera micrantha* (Saxifragaceae). *J. Heredity* **80,** 123–126.

Soltis, D.E., and Soltis, P.S. (1989d) Allopolyploid speciation in *Tragopogon*: insights from chloroplast DNA. *Amer. J. Bot.* **76,** 1119–1124.

Soltis, D.E., and Soltis, P.S. (1990a) Chloroplast DNA and nuclear rDNA variation: insights into autopolyploid and allopolyploid evolution. In: *Biological Approaches and Evolutionary Trends in Plants* (ed. S. Kawano), Academic Press, San Diego, pp. 97–117.

Soltis, D.E., and Soltis, P.S. (1990b) Isozyme evidence for ancient polyploidy in primitive angiosperms. *Syst. Bot.* **15,** 328–337.

Soltis, D.E., Soltis, P.S., and Ness, B.D. (1989a) Chloroplast DNA variation and multiple origins of autopolyploidy in *Heuchera micrantha* (Saxifragaceae). *Evolution* **43,** 650–656.

Soltis, D.E., Soltis, P.S., Ranker, T.A., and Ness, B.D. (1989b) Chloroplast DNA variation in a wild plant, *Tolmiea menziesii. Genetics* **121,** 819–826.

Soltis, P.S., and Soltis, D.E. (1986) Anthocyanin content in diploid and tetraploid cytotypes of *Tolmiea menziesii* (Saxifragaceae). *Syst. Bot.* **11,** 32–34.

Soltis, P.S., and Soltis, D.E. (1988) Electrophoretic evidence for genetic diploidy in *Psilotum nudum*. *Amer. J. Bot.* **75**, 1667–1671.

Soltis, P.S., and Soltis, D.E. (1991) Multiple origins of the allotetraploid *Tragopogon mirus* (Compositae): rDNA evidence. *Syst. Bot.* **16**, 407–413.

Soltis, P.S., Soltis, D.E., and Wolf, P.G. (1991) Allozymic and chloroplast DNA analyses of polyploidy in *Polystichum* (Dryopteridaceae). I. The origins of *P. californicum* and *P. scopulinum*. *Syst. Bot.* **16**, 245–256.

Stebbins, G.L. (1947) Types of polyploids: their classification and significance. *Adv. Genetics* **1**, 403–429.

Stebbins, G.L. (1971) *Chromosomal Evolution in Higher Plants*, Edward Arnold, London.

Szmidt, A.E., Alden, T., and Hallgren, J.-E. (1987) Paternal inheritance of chloroplast DNA in *Larix*. *Plant Mol. Biol.* **9**, 59–64.

Takhtajan, A. (1969) *Flowering Plants—Origin and Dispersal*, Smithsonian Institution Press, Washington, DC.

Takhtajan, A. (1980) Outline of the classification of flowering plants (Magnoliophyta). *Bot. Rev.* **46**, 225–359.

Thorne, R.F. (1983) Proposed new realignments in the angiosperms. *Nordic J. Bot.* **3**, 85–117.

Tindale, M.D., and Craven, L.A. (1988) Three new species of *Glycine* (Fabaceae: Phaseoleae) from north-western Australia, with notes on amphicarpy in the genus. *Austr. Syst. Bot.* **1**, 399–410.

Vida, G. (1976) The role of polyploidy in evolution. In: *Evolutionary Biology* (eds. V.J.A. Novak and Pacltova), Czechoslovak Academy of Sciences, Prague, pp. 267–304.

Wallace, H., and Langridge, W.H.R. (1971) Differential amphiplasty and the control of ribosomal RNA synthesis. *Heredity* **27**, 1–13.

Wendel, J.F. (1989) New World tetraploid cottons contain Old World cytoplasm. *Proc. Natl. Acad. Sci. USA* **86**, 4132–4136.

Werth, C.R., Guttman, S.I., and Eshbaugh, W.H. (1985a) Electrophoretic evidence of reticulate evolution in the Appalachian *Asplenium* complex. *Syst. Bot.* **10**, 184–192.

Werth, C.R., Guttman, S.I., and Eshbaugh, W.H. (1985b) Recurring origins of allopolyploid species in *Asplenium*. *Science* **228**, 731–733.

Wolf, P.G., Soltis, P.S., and Soltis, D.E. (1989) Tetrasomic inheritance and chromosome pairing behaviour in the naturally occurring autotetraploid *Heuchera grossulariifolia* (Saxifragaceae). *Genome* **32**, 655–659.

Wolf, P.G., Soltis, D.E., and Soltis, P.S. (1990) Chloroplast-DNA and electrophoretic variation in diploid and autotetraploid *Heuchera grossulariifolia*. *Amer. J. Bot.* **77**, 230–242.

Yatskievych, G., Stein, D.B., and Gastony, G.J. (1988) Chloroplast DNA evolution and systematics of *Phanerophlebia* (Dryopteridaceae) and related fern genera. *Proc. Natl. Acad. Sci. USA* **85**, 2589–2593.

Zimmer, E.A., Jupe, E.R., and Walbot, V. (1988) Ribosomal gene structure, variation and inheritance in maize and its ancestors. *Genetics* **120**, 1125–1136.

9

Molecular Systematics and Crop Evolution

John Doebley

During the past decade, the tools of molecular biology have been applied to plant systematics with remarkable success. New insights have been gained into such topics as phylogenetic reconstruction, introgression, genomic evolution, and levels of genetic variation in natural populations. Molecular methods have provided greater resolution than was previously possible with other approaches. Moreover, variation in DNA sequences is more readily subjected to statistical analysis than many previous types of data, and it can be less ambiguous, making interpretation of data more straightforward.

In this chapter, I review the application of molecular systematics to questions surrounding the origins and evolution of crop plants, specifically (1) identifying the progenitors of crops, (2) assaying levels of genetic variation in crops and their ancestors, and (3) determining the extent of introgression between crops and their relatives. Each of these issues has been addressed for several crops using molecular methods, and molecular evidence has provided definitive results in many cases. I will also outline issues in crop evolution that, although amenable to study with molecular methods, have received little or no study to date. These issues include (1) the genetic basis of morphologic changes induced by human selection and (2) the nature of molecular evolution under domestication.

Methodologic Considerations

Angiosperms possess three separate genomes: chloroplast DNA (cpDNA), mitochondrial DNA (mtDNA), and nuclear DNA (nDNA). Molecular analyses of each of these are applicable to the study of crop evolution, although their utility may vary depending on the question of interest. For studies of crop

Preparation of this chapter was supported in part by NSF grant number BSR-8806889. I thank Drs. K. Hosaka, K. Rasmusson, D. Spooner, and J. Wendel, and the editors for their many helpful suggestions which have greatly improved this chapter.

evolution, the most practical method of analysis of these genomes is restriction fragment length polymorphism (RFLP) analysis because it enables greater sampling of species and populations than the more labor-intensive method of nucleotide sequencing.

Chloroplast DNA is maternally inherited in most angiosperms, making it particularly suitable for identifying maternal parental species of polyploid crops. It also is typified by conservative rates of both structural rearrangements and sequence evolution (Palmer, 1987). This enables the identification of specific mutational events with great certainty, increasing confidence that restriction fragments, which appear to be the same, are in fact the same. Such unambiguous data are readily treated by statistical or phylogenetic analyses. The conservatism of cpDNA also results in generally low levels of intraspecific variation, often reducing its utility at this taxonomic level (but see Chapter 6, this volume).

The mitochondrial genome, like cpDNA, is usually maternally inherited in angiosperms. However, this genome is typified by a much higher rate of structural rearrangement (Palmer and Herbon, 1988). This feature generates higher levels of intraspecific RFLPs, but lowers confidence that two restriction fragments with the same electrophoretic mobility are indeed the same. Also, single structural changes may alter several restriction fragments, making accurate quantification of mtDNA data more difficult. Indeed, in the absence of detailed restriction mapping, it is practically impossible to sort out restriction site mutations and structural rearrangements in mtDNA.

The complexity of nDNA offers opportunity to study repetitive sequences, such as ribosomal genes (rDNA) (Doyle and Beachy, 1985; Zimmer et al., 1988), and single-copy or low-copy-number sequences or genes (Song et al., 1988a, b; Keim et al., 1989; Havey and Muehlbauer, 1989). Nuclear DNA sequences provide evidence for both the maternal and paternal lineages, and levels of polymorphism for nDNA are suitable for intraspecific analyses. Evolutionary rates for the nuclear genome vary depending upon the specific portion of the genome analyzed; however, in general, nDNA has a high rate of both sequence and structural evolution as compared to cpDNA. For nuclear low-copy-number sequences, detailed restriction maps provide the most precise data (Gepts and Clegg, 1989), but these can be assembled only through tedious laboratory analyses that, if performed, limit the number of accessions and genes that can be analyzed. Scoring presence/absence of specific low-copy-number restriction fragments is more practical; however, as with mtDNA, one can have only limited confidence that two restriction fragments with similar mobilities are indeed the same. Similarly, a single structural change in nDNA can produce several restriction fragment changes, creating uncertainties as how best to quantify the data.

Many questions surrounding crop evolution require quantification of molecular data, especially to obtain an accurate measure of genetic similarity between taxa or an estimate of the levels of variation. For cpDNA, crops and their relatives can be readily analyzed for the presence/absence of particular restriction sites and insertion/deletion events. These events can then be used to construct phylogenetic trees using parsimony (Palmer et al., 1985; Doebley et al., 1987; Ogihara and Tsunewaki, 1988). For nuclear and mitochondrial DNA, the higher levels of polymorphism create a more complex situation, especially if the molecular basis of the observed differences in restriction fragments is not understood. The best approach to mtDNA restriction fragment data may be to calculate distances based on the shared fragment method (McClean and Hanson, 1986; Graur et al., 1989). Because mtDNA RFLPs are apt to mix both restriction site changes and structural rearrangements, distances based upon them will not provide an accurate estimate of the proportion of substitutions per site, but should still provide a general measure of the degree of phenetic similarity. For rDNA, detailed restriction mapping has proven a practical approach (Zimmer et al., 1988). In the absence of detailed restriction mapping, low-copy-number nuclear sequence data are probably most accurately quantified when RFLPs are identified as alleles at a particular locus (Havey and Muehlbauer, 1989; Keim et al., 1989). This procedure increases confidence that structural rearrangements that alter several restriction fragments will not be weighted too heavily, and the data can be subjected to multivariate, phylogenetic, and statistical analyses, including measures of genetic distance and heterozygosity (Havey and Muehlbauer, 1989).

Origins of Crop Species

A principal concern of crop evolutionists has been the identification of the wild progenitors of domesticated species. This has often been difficult with classical taxonomic methods, because crops usually exhibit wide morphologic departures from their relatives. Genetic, cytogenetic, and biochemical analyses have contributed greatly to this subject (Simmonds, 1976; Doebley, 1989); however, much remains to be learned. The question of crop origins can be pursued at several different levels. First, some crops may have a clearly identified wild progenitor species, although the progenitor may be polymorphic and geographically widespread. Here, the question of interest may be to pinpoint the geographic region in which domestication occurred. Second, for some crops, there may be uncertainty as to which of several distinct wild taxa gave rise to the cultigen. Third, for polyploid crops, there is interest in identifying their diploid parental species. Finally, the question of whether a crop has been domesticated once or several times is often an issue.

Because domestication is a relatively recent event (the last 10,000 years), one expects crops to show relatively little genetic differentiation from their progenitors. This is true despite the often large differences in gross morphology between crop and progenitor because artificial (human) selection is likely to have acted primarily on the small suite of genes controlling morphologic traits of interest to humankind, leaving the vast majority of the genome to evolve at a much slower pace. Thus, for the conservatively evolving chloroplast genome, few (one or two) or no restriction site differences should be observed between crop and progenitor for studies using 20 to 25 restriction enzymes. Moreover, crop and progenitor should appear on the same (or very close) branchlet(s) of the cpDNA phylogenetic tree. For mtDNA and nDNA, which evolve more rapidly, greater differentiation is expected, although the progenitor should show greater similarity to the crop than to other congeneric wild species.

Barley

Clegg et al. (1984a) and Neale et al. (1988) examined cpDNA variation in barley (*Hordeum vulgare* ssp. *vulgare*) and its presumed progenitor (*H. vulgare* ssp. *spontaneum*). Each of these studies revealed a small number (3 to 5) of variable restriction sites that defined several cpDNA genotypes. These studies revealed that the most common cpDNA genotypes in cultivated barley were also present in wild barley, reflecting the close genetic relationship between these two subspecies. Further, this work showed greater cpDNA diversity in wild barley as compared to cultivated barley, which one would expect if the former were the progenitor species and the latter were the derivative.

Cotton

One of the most definitive molecular analyses of the origin of a crop has been that of cotton (Wendel, 1989). There are four separate species of cultivated cotton including two New World tetraploids, *Gossypium hirsutum* and *G. barbadense*. Cytogenetic analysis has shown that these tetraploids combine the A genome (of Asia-Africa) and the D genome (of the Americas). An unresolved question surrounding the origin of the New World cottons is when the A and D genome diploids hybridized to form the ancestor of the cultivated tetraploids. Various authors have proposed that hybridization took place (1) before Africa and South America were fully separated by continental drift (100 million or more years ago), (2) very recently when humans carried A genome species from Africa to the New World (during the past 15,000 years), or (3) within the past several million years after natural dispersal of an A genome species to the New World. The weight of the biosystematic evidence favors the third explanation; however, this topic is still controversial. Wendel (1989) showed that the degree of cpDNA divergence between the New World cottons and the A genome diploids indicates that the former arose 1–2 million years ago, clearly eliminating the first alternative and providing strong support for the third. Wendel (1989) also demonstrated that an Asian-African (A genome) species was the maternal parent of both New World tetraploids.

The "Irish" Potato

Solanum is a large, pantropical genus with between 1,400 and 2,000 species. Within it, the tuber-bearing species, including the "Irish" potato, belong to the subgenus *Potatoe* section *Petota*. Within this group, approximately six wild diploid species ($x = 12$) are taxonomically similar enough to domesticated potatoes to have potentially contributed to their origin (Hawkes et al., 1979; Hawkes, 1989). The cultivated potatoes include diploid, triploid, tetraploid, and pentaploid forms, which have been variously treated taxonomically. *Solanum goniocalyx, S. phureja,* and *S. stenotomum* are the primary cultivated diploids, and *S. tuberosum,* the potato of commerce, is a tetraploid. *Solanum tuberosum* includes two principal cultivated forms, ssp. *andigena* and ssp. *tuberosum*. The former subspecies arose in South America, and the latter is thought to have arisen twice by parallel evolution, once in southern Chile and once in Europe. Two questions surround the origin of cultivated potatoes: (1) which diploid(s) gave rise to the cultivated potatoes (both diploid and tetraploid) and (2) what is the relationship between ssp. *tuberosum* and ssp. *andigena*?

Hosaka (1986) and Hosaka and Hanneman (1988b) examined cpDNA variation in the cultivated potatoes and their relatives. These authors defined eight cpDNA types based on restriction site and insertion/deletion mutations. They also surveyed a large number of wild and cultivated accessions for these cpDNA variants (Hosaka and Hanneman, 1988b). Their work provides many valuable insights into the origin of the cultivated potatoes. First, spp. *andigena* possesses several cpDNA types that are also found in the cultivated diploids, strongly supporting the hypothesis that ssp. *andigena* was derived from the cultivated diploids. Second, the diversity of cpDNA types in ssp. *andigena,* along with other biosystematic data, suggests that ssp. *andigena* had multiple origins via repeated cycles of hybridization and introgression. Third, the occasional presence of wild-type cpDNAs (*S. sparsipilum* or *S. chacoense*) in ssp. *andigena* suggests occasional introgression from the wild forms into the domesticates. Fourth, the cpDNA data suggest a very close relationship among the three species of diploid cultivated potatoes (*S. goniocalyx, S. phureja,* and *S. stenotomum*).

Hosaka and Hanneman (1988a) also used cpDNA variation to assess the origin of ssp. *tuberosum*. They reported that ssp. *tuberosum* has a specific cpDNA type (T) not found in any diploid taxa, although it is found at low frequency in ssp. *andigena* from the Chilean-Argentine border. When combined with other data, cpDNA data provide strong support for the hypothesis that Chilean ssp. *tuberosum* was derived from Andean ssp. *andigena*. Chloroplast DNA data also indicate that the first potatoes introduced into Europe belonged to ssp. *andigena*. However, these apparently did not, for the most part, survive the potato blights of the 1840s and were replaced by the American cultivar "Rough Purple Chili," which belongs to ssp. *tuberosum*. Thus, molecular data argue against the hypothesis that ssp. *tuberosum* arose independently under domestication in Europe, but rather suggest that the modern European potato was derived from Chilean ssp. *tuberosum*.

Lentil

Havey and Muehlbauer (1989) examined RFLPs for low-copy-number nuclear sequences in the lentil (*Lens culinaris* ssp. *culinaris*) and its relatives. Treating the RFLPs as alleles at loci (probes), these authors calculated genetic distances among the species and subspecies. A UPGMA dendrogram based on these distances placed *Lens culinaris* ssp. *orientalis* closest to the cultivated lentil, in agreement with previous isozymic and morphologic data that indicated that this wild subspecies is the ancestor of the cultivated lentil. The RFLP data also demonstrated that the cultivated lentil possesses less genetic diversity than ssp. *orientalis*, consistent with the hypothesis that the former was a recent (during the past 10,000 years) derivative of the latter.

Maize

Timothy et al. (1979), Doebley et al. (1987), and Doebley (1990) examined cpDNA variation among *Zea mays* ssp. *mays* (maize) and its wild relatives (the teosintes). The teosinte species, *Z. diploperennis, Z. perennis,* and *Z. luxurians,* possessed cpDNAs that differed substantially (19 mutations) from that of maize. However, the four cpDNA genotypes found in maize were also present in the teosintes, *Z. mays* ssp. *mexicana* and ssp. *parviglumis*. These four cpDNA types are very similar to one another, each distinguished by a single restriction site variant or length mutation. These cpDNA results are consistent with previous cytologic and allozymic data that suggest one of these two wild subspecies was the progenitor of maize. The fact that four cpDNA types were found in both teosinte and maize implies either multiple domestications or introgression of wild cytoplasms into cultivated maize. A preponderance of other data suggests that the latter is the case (Doebley, 1990). The distribution of these cpDNA types in maize and teosinte was not sufficiently correlated with geography to enable the authors to discern a likely center for the origin of maize based on these data.

Further support that teosinte is ancestral to maize comes from restriction analysis of rDNA (Zimmer et al., 1988) and sequence analysis of heterochromatic DNA (Dennis and Peacock, 1984). Both of these studies showed that maize is essentially identical to Mexican annual teosinte (*Z. mays* ssp. *mexicana* or *parviglumis*), but distinct from the other teosinte species (*Z. diploperennis* and *Z. luxurians*). This result indicates that maize and Mexican annual teosinte share a more recent common ancestor than do Mexican annual teosinte and the other teosinte species.

Mustards and Cole Crops

The genus *Brassica* contains several important crop species, including the diploids, black mustard (*B. nigra*), turnip (*B. rapa*), and cabbage (*B. oleracea*), and all three possible allopolyploid hybrids between them, Abyssinian mustard (*B. carinata*), leaf mustard (*B. juncea*), and rape (*B. napus*). The relationships among these species have been established based on cytogenetic data (Fig. 9.1). Analysis of cpDNA showed that *B. nigra* was the maternal parent of *B. carinata*, whereas *B. rapa* was the maternal parent of both *B. napus* and *B. juncea* (Erickson

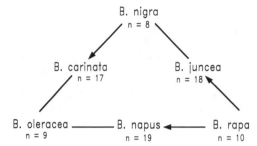

B. nigra
n = 8

B. carinata
n = 17

B. juncea
n = 18

B. oleracea ———— B. napus ◄——— B. rapa
n = 9   n = 19   n = 10

Figure 9.1. Phylogenetic relationships among the diploid and tetraploid species of *Brassica*. The arrows indicate the source of the cytoplasmic genomes for the amphidiploid species. Adapted from Palmer et al. (1983).

et al., 1983; Palmer et al., 1983). Phylogenetic analysis of low-copy-number nDNA supported the relationships outlined in Fig. 9.1, and indicated that the nuclear genomes of the amphidiploids are more similar to their maternal than their paternal parents (Song et al., 1988a). Detailed analysis of low-copy-number nDNA sequences of *B. rapa* indicated that this species consists of two well-defined groups, European and east Asian, which may represent two separate domestications from separate wild populations (Song et al., 1988b).

Pea

Palmer et al. (1985) analyzed cpDNA variation in *Pisum sativum* (pea) and its wild relatives, *P. humile, P. elatius,* and *P. fulvum*. Among 13 accessions of cultivated pea, they found two cpDNA types. The more common (12 of the 13 accessions) of these cpDNA types was similar to that of an accession of *P. humile* from northern Israel, where cytologic and morphologic data place the origin of domestication for pea. The second cpDNA type, which was found in a single accession of cultivated pea, differed from the first type by five mutations. As noted by the authors, it represents either a separate domestication event or introgression of a foreign cytoplasm into cultivated pea.

Soybean

Relatively few comparative studies of nDNA in crops and their progenitors have been conducted; however, soybean (*Glycine max*) and its presumed progenitor (*G. soja*) have been the subject of several such studies. Doyle and Beachy (1985) examined the 18S–25S ribosomal genes of over 40 diverse accessions of wild and cultivated soybean and failed to find a single variant within or between these species. Similarly, Doyle (1988) found only a single repeat length size (345 bp) for the 5S ribosomal genes of soybean. Both this repeat size plus a smaller variant (334 bp) were found in the 5S ribosomal genes of *G. soja*. Data from both ribosomal genes suggest a very close relationship between soybean and its presumed progenitor.

Keim et al. (1989) examined RFLPs of 17 low-copy-number nuclear sequences

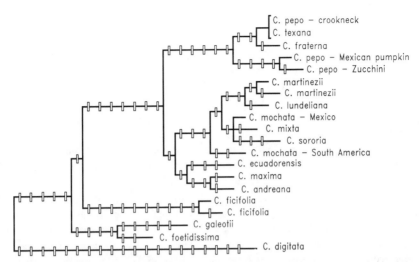

Figure 9.2. Wagner parsimony phylogenetic tree for 15 species of the genus *Cucurbita* based on cpDNA restriction site presence/absence. Cross-bars along the branches represent individual restriction site losses/gains. The tree required 95 steps to account for 86 restriction site changes. The restriction site characters were polarized using *Lagenaria siceraria* and *Luffa aegyptiaca* as outgroups. Data are from Doebley et al. (unpublished data).

in cultivated and wild soybean (*G. soja*). This study uncovered several facts that appear inconsistent with the simple scenario that *G. max* is a recent, cultivated derivative of *G. soja*. First, there is greater genetic diversity within soybean than within *G. soja*. Second, a principal component analysis of the RFLP data revealed little overlap between these two species, suggesting that most soybean accessions have become differentiated from their presumed progenitor for these 17 nuclear sequences. Third, one RFLP, present in 90% of all soybean accessions, was absent from *G. soja*, a pattern that would not be anticipated if the former were recently derived from the latter. These results contrast with typical observations for allozymic variation between a crop and its progenitor for which one normally finds (1) the same common alleles in the domesticate and its ancestor and (2) greater genetic variation in the progenitor (Doebley, 1989). Too little is known about RFLPs of low-copy-number nDNA sequences among species to draw hard conclusions from the results of Keim et al. (1989). It will be of interest to see data of this nature accumulate for other crops and their wild relatives.

Squash, Pumpkin, and Ornamental Gourds

Chloroplast DNA variation was analyzed in *Cucurbita*, including five cultivated and ten wild species (Fig. 9.2) (Doebley et al., unpublished data). These analyses revealed a close association (two restriction site differences) between the cultivated hubbard squash (*C. maxima*) and its presumed progenitor, *C.*

andreana. They also revealed a close association (one restriction site difference) between the cultigens cushaw (*C. mixta*) and butternut squash (*C. mochata* from Mexico), suggesting that these two cultigens may share the same maternal progenitor species. *Cucurbita sororia* (four accessions), thought to be the progenitor of cushaw, differed from it by four restriction site mutations. This is a somewhat greater degree of divergence than one would expect between a crop and its progenitor, although a form of *C. sororia* more similar to cushaw may be found among additional accessions of this wild species. Another anomalous result was that *C. mochata* from South America differed from Mexican varieties of this species by four restriction sites, somewhat more than anticipated among varieties of a cultigen. This suggests separate origins for this domesticate in Mexico and South America or perhaps introgression. Finally, cpDNA analyses revealed two distinct cpDNA forms in the cultivated species, *C. pepo* (squash, ornamental gourd, and pumpkin). The first of these, which was found in ornamental gourd and crookneck squash, was identical to that of the wild species *C. texana*, agreeing with recent allozymic data that suggest this wild species had a role in the origin of *C. pepo* (Decker and Wilson, 1987; Decker, 1988). The second cpDNA form, which was found in zucchini and Mexican pumpkin, differed from the first by six restriction site mutations. This cpDNA type is unknown in any wild form, including a suspected progenitor, *C. fraterna*. This too agrees with previous allozymic data (Wilson, 1989) and suggests that the wild progenitor of pumpkin has yet to be discovered.

Sunflower

Rieseberg and Seiler (1990) examined cpDNA variation in cultivated and wild accessions of *Helianthus annuus* (sunflower) and found four closely related cpDNA genotypes among 11 accessions of wild/weedy sunflower. The cpDNAs of 23 diverse accessions of cultivated sunflower were all the same, and they were identical to one of the four wild-type cpDNAs. The cpDNA type associated with the cultigen is widespread in the wild sunflower (Missouri to California), and thus these data do not help pinpoint the geographic origin of the sunflower. However, when combined with other data, the lack of polymorphism in the cpDNA of the sunflower suggests that this crop was domesticated only once.

Wheat

The cultivated wheats are part of a polyploid complex in the genus *Triticum*. There are four primary taxa of cultivated wheat, each of which has its own presumed ancestor: the diploid *T. monococcum* var. *monococcum* (AA genome) derived from the wild-type *T. monococcum* var. *boeoticum* (AA); the tetraploid *T. turgidum* var. *dicoccum* (AABB) derived from the wild-type *T. turgidum* var. *dicoccoides* (AABB); the tetraploid *T. timopheevii* var. *timopheevii* (AAGG) derived from the wild-type *T. timopheevii* var. *araraticum* (AAGG); and the hexaploid *T. aestivum* (AABBDD) derived from a cross of the cultivated *T. turgidum* (AABB) and the wild *T. tauschii* (DD) (Feldman, 1976). As indicated, each of the cultivated polyploids has a wild ancestral form of the same ploidal level with the exception of *T. aestivum,* which presumably arose by polyploidy under domestication.

The origin of the A, B, and D genomes of hexaploid wheat (*T. aestivum*) has not been fully resolved. Based on cytogenetic analyses, two of the genomes can be clearly associated with specific, wild, diploid taxa: (1) *T. monococcum* var. *boeoticum* for the A genome, and (2) *T. tauschii* for the D genome. The donor of the B genome is uncertain; however, several species of *Triticum* have been favored based on cytogenetic, taxonomic, and biochemical data, including *T. speltoides, T. longissimum,* and *T. bicorne*. The donor of the G genome of *T. timopheevii* is also uncertain, although *T. speltoides* has been suggested by some authors (see Feldman, 1976).

Identification of the wild species that contributed the B genome to the wheats has been attempted with cpDNA and mtDNA analyses. This is possible because it was the ancestral B genome species that was the maternal (cytoplasmic) parent of *T. turgidum* and *T. aestivum*. Thus, these species may be said to contain B cytoplasm. Tsunewaki and Ogihara (1983) examined cpDNA of the wheats and concluded that the maternal or B genome parent was *T. longissimum;* however, Bowman et al. (1983) were unable to confirm this result. Subsequent analyses suggested that the *T. longissimum* cytoplasm examined by Tsunewaki and Ogihara (1983) had been mislabeled (Ogihara and Tsunewaki, 1988). Recent comprehensive analyses revealed the cpDNA of *T. turgidum* and *T. aestivum* to be distinct (four restriction site mutations and two insertion differences from *T. speltoides*) within the genus (Ogihara and Tsunewaki, 1988). Analysis of mtDNA of the cultivated wheats and their relatives also demonstrated the mtDNA of the B cytoplasm to be unique within the genus (Graur et al., 1989). This is a curious result because it shows that the cytoplasmic genome of the tetraploid, *T. turgidum* (AABB), and the hexaploid, *T. aestivum* (AABBDD), is unknown in any diploid species. Either there are additional species or populations that have not been discovered and analyzed, or the donor of the B genome and B cytoplasm is extinct. The fact that considerable heterogeneity has been found among the mtDNA of *T. speltoides* suggests that more comprehensive sampling would be informative (Breiman, 1987).

Determination of other progenitor species of the wheats has been much more straightforward. Ogihara and Tsunewaki (1988) analyzed cpDNA of *T. timopheevii* and found it to be identical to that of *Aegilops aucheri* (a form of *T. speltoides*), its presumed maternal progenitor. Analysis of mtDNA (Graur et al., 1989) and cpDNA (Ogihara and Tsunewaki, 1988) support the hypothesis that *T. turgidum* var. *dicoccoides* was the maternal progenitor of *T. turgidum* var. *turgidum,* which in turn was the maternal progenitor of *T. aestivum*.

Other Crops

Several other studies have examined organellar DNA in crops and their relatives. Clegg et al. (1984b) and Gepts and Clegg (1989) found no restriction site differences between cpDNAs of pearl millet (*Pennisetum glaucum* ssp. *glaucum*) and its presumed wild progenitor (ssp. *monodii*). Perl-Treves and Galun (1985) reported no restriction site or other differences between cpDNAs of *Cucumis sativus* (cucumber) and its presumed wild progenitor, *C. hardwickii*. These same authors also examined cpDNA of melon (*Cucumis melo* var. *melo*) and its presumed progenitor (*C. melo* var. *agrestis*) and found only a single, small deletion to differentiate them. Duvall and Doebley (1990) found little or no differentiation (0 to 2 mutations) between the cpDNAs of accessions of sorghum (*Sorghum bicolor* ssp. *bicolor*) and its presumed progenitor (*S. bicolor* ssp. *arundinaceum*). McClean and Hanson (1986) studied mtDNA of tomato (*Lycopersicon esculentum* var. *esculentum*) and its wild relatives. Based on the shared restriction fragment method, they calculated divergence values among the taxa. These revealed *L. esculentum* var. *cerasiforme* to be most similar to tomato, supporting the hypothesis that it is the progenitor of tomato (Rick and Fobes, 1975).

The origins of several polyploid crops have been subjected to molecular analyses. First, tobacco (*Nicotiana tabacum*) is a tetraploid known only in cultivation. Cytotaxonomic evidence suggests that *N. sylvestris* and *N. tomentosiformis* are its parental diploids (Gerstel, 1976). Analysis of both cpDNA (Kung et al., 1982) and mtDNA (Bland et al., 1985) demonstrated that tobacco inherited these cytoplasmic genomes from *N. sylvestris*. Second, finger millet (*Eleusine coracana*) is a tetraploid cereal crop of Africa. Analysis of cpDNA restriction patterns indicated that finger millet inherited its chloroplast genome from *E. indica* (Hilu, 1988). Third, the evolution of citrus fruits is known to have involved frequent hybridizations. Analysis of cpDNA revealed that sour orange (*Citrus aurantium*), grapefruit (*C. x paradisi*), lemon (*C. limon*), and sweet orange (*C. sinensis*) all share the same cpDNA genotype (Green et al., 1986). This cpDNA type was the same as that of *C. grandis*, the presumed maternal progenitor of these cultivated fruit species. Finally, coffee (*Coffea arabica*) is a tetraploid tree species of Africa. Molecular analyses revealed no differences between the mtDNA and cpDNA of coffee and *C. eugenioides,* a diploid previously thought to be one of the ancestors of coffee (Berthou et al., 1983).

In summary, it is clear that molecular analyses have been successfully applied to many questions surrounding the origins of crop plants. Our knowledge of the ancestry of cotton, the cole crops, sunflower, and pea have all been enhanced substantially. In other cases, such as the origin of the B genome of hexaploid wheat and the origin of pumpkin, molecular analyses have not identified any likely progenitor species. In these cases, it appears that all relevant wild species have yet to be discovered, or perhaps that they exist in germplasm collections but have yet to be analyzed. This situation calls attention to the need for more field work, taxonomic studies, and evaluation of germplasm collections.

Genetic Variation in Crop Plants and their Wild Ancestors

Although there is little hard evidence concerning the beginnings of plant domestication, it seems a reasonable assumption that the first farmers experimented with only a small fraction of the variation present within the progenitor species of today's crops. Further, as the domestication process proceeded, these farmers probably selected only the best phenotypes when setting seed aside for sowing the following year. For these reasons, one expects a significant loss of genetic variation over the six to ten millennia since most crops were first domesticated. Competing forces such as introgression from wild relatives and selection against loss of fitness due to inbreeding have probably counterbalanced the expected loss of genetic variation.

Many authors investigating molecular variation in crops and their progenitors have addressed the above issue. In general, these authors report lower levels of genetic variation in crops than in their progenitors. Studies of cpDNA in barley (Clegg et al., 1984a; Neale et al., 1988) and sunflower (Rieseberg and Seiler, 1990) have found greater variation in the progenitors of these crops than in the crops themselves. Analyses of ribosomal genes in soybean (Doyle, 1988), barley (Allard, 1988), and pearl millet (Gepts and Clegg, 1989) also suggest that greater variation is found in the progenitor than in its associated crop. Examination of low-copy-number nDNA sequences in lentil shows considerably less total genetic diversity (H_t) in lentil (0.18) than in its progenitor (0.27) (Havey and Muehlbauer, 1989).

A few studies reached different conclusions concerning this issue. First, equivalent levels of polymorphism for mtDNA and cpDNA were reported in wild and cultivated barley (Holwerda et al., 1986), in contrast with the two analyses of barley cpDNA cited above. Second, Gepts and Clegg (1989) reported equivalent levels of restriction fragment variation at the alcohol dehydrogenase loci in wild and cultivated pearl millet in contrast with the results of these same authors for rDNA in pearl millet. Third, Keim et al. (1989) showed greater diversity for low-copy-number nDNA sequences in soybean (diversity = 0.37) than in its presumed progenitor, *G. soja* (0.22). They obtained this result despite equivalent sample sizes (ten accessions of *G. max* plant introductions; eight accessions of *G. soja*). This result agrees with the results of Shoemaker et al. (1986) who found three cpDNA types in soybean but only one in *G. soja*.

Table 9.1 summarizes the molecular evidence for loss of genetic variation in the cpDNAs of crops. This table includes only those studies for which the mutational basis (restriction site change versus structural change) of the observed

Table 9.1.   A comparison of cpDNA variation in crops and their progenitors.*

Crop Progenitor	D	TS	P	N	Reference
Hordeum vulgare ssp. *vulgare*	0.08	5	0.20	9	Clegg et al. (1984a)
H. vulgare ssp. *spontaneum*	0.43	5	1.00	11	Clegg et al. (1984a)
Hordeum vulgare ssp. *vulgare*	0.03	3	0.33	51	Neale et al. (1988)
H. vulgare ssp. *spontaneum*	0.45	3	1.00	30	Neale et al. (1988)
Pisum sativum	0.10	6	0.50	13	Palmer et al. (1985)
P. humilis	0.43	6	0.67	4	Palmer et al. (1985)
Sorghum bicolor ssp. *bicolor*	0.06	8	0.13	3	Duvall and Doebley (1990)
S. bicolor ssp. *arundinaceum*	0.35	8	1.00	6	Duvall and Doebley (1990)
Helianthus annuus cultivated	0.00	4	0.00	23	Rieseberg and Seiler (1990)
H. annuus wild/weedy	0.25	4	1.00	11	Rieseberg and Seiler (1990)
Zea mays ssp. *mays*	0.22	3	0.67	80	Doebley (1990)
Z. mays ssp. *mays*	0.38	3	1.00	31	Doebley (1990)
Glycine max	0.20	2	1.00	46	Close et al. (1989)
G. soja	0.47	2	1.00	8	Close et al. (1989)

*Diversity (*D*), total number of polymorphic sites in crop and its progenitor (TS), proportion of the total sites that are polymorphic (*P*), and number of accessions analyzed (*N*) for pairs of crops and their presumed progenitors. Diversity is defined as the average probability that two cpDNAs will differ at a polymorphic site.

differences had been determined. The cpDNA data were used to calculate diversity (*D*) as the average probability per variable site that two accessions will differ (Clegg et al., 1984a). Also presented is the proportion of all polymorphic sites (i.e., sites polymorphic in either crop or progenitor) that were polymorphic in the crop and in the progenitor (*P*). The data for seven studies (six crops) show consistently lower levels of cpDNA variation in crops as compared to their progenitors (Table 9.1). Furthermore, these data agree with a previous review of allozymic variation, which suggested that crops possess less genetic variation than their progenitors (Doebley, 1989). Nevertheless, the cpDNA data should be viewed cautiously as sample sizes are generally small, and sampling bias is a possibility. Finally, although these data on cpDNA are of interest, more detailed studies of the levels of genetic variation in the nuclear genome are needed. Studies of the nuclear genome such as that of Havey and Muehlbauer (1989) have the potential of giving a very precise assessment of the effects of domestication on genetic variation within crops.

Introgression between Crops and their Relatives

Introgression between crops and their relatives is a critical concern of crop evolutionists, because it is a potential source of new variation during crop evolution and it may give rise to weedy intermediates between crops and their relatives. Ample opportunity exists for such introgression, because crops and their wild relatives usually grow sympatrically and often lack barriers to hybridization. Thus, one would expect introgression to be a common phenomenon. However, disruptive selection may act to restrict introgression. Crops are adapted to human needs and their wild relatives to survival in nature. Hybrids between the two satisfy the demands of neither nature nor humankind, and thus, will be strongly selected against. Because crops and their progenitors differ for many traits controlled by at least dozens of genes, a large portion of their genomes will be linked to one or more of these genes. Thus, large portions of their genomes may be hindered from introgressing freely because of such linkage.

Molecular and biochemical techniques offer a powerful means of detecting introgression and distinguishing it from other phenomena, such as joint retention of the ancestral condition, clinal variation, and convergence. Nevertheless, I am unaware of any studies explicitly designed to determine the extent of introgression between crops and their relatives using molecular data. In fact, with the exception of several studies that employed isozyme analysis (see Doebley, 1989, for a review), reports of introgression between crops and their relatives have been based almost exclusively on field observations of morphology. A few cases of introgression between crops and their relatives have been discovered during studies whose primary intention was to examine molecular variation in a crop or to elucidate its phylogeny.

Palmer et al. (1983) found two accessions of rape (*B. napus*) that possessed a foreign cpDNA type as compared with other accessions of this species. Analysis of rDNA of these two accessions indicated that their nuclear genomes were not different from other *B. napus*. This strongly indicates introgression of a foreign cytoplasm into the nuclear background of *B. napus*. Doebley and Sisco (1989) reported a similar situation in maize where a form of cytoplasmic male sterile maize (CMS-S) possesses the cytoplasm of the teosinte *Zea mays* ssp. *mexicana* as the result of introgression. Hosaka and Hanneman (1988b) found foreign (wild-type) cpDNAs in nine accessions of the potato out of more than 140 accessions analyzed. This suggests introgression of the cytoplasms of wild species into the nuclear background of potato, although, in the absence of analysis of the nuclear genome, multiple domestications cannot be conclusively rejected. A situation similar to that with potato was described by Palmer et al. (1985) for pea. Doebley et al. (1987) and Doebley (1990) demonstrated that maize and the Mexican annual teosintes share four distinct cpDNA genotypes. Because evidence from the nuclear genome indicates that maize was domesticated only once (Doebley, 1990), the shared presence of these four cpDNA types probably represents introgression of the cytoplasms of teosinte into maize or the reverse.

All of the above evidence for introgression between crops and their relatives involves the introgression of a foreign cytoplasm into a distinct nuclear background. Introgression of this nature is relatively easy to document because cytoplasmic genomes can provide several linked markers (restriction site mutations) that, unlike markers in the nuclear genome, cannot recombine. A more difficult task is the documentation of introgression for the nuclear genome, although this is possible through the analysis of RFLPs of low-copy-number nuclear sequences or rDNA. In such studies, it will be important to eliminate explanations other than introgression, such as convergence, joint retention of the ancestral condition (allele), and multiple domestications from distinct wild types.

Future Prospects

To understand fully the domestication process, one must explain the genetic basis of the morphologic and physiologic changes that differentiate crops from their progenitors. These changes are many, including retention of the seed/fruit on the plant at maturity, loss of germination inhibitors, increase in seed/fruit size, changes in starch, sugar, and protein content of the seed/fruit, loss of bitter substances in the seed/fruit, increase in vegetative vigor, increase in apical dominance, restoration of fertility to sterile floral parts, synchrony of flowering, more determinate growth, and a smaller number of larger fruits or inflorescences (Harlan, 1975; Simmonds, 1979). For some of these traits, such as retention of seed, genetic analyses indicate that they are under simple genetic control, often involving one or two genes (Harlan et al., 1973). However, there is little evidence concerning genetic control of most traits distinguishing crops from their progenitors. Harlan (1975) stated that the number of genes controlling wild versus domesticated morphologies is often rather small. Evidence for this point of view comes from studies of F_2 populations derived from wild–domesticate crosses. In such populations, plants resembling the parental types can be recovered at relatively high frequencies (Beadle, 1972).

The joining of molecular biology and quantitative genetics offers crop evolutionists a powerful opportunity to investigate genetic control of the morphologic evolution of crop species. Through the use of RFLPs as markers to locate quantitative trait loci, the minimum number of genes controlling a trait and the chromosomal locations of these genes can be discerned (Edwards et al., 1987; Paterson et al., 1988). Moreover, the relative contributions of these genes to the total variance for a trait can be estimated, and thus, major loci distinguished from modifier loci. Thus, it will be possible to construct a detailed picture of the genetic basis of the morphologic differences between crops and their progenitors.

This type of analysis has been applied to the origin of maize from its progenitor teosinte (Doebley et al., 1990). The results indicated that each morphologic trait distinguishing maize from teosinte is under the control of four or more chromosomal regions. However, the chromosomal regions affecting any one morphological trait tend to differ markedly in the magnitude of their effects. Regions with major effects on the traits that distinguish maize and teosinte are restricted to five of the ten chromosomes (Fig. 9.3), whereas regions with small

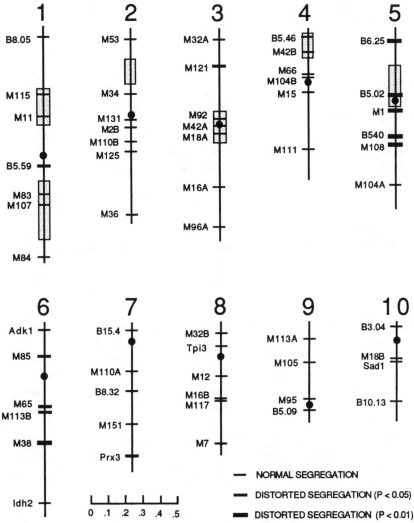

Figure 9.3. Schematic diagram of the 10 chromosomes of maize and teosinte showing the locations of regions with large effects on the morphologic traits that distinguish maize from its progenitor, teosinte (stippled blocks). Cross-bars indicate the locations of molecular marker loci (RFLPs or isozymes) used to detect the presence of morphologic trait loci. Thickness of the cross-bars represents the probability that segregation at a particular marker locus fit Mendelian expectations. The scale indicates the recombination fraction (r) between marker loci (Doebley et al., 1990).

effects are found throughout the entire genome. This explains why maize-like and teosinte-like plants can be recovered in relatively high frequencies in F_2 populations derived from maize–teosinte hybrids (Beadle, 1972), despite the fact that a large number (20 or more) of loci were involved in the domestication of maize.

Harlan (1975, p. 140) comments that one should expect considerable linkage of genes controlling the differences between wild and cultivated morphologies. This opinion is based largely on theoretical arguments that groups of genes controlling a trait will respond more readily to selection when they are linked. Proposed cases of this phenomenon include wheat (see Harlan, 1975) and maize (see Beadle, 1972). In the case of maize, the genes distinguishing maize from teosinte (its progenitor) are said to be assembled in five or six blocks (Beadle, 1972). Galinat (1988) speculates that cryptic structural changes within these blocks suppress crossing over, and thereby, act to preserve the integrity of the blocks. Thus, when a maize–teosinte hybrid is formed, parental types will be recovered among its progeny at higher than expected frequencies because of suppression of crossing over within the blocks. These ideas on linkage relationships can be readily tested through the analysis of linkage for low-copy-number nDNA sequences and quantitative trait loci controlling the differences between crops and their progenitors. In a maize–teosinte F_2 population, restriction to recombination was observed throughout the entire genome, as opposed to being confined to regions containing the morphologic trait loci (Doebley et al., 1990).

With the start of domestication, incipient crop species were placed under dramatically different selective regimes from those which they experienced in nature. Therefore, it is possible that the rate of evolution for some genes may differ under domestication as compared to their rate under natural selection. Furthermore, the domestication process may have subjected incipient cultigens to periods of intense stress, either through inbreeding, hybridization, or other means. During these episodes, transposable elements may have been mobilized, causing both restructuring of the genome and elevated mutation rates (McClintock, 1984). One wonders if the Northern Flint Corns, which differ so dramatically from all other types of corn, are not the result of some such process (Doebley et al., 1986). Molecular analyses of the genomes of crops and their progenitors are certain to provide new insights into these questions.

Conclusions

Molecular systematics has contributed substantially to our understanding of crop evolution in several capacities. First, it has clarified the phylogenies and origins of many crop species, including cotton and potato, whose origins have been surrounded by uncertainty. Second, molecular analyses have provided support for previous allozymic evidence that crops possess lower levels of genetic diversity than their progenitors, although more studies addressing this issue are clearly needed. Third, molecular evidence has revealed several cases in which the cytoplasmic genomes of cultigens have been replaced by those of their relatives through introgressive hybridization. Molecular analysis of introgression of nDNA between crops and their relatives promises to provide new insights into this critical issue in crop evolution. Finally, molecular approaches will play an important role in understanding the morphologic evolution of crops and the evolution of the genomes of crop species under domestication.

References

Allard, R.W. (1988) Genetic changes associated with the evolution of adaptedness in cultivated plants and their wild relatives. *J. Hered.* **79,** 225–238.

Beadle, G.W. (1972) The mystery of maize. *Field Mus. Nat. Hist. Bull.* **43,** 2–11.

Berthou, F., Mathieu, C., and Vedel, F. (1983) Chloroplast and mitochondrial DNA variation as an indicator of phylogenetic relationships in the genus *Coffea* L. *Theor. Appl. Genet.* **65,** 77–84.

Bland, M.M., Matzinger, D.F., and Levings, C.S. (1985) Comparison of the mitochondrial genome of *Nicotiana tabacum* with its progenitor species. *Theor. Appl. Genet.* **69,** 535–541.

Bowman, C.M., Bonnard, G., and Dyer, T.A. (1983) Chloroplast DNA variation between species of *Triticum* and *Aegilops*. Location of the variation on the chloroplast genome and its relevance to the inheritance and classification of the cytoplasm. *Theor. Appl. Genet.* **65,** 247–262.

Brieman, A. (1987) Mitochondrial DNA diversity in the genera of *Triticum* and *Aegilops* revealed by Southern blot hybridization. *Theor. Appl. Genet.* **73,** 563–570.

Clegg, M.T., Brown, A.H.D., and Whitfeld, P.R. (1984a) Chloroplast DNA diversity in wild and cultivated barley: implications for genetic conservation. *Genet. Res.* **43,** 339–343.

Clegg, M.T., Rawson, J.R., and Thomas, K. (1984b) Chloroplast DNA variation in pearl millet and related species. *Genetics* **106,** 449–461.

Close, P.S., Shoemaker, R.C., and Keim, P. (1989) Distribution of restriction site polymorphism within the chloroplast genome of the genus *Glycine,* subgenus *Soya. Theor. Appl. Genet.* **77,** 768–776.

Decker, D.S. (1988) Origin(s), evolution, and systematics of *Cucurbita pepo* (Cucurbitaceae). *Econ. Bot.* **42,** 4–15.

Decker, D.S., and Wilson, H.D. (1987) Allozyme variation in the *Cucurbita pepo* complex: *C. pepo* var. *ovifera* vs. *C. texana. Syst. Bot.* **12,** 263–273.

Dennis, E.S., and Peacock, W.J. (1984) Knob heterochromatin homology in maize and its relatives. *J. Molec. Evol.* **20,** 341–350.

Doebley, J.F. (1989) Isozymic evidence and the evolution of crop plants. In: *Isozymes in Plant Biology* (eds. D. Soltis and P. Soltis), Dioscorides Press, Portland, OR, pp. 165–191.

Doebley, J.F. (1990) Molecular evidence and the evolution of maize. *Econ. Bot.* **445**, 6–27.

Doebley, J.F., and Sisco, P.H. (1989). On the origin of the maize male sterile cytoplasms: it's completely unimportant, that's why it's so interesting. *Maize Genet. Newsl.* **63**: 108–109.

Doebley, J.F., Goodman, M.M., and Stuber, C.W. (1986) Exceptional genetic divergence of Northern Flint Corn. *Amer. J. Bot.* **73**, 64–69.

Doebley, J., Renfroe, W., and Blanton, A. (1987) Restriction site variation in the *Zea* chloroplast genome. *Genetics* **117**, 139–147.

Doebley, J., Stec, A., Wendel, J., and Edwards, M. (1990) Genetic and morphological analysis of a maize-teosinte F_2 population: implications for the origin of maize. *Proc. Natl. Acad. Sci. USA* **87**, 9888–9892.

Doyle, J.J. (1988) 5S ribosomal gene variation in the soybean and its progenitor. *Theor. Appl. Genet.* **75**, 621–624.

Doyle, J.J., and Beachy, R.N. (1985) Ribosomal gene variation in soybean (*Glycine*) and its relatives. *Theor. Appl. Genet.* **70**, 369–376.

Duvall, M.R., and Doebley, J. (1990) Restriction site variation in the chloroplast genome of *Sorghum* (Poaceae). *Syst. Bot.* **15**, 472–480.

Edwards, M.D., Stuber, C.W., and Wendel, J.F. (1987) Molecular marker-facilitated investigations of quantitative-trait loci in maize. I. Numbers, genomic distribution and types of gene action. *Genetics* **116**, 113–125.

Erickson, L.R., Strauss, N.A., and Beversdorf, W.D. (1983) Restriction patterns reveal origins of chloroplast genomes in *Brassica* amphiploids. *Theor. Appl. Genet.* **65**, 201–206.

Feldman, M. (1976) Wheats. *Triticum* spp. (Gramineae-Triticinae). In: *Evolution of Crop Plants* (ed. N.W. Simmonds), Longman, New York, pp. 120–127.

Galinat, W.C. (1988) The origin of corn. In: *Corn and Corn Improvement* (ed. B. Walden), Crop Science Society of America, Madison, WI, pp. 1–31.

Gepts, P., and Clegg, M.T. (1989) Genetic diversity in pearl millet (*Pennisetum glaucum* (L.) R. Br.) at the DNA sequence level. *J. Hered.* **80**, 203–208.

Gerstel, D.U. (1976) Tobacco. In: *Evolution of Crop Plants* (ed. N.W. Simmonds), Longman, New York, pp. 273–277.

Graur, D., Bogher, M., and Brieman, A. (1989) Restriction endonuclease profiles of mitochondrial DNA and the origin of the B genome of bread wheat, *Triticum aestivum*. *Heredity* **62**, 335–342.

Green, R.M., Vardi, A., and Galun, E. (1986) The plastome of *Citrus*. Physical map, variation among *Citrus* cultivars and species and comparison with related genera. *Theor. Appl. Genet.* **72**, 170–177.

Harlan, J.R. (1975) *Crops and Man,* Crop Science Society of America, Madison, WI.

Harlan, J.R., de Wet, J.M.J., and Price, E.G. (1973) Comparative evolution of cereals. *Evolution* **27**, 311–325.

Havey, M.J., and Muehlbauer, F.J. (1989) Variability for restriction lengths and phylogenies in lentil. *Theor. Appl. Genet.* **77,** 839–843.

Hawkes, J.G. (1989) Nomenclatural and taxonomic note on infrageneric taxa of the tuber-bearing Solanums (Solanaceae). *Taxon* **38,** 489–492.

Hawkes, J.G., Lester, R.N., and Skelding, A.D. (1979) *The Biology and Taxonomy of the Solanaceae,* Academic Press, New York.

Hilu, K.W. (1988) Identification of the "A" genome of finger millet using chloroplast DNA. *Genetics* **118,** 163–167.

Holwerda, B.C., Jana, S., and Crosby, W.L. (1986) Chloroplast and mitochondrial DNA variation in *Hordeum vulgare* and *Hordeum spontaneum. Genetics* **114,** 1271–1291.

Hosaka, K. (1986) Who is the mother of the potato?—restriction endonuclease analysis of chloroplast DNA of cultivated potatoes. *Theor. Appl. Genet.* **72,** 606–618.

Hosaka, K., and Hanneman, R.E. (1988a) The origin of the cultivated tetraploid potato based on chloroplast DNA. *Theor. Appl. Genet.* **76,** 172–176.

Hosaka, K., and Hanneman, R.E. (1988b) Origin of chloroplast DNA diversity in the Andean potatoes. *Theor. Appl. Genet.* **76,** 333–340.

Keim, P., Shoemaker, R.C., and Palmer, R.G. (1989) Restriction fragment length polymorphism diversity in soybean. *Theor. Appl. Genet.* **77,** 786–792.

Kung, S.D., Zhu, Y.S., and Shen, G.F. (1982) *Nicotiana* chloroplast genome. III. Chloroplast DNA evolution. *Theor. Appl. Genet.* **61,** 73–79.

McClean, P.E., and Hanson, M.R. (1986) Mitochondrial DNA sequence divergence among *Lycopersicon* and related *Solanum* species. *Genetics* **112,** 649–667.

McClintock, B. (1984) The significance of the response of the genome to challenge. *Science* **226,** 792–801.

Neale, D.B., Saghai-Maroof, M.A., Allard, R.W., Zhang, Q., and Jorgensen, R.A. (1988) Chloroplast DNA diversity in populations of wild and cultivated barley. *Genetics* **120,** 1105–1110.

Ogihara, Y., and Tsunewaki, K. (1988) Diversity and evolution of chloroplast DNA in *Triticum* and *Aegilops* as revealed by restriction fragment analysis. *Theor. Appl. Genet.* **76,** 321–332.

Palmer, J.D. (1987) Chloroplast DNA evolution and biosystematic uses of chloroplast DNA variation. *Amer. Natur.* **130,** S6–S29.

Palmer, J.D., and Herbon, L.A., (1988) Plant mitochondrial DNA evolves rapidly in structure, but slowly in sequence. *J. Mol. Evol.* **28,** 87–97.

Palmer, J.D., Shields, C.R., Cohen, D.B., and Orton, T.J. (1983) Chloroplast DNA evolution and the origin of amphidiploid *Brassica* species. *Theor. Appl. Genet.* **65,** 181–189.

Palmer, J.D., Jorgensen, R.A., and Thompson, W.F. (1985) Chloroplast DNA variation and evolution in *Pisum*: patterns of change and phylogenetic analysis. *Genetics* **109,** 195–213.

Paterson, A.H., Lander, E.S., Hewitt, J.D., Peterson, S., Lincoln, S.E., and Tanksley,

S.D. (1988) Resolution of quantitative traits into Mendelian factors by using a complete linkage map of restriction fragment length polymorphisms. *Nature* **335**, 721–726.

Perl-Treves, R., and Galun, E. (1985) The *Cucumis* plastome: physical map, intrageneric variation and phylogenetic relationships. *Theor. Appl. Genet.* **71**, 417–429.

Rick, C.M., and Fobes, J.F. (1975) Allozyme variation in the cultivated tomato and closely related species. *Bull. Torrey Bot. Club* **102**, 376–384.

Rieseberg, L.H., and Seiler, G. (1990) Molecular evidence and the origin and development of the domesticated sunflower (*Helianthus annuus* L.). *Econ. Bot.* **44S**, 79–91.

Shoemaker, R.C., Hatfield, P.M., Palmer, R.G., and Atherly, A.G. (1986) Chloroplast DNA variation in the genus *Glycine* subgenus *Soja*. *J. Hered.* **77**, 26–30.

Simmonds, N.W. (1976) *Evolution of Crop Plants,* Longman, New York.

Simmonds, N.W. (1979) *Principles of Crop Improvement,* Longman, New York.

Song, K.M., Osborn, T.C., and Williams, P.H. (1988a) *Brassica* taxonomy based on nuclear restriction fragment length polymorphisms (RFLPs) 1. Genome evolution of diploid and amphidiploid species. *Theor. Appl. Genet.* **75**, 784–794.

Song, K.M., Osborn, T.C., and Williams, P.H. (1988b) *Brassica* taxonomy based on nuclear restriction fragment length polymorphisms (RFLPs). 2. Preliminary analysis of subspecies within *B. rapa* and *B. oleracea*. *Theor. Appl. Genet.* **76**, 593–600.

Timothy, D.H., Levings, C.S., Pring, D.R., Conde, M.F., and Kermicle, J.L. (1979) Organelle DNA variation and systematic relationships in the genus *Zea*: teosinte. *Proc. Natl. Acad. Sci. USA* **76**, 4220–4224.

Tsunewaki, K., and Ogihara, Y. (1983) The molecular basis of genetic diversity among cytoplasms of *Triticum* and *Aegilops* species. II. On the origin of the polyploid wheat cytoplasms as suggested by chloroplast DNA restriction fragment patterns. *Genetics* **104**, 155–171.

Wilson, H.D. (1989) Discordant patterns of allozyme and morphological variation in Mexican *Cucurbita*. *Syst. Bot.* **14**, 612–623.

Wendel, J.F. (1989) New World tetraploid cottons contain Old World cytoplasms. *Proc. Natl. Acad. Sci. USA* **86**, 4132–4136.

Zimmer, E.A., Jupe, E.R., and Walbot, V. (1988) Ribosomal gene structure, variation and inheritance in maize and its ancestors, *Genetics* **120**, 1125–1136.

10

Contributions of Molecular Data to Papilionoid Legume Systematics

Jeff J. Doyle, Matt Lavin, and *Anne Bruneau*

The Leguminosae, with some 650 genera and over 18,000 species, is the third largest family of flowering plants (Polhill et al., 1981). Of the three subfamilies of the Leguminosae, the Papilionoideae is the largest (450 genera, 12,000 species; Polhill, 1981), most diverse, and by far the most important economically. In part because of its size and diversity, and because of reliance on a small number of "key" taxonomic characters, phylogenetic relationships at the subtribal and tribal levels in the subfamily have remained unclear. A concerted effort has been made in recent years to remedy this problem, by identifying new systematic characters, reevaluating traditional characters, and applying rigorous phylogenetic analytical methods to questions at all taxonomic levels in the family. This has resulted in several volumes, some based on international symposia, in which information from a diversity of sources and disciplines has been compiled in an effort to improve our understanding of the family (Summerfield and Bunting, 1980; Polhill and Raven, 1981; Stirton, 1987).

Among the most promising sources of phylogenetic characters, as in other families, are DNA sequences. Given the economic importance of many legumes and the role that others, such as the garden pea, occupy as taxa of choice for a wide variety of biological disciplines, it is not surprising that this family should be the focus of both strictly molecular and molecular systematic studies as the techniques of recombinant DNA technology have developed. Molecular systematic studies in the family through the first half of the 1980s were reviewed relatively recently in a paper that focused largely on the potential of molecular characters (Doyle, 1987). Since 1987, considerable progress has been made in realizing that potential. In the previous review, the utility of chloroplast,

We wish to thank the many people who have been associated with the various projects described herein. Principal among these are Tony Brown, Jane Doyle, Dave Garvin, Jeff Palmer, Dan Potter, and Norm Weeden. The authors received support from NSF grants BSR-8805630 and BSR-8516630 to JJD, BSR-8900151 and an NSERC scholarship to AB, and an NSF postdoctoral fellowship to ML.

mitochondrial, and nuclear DNA characters was assessed, and considerable possibilities were seen for all three genomes, although less so for the mitochondrial genome. In the legumes, as is generally true throughout the plant kingdom, the mitochondrial genome remains underexploited as a source of phylogenetic characters, mostly because of its large and variable size, extensive intramolecular recombination leading to a diversity of subgenomic molecules, and the presence of promiscuous DNA sequences (see Chapter 3, this volume). This remains true despite the potential utility of both individual gene sequences and of genomic restriction mapping at higher taxonomic levels, a utility suggested by the recently documented slow rate of plant mtDNA sequence evolution (Wolfe et al., 1987; Palmer and Herbon, 1988). At lower taxonomic levels, mitochondrial DNA restriction fragment patterns continue to find some use as a means of characterizing germplasm in plants such as soybean (e.g., Grabau et al., 1989).

The Nuclear Genome

Much progress has been made in the isolation and sequencing of low-copy nuclear genes of leguminous genera, with continued emphasis on the small number of taxa that are well-characterized "systems" for plant genetics, developmental biology, and molecular studies. At least some nuclear genes have been studied from such taxonomically diverse and economically important genera as pea (*Pisum*), beans (*Phaseolus*), soybean (*Glycine*), broad bean (*Vicia*), alfalfa (*Medicago*), lupine (*Lupinus*), guar (*Cyamopsis*), mung bean (*Vigna*), winged bean (*Psophocarpus*), and jackbean (*Canavalia*).

DNA sequences have not been obtained for the purpose of making phylogenetic comparisons, however, but rather for studying gene structure, function, and regulation. The sequence data for any particular gene seldom are compared across taxa, except to report percent sequence similarity values in selected cases, often in an effort to identify putatively functionally conserved regions. Legume nuclear gene sequences therefore represent an untapped resource for the molecular systematist willing to glean such data from the literature and perform comparisons in a rigorous manner. The taxonomic distribution of the genera listed above is such that, were sequences for any particular gene available for most of them, it would be feasible to test current concepts of papilionoid legume phylogeny, as represented, for example, in the scheme outlined by Polhill (1981; see also Figs. 10.1 and 10.2).Conversely, our improving understanding of legume phylogeny is providing a framework for studying the complexities of nuclear gene evolution.

A major concern in investigations involving nuclear genes is the effect of gene duplication on phylogeny reconstruction. Duplication followed by divergence leads to greater similarity between some members of a multigene family across species than within the multigene family of the same species. In using such multigene families as a source of organismal evolutionary data, many of the possible comparisons are phylogenetically illegitimate, paralogous comparisons

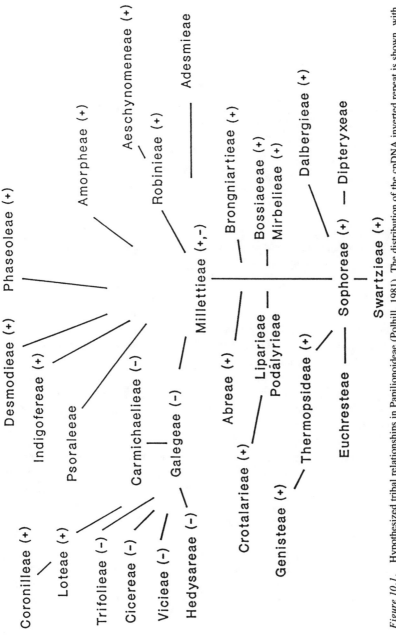

Figure 10.1. Hypothesized tribal relationships in Papilionoideae (Polhill, 1981). The distribution of the cpDNA inverted repeat is shown, with "+" indicating presence of the repeat or "—" indicating its absence (data from Lavin et al., 1990).

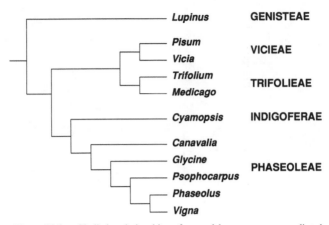

Figure 10.2. Cladistic relationships of several legume genera predicted from current hypotheses of legume phylogeny (as represented in Fig. 10.1). Genera are those for which at least some sequences of nuclearly encoded genes are available (see text for details).

(Fitch, 1970). Thus, random sampling of the multigene families from each of several taxa—either by the systematist or by the evolutionary process itself—can produce a gene phylogeny that does not reflect organismal evolutionary relationships. The problem facing the systematist in such cases is to determine which are the orthologous genes, so that legitimate comparisons can be made. Clearly, evidence of many kinds may be utilized in such a determination, including chromosomal position if a linkage map is known, structural features such as number and position of introns, and functional characteristics such as tissue-specific expression patterns.

One gene family for which regulatory information is available encodes glutamine synthetase (GS), an important enzyme in the nitrogen metabolism of microorganisms and higher plants. Genomic or complementary DNA (cDNA) sequences have been described from *Lupinus* (Grant et al., 1989), *Medicago* (Tischer et al., 1986), *Phaseolus* (Gebhardt et al., 1986; Lightfoot et al., 1988; Bennett et al., 1989), *Pisum* (Tingey et al., 1987, 1988), *Nicotiana* (tobacco, Tingey and Coruzzi, 1987), *Hordeum* (barley, Freeman et al., 1990), *Oryza* (rice, Sakamoto et al., 1989), and *Zea* (maize, Snustad et al., 1988). In each case GS is encoded by a small multigene family, with the products of different genes expressed differentially in various cellular compartments. In leaves, two GS isoforms occur, one cytosolic and one associated with the chloroplast, whereas in roots only a single isoform is found, which is similar to the leaf cytosolic enzyme. In legumes a third major site of GS expression is the nodule, where novel GS isoforms predominate. Both the cytosolic and chloroplast proteins appear to be eukaryotic; the chloroplast protein is far more similar in structure

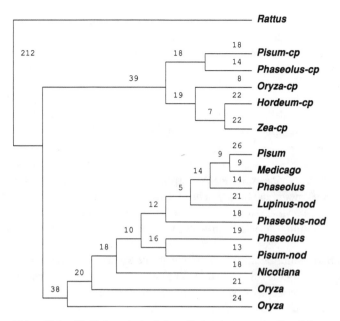

Figure 10.3. Cladistic analysis of glutamine synthetase genes. One of three shortest Wagner parsimony trees (704 steps; consistency index 0.555 without uninformative characters) for a data set in which third codon positions were excluded. Genes most strongly expressed in chloroplasts (cp) or nodules (nod) are indicated; other plant genes are cytosolic. The number of characters is given over each internode. 212 unpolarized steps separate the outgroup (rat) from all plant genes. The other trees at 704 steps differ only in the placements of *Phaseolus*, *Phaseolus*-nod, and *Lupinus*-nod relative to one another in the terminal clade of cytosolic genes.

and sequence to plant cytosolic GS and animal GS than it is to prokaryotic sequences.

Cladistic analyses of these sequences (Doyle, unpublished data) suggest that information on expression of GS is of varying utility in identifying orthologous sequences. In all analyses, the chloroplastic and cytosolic (including nodule) GS genes form separate clades when polarized using the rat (van de Zande et al., 1988) GS sequence (Fig. 10.3). Thus, identifying orthologous groups of genes at this level appears straightforward, because the chloroplastic genes also have diagnostic structural features (transit peptides and 3′ extensions). The topology of the cytosolic genes is less stable in cladistic analyses. In most analyses there is support for a dicot clade, and for a legume clade, but within the legume clade there is little apparent agreement with taxonomic concepts. Moreover, hypothesizing that similarly regulated genes are orthologous does nothing to improve the situation, because in no case do nodule sequences form a clade (Fig.

10.3). Thus, although the smaller chloroplast gene family may be developed as a useful phylogenetic tool with the addition of new sequences, the cytosolic sequences must await better means of determining orthology. It is possible that such evidence will be provided when gene sequences, including introns, become available.

The problem of mixing orthologous and paralogous comparisons would appear to be overcome when concerted evolution operates to "homogenize" the members of a multigene family. Concerted evolution can produce situations in which all genes within a species are more closely related to one another than to any genes from another species. A classic and familiar example of this phenomenon is the nuclear 18S–25S ribosomal RNA cistron (rDNA), which exists as a large but homogeneous multigene family in most plants (see Chapters 4 and 5, this volume, for reviews). To what extent the opposed processes of functional divergence and concerted evolution operate in low-copy, protein-coding gene families is largely unknown but may depend both on the gene and on the taxon.

Concerted evolution appears to occur in the small (< 20-member) multigene families encoding the 7S seed storage proteins, albeit not to the extent observed for rDNA. This conclusion is based on the observation that the several sequenced 7S genes of *Phaseolus* and *Glycine* are all more similar within each genus than to any members from the other genus (Doyle et al., 1986), despite the likelihood that common ancestor of these related plants possessed a 7S multigene family. The most parsimonious explanation for this observation is concerted evolution, despite the suggestion that the similarity within a taxon is due to amplification alone (Gibbs et al., 1989).

The potential of the 7S seed storage proteins as a source of systematic characters has been reviewed previously (Doyle, 1987). Nucleotide sequences from one or more 7S genes or cDNAs have been reported from *Pisum* (Lycett et al., 1983; Casey et al., 1984; Higgins et al., 1988), *Vicia* (Bassuner et al., 1987; Weschke et al., 1987), *Glycine* (Doyle et al., 1986; Harada et al., 1989), *Phaseolus* (Slightom et al., 1983), and *Canavalia* (Yamauchi et al., 1988). Chlan et al. (1987) have also reported the sequence of a *Gossypium* (cotton) gene encoding a protein with considerable nucleotide and amino acid similarity to the legume genes and proteins, providing an outgroup for cladistic comparisons of the legume sequences.

Such cladistic analyses support the hypothesis of concerted evolution among the 7S genes of *Phaseolus* and *Glycine* (Doyle, unpublished data; Fig. 10.4). The situation in *Pisum* and *Vicia* is less clear, with the single *Vicia* sequence consistently joining a particular *Pisum* gene, suggesting a retention of characters supporting orthologous relationships across generic boundaries. The grouping together of *Phaseolus* and *Glycine* sequences in one clade, and *Vicia* and *Pisum* in another, is in good agreement with tribal taxonomic classifications. *Canavalia*, as a member of the Phaseoleae, would be expected to have 7S genes that would form a clade with *Phaseolus* and *Glycine*. In analyses excluding third codon

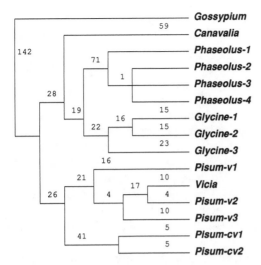

Figure 10.4. Cladistic analysis of 7S seed storage protein genes. One of two shortest Wagner parsimony trees (577 steps; consistency index 0.71 without uninformative characters) for a data set in which third codon positions were excluded. Numbers following generic names refer to different sequences; *Pisum* genes encode either approximately 50-kda vicilin proteins (v) or 70-kda convicilin proteins (cv). Numbers above internodes refer to the number of supporting characters. 142 unpolarized changes separate the legume sequences from the outgroup, cotton. The other tree at 577 steps differs only in the resolution within the *Phaseolus* clade.

positions this expectation is realized (Fig. 10.4), whereas with other subsets of the data different topologies are supported. Furthermore, even for the data excluding third codon positions, trees only slightly longer than the most parsimonious solution place the *Canavalia* gene basal to the remainder of the legumes, a result also found by Gibbs et al. (1989) in their analysis of amino acid data. The lack of strong support for any topology involving *Canavalia*, taken together with the possibility that concerted evolution has acted to varying degrees in even the small number of legume taxa thus far examined, suggests that caution be exercised in using these nuclear genes as a source of organismal phylogenetic characters.

These rather ambiguous results are reminiscent of plant amino acid sequence phylogenies, several of which show the legumes paired unexpectedly with Brassicaceae (Martin et al., 1985). Archie (1989) has suggested that the phylogenetic information contained in these amino acid sequence data does not differ significantly from that contained in a random data set. Bremer (1988) showed that strict consensus analysis of trees only a few steps longer than the shortest trees resulted in complete loss of phylogenetic resolution among angiosperm families for the same data set. Unexpected phylogenetic relationships seen in distance analysis of ribulose bisphosphate carboxylase genes (Meagher et al., 1989) are also reflected in cladistic analyses of the same data and appear to be due to limited resolving power of these sequences (Doyle and Meagher, unpublished data). It is possible that the source of some homoplasy, and hence of analytical ambiguities, may be the process of concerted evolution itself. This is because recombination among genes is formally analogous, at the molecular level, to hybridization among species; in both cases the resulting reticulation is expected to produce homoplasy. Simulations of this concerted evolution that result in taxon-specific gene families also increase homoplasy (Doyle and Sanderson, unpublished data), suggesting that, even though concerted evolution may be beneficial in reducing the need for comprehensive sampling, it may do so at the expense of phylogenetic resolution. In any case, these results are in keeping with previous indications that the mere existence of large bodies of sequence data does not guarantee phylogenetic resolution.

Because of their abundance and scientific interest, most of the low-copy-number nuclear genes thus far isolated and sequenced are members of small to medium multigene families. However, it is thought that the slowly reannealing fraction of the plant nuclear genome is largely composed of true "single-copy" genes (Bernatzky and Tanksley, 1986). If concerted evolution is a source of homoplasy, perhaps such genes will yield more reliable phylogenetic data than will members of multigene families. Unique sequences also provide a further source of characters when rare duplications increase their copy number. The duplication of glucosephosphate isomerase (GPI) in *Clarkia* (Onagraceae) has long been exploited at the protein level and more recently has been studied at the DNA sequence level (Gottlieb, 1988). Glucosephosphate isomerase also appears to have been duplicated in the legumes, presumably in a basal group of the subfamily Papilionoideae, followed by numerous silencings in a diversity of unrelated groups (Weeden et al., 1989). Detailed studies of the DNA sequence may eventually provide characters defining monophyletic groups of legume taxa; these studies are beginning with the isolation and characterization of duplicated GPI genes of *Phaseolus* (Garvin et al., unpublished data).

Restriction fragment length polymorphism (RFLP) studies of single-copy nuclear genes are being used, along with isozymes and morphologic markers, to construct saturated linkage maps in a variety of legume taxa, such as *Pisum, Lens* (lentil), and *Cicer* (chickpea; Weeden and Wolko, 1990), *Medicago* (alfalfa; Brummer et al., 1989; Echt and McCoy, 1989), and *Glycine* (Diers et al., 1989; Keim et al., 1989; Tingey et al., 1989). Eventually, such studies should provide information on the structural evolution of legume nuclear genomes and have the potential to provide informative characters for phylogenetic analysis. Current results indicate, for example, that many features of the nuclear map are conserved. between *Pisum* and *Lens,* and to a lesser degree between these genera and *Cicer* (Weeden, unpublished data). Many of the sequences used in such studies are selected at random from cDNA or genomic libraries, and it is unknown what they encode—if anything, in the case of genomic libraries. Nevertheless, such true single-copy sequences may prove useful in future DNA sequencing studies.

Such random low-copy-number sequences are selected for use in mapping studies precisely because they exhibit restriction pattern polymorphism between the often relatively closely related parents used in segregational analysis. Thus it is to be expected that these same markers will show variation in surveys of additional species. Analysis of this type of variation is complicated because of the short length of the probes used, which makes assumptions about fragment homologies tenuous. Nevertheless, it is to be expected that, as more single-copy gene libraries become available for more taxa, there will be increasing motivation to use such RFLP variation as a source of phylogenetic characters, as has been done in *Brassica* (Song et al., 1988). In the legumes, RFLP markers from *Glycine max,* the annual cultivated soybean (Apuya et al., 1988; Keim et al., 1989), are being used to study the wild perennial taxa of subgenus *Glycine;* preliminary results suggest at least partial agreement with other sources of phylogenetic data (Menancio et al., 1990; Hannah et al., unpublished data).

The more highly repeated genes encoding the nuclear ribosomal RNAs continue to be useful in systematic studies. The potential utility of the various regions of the genes encoding the 18S–5.8S–25S rRNAs in legume systematics has been reviewed recently by Jorgensen and Cluster (1988). The highly conserved sequences of nuclear ribosomal RNAs have not yet been exploited as a source of phylogenetic data in legumes. However, the more variable regions of the rDNA cistron continue to provide characters for study at lower taxonomic levels. These genes, as well as the genes encoding 5S nuclear RNA, have been used in several studies of *Glycine,* including a comparison of the cultivated soybean and its progenitor (Doyle, 1988) and as a source of fixed hybrid markers for tracing the origins of allopolyploids in subgenus *Glycine* (Doyle and Brown, 1989; see also Chapter 8, this volume).

The Chloroplast Genome

Both the mitochondrial and nuclear genomes, although potential sources of phylogenetic characters, generally remain underutilized in the legumes and elsewhere in the angiosperms. In contrast, as is obvious from this volume, chloroplast DNA (cpDNA) has become the tool of choice for the plant molecular systematist. The previous review of legume molecular systematics (Doyle, 1987) suggested that cpDNA had great potential at a wide variety of taxonomic levels in the family. This remains true, and the increasing evidence of intraspecific cpDNA polymorphism in both leguminous and other taxa, although raising a cautionary note (see Chapter 6, this volume), only adds to the versatility of this remarkable source of phylogenetic characters. Two major classes of cpDNA characters are described below: structural rearrangements and restriction site variation.

Structural Mutations as a Source of
Phylogenetic Information

The structurally conservative chloroplast genome has undergone several major modifications among legume taxa. A 50-kb inversion of the large single-copy (LSC) region relative to the common angiosperm chloroplast genome has been identified in all legumes for which detailed mapping has been performed. Although this remains a small number of genera, and no such maps have been published for members of Mimosoideae or Caesalpinioideae, current evidence suggests that this inversion is present in all three subfamilies (Palmer et al., 1988b). Thus, as suggested previously (Doyle, 1987), this character may be of utility in identifying a sister group to the Leguminosae. Studies of such families as Sapindaceae, Connaraceae, and Chrysobalanaceae are in progress (Palmer and Doyle, unpublished data).

A second character that may have phylogenetic utility at a similarly broad level is the loss of the chloroplast gene *rpl22*. This gene is missing from the chloroplast genomes of members of all three subfamilies, but is present in Sapindaceae and Rosaceae (Palmer and Doyle, unpublished data cited in Palmer et al., 1988a, and in Chapter 2, this volume).

Neither of these structural mutations has as yet provided new insights into legume phylogeny. In contrast, two other mutations, the loss of the inverted repeat (IR) and a 78-kb inversion of the LSC region, have proven useful as characters at the tribal and subtribal levels.

Loss of the Chloroplast DNA Inverted Repeat and Tribal Relationships in the Papilionoideae

A large inverted repeat (IR) that includes the 16S and 23S ribosomal RNA genes is a common feature of the chloroplast genomes of most land plants, from liverworts to angiosperms. Chloroplast genomes lacking the inverted repeat were known only from the Leguminosae until recent studies also found only a single copy of this region in conifer cpDNAs (Strauss et al., 1988; Raubeson and Jansen, 1989). In legumes, the lack of an IR was first noted in the chloroplast genomes of *Pisum* (Kolodner and Tewari, 1979; Chu et al., 1981; Palmer and Thompson, 1981), *Vicia* (Koller and Delius, 1980), and *Cicer* (Chu and Tewari, 1982). Both the *Pisum* and *Vicia* genomes are substantially rearranged relative to legumes possessing the inverted repeat (Palmer and Thompson, 1981, 1982; Ko et al., 1987; Michalowski et al., 1987; Palmer et al., 1987b, 1988b). These observations suggested to Palmer and Thompson (1982) the possibility that the presence of the IR could play a role in the observed structural stability of chloroplast genomes. Later studies shed new light on these speculations. At least one species that lacks the IR and is highly rearranged, *Trifolium subterraneum,* also possesses dispersed repeated sequences that could have played a role in rearrangement (Palmer et al., 1987b; Milligan et al., 1989), much as similar recombination among repeats has apparently led to rearrangement of the *Pelargonium* (Geraniaceae) genome, which retains the inverted repeat structure (Palmer et al., 1987a). In other legumes with rearranged chloroplast genomes, however, repeats are either absent, or if present they are not implicated directly in rearrangement (Wolfe, 1988; Palmer et al., 1988a).

Arguing against the loss of the IR as a causative agent in chloroplast genome rearrangement is the observation that the genomes of *Medicago* and *Wisteria* both lack the IR, but retained the gene order common to legumes such as *Lupinus, Glycine, Phaseolus,* and *Vigna* that had IRs (Palmer et al., 1987b). Recent findings that conifers lacking the IR also have rearranged chloroplast genomes

(Strauss et al., 1988) again have raised the issue of the role of the IR in chloroplast genomic stability.

The potential of this structural mutation for phylogenetic studies was recognized almost immediately (Palmer and Thompson, 1982), and was further suggested by later findings that it was in all cases the same copy of the IR that had been lost in legumes (Palmer et al., 1987b). These later studies, although more concerned with evolution of the chloroplast molecule than with legume phylogeny, included taxa that further indicated its utility (Palmer et al., 1987b). The lack of the IR in genera of Vicieae, Cicereae, and Trifolieae led to the prediction that the IR would be absent also in members of the remainder of the so-called "temperate herbaceous group" of tribes, Galegeae, Hedysareae, Carmichaelieae, Loteae, and Coronilleae. Similarly, the observation that *Wisteria* (Millettieae) lacked the IR, whereas genera of the Phaseoleae, a tribe thought to have been derived from millettioid ancestors, possessed the IR, suggested that this character might be useful at the generic level in Millettieae (Doyle, 1987).

These predictions were tested in a survey in which 25 of the 31 tribes of Papilionoideae were included, as well as representatives of Caesalpinioideae and Mimosoideae (Lavin et al., 1990). All members of Vicieae, Cicereae, Trifolieae, Galegeae, Hedysareae, and Carmichaelieae lack the IR (Fig. 10.1). These tribes share numerous characters and trends that suggest their monophyly, including a predominantly herbaceous growth habit, centers of diversity in the temperate regions of the Old World, epulvinate leaves, and stipules adnate to petioles. Thus the distribution of the IR loss character is basically compatible with current concepts of papilionoid phylogeny. However, the sister tribes Loteae and Coronilleae retain the IR, an unexpected result as the genera of these tribes are both temperate in distribution and herbaceous in habit and have long been considered related to the tribes now known to lack the IR. The assumption that Loteae and Coronilleae belong to the temperate herbaceous group has led to the interpretation of numerous characters shared by these two tribes as derived features within the temperate clade. However, some of these "derived" features may be viewed as being shared with predominantly tropical tribes such as Phaseoleae and Millettieae (Lavin et al., 1990). Such characters as pollen ultrastructure, root nodule morphology, inflorescence type, presence of benzofurans, and stamen filament morphology are characters that may, when homologies are fully analyzed, indicate phylogenetic relationships between Loteae/Coronilleae and tropical groups. Thus it may be possible to reconcile traditional characters, if not traditional taxonomy, with the new cpDNA character.

The temperate herbaceous group is thought to be derived from woody tropical elements similar to modern Millettieae, the tribe to which *Wisteria* belongs (Polhill, 1981; Fig. 10.1). It has been suggested that this tribe, which is considered basal to so many other papilionoid tribes, is paraphyletic (Geesink, 1984; van de Zande and Geesink, 1987). The observation that *Wisteria* lacks the IR, but other members of the tribe retain it, is thus perhaps not unexpected and indeed may be paralleled by other characters. *Wisteria* has features typical of Millettieae, such as a woody liana habit, pulvinate leaves, and hypogeal germination. However, it also possesses several characters, such as its temperate distribution and base chromosome number of $x = 8$, that are unusual for Millettieae, but which could represent synapomorphies with Galegeae of the temperate herbaceous group. Furthermore, other characters that place *Wisteria* in Millettieae are not totally anomalous in Galegeae and its allies. Woodiness is known from members of Galegeae and Carmichaelieae, Cicereae and Vicieae have hypogeal germination, and *Glycyrrhiza* of the Galegeae has pulvinate leaves. Thus, the placement of *Wisteria* as a derived member of Millettieae, but basal to the temperate herbaceous group as suggested by the loss of the IR, is not inconsistent with morphologic data.

The distribution of cpDNA transmission types among papilionoid genera shows some correlation with loss of the IR. Among the Leguminosae, as in most angiosperm families, maternal inheritance predominates, but the potential for biparental cpDNA transmission occurs in a number of legume genera (Corriveau and Coleman, 1988), and biparental or even predominantly paternal inheritance of cpDNA has been described in *Medicago* (Lee et al., 1988; Johnson and Palmer, 1989; Schumann and Hancock, 1989). The distribution of inheritance types in the subfamily (Corriveau and Coleman, 1988) is such that all of the genera with the potential for biparental plastid transmission are members of the temperate herbaceous tribes, with one exception—*Wisteria*. *Lotus* and *Coronilla*, in contrast, show strictly maternal transmission. The correlation is not perfect, however, as not all members of the tribes that lack the IR possess the capacity for paternal plastid transmission. Thus in the Vicieae, *Pisum* and *Lathyrus* are listed by Corriveau and Coleman (1988) as having biparental cpdna inheritance, whereas *Vicia* and *Lens* have maternal, and in Trifolieae *Medicago* and *Melilotus* have biparental but *Trifolium* has maternal transmission. Elsewhere in the temperate herbaceous group, *Cicer* (Cicereae) has biparental plastid inheritance, whereas an *Astragalus* species (Galegeae) has maternal transmission (Corriveau and Coleman, 1988).

The distribution of the loss of the chloroplast IR thus has provided an important character for assessing phylogenetic relationships in Papilionoideae. However, most of the affinities thus revealed were not unexpected, as they were largely concordant with traditional taxonomic placements based on morphologic data. Where apparent disagreements were observed, it appears quite likely that careful analysis of morphologic, chemical, and developmental characters will reveal the overall congruence of the various types of data.

The Distribution of a 78-kb Chloroplast DNA Inversion in the Tribe Phaseoleae

Detailed restriction site mapping has revealed the presence of a gene rearrangement interpreted as a 78-kb inversion in the chloroplast genomes of

Phaseolus and *Vigna* (Palmer et al., 1988b). This rearrangement is absent in other legumes, notably *Glycine,* which is a member of the same tribe, Phaseoleae, as the genera containing the inversion. The differential presence of the inversion within Phaseoleae suggested that the character might have taxonomic utility at the generic or subtribal level in this large and economically important tribe. Representatives of 45 of the 84 genera of Phaseoleae were sampled for the presence of the inversion (Bruneau et al., 1990), including members of all eight subtribes recognized by Lackey (1981). Nine genera of Millettieae, the tribe considered basal to Phaseoleae, were also included.

Restriction fragment patterns consistent with the presence of the inversion were observed only in representatives of 10 genera, all of which are members of the subtribe Phaseolinae (Bruneau et al., 1990). Although both its name and its precise composition have varied, the Phaseolinae (*sensu* Lackey, 1981) consistently has been recognized as natural by legume taxonomists for over 100 years (Bentham, 1865; Taubert, 1894; Hutchinson, 1964; Baudet, 1978). Lackey (1981) emphasizes the uniqueness of several characters uniting the subtribe including the presence of a seed epihilum and a number of specialized floral features associated with entomophily. Both the 78-kb inversion and these putative subtribal synapomorphies are present in *Macrotyloma,* which because of its tuberculate pollen and other unique characters has sometimes been excluded from Phaseolinae (Baudet, 1978). In contrast, such genera as *Pachyrhizus* (jicama) and *Psophocarpus* (winged bean) lack the 78-kb inversion and one or more of the morphologic characters by which a core Phaseolinae may be defined. Thus, as with the IR study, the results were generally congruent both with current phylogenetic concepts and with the nonmolecular characters that support these hypotheses.

Restriction Site Variation and
Legume Phylogeny

Unique structural mutations of the chloroplast genome are useful as systematic characters, in part because their rarity suggests phylogenetic stability. However, because of this same scarcity, such characters do not generally afford much resolution, permitting only broad groupings of taxa. In contrast, restriction site variation of cpDNA provides one of the simplest and most effective sources of data for addressing phylogenetic questions in plants. Despite the general versatility and proven effectiveness of this approach, the presence of either too much or too little variation within or among the taxa in question can be limiting. At one extreme, surprisingly low levels of divergence have been observed in several woody families (e.g., Juglandaceae, Fagaceae, Palmae), whereas at the other, levels of variation too great for ready interpretation of fragment patterns may be encountered when genera or even species of herbaceous groups are compared.

Early studies in the legumes suggested that the latter problem might be prevalent. An intergeneric estimate of cpDNA divergence among *Pisum, Vicia,* and *Vigna* using reassociation kinetic measurements (Palmer and Thompson, 1982) revealed a high degree of cpDNA differentiation in the family. Palmer et al. (1983) used restriction site variation to estimate cpDNA divergence levels for three members of the Phaseoleae (*Glycine, Phaseolus,* and *Vigna*) and found greater differentiation among these relatively closely related herbaceous genera than had been expected from studies of other angiosperm families. However, manageably low levels of restriction site variation were found in a cpDNA phylogenetic study of *Pisum* (Palmer et al., 1985). Recent studies in our laboratories have included intra- and intergeneric comparisons of both herbaceous and woody genera of papilionoid legumes. The general pattern that emerges is that the level of cpDNA divergence among legumes, while certainly greater than in some groups of angiosperms, is not so extraordinarily great as to preclude the use of the restriction fragment approach as a phylogenetic tool at a diversity of taxonomic levels in the family.

Chloroplast DNA Variation in Woody Papilionoideae

The slow rate of cpDNA evolution that appears to typify many woody plant families also is found among woody papilionoids. Although a low rate of differentiation may be a liability for studies within a genus, such sequence conservation has meant that the restriction site approach can be applied at higher taxonomic levels in the family. In the tropical woody papilionoid tribes Millettieae and Robinieae, the slowly evolving IR and the small single-copy region have provided restriction map characters useful at the tribal level, and even the more labile large single-copy region has proven useful as a source of characters for the reconstruction of generic phylogenies within tribes (Lavin and Doyle, unpublished data).

Restriction site studies suggest that Robinieae is a monophyletic tribe comprising two distinct lineages that agree well with the traditionally recognized subtribes Sesbaniinae and Robiniinae (Fig. 10.5). These cpDNA data are more consistent with the classification of Rydberg (1924) than with the cladistic treatment of Lavin (1987), which depicted Sesbaniinae as closely related to derived elements of Robiniinae, notably the genus *Sabinea.* This has prompted a subsequent reevaluation of the morphology of Sesbaniinae, in which a different interpretation of polarity in two morphologic characters (leaflet nyctinasty and bracteolate flowers) produces a most parsimonious topology, which agrees with the cpDNA data in showing a sister-group relationship between the two subtribes (Lavin, unpublished data). There thus appears to be little conflict between the two data sets.

The monophyly of Millettieae has been questioned (Geesink, 1981, 1984). As already noted, *Wisteria* lacks the inverted repeat, making direct comparison with

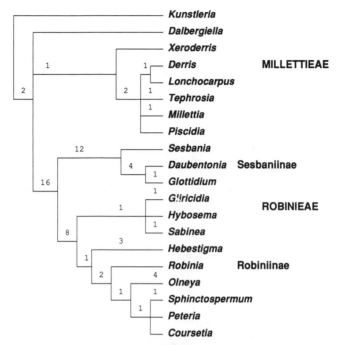

Figure 10.5. Cladistic relationships (Wagner parsimony) among inverted repeat and small single-copy regions of Robinieae chloroplast genomes (Lavin and Doyle, 1991). Restriction site characters were polarized using chloroplast genomes of members of tribe Millettieae (primarily, *Kunstleria*, as shown here). Branch lengths are indicated; consistency index of this tree is 0.85 without autapomorphies. Tribal and subtribal classifications are shown to right of genera.

other Millettieae difficult. The single remaining copy of this region in *Wisteria* does not appear to share restriction sites synapomorphic for the remaining Millettieae, which could be taken as further evidence that the closest affinities of this genus are with other taxa lacking the IR. The remainder of the tribe, although relatively homogeneous with respect to their cpDNA restriction maps in the IR and small single-copy regions, is also likely to be para- or polyphyletic. Our current polarization of these restriction site data suggests that most of these sites shared among the genera studied are plesiomorphic (Lavin and Doyle, 1991), and there is at this stage no support for a monophyletic Millettieae.

The numerous restriction site characters distinguishing Robinieae and Millettieae provide a means of assessing the tribal relationships of certain problematic genera. For example, *Tephrosia* has been considered most closely related to Robinieae by Sousa and de Sousa (1981) but to genera of Millettieae by others (Geesink, 1984; Lavin, 1987); cpDNA data support the latter conclusion (Fig. 10.5). Similarly, the affinities of the vegetatively and florally reduced *Sphinc-*

tospermum have been debated, with either *Tephrosia* of the Millettieae (Lavin, 1987) or *Coursetia* of the Robinieae (Wood, 1949; Polhill and Sousa, 1981) suggested as possible relatives. Both the cpDNA data and a recent cladistic analysis of morphologic characters strongly support a relationship of this genus with Robinieae (Lavin, 1990; Fig. 10.5).

Low levels of cpDNA variation within genera are not unique to these two woody tribes. *Erythrina* (Phaseoleae), a pantropical genus with over 100 species of shrubs and trees, has been the focus of a restriction site survey. Species of *Erythrina* all have very similar cpDNA restriction maps, although some groups are clearly resolved. In contrast, the chloroplast genomes of the genus are markedly divergent from other Phaseoleae thus far studied (Bruneau et al., 1988; Bruneau, unpublished data). This is consistent with the morphologic and chemical uniqueness of the genus.

Chloroplast DNA Restriction Site Phylogenetic Studies in Genera of Herbaceous Papilionoideae

Several studies have characterized cpDNA variation among taxa of *Glycine* subgenus *Glycine* (Phaseoleae), the wild perennial relatives of the soybean (Doyle et al., 1990a–e). Results of these studies have been useful not only as a source of data for phylogenetic hypotheses, but also in investigating the origins of polyploid taxa, as discussed in Chapter 8 (Soltis et al., this volume). An initial survey of cpDNA variation in the subgenus utilized over 150 restriction site characters to construct a cladogram with a consistency index (0.91) typical for interspecific studies of angiosperm genera in which relatively small numbers of taxa are included (Doyle et al., 1990a). Although low levels of divergence in some cases precluded formulation of robust hypotheses of interspecific relationships, chloroplast genomes were grouped into three well-resolved groups (Fig. 10.6). These major cpDNA clades were largely congruent with groupings of species based on crossing relationships and morphologic variation (Putievsky and Broue, 1979; Newell and Hymowitz, 1983; Grant et al., 1984b, 1986; Singh and Hymowitz, 1985a, b; Singh et al., 1987, 1988).

Two apparent exceptions to this pattern highlight the inherent difficulties of comparing cladistically determined cpDNA phylogenetic hypotheses with relatedness estimates based on other characters, particularly hybrid fertility. The first involves *G. tomentella*, a polyploid complex with several morphologically, geographically, isozymically, and genetically divergent elements (Grant et al., 1984a; M. Doyle and Brown, 1985; M. Doyle et al., 1986) that also exhibit cpDNA polymorphism (Doyle et al., 1990a; Doyle, unpublished data). All chloroplast genomes occurring in members of this complex belong to the core clade of the A plastome group despite the fact that fertility barriers exist not only within this polyploid complex but also between its various races and the other species having A-type plastomes (Doyle et al., 1990a; Doyle, unpublished data). How-

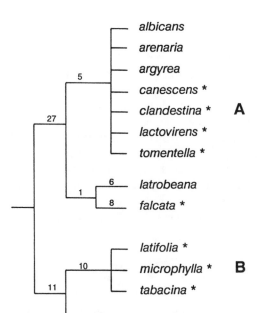

Figure 10.6. Chloroplast DNA phylogeny for diploid members of *Glycine* subgenus *Glycine,* rooted using *G. soja* of subg. *Soja* (Doyle et al., 1990a, b). Consistency index of this tree is 0.91. Plastome groups (A, B, or C) are indicated by capital letters to right of species. Species in which intraspecific variation has been observed are indicated with stars. Branch lengths are shown; autapomorphies are not given for taxa of the A or B clades because of the complexity of intraspecific variation (however, within the A group, no more than three autapomorphies have been described for any accession).

ever, morphologic characters agree with cpDNA in linking *G. tomentella* to the other species of the A group. The second discrepancy between cpDNA and current concepts of the subgenus involves *G. falcata,* which not only is reproductively isolated from all other members of the subgenus but also has a highly divergent morphology, chemistry, and ecology. Despite this uniqueness, however, *G. falcata* possesses a chloroplast genome belonging to the A cpDNA clade.

In each of these cases, it is possible to reconcile the cpDNA phylogeny with nonmolecular data. In both *G. tomentella* and *G. falcata,* inability to form fertile hybrids with species of the same cpDNA clade may be due to novel mutations, and such autapomorphies do not conflict with cpDNA synapomorphies supporting the inclusion of these taxa in a plastome clade. Similarly, the numerous other unique features of *G. falcata* are also autapomorphies; these characters have no bearing on the cladistic placement of the species within the subgenus and do not conflict with the cpDNA results. Our findings, however, do indicate extremely different levels of differentiation between the chloroplast genome and such characters as morphology, soluble proteins, physiology, and chromosome pairing. Species belonging to the A plastome group, in particular, although diverse ecologically and morphologically, have cpDNAs that in many cases are only slightly diverged from one another (Fig. 10.6).

A number of other herbaceous papilionoid taxa are currently under investigation using the restriction site approach. Preliminary results from a survey of New

World species belonging to several genera of Phaseoleae subtribe Phaseolinae suggest good agreement between cpDNA data and recent taxonomic treatments (Delgado et al., unpublished data). Morphologic and cpDNA phylogenies also appear congruent in the genus *Sphenostylis*, where the origin of a cultivated species, the African yam bean (*S. stenocarpa*), is being investigated (Potter and Doyle, 1989). The utility of cpDNA restriction data in *Trifolium* is described elsewhere in this volume (see Chapter 6, this volume), whereas studies on the largest legume genus, *Astragalus*, and its allies are in progress in several different laboratories (Lavin, unpublished data; Liston, personal communication; Sanderson, personal communication). Mapping of the slowly evolving inverted repeat regions is underway for a large sample of herbaceous Phaseoleae, and preliminary data suggest that this approach will help delimit natural groups of chloroplast genomes—and, hopefully, genera—for more detailed study (Doyle and Doyle, unpublished data).

Populational Variation and
Species Phylogenies

Intraspecific variation of cpDNA has been discovered in several papilionoid legume genera. One of the first studies specifically aimed at determining the extent of cpDNA polymorphism in natural plant populations was conducted on a papilionoid legume, *Lupinus texensis* (Banks and Birky, 1985). Both size and restriction site variants were found in this Texas endemic, yet ironically this study has been cited as evidence for a lack of intraspecific cpDNA variation in natural populations, and thus used to suggest that sampling is unimportant. As in *L. texensis*, cpDNA variation also was found in the phylogenetic study of *Pisum* species, despite a relatively small sample (Palmer et al., 1985), and polymorphism was observed in a survey of the cultivated soybean and its wild progenitor (Shoemaker et al., 1986). Populational variation of cpDNA also occurs in *Astragalus* and *Gliricidia* (Lavin, unpublished data), *Erythrina* (Bruneau, unpublished data), *Sphenostylis* (Potter and Doyle, 1989), and *Trifolium* (Milligan, unpublished data, discussed in Soltis et al., Chapter 6, this volume). It is clear from these studies, and from investigations throughout the angiosperms, that intraspecific cpDNA variation is far more prevalent than has been suggested previously (Birky, 1988; see Chapter 6, this volume).

In subgenus *Glycine*, the initial cpDNA phylogenetic study revealed several cases of cpDNA polymorphism, but no instances in which accessions of a species possessed plastomes not placed in the same major cpDNA group as conspecific accessions (Doyle et al., 1990a). Subsequent studies have confirmed these findings. Polymorphism is the rule in subgenus *Glycine* (Fig. 10.6), and to date the only species in which intraspecific cpDNA variation has not been encountered are those for which little or no sampling has been attempted. Recently, three new species were described in the subgenus (Tindale and Craven, 1988), and their chloroplast genomes were surveyed for a small subset of the characters used for the full study of the subgenus. In *G. lactovirens,* all three accessions studied showed restriction site differences, and in *G. hirticaulis,* two variants were found among three accessions; the third species, *G. albicans,* was represented by only a single accession (Doyle et al., 1990b).

One of the groups in which restriction site polymorphism was observed in the initial survey was the B plastome group, which included three species whose cladistic relationships were unresolved (Fig. 10.6). Extensive sampling of these taxa identified 27 plastome types among 75 accessions studied (Doyle et al., 1990c). Phylogenies of these plastome types reconstructed from restriction site data contained substantially more homoplasy than was observed in the study of the entire subgenus. Perhaps more importantly, although all of the accessions were shown to belong to the B clade, as predicted, there were several instances of apparent incongruence between the cpDNA tree and taxonomic, morphologic, or isozyme groupings (Doyle et al., 1990c). In some cases, this is doubtless an artifact of the state of taxonomic (and morphologic) understanding of this group of species, as it is already clear that the current delimitation of only three species has created some very arbitrary taxa even when only morphology or isozyme profiles are concerned. However, numerous instances were encountered in which morphologically, geographically, and isozymically uniform groups of accessions were polymorphic for cpDNA, and, moreover, these polymorphisms clearly overlapped with other well-differentiated taxa. For example, most accessions of *G. microphylla,* a species that is relatively uniform for characters other than cpDNA, belonged to a single major cpDNA clade comprising several plastome types; however, this clade also included at least one accession that was clearly not *G. microphylla.* Additionally, morphologically and isozymically typical *G. microphylla* accessions possessed plastomes belonging to other clades that included accessions of other taxa (Fig. 10.7). In one case, two morphologically "good" *G. microphylla* plants with identical isozyme profiles collected from the same population were found to have plastomes belonging to different cpDNA clades, one of which was more closely related to plastomes found in other B genome species than to the main group of *G. microphylla* accessions.

Possible explanations for the apparent disagreement of cpDNA data with groupings based on nonmolecular characters include hybridization (including introgressive cases), as well as "lineage sorting," in which stochastic partitioning of ancestral polymorphisms into derivative species leads to gene trees that do not reflect species relationships (Neigel and Avise, 1986; Nei, 1987; Pamilo and Nei, 1988). Partly fertile artificial hybrids have been produced from among these taxa, suggesting that gene flow is, or at least has been, possible. Lineage sorting requires a polymorphic ancestor, which appears likely given the degree of variation observed among these taxa, and is promoted by recency of divergence, which is suggested by much circumstantial evidence.

It is clear from these studies that cpDNA results alone do not always resolve

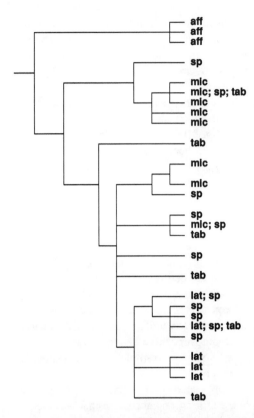

Figure 10.7. Chloroplast DNA polymorphism and phylogeny in the B diploid genome of *Glycine* subg. *Glycine*. Strict consensus tree of most parsimonious cladograms for 74 accessions, modified from Doyle et al. (1990c); consistency index is 0.81. Abbreviations: lat, *G. latifolia;* mic, *G. microphylla;* tab, *G. tabacina;* aff, *G.* sp. aff. *tabacina;* sp, any of several undescribed B genome diploid taxa. Each of the 27 chloroplast types (plastomes) shown possessed synapomorphies for the B chloroplast clade (Fig. 10.6). Some plastome types were found in more than one taxon and numerous discrepancies occur between the topology of this tree and taxonomic designations.

issues of phylogenetic relationships. In the face of gene flow or lineage sorting, cpDNA becomes a single character, moving as a unit from taxon to taxon, and may therefore be taken as a point estimate of relationship. In the case of the B genome species, it may prove virtually impossible to resolve phylogenetic relationships, perhaps for the simple reason that the populations studied are simply at a stage of their evolution in which there are no "phylogenies" at all, in the Hennigian sense, but only a tokogenetic, reticulate history.

Conclusions

In contrast to the mitochondrial and nuclear genomes, cpDNA characters have provided significant insights into phylogenetic patterns at various taxonomic levels in the Papilionoideae. Major structural mutations provide only single cladistic characters, but their apparent rarity makes them very informative, in some cases at taxonomic levels beyond which restriction mapping (but perhaps not DNA sequencing) can be easily applied. As with any taxonomic character, our confidence in the stability of these structural mutations is greatly enhanced by their general agreement with our concepts of relationship based on other characters, including morphology and chemistry. Furthermore, and again like other good characters, the novel information provided by these structural changes points to areas in which careful reevaluation of other data is likely to produce significant results, as in the case of the tribes Coronilleae/Loteae or genera such as *Wisteria* and *Psophocarpus*. The addition of new characters, combined with more intensive and methodologically rigorous examination of older data, is by no means confined, in the legumes, to molecular characters. In this sense the two processes involved, of gathering new data and of explicit analysis, transcend any one class of information, as is readily apparent in Stirton (1987). One might even claim that the true revolution embodied by molecular systematics is only partly due to the molecular data themselves, and is at least equally a consequence of precise definitions of homology and the application of rigorous analytic methods. That molecular data lend themselves to these procedures, however, cannot be denied.

At the lower taxonomic levels, the now-standard methods of restriction endonuclease-based phylogeny reconstruction are making the same impact in the legumes as in other plant groups. Indeed, one of the earlier applications of these procedures was in *Pisum* (Palmer et al., 1985), a work that is in many ways a classic for its combination of molecular and nonmolecular data, its finding of intraspecific cpDNA variation, and its suggestion that inconsistencies between cpDNA trees and traditional characters need not be resolved in favor of the molecular data. In the legumes as elsewhere in the flowering plants, the level at which restriction endonuclease methods can be fruitfully applied may be expected to vary with the group at hand. Correlations between levels of cpDNA variation and life history traits such as woodiness or generation time are becoming apparent, and more studies should provide a clearer picture. With adequate sampling, it is expected that it will also become apparent that polymorphism within species is the rule, rather than the exception, and that significant levels of polymorphism may occur in the absence of major discontinuities between related species.

As for the question of congruence, it seems likely at this point in the history of systematics that the limiting factor in making reasonable comparisons between morphologic and molecular data will be the availability of morphologic cladistic hypotheses. The ability of molecular systematists to generate explicit and well-supported hypotheses of chloroplast genealogy, at least at some taxonomic levels, has been proven beyond any reasonable doubt. Furthermore, restriction fragment mobility and hybridization probe specificity are easier criteria from which to hypothesize homology than are many of the criteria available to the morphologic systematist, which accounts for some of the claims for the superiority of molecular data. For whatever reason, too few comparable studies exist with which to do justice to questions about congruence among different data sets. Far too many studies that claim to have detected incongruence are comparing cladistically analyzed molecular data with traditional taxonomic treatments of the sort whose lack of a sound theoretical basis have long been decried by those demanding objectivity in systematics. Even so, all indications seem to suggest that, as for the major cpDNA mutations, cpDNA variation patterns at the lower taxonomic levels are generally congruent with morphology, crossability, and chemistry, in the legumes and elsewhere. Departures from this rule suggest fertile avenues for further investigation, and the precision of the hypotheses formulated from cpDNA data may point more clearly to apparent discrepancies.

Yet it is also clear that when incongruence is observed, the first thought of the systematist should not always be, "Why are all the morphologic characters wrong?" An equally suspicious eye should be turned to the molecular data, and the taxa at hand considered carefully for factors likely to influence the assumptions of cpDNA analysis. In general, it should be realized by all systematists, molecular and otherwise, that no matter how internally consistent cpDNA phylogenetic trees are, it still requires a leap of faith to infer a species phylogeny from a cpDNA gene tree. And blind faith in any character, as evidenced repeatedly in the history of taxonomy, is not sound systematic philosophy.

References

Apuya, N., Frazier, B., Keim, P., Roth, E.J., and Lark, K.G. (1988) Restriction fragment polymorphisms as genetic markers in soybean. *Theor. Appl. Genet.* **75**, 889–901.

Archie, J.W. (1989) Phylogenies of plant families: a demonstration of phylogenetic randomness in DNA sequence data derived from proteins. *Evolution* **43**, 1796–1800.

Banks, J.A., and Birky, C.W., Jr. (1985) Chloroplast DNA diversity is low in a wild plant, *Lupinus texensis. Proc. Natl. Acad. Sci. USA* **82**, 6950–6954.

Bassuner, R., Hai, N.V., Jung, R., Saalbach, G., and Muntz, K. (1987) The primary structure of the predominating vicilin storage protein subunit from field bean seeds (*Vicia faba* L. var. *minor* cv. Fribo). *Nucleic Acids Res.* **15**, 9609.

Baudet, J.C. (1978) Prodrome d'une classification générique des Papilionaceae-Phaseoleae. *Bull. Jard. Bot. Natl. Belg.* **48**, 183–220.

Bennett, M.J., Lightfoot, D.A., and Cullinore, J.V. (1989) cDNA sequence and differential expression of the gene encoding the glutamine synthetase polypeptide of *Phaseolus vulgaris. Plant Mol. Biol.* **12**, 553–565.

Bentham, G. (1865) Phaseoleae. In: *Genera Plantarum* (eds. G. Bentham and J.D. Hooker), Reeve and Co., London, pp. 451–454.

Bernatzky, R., and Tanksley, S.D. (1986) Majority of random cDNA clones correspond to single loci in the tomato genome. *Mol. Gen. Genet.* **203**, 8–14.

Bremer, K. (1988) The limits of amino acid sequence data in Angiosperm phylogenetic reconstruction. *Evolution* **42**, 795–803.

Brummer, E.C., Kochert, G.D., and Bouton, J.H. (1989) Restriction fragment length polymorphism in diploid and tetraploid alfalfa (*Medicago sativa* L.). *General Index to 1989 Agronomy Abstracts,* 81st Annual Meetings, p. 174.

Bruneau, A., Doyle, J.J., and Neill, D.A. (1988) Chloroplast DNA structure, inheritance, variation, and phylogeny in *Erythrina* (Leguminosae: Phaseoleae). *Amer. J. Bot.* **75,** s162.

Bruneau, A., Doyle, J.J., and Palmer, J.D. (1990) A chloroplast DNA inversion as a subtribal character in the Phaseoleae (Leguminosae). *Syst. Bot.* **15,** 378–386.

Casey, R., Domoney, C., and Stanley, J. (1984) Convicilin mRNA from pea (*Pisum sativum* L.) has sequence homology with other legume 7S storage protein mRNA species. *Biochem. J.* **224,** 661–666.

Chlan, C.A., Pyle, J.B., Legocki, A.B., and Dure, L., III (1987) Developmental biochemistry of cottonseed embryogenesis and germination. XVIII. cDNA and amino acid sequences of members of the storage protein families. *Plant Mol. Biol.* **7,** 475–489.

Chu, N.M., and Tewari, K.K. (1982) Arrangement of the ribosomal RNA genes in chloroplast DNA of Leguminosae. *Mol. Gen. Genet.* **186,** 23–32.

Chu, N.M., Oishi, K.K., and Tewari, K.K. (1981) Physical mapping of the pea chloroplast DNA and localization of the ribosomal RNA genes. *Plasmid* **6,** 279–292.

Corriveau, J.L., and Coleman, A.W. (1988) Rapid screening method to detect potential biparental inheritance of plastid DNA and results for over 200 angiosperm species. *Amer. J. Bot.* **75,** 1443–1458.

Diers, B.W., Keim, P., and Shoemaker, R.C. (1989) Restriction fragment length polymorphism analysis of soybean seed size, oil and protein. *General Index to 1989 Agronomy Abstracts,* 81st Annual Meetings, p. 175.

Doyle, J.J. (1987) Variation at the DNA level: uses and potential in legume systematics. In: *Advances in Legume Systematics,* Part 3 (ed. C.H. Stirton), Royal Botanic Gardens, Kew, pp. 1–30.

Doyle, J.J. (1988) 5S ribosomal gene variation in the soybean and its progenitor. *Theor. Appl. Genet.* **75,** 621–624.

Doyle, J.J., and Brown, A.H.D. (1989) 5S nuclear ribosomal gene variation in the *Glycine tomentella* polyploid complex (Leguminosae). *Syst. Bot.* **14,** 398–407.

Doyle, J.J., Schuler, M.A., Godette, W.D., Zenger, V., and Beachy, R.N. (1986) The glycosylated seed storage proteins of *Glycine max* and *Phaseolus vulgaris*. *J. Biol. Chem.* **261,** 9228–9238.

Doyle, J.J., Doyle, J.L., and Brown, A.H.D. (1990a) A chloroplast DNA phylogeny of the wild perennial relatives of the soybean (*Glycine* subgenus *Glycine*): congruence with morphological and crossing groups. *Evolution* **44,** 371–389.

Doyle, J.J., Doyle, J.L., and Brown, A.H.D. (1990b) Chloroplast DNA phylogenetic affinities of newly discovered species in *Glycine* (Leguminosae: Phaseoleae). *Syst. Bot.* **15,** 466–471.

Doyle, J.J., Doyle, J.L., and Brown, A.H.D. (1990c) Chloroplast DNA polymorphism and phylogeny in the B genome of *Glycine* subgenus *Glycine* (Leguminosae). *Amer. J. Bot.* **77,** 772–782.

Doyle, J.J., Doyle, J.L., Brown, A.H.D., and Grace, J.P. (1990d) Multiple origins of polyploids in the *Glycine tabacina* complex inferred from chloroplast DNA polymorphism. *Proc. Natl. Acad. Sci. USA* **87**, 714–717.

Doyle, J.J., Doyle, J.L., Grace, J.P., and Brown, A.H.D. (1990e) Reproductively isolated polyploid races of *Glycine tabacina* (Leguminosae) had different chloroplast genome donors. *Syst. Bot.* **15**, 173–181.

Doyle, M.J., and Brown, A.H.D. (1985) Numerical analysis of isozyme variation in *Glycine tomentella*. *Biochem. Syst. Ecol.* **13**, 413–419.

Doyle, M.J., Grant, J.E., and Brown, A.H.D. (1986) Reproductive isolation between isozyme groups of *Glycine tomentella* (Leguminosae), and spontaneous doubling in their hybrids. *Austral. J. Bot.* **34**, 523–535.

Echt, C.S., and McCoy, T.J. (1989) Using RFLP markers in genetic analysis of alfalfa species. *General Index to 1989 Agronomy Abstracts*, 81st Annual Meetings, p. 175.

Fitch, W.M. (1970) Distinguishing homologous from analogous proteins. *Syst. Zool.* **19**, 99–113.

Freeman, J., Marquez, A.J., Wallsgrove, R.M., Saarelainen, R., and Forde, B.G. (1990). Molecular analysis of barley mutants deficient in chloroplast glutamine synthetase. *Plant Mol. Biol.* **14**, 297–311.

Gebhardt, C., Oliver, J.E., Forde, B.G., Saarelainen, R., and Miflin, B.J. (1986) Primary structure and differential expression of glutamine synthetase genes in nodules, roots and leaves of *Phaseolus vulgaris*. *EMBO J.* **5**, 1429–1435.

Geesink, R. (1981) Tephrosieae. In: *Advances in Legume Systematics*, Part 1 (eds. R.M. Polhill and P.H. Raven), Royal Botanic Gardens, Kew, pp. 245–260.

Geesink, R. (1984) *Scala Millettiarum*, Leiden Botanical Series 8, E.J. Brill/Leiden University Press, Leiden.

Gibbs, P.E., Strongin, K.B., and McPherson, A. (1989) Evolution of legume seed storage proteins—a domain common to legumins and vicilins is duplicated in vicilins. *Mol. Biol. Evol.* **6**, 614–623.

Gottlieb, L.D. (1988) Towards molecular genetics in *Clarkia*: gene duplications and molecular characterization of PGI genes. *Ann. Missouri Bot. Gard.* **75**, 1169–1179.

Grabau, E.A., Davis, W.H., and Gengenbach, B.G. (1989) Restriction fragment length polymorphism in a subclass of the 'Mandarin' soybean cytoplasm. *Crop Sci.* **29**, 1554–1559.

Grant, J.E., Brown, A.H.D., and Grace, J.P. (1984a) Cytological and isozyme diversity in *Glycine tomentella* Hayata (Leguminosae). *Austral. J. Bot.* **32**: 665–677.

Grant, J.E., Grace, J.P., Brown, A.H.D., and Putievsky, E. (1984b). Interspecific hybridization in *Glycine* Willd. subgenus *Glycine* (Leguminosae). *Austral. J. Bot.* **32**, 655–663.

Grant, J.E., Pullen, R., Brown, A.H.D., Grace, J.P., and Gresshoff, P.M. (1986) Cytogenetic affinity between the new species *Glycine argyrea* and its congeners. *J. Hered.* **77**, 423–426.

Grant, M.R., Carne, A., Hill, D.F., and Farnden, K.J.F. (1989) The isolation and characterization of a cDNA clone encoding *Lupinus angustifolius* root nodule glutamine synthetase. *Plant Mol. Biol.* **13**, 481–490.

Harada, J.J., Barker, S.J., and Goldberg, R.B. (1989) Soybean β-conglycinin genes are clustered in several DNA regions and are regulated by transcriptional and posttranscriptional processes. *Plant Cell* **1**, 415–425.

Higgins, T.J.V., Newbigin, E.J., Spencer, D., Llewellyn, D.J., and Craig, S. (1988) The sequence of a pea vicilin gene and its expression in transgenic tobacco plants. *Plant Mol. Biol.* **11**, 683–695.

Hutchinson, J. (1964) *The Genera of Flowering Plants,* Vol. I, Oxford University Press, London.

Johnson, L.B., and Palmer, J.D. (1989) Heteroplasmy in chloroplast DNA of *Medicago*. *Plant Mol. Biol.* **12**, 3–11.

Jorgensen, R.A., and Cluster, P.D. (1988) Modes and tempos in the evolution of nuclear ribosomal DNA: new characters for evolutionary studies and new markers for genetic and population studies. *Ann. Missouri Bot. Gard.* **75**, 1238–1247.

Keim, P., Shoemaker, R.C., and Palmer, R.G. (1989) Restriction fragment length polymorphism diversity in soybean. *Theor. Appl. Genet.* **77**, 786–792.

Ko, K., Orfanides, A.G., and Straus, N.A. (1987) A model for the evolution of the *Vicia faba* chloroplast genome. *Theor. Appl. Genet.* **74**, 125–139.

Koller, B., and Delius, H. (1980). *Vicia faba* chloroplast DNA has only one set of ribosomal RNA genes as shown by partial denaturation mapping and R-loop analysis. *Mol. Gen. Genet.* **178**, 261–269.

Kolodner, R., and Tewari, K.K. (1979) Inverted repeats in chloroplast DNA from higher plants. *Proc. Natl. Acad. Sci. USA* **76**, 41–45.

Lackey, J.A. (1981) Phaseoleae. In: *Advances in Legume Systematics,* Part 1 (eds. R.M. Polhill and P.H. Raven), Royal Botanic Gardens, Kew, pp. 301–327.

Lavin, M. (1987) A cladistic analysis of the tribe Robinieae (Papilionoideae; Leguminosae). In: *Advances in Legume Systematics,* Part 3 (ed. C.H. Stirton), Royal Botanic Gardens, Kew, pp. 31–64.

Lavin, M. (1990) The genus *Sphinctospermum:* taxonomy and tribal relationships as inferred from a cladistic analysis of traditional data. *Syst. Bot.* **15**, 544–559.

Lavin, M., and Doyle, J.J. (1991) Tribal relationships of *Sphinctospermum* (Leguminosae): integration of traditional and chloroplast DNA data. *Syst. Bot.* **16**, 162–172.

Lavin, M., Doyle, J.J., and Palmer, J.D. (1990) Evolutionary significance of the loss of the chloroplast DNA inverted repeat in the Leguminosae subfamily Papilionoideae. *Evolution* **44**, 390–402.

Lee, D.J., Blake, T.K., and Smith, S.E. (1988) Biparental inheritance of chloroplast DNA and the existence of heteroplasmic cells in alfalfa. *Theor. Appl. Genet.* **76**, 545–549.

Lightfoot, D.A., Green, N.K., and Cullinore, J.V. (1988) The chloroplast-located glutamine synthetase of *Phaseolus vulgaris* L.: nucleotide sequence, expression in different organs and uptake into isolated chloroplasts. *Plant Mol. Biol.* **11**, 191–202.

Lycett, G.W., Delauney, A.J., Gatehouse, J.A., Gilroy, J., Croy, R.R.D., and Boulter, D. (1983). The vicilin gene family of pea (*Pisum sativum* L.): a complete cDNA coding sequence for preprovicilin. *Nucleic Acids Res.* **8**, 2367–2380.

Martin, P.G., Boulter, D., and Penny, D. (1985) Angiosperm phylogeny studied using sequences of five macromolecules. *Taxon* **34**, 393–400.

Meagher, R.B., Berry-Lowe, S., and Rice, K. (1989) Molecular evolution of the small subunit of ribulose bisphosphate carboxylase: nucleotide substitution and gene conversion. *Genetics* **123**, 845–863.

Menancio, D.I., Hepburn, A.G., and Hymowitz, T. (1990) Restriction fragment length polymorphism (RFLP) of wild perennial relatives of soybean. *Theor. Appl. Genet.* **79**, 235–240.

Michalowski, C., Breunig, K., and Bohnert, H.J. (1987) Points of rearrangement between plastid chromosomes: location of protein coding regions on broad bean chloroplast DNA. *Curr. Genet.* **11**, 265–274.

Milligan, B.G., Hampton, J.N., and Palmer, J.D. (1989) Dispersed repeats and structural reorganization in subclover chloroplast DNA. *Mol. Biol. Evol.* **6**, 355–368.

Nei, M. (1987) *Molecular Evolutionary Genetics*. Columbia University Press, New York.

Neigel, J.E., and Avise, J.C. (1986) Phylogenetic relationships of mitochondrial DNA under various demographic models of speciation. In: *Evolutionary Processes and Theory* (eds. S. Karlin and E. Nevo), Academic Press, New York, pp. 515–534.

Newell, C.A., and Hymowitz, T. (1983) Hybridization in the genus *Glycine* subgenus *Glycine*. *Amer. J. Bot.* **70**, 334–348.

Palmer, J.D., and Herbon, L.A. (1988) Plant mitochondrial DNA evolves rapidly in structure but slowly in sequence. *J. Mol. Evol.* **28**, 87–97.

Palmer, J.D., and Thompson, W.F. (1981) Rearrangements in the chloroplast genomes of mung bean and pea. *Proc. Natl. Acad. Sci. USA* **78**, 5533–5537.

Palmer, J.D., and Thompson, W.F. (1982) Chloroplast DNA rearrangements are more frequent when a large inverted repeat sequence is lost. *Cell* **29**, 537–550.

Palmer, J.D., Singh, G.P., and Pillay, D.T.N. (1983) Structure and sequence evolution of three legume chloroplast DNAs. *Mol. Gen. Genet.* **190**, 13–19.

Palmer, J.D., Jorgensen, R.A., and Thompson, W.F. (1985) Chloroplast dna variation and evolution in *Pisum*: patterns of change and phylogenetic analysis. *Genetics* **109**, 195–213.

Palmer, J.D., Nugent, J.M., and Herbon, L.A. (1987a) Unusual structure of geranium chloroplast DNA: a triple-sized inverted repeat, extensive gene duplications, multiple inversions, and two repeat families. *Proc. Natl. Acad. Sci. USA* **84**, 769–773.

Palmer, J.D., Osorio, B., Aldrich, J., and Thompson, W.F. (1987b) Chloroplast DNA evolution among legumes: loss of a large inverted repeat occurred prior to other sequence rearrangements. *Curr. Gen.* **11**, 275–286.

Palmer, J.D., Jansen, R.K., Michaels, H.J., Chase, M.W., and Manhart, J.R. (1988a) Chloroplast DNA variation and plant phylogeny. *Ann. Missouri Bot. Gard.* **75**, 1180–1206.

Palmer, J.D., Osorio, B., and Thompson, W.F. (1988b). Evolutionary significance of inversions in legume chloroplast DNAs. *Curr. Genet.* **14,** 65–74.

Pamilo, P., and Nei, M. (1988) Relationships between gene trees and species trees. *Mol. Biol. Evol.* **5,** 568–583.

Polhill, R.M. (1981) Papilionoideae. In: *Advances in Legume Systematics,* Part 1, (eds. R.M. Polhill and P.H. Raven), Royal Botanic Gardens, Kew, pp. 191–208.

Polhill, R.M., and Raven, P.H. (1981) *Advances in Legume Systematics,* Part 1, Royal Botanic Gardens, Kew.

Polhill, R.M., and Sousa, M. (1981) Robinieae. In: *Advances in Legume Systematics,* Part 1 (eds. R.M. Polhill and P.H. Raven), Royal Botanic Gardens, Kew, pp. 283–288.

Polhill, R.M., Raven, P.H., and Stirton, C.H. (1981) Evolution and systematics of the Leguminosae. In: *Advances in Legume Systematics,* Part 1 (eds. R.M. Polhill and P.H. Raven), Royal Botanic Gardens, Kew, pp. 1–26.

Potter, D., and Doyle, J.J. (1989) Chloroplast DNA systematics of *Sphenostylis* (Leguminosae: Phaseoleae). *Amer. J. Bot.* **76,** s266.

Putievsky, E., and Broue, P. (1979) Cytogenetics of hybrids among perennial species of *Glycine* subgenus *Glycine* Willd. (Leguminosae: Papilionoideae). *Austral. J. Bot.* **27,** 713–723.

Raubeson, L.A., and Jansen, R.K. (1989) Molecular evidence on conifer phylogeny: structural variation in the chloroplast genome. *Amer. J. Bot.* **76,** s222.

Rydberg, P.A. (1924) Genera of North American Fabaceae II. *Amer. J. Bot.* **11,** 470–482.

Sakamoto, A., Ogawa, M., Masumura, T., Shibata, D., Takeba,G., Tanaka, K., and Fujii, S. (1989) Three cDNA sequences coding for glutamine synthetase polypeptides in *Oryza sativa* L. *Plant Mol. Biol.* **13,** 611–614.

Schumann, C.M., and Hancock, J.F. (1989) Paternal inheritance of plastids in *Medicago sativa. Theor. Appl. Genet.* **78,** 863–866.

Shoemaker, R.C., Hatfield, P.M., Palmer, R.G., and Atherly, A.G. (1986) Chloroplast DNA variation in the genus *Glycine* subgenus *Soja. J. Hered.* **77,** 26–30.

Singh, R.J., and Hymowitz, T. (1985a) Intra- and interspecific hybridization in the genus *Glycine,* subgenus *Glycine* Willd.: chromosome pairing and genome relationships. *Z. Pflanzenzucht.* **95,** 289–310.

Singh, R.J., and Hymowitz, T. (1985b) The genomic relationships among six wild perennial species of the genus *Glycine* subgenus *Glycine* Willd. *Theor. Appl. Genet.* **71,** 221–230.

Singh, R.J., Kollipara, K.P., and Hymowitz, T. (1987) Polyploid complexes of *Glycine tabacina* (Labill.) Benth. and *G. tomentella* Hayata revealed by cytogenetic analysis. *Genome* **29,** 490–497.

Singh, R.J., Kollipara, K.P., and Hymowitz, T. (1988) Further data on the genomic relationships among wild perennial species ($2n = 40$) of the genus *Glycine* Willd. *Genome* **30,** 166–176.

Slightom, J.L., Sun, S.M., and Hall, T.C. (1983) Complete nucleotide sequence of a French bean storage protein gene: phaseolin. *Proc. Natl. Acad. Sci. USA* **80,** 1897–1901.

Snustad, D.P., Hunsperger, J.P., Chereskin, B.M., and Messing, J. (1988) Maize glutamine synthetase cDNAs: isolation by direct genetic selection in *Escherichia coli. Genetics* **120,** 1111–1124.

Song, K.M., Osborn, T.C., and Williams, P.H. (1988) *Brassica* taxonomy based on nuclear restriction fragment length polymorphisms. 1. Genome evolution of diploid and amphidiploid species. *Theor. Appl. Genet.* **76,** 784–794.

Sousa, M., and de Sousa, M.P. (1981) New World Lonchocarpinae. In: *Advances in Legume Systematics* Part 1 (eds. R.M. Polhill and P.H. Raven), Royal Botanic Gardens, Kew, pp. 261–281.

Stirton, C.H. (1987) *Advances in Legume Systematics,* Part 3. Royal Botanic Gardens, Kew.

Strauss, S.H., Palmer, J.D., Howe, G.T., and Doerksen, A.H. (1988) Chloroplast genomes of two conifers lack a large inverted repeat and are extensively rearranged. *Proc. Natl. Acad. Sci. USA* **85,** 3898–3902.

Summerfield, R.J., and Bunting, A.H. (1980) *Advances in Legume Science.* Royal Botanic Gardens, Kew.

Taubert, P. (1894) Leguminosae. In: *Die Natürlichen Pflanzenfamilien* III(3) (eds. A. Engler and K. Prantl), Engelmann, Leipzig, pp. 70–388.

Tindale, M.D., and Craven, L.A. (1988) Three new species of *Glycine* (Fabaceae: Phaseoleae) from north-western Australia, with notes on amphicarpy in the genus. *Austr. Syst. Bot.* **1,** 399–410.

Tingey, S., and Coruzzi, G.M. (1987) Glutamine synthetase of *Nicotiana plumbaginifolia. Plant Physiol.* **84,** 366–373.

Tingey, S.V., Walker, E.L., and Coruzzi, G.M. (1987) Glutamine synthetase genes of pea encode distinct polypeptides which are differentially expressed in leaves, roots and nodules. *EMBO J.* **6,** 1–9.

Tingey, S.V., Tsai, F.-Y., Edwards, J.W., Walker, E.L., and Coruzzi, G.M. (1988) Chloroplast and cytosolic glutamine synthetase are encoded by homologous nuclear genes which are differentially expressed *in vivo. J. Biol. Chem.* **263,** 9651–9657.

Tingey, S., Sebastian, S., and Rafalski, J.A. (1989) A RFLP map of the soybean genome. *General Index to 1989 Agronomy Abstracts,* 81st Annual Meetings, p. 180.

Tischer, E., DasSarma, S., and Goodman, H.M. (1986) Nucleotide sequence of an alfalfa glutamine synthetase gene. *Mol. Gen. Genet.* **203,** 221–229.

van de Zande, L., Labruyere, W.T., Smaling, M.M., Moorman, A.F.M., Wilson, R.H., Charles, R., and Lamers, W.H. (1988) Nucleotide sequence of rat glutamine synthetase mRNA. *Nucleic Acids Res.* **16,** 7726.

Weeden, N.F., and Wolko, B. (1990) Linkage map for the garden pea. In: *Genetic Maps* (ed. S. O'Brien), Cold Spring Harbor, New York, pp. 6.106–6.112.

Weeden, N.F., Doyle, J.J., and Lavin, M. (1989) Distribution and evolution of a glucose-phosphate isomerase duplication in the Leguminosae. *Evolution* **43,** 1637–1651.

Weschke, W., Baumlein, H., and Wobus, U. (1987) Nucleotide sequence of a field bean (*Vicia faba* L. var. *minor*) vicilin gene. *Nucleic Acids Res.* **15,** 10065.

Wolfe, K.H. (1988) The site of deletion of the inverted repeat in pea chloroplast DNA contains duplicated gene fragments. *Curr. Genet.* **13,** 97–99.

Wolfe, K.H., Li, W.-H., and Sharp, P.M. (1987) Rates of nucleotide substitution vary greatly among plant mitochondrial, chloroplast and nuclear DNAs. *Proc. Natl. Acad. Sci. USA* **84,** 9045–9058.

Wood, C.E. (1949) The American barbistyled species of *Tephrosia* (Leguminosae). *Contr. Gray Herb.* **170,** 193–384.

Yamauchi, D., Nakamura, K., Asahi, T., and Minamikawa, T. (1988) cDNAs for canavalin and concanavalin A from *Canavalia gladiata* seeds. *Eur. J. Biochem.* **170,** 515–520.

Zandee, M., and Geesink, R. (1987) Phylogenetics and legumes: a desire for the impossible? In: *Advances in Legume Systematics,* Part 3 (ed. C.H. Stirton), Royal Botanic Gardens, Kew, pp. 131–167.

11

Chloroplast DNA Variation in the Asteraceae: Phylogenetic and Evolutionary Implications

Robert K. Jansen, Helen J. Michaels,
Robert S. Wallace, Ki-Joong Kim, Sterling C. Keeley,
Linda E. Watson, and Jeffrey D. Palmer

The sunflower family (Asteraceae) is one of the largest families of flowering plants, consisting of approximately 1,100 genera and 25,000 species (Jeffrey, 1978). The family has been variously divided into two or three subfamilies and 10 to 17 tribes (reviewed in Jansen et al., 1991a; Bremer, 1987). In addition to the disagreement concerning intrafamilial relationships, the phylogenetic affinities of the Asteraceae to other angiosperm families have been the topic of considerable debate (Cronquist, 1955, 1977, 1981; Carlquist, 1976; Wagenitz, 1976; Stebbins, 1977; Turner, 1977a; Jeffrey, 1977; Bremer, 1987). Until recently, systematic studies in the family relied on more traditional taxonomic characters and less rigorous phylogenetic methods. During the past five years we have been examining chloroplast DNA (cpDNA) variation in the Asteraceae to evaluate phylogenetic relationships at a wide range of taxonomic levels using several cladistic methods. We have accumulated restriction site data for approximately 250 genera in the family from all currently recognized tribes. In addition, we have sequenced the gene encoding the large subunit of ribulose-1,5-bisphosphate carboxylase (*rbc*L) from representative species in the Asteraceae and putatively related families. Our studies, which have produced the largest molecular data set for any plant family, have allowed us to perform phylogenetic comparisons from the intraspecific to the interfamilial levels.

In this chapter, we summarize the results of phylogenetic comparisons of both restriction site and gene sequence data. We show that cpDNA is useful for resolving systematic relationships in the Asteraceae primarily because of its generally conservative evolution. However, it is also clear that rates of cpDNA

We thank K. Bremer and L. King for providing unpublished data and J. Johansson, J. Whitton, and Y.-D. Kim for critically reading the manuscript. The chloroplast DNA research has been supported by grants from the NSF to JDP and RKJ (BSR-8415934), RKJ (BSR 87-08246), HJM (BSR-8700195), JDP (BSR-8717600), SCK (RII-8800730), LEW (BSR 87-00977), and two University of Connecticut Research Foundation Grants to RKJ.

evolution are variable in the family and that certain regions of the genome are hotspots for both restriction site and length mutations. The availability of extensive data on the spatial distribution of mutations provides valuable information for planning future systematic and evolutionary investigations.

Phylogenetic Relationships at Higher Taxonomic Levels

Gene Sequencing: Interfamilial Relationships

The interfamilial relationships of the Asteraceae have been debated for many years. A total of 13 different families have been proposed as closest relatives by various authors (reviewed in Bremer, 1987). Based largely on morphologic evidence suggesting the Heliantheae as the basal tribe, Conquist (1955, 1977, 1981) has advocated that the Asteraceae are most closely related to the Rubiaceae. Others have proposed alliances to the Campanulaceae (Stebbins, 1977; Jeffrey, 1977), Lobeliaceae (Bremer, 1987), and the Calyceraceae (Jeffrey, 1977; Turner, 1977a). To elucidate interfamilial relationships, representative taxa from the Asteraceae and 10 putatively related families were examined using comparative sequencing of the *rbc*L gene (Michaels et al., unpublished data).

Our preliminary analyses of nearly complete *rbc*L data from 17 taxa included published sequences from *Spinacia* (Chenopodiaceae; Zurawski et al., 1981), *Nicotiana* (Solanaceae; Shinozaki and Sugiura, 1982), and *Flaveria* (Asteraceae: Heliantheae; Hudson et al., 1990). Alignments of these sequences identified 398 variable nucleotide positions, 216 of which were phylogenetically informative. Phylogenetic analyses using *Spinacia* as an outgroup generated two equally parsimonious trees with a length of 844 steps (including autapomorphies). The two trees showed only minor differences in the placement of *Viburnum* (Caprifoliaceae). The majority-rule consensus tree from a bootstrap analysis has the same topology as one of the shortest trees (Fig. 11.1). The *rbc*L data clearly indicate that the families Calyceraceae and Goodeniaceae are the closest relatives of the Asteraceae, whereas the Campanulaceae and Lobeliaceae form the next nearest sister group. The data imply a much more distant relationship between the Asteraceae and Rubiaceae. The position of the Calyceraceae and Goodeniaceae as the sister group to the Asteraceae is quite stable, as this topology occurs in both of the shortest trees and all trees up to three steps longer and it has strong statistical support from the bootstrap analysis (Fig. 11.1, 92% confidence interval).

Gene Order and Restriction Site Comparisons: Subfamilial and Tribal Relationships

The intrafamilial systematics of the Asteraceae has been extensively studied during the past 150 years, including a symposium devoted to the family (Heywood

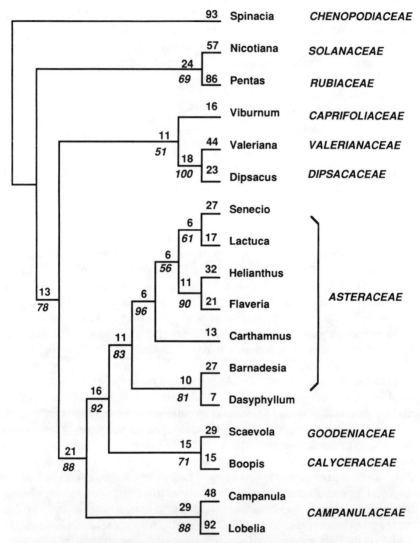

Figure 11.1. Chloroplast DNA tree of 17 species from the Asteraceae and putatively related families in the Asteridae based on *rbc*L sequence data. From Michaels et al. (unpublished data). The tree was produced by Wagner parsimony and has a total length of 844 steps (including 187 autapomorphies) and a consistency index of 0.51. The numbers above and below the nodes indicate the number of nucleotide substitutions and the number of times that a monophyletic group occurred in 100 bootstrap replicates, respectively.

Table 11.1.   Genera examined for chloroplast DNA variation in the Asteraceae.

Subfamily Tribe	No. Examined/ Total in Tribe[a]	Reference[c]	Subfamily Tribe	No. Examined/ Total in Tribe	Reference
Lactucoideae			Asteroideae		
Mutisieae[b]	13/89*	1	Astereae	70/130*	3,6,7,8
Tarchonantheae	2/2	2	Anthemideae	5/108	3
Cardueae	5/79	3	Inuleae	4/193	3
Vernonieae	11/81*	3,4	Senecioneae	5/114	3
Liabeae	3/15	3,4	Calenduleae	3/8	3
Lactuceae	45/70*	3,5	Eupatorieae	16/180*	3,9,10
Arctoteae	3/16	3	Heliantheae	72/240*	3,9,10,11
			Coreopsideae	5/24	3,10
			Tageteae	10/16*	3,10

[a]The first number indicates the number of genera surveyed (at least one species per genus) and the second gives the approximate number of genera in each tribe. Asterisks indicate those tribes that have been or are currently being examined in detail (see section on Generic and Interspecific Relationships).

[b]The Mutisieae includes the subtribe Barnadesiinae, which is currently being elevated to subfamilial status (Bremer and Jansen, unpublished data).

[c]1—Jansen and Palmer (1988); 2—Keeley and Jansen (1991); 3—Jansen et al. (1990, 1991a); 4—Keeley and Jansen (unpublished data); 5—Jansen et al. (1990c), Wallace and Jansen (unpublished data); 6—Suh and Simpson (1990); 7—Morgan and Simpson (unpublished data); 8—Lane and Jansen (unpublished data); 9—Watson et al. (1991b); 10—Kim et al. (unpublished data); 11—Baldwin et al. (1990).

et al., 1977). In spite of these intensive efforts using a broad diversity of systematic approaches and characters, considerable disagreement still exists concerning subfamilial and tribal relationships and limits. The degree of controversy is obvious from the fact that eight markedly different schemes of phylogenetic relationships among the subfamilies and tribes of Asteraceae have been proposed during the past 30 years (Cronquist, 1955; Poljakov, 1967; Carlquist, 1976; Wagenitz, 1976; Jeffrey, 1978; Robinson, 1981, 1983; Thorne, 1983; Bremer, 1987). Most of these workers agree that two subfamilies (Asteroideae and Lactucoideae) should be recognized; however, there is no consensus concerning their circumscription. The only cladistic analysis of the family based on morphologic characters (Bremer, 1987) supported the monophyly of the Asteroideae (eight tribes) but indicated that the Lactucoideae (six tribes) are paraphyletic. In addition to the lack of agreement concerning subfamilial circumscription, relationships among the 12 to 17 recognized tribes are also controversial.

   Our initial studies of cpDNA structure in the Asteraceae (Jansen and Palmer, 1987a,b) revealed two chloroplast genome arrangements in the family. Chloroplast DNAs from species in the subtribe Barnadesiinae (Mutisieae) are colinear with the consensus gene order for land plants, whereas all other Asteraceae share a 22-kb inversion. This single structural mutation, which has now been found in 250 genera from all tribes of Asteraceae (Table 11.1), defines an ancient evolutionary split in the family. The Barnadesiinae, which lack this inversion, are clearly placed as the sister group to the rest of the Asteraceae. Subsequent studies using restriction site comparisons (Jansen and Palmer, 1988), morphology (Bremer, 1987), and DNA sequencing of the *rbc*L gene (Fig. 11.1, see above) have confirmed this ancient dichotomy in the family. The molecular data provide overwhelming support for the recognition of the Barnadesiinae as a new subfamily. The identification of the Barnadesiinae as the sister group to the rest of the Asteraceae also has potential implications concerning the place of origin of the family. The center of diversity for this subtribe is the northern Andes (Cabrera, 1977), which supports previous suggestions that the family originated in montane South America (Bentham, 1873; Small, 1919; Raven and Axelrod, 1974; Turner, 1977b). Another important implication is that we can now use the Barnadesiinae as an outgroup to polarize cpDNA mutations for phylogenetic analyses within the Asteraceae.

In two previous studies (Jansen et al., 1990, 1991a), we performed phylogenetic analyses of 328 restriction site mutations from 57 species representing all of the currently recognized tribes in the Asteraceae (sensu Jeffrey, 1978). The results are shown in Figs. 11.2 and 11.3, which illustrate both Dollo and Wagner majority-rule consensus trees based on a bootstrap analysis (Felsenstein, 1985). The Wagner tree (Fig. 11.3B) is reduced to show only the tribes since there were only minor differences between the two parsimony algorithms at terminal nodes (Jansen et al., 1990). The cpDNA phylogeny provides strong support for the monophyly of the subfamily Asteroideae, including the nine tribes Tageteae, Coreopsideae, Heliantheae, Eupatorieae, Calenduleae, Senecioneae, Inuleae, Anthemideae, and Astereae. The composition of this subfamily in our phylogeny is in agreement with the circumscription of the Asteroideae by Robinson (1981), Thorne (1983, Fig. 11.4A), and Bremer (1987, Fig. 11.4C), but it differs from all other subfamilial classifications of the Asteraceae. The cpDNA data provide only moderate support for the monophyly of the group consisting of the remaining seven tribes, which have been placed in the Lactucoideae by several recent classifications (Poljakov, 1967; Carlquist, 1976; Jeffrey, 1978; Thorne, 1983; Robinson, 1983). The small number of mutations (two in Fig. 11.2) supporting the monophyly of this subfamily and the lack of support for its monophyly in both morphological studies (Fig. 11.4C, Bremer, 1987; Karis et al., unpublished data) and Wagner parsimony analyses of the same cpDNA data (Fig. 11.3B) suggest that further investigations of the Lactucoideae are needed. Certainly there is no support for previous suggestions (Carlquist, 1976; Cronquist, 1955, 1977) that the Lactuceae represent a separate subfamily.

The cpDNA phylogeny supports the recognition of 13 monophyletic tribes [Mutisieae (excluding the subtribe Barnadesiinae), Cardueae, Lactuceae, Liabeae, Vernonieae, Arctoteae, Astereae, Calenduleae, Anthemideae, Senecioneae, Inuleae, Coreopsideae, and Eupatorieae]. The tribes Tageteae and Heliantheae are paraphyletic. The monophyly of most of these tribes is strongly supported by numerous synapomorphic mutations (Fig. 11.2) and by high confidence intervals in bootstrap

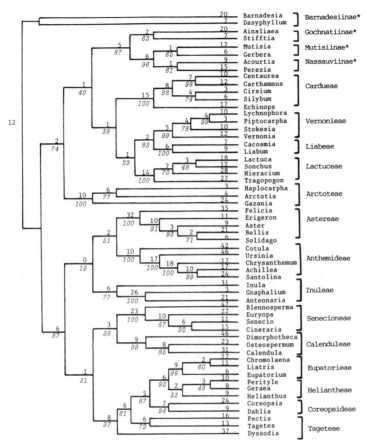

Figure 11.2. Dollo tree summarizing phylogenetic relationships in the Asteraceae using 328 phylogenetically informative cpDNA restriction site mutations. From Jansen et al. (1991a). This is a majority-rule consensus tree based on a bootstrap analysis; it has a total of 1,345 steps (including autapomorphies) and a consistency index of 0.44 (excluding autapomorphies). The numbers above (arabic) and below (italics) the nodes indicate the number of restriction site changes and the number of times that a monophyletic group occurred in 100 bootstrap replicates, respectively. Brackets show the current circumscription of 15 tribes, and the four subtribes of the Mutisieae (*sensu* Cabrera, 1977) are indicated by asterisks.

analyses (Figs. 11.2–11.3). The paraphyly of the Heliantheae is not unexpected because of the many recent changes in the circumscription of this tribe (Stuessy, 1977; Robinson, 1981; Turner and Powell, 1977). However, the paraphyly of the Tageteae is surprising in view of the previous consensus that this tribe represents a monophyletic group (Strother, 1977; Bremer, 1987).

Relationships among tribes of Asteraceae are not fully resolved in the cpDNA phylogeny (Figs. 11.2–11.3). In the subfamily Lactucoideae, a close phylogenetic

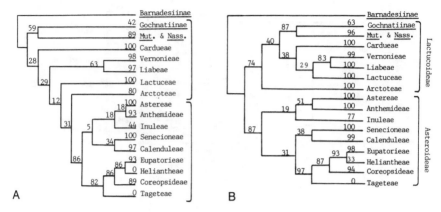

Figure 11.3. Dollo (A) and Wagner (B) trees summarizing phylogenetic relationships in the Astera-ceae using 328 phylogenetically informative cpDNA restriction site mutations. From Jansen et al. (1990). Both trees are majority-rule consensus trees based on a bootstrap analysis. The trees have been reduced to show the tribes of Asteraceae only except for the four subtribes of the Mutisieae (underlined). Numbers indicate how many times out of 100 bootstrap replicates that a monophyletic group occurred. Brackets show the circumscription of the subfamilies according to Thorne (1983).

relationship is supported only between the tribes Liabeae and Vernonieae. Histori-cally, the Liabeae have been treated as a subtribe of the Senecioneae (Bentham, 1873), but more recently this group has been accorded separate tribal status close to the Vernonieae (Robinson and Brettell, 1973a; Robinson, 1977, 1983; Bremer, 1987). The cpDNA phylogeny provides better resolution of relationships among the tribes in the subfamily Asteroideae (Figs. 11.2–11.3). The strongest clade contains the tribes Tageteae, Coreopsideae, Heliantheae, and Eupatorieae. The close relationship of the Tageteae, Coreopsideae, and Heliantheae has long been recognized (Cronquist, 1955; Stuessy, 1977; Robinson, 1981; Bremer, 1987). However, the inclusion of the Eupatorieae in this group disagrees with Bremer's (1987) suggestion that this tribe is closely allied to the Astereae (Fig. 11.4C). A number of morphologic characters, including carbonized achenes and opposite, trinerved leaves, also support a close relationship of the Eupatorieae and Helian-theae.

Tribal Placement of Morphologically Anomalous Genera

Several genera of Asteraceae have eluded tribal placement because of their anomalous morphology (reviewed in Carlquist, 1976; Wagenitz, 1976). Strong support for the monophyly of most tribes of Asteraceae in the cpDNA phylogenies (Figs. 11.2–11.3) provides the phylogenetic background for resolving tribal affinities of these problematic genera. Our studies have included nine genera whose tribal status has been debated. Five of these (*Blennosperma, Cotula,*

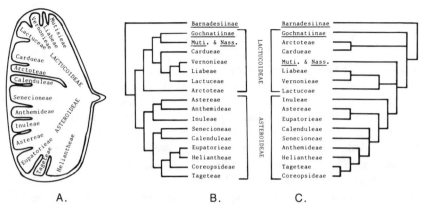

A.   B.   C.

Figure 11.4. Comparison of molecular and morphological phylogenies of the Asteraceae. From Jansen et al. (1991a). A.—Thorne's (1983) phyletic tree of relationships among the tribes. B.— Chloroplast DNA phylogeny reduced from Fig. 11.2 to show only the tribes. C.—Bremer's (modified from Fig. 6, 1987) cladogram reduced to show the same tribes indicated in the cpDNA phylogeny.

Echinops, Stokesia, and *Ursinia*) were included in our initial survey (Fig. 11.2), and four additional genera (*Brachylaena, Marshallia, Pluchea,* and *Tarchonanthus*) whose tribal status is even more uncertain have recently been added (Keeley and Jansen, 1991; Watson et al., 1991). The reliability of cpDNA data for resolving the tribal placement of these controversial genera demonstrates the potential of the chloroplast genome for assessing tribal affinities of other morphologically anomalous genera of Asteraceae.

The genera *Cotula* and *Ursinia* have been placed in three different tribes or elevated to tribal status (Robinson and Brettell, 1973b) in the subfamily Asteroideae (reviewed in Heywood and Humphries, 1977). Both genera are clearly placed in the Anthemideae in the cpDNA phylogeny. The grouping of *Echinops* with core genera of the Cardueae strongly supports its inclusion in this tribe. *Stokesia,* although previously placed in the Vernonieae, is anomalous in this tribe because it has heads with ligulate and deeply lobed outer flowers, resembling some Mutisieae. The genus is nested well within the Vernonieae based on cpDNA data (Fig. 11.2). *Blennosperma* has been placed in four different tribes (reviewed in Ornduff et al., 1973), whereas the cpDNA data clearly show that the genus belongs in the Senecioneae.

Four of the most controversial genera in terms of tribal placement are *Brachylaena, Tarchonanthus, Pluchea,* and *Marshallia.* Historically, *Brachylaena* and *Tarchonanthus* have been grouped together because they share an unusual combination of morphologic characters, including filiform corollas, shrub or tree growth form, and a dioecious sexual system. Various authorities have placed the genera in six different tribes in both the subfamilies Asteroideae and Lactucoideae (Keeley and Jansen, 1991). The cpDNA data confirm the close phylogenetic relationship of *Brachylaena* and *Tarchonanthus,* but the genera do not group consistently with any of the currently recognized tribes (Figs. 1 and 2 in Keeley and Jansen, 1991). This result, in combination with the anomalous morphology of these genera, provides compelling evidence for recognizing this group as a distinct tribe. *Pluchea* has been placed in three different tribes in both subfamilies of the Asteraceae. The cpDNA phylogeny clearly positions the genus in the Inuleae *sensu lato* (Fig. 1 in Keeley and Jansen, 1991), in agreement with recent morphologically based cladistic studies by Bremer (1987) and Anderberg (1989). *Marshallia* at various times has been placed in the tribes Eupatorieae, Heliantheae, Vernonieae, and Inuleae. The genus is anomalous in possessing an unusual combination of morphologic features, including discoid heads, anthocyanic corollas with spirally twisted lobes, a paleaceous receptacle, and alternate leaves. The cpDNA data clearly group *Marshallia* in the paraphyletic tribe Heliantheae (Fig. 1 in Watson et al., 1991a).

Comparison of Molecular and
Morphologic Phylogenies

Comparison of our cpDNA phylogeny with the two most recent morphologic schemes of relationships in the Asteraceae reveals several similarities and differences (Fig. 11.4A–C). The Dollo tree (Fig. 11.2 and reduced to show only tribes in Fig. 11.3A) is used in these comparisons, because this topology is favored in phylogenetic analyses of restriction site data (Jansen et al., 1990). The cpDNA phylogeny shows excellent congruence with the less explicit scheme of phyletic relationships proposed by Thorne (1983; Fig. 11.4A). Similarities include the monophyly of both the subfamilies Lactucoideae and Asteroideae, the inclusion of the same tribes in both subfamilies, and a close relationship between the Eupatorieae and Heliantheae. The morphologically based tree of Bremer (1987) enables a more reliable comparison of morphologic and molecular data, because this tree was produced by cladistic methods (Figs. 11.4B and 11.4C). Considerable incongruence exists concerning the monophyly of the subfamily Lactucoideae and the tribes Mutisieae (excluding the subtribe Barnadesiinae) and Inuleae, as well as relationships among several of the tribes, especially the Eupatorieae, Heliantheae, and Astereae.

Character Evolution

Previous hypotheses of character evolution in the Asteraceae have been clouded by circularity because the characters being discussed were used to construct phylogenetic trees. Our cpDNA phylogeny allows an independent assessment of the patterns of morphologic, chromosomal, and chemical evolution in the family. To assess the patterns of character evolution in the Asteraceae, the distribution of selected characters has been plotted on the cpDNA phylogeny (Fig. 11.5A–B, re-

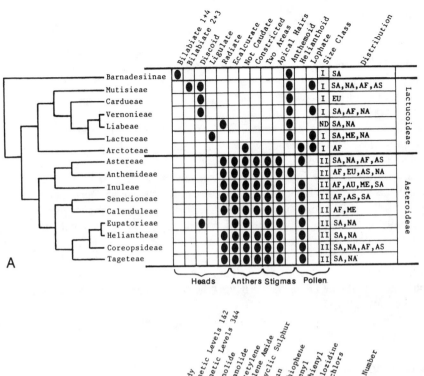

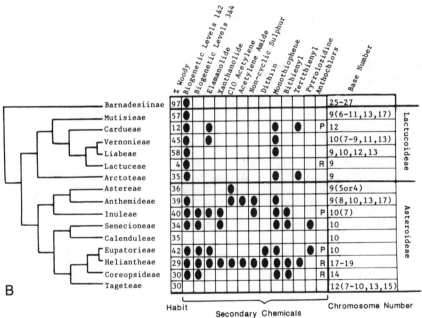

Figure 11.5. Distribution of selected nonmolecular characters on the chloroplast DNA restriction site phylogeny of the Asteraceae (reduced from Fig. 11.2 to show only tribes). From Jansen et al. (1991a). The predominant condition of the characters is listed for each tribe. "●" denotes the presence of the character indicated; ND indicates that no data are available for the character. Character information was obtained from the references given in Jansen et al. (1991a). A.— Distribution of geographic, floral, and pollen characters. Abbreviations for geographic distributions: AF, Africa; AS, Asia; AU, Australia; EU, Eurasia; ME, Mediterranean; NA, North America; SA, South America. B.—Distribution of growth habit, chemical, and chromosomal characters. Type P and R anthochlors refer to the phloroglucinol and resorcinol biosynthetic pathways, respectively.

duced to show only the tribes). The occurrence of woody habit, bilabiate flowers, and anthemoid pollen in the Barnadesiinae suggests that these characters may be ancestral within the Asteraceae. A more detailed discussion of character evolution is given in Jansen et al. (1991a); only a brief description of the general trends is summarized here.

One interesting result is the low level of homoplasy in morphologic features (Figs. 11.5A–B). If one ignores occasional variation within tribes, only a single origin is required to explain the phylogenetic distribution of many of the floral characters, especially features of the anthers and stigmas (Fig. 11.5A). The low incidence of homoplasy in floral characters is not too surprising because these features have been used historically for tribal and subfamilial circumscription. Several floral traits provide morphologic support for the monophyly of the subfamily Asteroideae, including radiate heads, ecalcurate and constricted anthers, and styles with two stigmatic areas and apical hairs. In most instances, pollen wall ultrastructure also supports the subfamilial groupings shown by the cpDNA phylogeny, with the Lactucoideae having the anthemoid type and the Asteroideae the helianthoid type. In contrast to morphologic characteristics, the basic chromosome number and major classes of secondary chemicals show a higher degree of homoplasy (Fig. 11.5B).

Phylogenetic Relationships at Lower Taxonomic Levels

Generic and Interspecific Relationships

Chloropast DNA variation in the Asteraceae has been most widely used for comparisons among species and genera. At these taxonomic levels, there is usually enough restriction site variation to provide a sufficient number of phylogenetically informative mutations, and the levels of homoplasy are low, usually ranging from 0 to 30%. Thus it is possible to generate one or only a few equally parsimonious trees in which there is strong statistical support for the included monophyletic groups. Numerous studies at lower taxonomic levels in at least seven different tribes of the Asteraceae have been completed or are in progress (Table 11.1, asterisks). In this chapter, we demonstrate the utility of cpDNA restriction site comparisons at the generic and specific levels by summarizing the results in the tribe Lactuceae. The examples are selected to illustrate both the power and limitations of using the chloroplast genome for systematic and evolutionary studies.

The tribe Lactuceae has a cosmopolitan distribution and consists of approximately 2,300 species and 70 genera (Stebbins, 1953; Tomb, 1977). During the past three years, cpDNA variation within the tribe has been examined to resolve phylogenetic relationships among genera, as well as interspecific relationships within several selected genera. Our most extensive studies have been completed

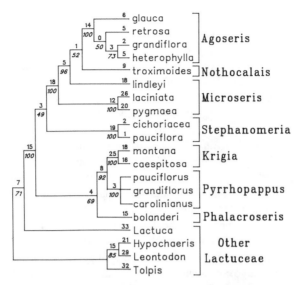

Figure 11.6   Dollo tree summarizing phylogenetic relationships in the subtribe Microseridinae (Lactuceae) using 180 phylogenetically informative cpDNA restriction site mutations. From Jansen et al. (1991b). This is a majority-rule consensus tree based on a bootstrap analysis; it has a total of 468 steps (including autapomorphies) and a consistency index of 0.65 (excluding autapomorphies). The numbers above (arabic) and below (italics) the nodes indicate the number of restriction site changes and the number of times that a monophyletic group occurred in 100 bootstrap replicates, respectively. Brackets show the current circumscription of the genera.

in the subtribe Microseridinae (Wallace and Jansen, 1990; Kim and Mabry, 1991; Kim et al., 1991a, b; Jansen et al., 1991b), which includes *Agoseris, Krigia, Microseris, Nothocalais, Phalacroseris, Pyrrhopappus,* and *Picrosia.* Below we briefly summarize the results of two studies, one at the generic level in the Microseridinae, and a second at the interspecific level within *Microseris.*

Phylogenetic relationships among 35 of the 46 species from six of the seven genera of the Microseridinae were assessed by comparative restriction site mapping of 17 enzymes (Jansen et al., 1991b). The 180 informative cpDNA mutations were used to generate phylogenetic trees using both Dollo and Wagner parsimony, and the resulting monophyletic groups were evaluated by the bootstrap method (Felsenstein, 1985). Only the results of the Dollo analyses are presented for the reasons noted earlier. The majority-rule consensus tree from the bootstrap analysis (Fig. 11.6) has the same topology as one of the four equally parsimonious Dollo trees (strict consensus trees are given in Jansen et al., 1991b). The results indicate that the subtribe Microseridinae is not monophyletic since *Stephanomeria* (subtribe Stephanomerinae) is nested within the Microseridinae. Any taxonomic changes at the subtribal level will have to await completion of our broader studies of cpDNA variation in the Lactuceae.

Five of the genera of the Microseridinae are monophyletic, whereas *Microseris* is paraphyletic because one of its species, *M. lindleyi*, is more closely related to *Agoseris* and *Nothocalais* (Jansen et al., 1991b; Fig. 11.6). The distinctiveness of *M. lindleyi* has long been recognized. It is anomalous in many of its morphologic features, DNA content (Price and Bachmann, 1975), allozymes (Irmler et al., 1982), and cpDNA (Wallace and Jansen, 1990). Chambers (1955) placed *M. lindleyi* in subgenus *Microseris* because of its annual habit, ability to produce vigorous but sterile hybrids with some other annual species, and involvement in the origin of two allotetraploids in the same subgenus. The results of our cpDNA studies, combined with the other data sets noted above provide compelling evidence for recognizing *M. lindleyi* as a distinct genus. This study demonstrates the power of cpDNA for providing the phylogenetic framework to clarify generic relationships and limits in the Asteraceae. This is particularly significant because generic delimitation is problematic in many parts of the family, especially in those tribes that have one large core genus, such as *Eupatorium, Senecio, Vernonia*, and *Viguiera*. Recent studies in *Viguiera* (Schilling and Jansen, 1989) and *Vernonia* (Keeley and Jansen, unpublished data) demonstrate the utility of cpDNA restriction site analysis for carving out generic limits in large core genera.

At the interspecific level we summarize a recent study of *Microseris* (Wallace and Jansen, 1990), which has important implications relating to the utility of cpDNA restriction site studies at this taxonomic level. The genus consists of 16 species of annual and perennial herbs, native primarily to the western United States, with one disjunct species in Australia/New Zealand and a second in Chile. *Microseris* has been the focus of intensive systematic and evolutionary studies using a wide diversity of approaches, including morphology (Chambers, 1955; 1957), experimental hybridization (Chambers, 1955), allozymes (Irmler et al., 1982), and DNA content (Bachmann et al., 1979; Price et al., 1983). In spite of these intensive efforts, many questions remain unanswered concerning phylogenetic relationships among the major species groups proposed by Chambers (1955; see Figure 11.7).

Our cpDNA studies of *Microseris* (Wallace and Jansen, 1990) identified 115 restriction site mutations among the 16 species, 78 of which were phylogenetically informative. The data were free of homoplasy, permitting the construction of a single most parsimonious tree (Fig. 11.7). The tree identified three distinct clades in *Microseris:* one including *M. lindleyi;* the second clade consisting of all perennial species of *Microseris* except the disjunct allotetraploid *M. scapigera* from Australia and New Zealand; and the third group including all other annual species and *M. scapigera*. The overall topology of the *Microseris* tree is characteristic of the tree topology often obtained in interspecific studies in the Asteraceae, one in which the major groups are well resolved but in which there are insufficient mutations to clarify relationships among closely related species within these

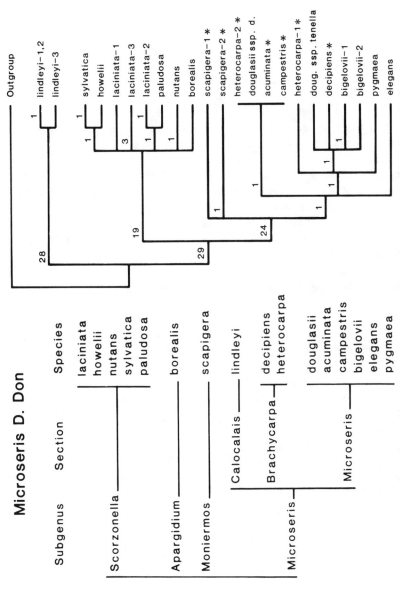

Figure 11.7. Left—systematic relationships within the genus *Microseris* as proposed by Chambers (1955). Right—Single most parsimonious cladogram showing the phylogenetic relationships of *Microseris* based on 115 chloroplast DNA restriction site mutations. From Wallace and Jansen (1990). The tree has a consistency index of 1.0. Numbers at nodes and along each lineage indicate the number of mutational changes. Asterisks indicate tetraploid species; all others are diploid.

groups. Other examples include *Coreopsis* (Crawford et al., 1990), *Krigia* (Kim et al., 1991a), and *Malacothrix* (Whitton et al., unpublished data).

The cpDNA phylogeny for *Microseris* shows considerable incongruence with Chambers' (1955) subgeneric classification (Figs. 11.7A–B). Except for the placement of *M. lindleyi*, the lack of congruence between morphology and cpDNA is due entirely to the maternal mode of inheritance of the chloroplast characters. The inclusion of *M. scapigera* in the annual clade is due to the involvement of an annual species in the hybrid origin of this allotetraploid (Chambers, 1955). A similar explanation applies to the position of two populations of the allotetraploid *M. heterocarpa*, which has had multiple origins with different maternal parents (Wallace and Jansen, unpublished data; see next section). These results clearly emphasize that one must be cautious when using cpDNA data to construct phylogenies in groups that have undergone reticulate evolution. On the other hand, with a knowledge of hybrid speciation in a group, cpDNA data can resolve the maternal parentage of hybrid taxa (see also Chapters 7 and 8, this volume).

Origin and Evolution of Polyploids

Polyploidy, which has been estimated to occur in 47% of the species of angiosperms (Grant, 1981), is a very common mode of speciation in the Asteraceae. Many of the classic studies of polyploid evolution have involved genera in this family, such as *Crepis* (Babcock and Stebbins, 1938) and *Tragopogon* (Ownbey, 1950; Roose and Gottlieb, 1976; Soltis and Soltis, 1989). The advent of DNA techniques has provided a new set of powerful tools to refine further our understanding of polyploidy in the Asteraceae. The combined use of restriction site comparisons of chloroplast and nuclear ribosomal DNA (rDNA) has already provided new insights into the origin and evolution of both auto- and allopolyploids in the genera *Krigia* (Kim and Mabry, 1991; Kim et al., 1991a, b), *Microseris* (Wallace and Jansen, 1990), and *Tragopogon* (Soltis and Soltis, 1989, 1991). In this chapter, we summarize the results of studies in the former two genera; a complete review of the application of DNA methods in polyploid groups is provided by Soltis et al. in Chapter 8 (this volume).

Microseris contains five tetraploid species, four of which are placed in the annual subgenus *Microseris* (Chambers, 1955). Chambers used morphologic characters and experimental hybridizations to propose the parentage of the four annual tetraploids (Fig. 11.8). Examination of chloroplast and nuclear DNA variation among 70 populations of these polyploids and their putative diploid progenitors has confirmed the parentage of two of these tetraploids, *M. heterocarpa* and *M. decipiens* (Wallace and Jansen, unpublished data). The paternal parent of both is *M. lindleyi*, a species that has recently been segregated into the monotypic genus *Uropappus*, because it is more closely related to the genera *Nothocalais* and *Agoseris* (see above and Jansen et al., 1991b). Artificial crosses between *M. lindleyi* and *M. douglasii* or *M. bigelovii* produced vigorous but sterile

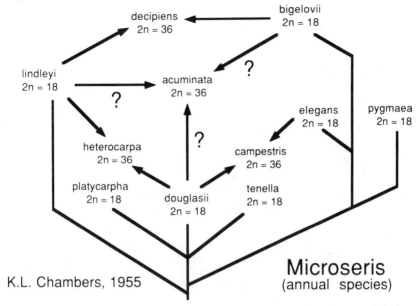

Figure 11.8. Proposed parentage of allotetraploids in *Microseris* based on morphologic characters and experimental crosses. Redrawn from Chambers (1955).

hybrids (Chambers, 1955; Irmler et al., 1982). Thus, the involvement of the diploid *M. lindleyi* in the origin of these alloploids is enigmatic in view of its distant relationship to the other diploid parents. Another exciting result is that *M. heterocarpa* has had multiple origins with the maternal parent being *M. douglasii* and the paternal parent *M. lindleyi*. This is one of several studies (Wyatt et al., 1988; Soltis and Soltis, 1989, 1991, Soltis et al., 1989b; Doyle et al., 1990a; Wolf et al., 1990) demonstrating the power of cpDNA for documenting multiple origins of polyploids.

Krigia comprises seven annual and perennial species distributed in eastern and central North America (Kim, 1989). The genus exhibits a broad range of chromosome numbers, including $n = 4, 5, 6, 9, 10, 15,$ and 60. Recent molecular studies have helped resolve the origin and evolution of both auto- and allopolyploids (Kim and Mabry, 1991; Kim et al., 1991a, b). A cpDNA phylogeny of 11 diploid and tetraploid populations of *K. virginica* indicates that the autoploid populations of this species had a single origin (Kim et al., 1991a, Fig. 11.9).

Evolutionary Implications of
Intraspecific Variation

Low levels of sequence divergence in the chloroplast genome have limited its usefulness in evolutionary studies at the population level. Intraspecific cpDNA polymorphisms have been documented in studies of some cultivated plants (Clegg et al., 1984; Holwerda et al., 1986; Neale et al., 1986; Palmer et al., 1983; 1985;

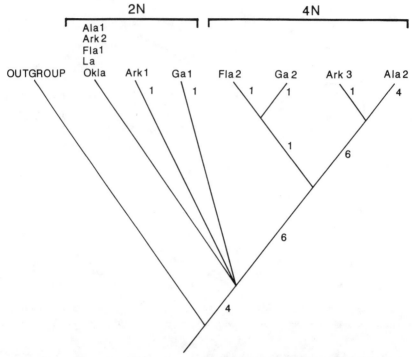

Figure 11.9. Chloroplast DNA restriction site phylogeny showing relationships among 11 populations of *Krigia virginica*. From Kim et al. (1991a). Numbers at nodes and along each lineage indicate the number of mutational changes. Abbreviations indicate geographic location (Ala—Alabama, Ark—Arkansas, Fla—Florida, Ga—Georgia, La—Louisiana) and number of population examined from the same state. Brackets indicate tetraploid and diploid populations.

Timothy et al., 1979), as well as in several wild species (Palmer and Zamir, 1982; Banks and Birky, 1985; Rieseberg et al., 1988; Soltis et al., 1989a, b). This type of variation is generally rare, but, when found, it provides important insights into the origin of crop plants, hybrids, and polyploids. Although Chapter 6 of this book is devoted to intraspecific variation in cpDNA, we feel it is important to discuss variation at this taxonomic level because so many systematists are currently using this genome for phylogenetic studies in the Asteraceae. From the discussions below, it should be obvious that multiple populations should be examined in future cpDNA studies in the family.

The prevailing dogma in the Asteraceae has been that intraspecific chloroplast DNA variation is extremely rare, with only three published examples in the genera *Helianthus* (Rieseberg et al., 1988): *Microseris* (Wallace and Jansen, 1990), and *Tragopogon* (Soltis and Soltis, 1989). One reason for this viewpoint is that few studies of cpDNA variation in the Asteraceae have examined multiple populations. More recent investigations have performed extensive population

Table 11.2.   Levels of intraspecific cpDNA variation in the Asteraceae and other selected plant groups.

Species (ploidal level)	Number of Populations	Number of Sites	Number of Enzymes	Mutations Site	Length	Reference[a]
Asteraceae						
Antennaria plantaginifolia (2N)	13	1,000	12	10	0	1
Antennaria parlinii (4N)	34	1,000	12	9	1	1
Coreopsis grandiflora (2N,6N)	3	1,060	26	0	0	2
Helianthus annuus (2N)	9	507	36	4	0	3
Helianthus bolanderi (2N)	9	507	36	9	0	3
Krigia biflora (2N,4N)	4	1,100	22	8	1	4
Krigia dandelion (8N)	7	1,100	22	0	0	4
Krigia virginica (2N,4N)	11	1,100	22	22	0	4
Marshallia caespitosa (2N,4N)	3	1,100	22	9	0	5
Microseris heterocarpa (4N)	2	1,100	22	2	0	6
Microseris laciniata (2N)	3	1,100	22	5	0	6
Microseris lindleyi (2N)	3	1,100	22	1	0	6
Taraxacum officinale (2N–6N)	19	64	6	11	11	7
Viguiera dentata (2N)	3	1,060	16	1	0	8
Other Plants						
Brassica compestris (2N)	3	520	28	1	nd[b]	9
Hordeum vulgare (2N)	15	380	17	8	5	10
Heuchera micrantha (2N,4N)	28	292	19	14	3	11
Lupinus texensis (2N)	21	134	7	2	1	12
Lycopersicon peruvianum (2N)	6	484	25	2	0	13
Pinus banksiana (2N)	68	26	2	0	6	14
Pisum sativum (2N)	12	132	11	1	5	15
Tolmiea menzieii (2N,4N)	37	352	19	7	5	16
Zea mays (2N)	13	580	21	2	1	17

[a]1—Michaels and Palmer (unpublished data); 2—Crawford et al. (1990); 3—Rieseberg et al. (1988); 4—Kim et al. (1991a); 5—Watson et al. (unpublished data); 6—Wallace and Jansen (1990); 7—King (unpublished data); 8—Schilling and Jansen (1989); 9—Palmer et al. (1983); 10—Holwerda et al. (1986); 11—Soltis et al. (1989b); 12—Banks and Birky (1985); 13—Palmer and Zamir (1982); 14—Wagner et al. (1987); 15—Palmer et al. (1985); 16—Soltis et al. (1989a); 17—Doebley et al. (1987).

[b]nd indicates no data.

sampling within several genera from phylogenetically distant tribes (summarized in Table 11.2 and compared with representative studies in other angiosperm families). In comparing the levels of intraspecific variation in these genera, it is evident that a wide range of variation patterns exists, with some groups exhibiting few, if any, polymorphisms (e.g., *Antennaria, Coreopsis,* and *Marshallia*), whereas others (e.g., *Krigia*) have some of the highest levels of intraspecific variation documented so far in angiosperms. *Krigia* exhibits both extremes (Table 11.2), with some invariant species and others that have many intraspecific cpDNA polymorphisms. Sequence divergence among populations ranged from 0 to 0.26%, with an average value of 0.11%. This level of interpopulational cpDNA sequence divergence exceeds all previous estimates for species from the Astera-ceae, as well as most other angiosperm families. Furthermore, the levels of intraspecific cpDNA divergence in *Krigia* are higher than those found at the interspecific level in many other genera of Asteraceae, including *Coreopsis* (Crawford et al., 1990), *Marshallia* (Watson et al., unpublished data), and *Antennaria* (Michaels and Palmer, unpublished data). The different patterns of intraspecific variation could be due to different rates of cpDNA mutations. Several studies have clearly demonstrated unequal rates of restriction site mutations in the family (see below). In *Krigia,* there is a correlation between rates of cpDNA evolution and increasing ploidal level at both the intra- and interspecific levels (Kim et al., 1991a, b). It is noteworthy that many other angiosperm species that exhibit high levels of intraspecific cpDNA variation often include polyploid populations (Table 11.2).

When intraspecific cpDNA variation is encountered, it can provide exciting evolutionary insights. In the 11 populations of *Krigia virginica* examined, 22 restriction site differences were detected, 20 of which were found only in one or more of the four examined tetraploid populations. Thirteen of these cpDNA mutations were shared by more than one population, allowing the construction of a single most parsimonious tree for the 11 populations (Fig. 11.9). The implications of this population phylogeny are that the diploids and tetraploids from the same geographic region are not grouping together and that the autotetraploid populations had a single origin.

Relative Rates and Locations of Chloroplast DNA Mutations

Several studies in the Asteraceae have examined rates of DNA evolution using the two-tailed Wilcoxon matched-pair signed rank test developed for restriction site data (Templeton, 1983). This test, which examines relative rates by compar-ing the number of restriction site differences along two lineages that share a common ancestor, has been used to demonstrate rate inequities in comparisons between some lineages in two other angiosperm families (*Clarkia,* Sytsma and Gottlieb, 1986; *Glycine,* Doyle et al., 1990b). Studies in the Asteraceae have found rate inequities in comparisons between some lineages, whereas a molecular clock could not be rejected in comparisons between others. Schilling and Jansen (1989) suggested that different evolutionary rates within *Viguiera* and related genera of the tribe Heliantheae may be correlated with generation time. Studies of animal mitochondrial DNA have implicated generation time in affecting evolu-tionary rates in this genome (Easteal, 1985; Wu and Li, 1985). Unequal rates of cpDNA evolution among genera and species in the Microseridinae (tribe Lactu-ceae) are not always correlated with differences in the annual or perennial growth habit (Wallace and Jansen, 1990; Jansen et al., 1991b, c; Kim et al., 1991b). In the genus *Krigia,* a strong correlation exists between rates of cpDNA evolution

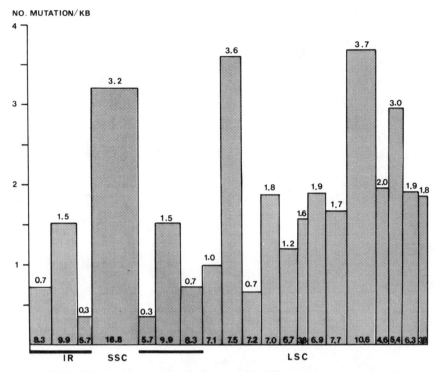

NO. MUTATION/KB

Figure 11.10. Histogram summarizing the location of the 252 restriction site mutations detected in different regions of the *Krigia* chloroplast genome. From Kim et al. (1991b). The regions are indicated by the cloned *Sac*I restriction fragments (Jansen and Palmer, 1987a; Fig. 11.11). The solid bar indicates the position and extent of the inverted repeat in lettuce. The numbers at the top of each bar indicate the average number of restriction site mutations per kilobase of the genome.

and differences in chromosome base number and ploidy (Kim et al., 1991a, b). The cause of this correlation is unknown, although one possible hypothesis is that chromosomal changes during the evolutionary process may alter the interaction between the nuclear and chloroplast genome, thereby inducing a more rapid evolution of the chloroplast genome (Kim et al., 1991b).

We have also examined the frequency of restriction site and length mutations in different regions of the chloroplast genome. Jansen and Palmer (1987a) suggested that the inverted repeat is the most highly conserved portion of the chloroplast genome in the Asteraceae. Sequence comparisons (Wolfe et al., 1987) have confirmed the slower rate of evolution of the inverted repeat. Certain portions of the large and small single-copy regions of the chloroplast genome in the Asteraceae accumulate mutations more rapidly (Jansen and Palmer, 1987a). A detailed comparison in *Krigia* (Kim et al., 1991b) identified four mutational hotspots, defined by more than three mutations per kilobase (Fig. 11.10). These areas correspond to the 10.6-, 7.5-, 18.8-, and 5.4-kb *Sac*I fragments of lettuce cpDNA

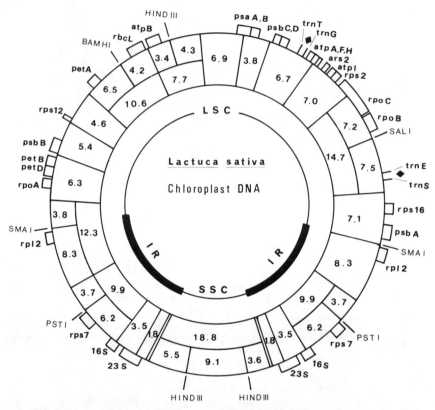

Figure 11.11. Restriction map of lettuce chloroplast DNA showing the location of 22 cloned restriction fragments and genes. Redrawn from Jansen and Palmer (1987a, 1988). All restriction sites are for *Sac*I, except where indicated. The two heavy lines indicate the extent of the inverted repeat (IR) and LSC and SSC indicate the large and small single-copy regions, respectively. Solid diamonds within the 6.7- and 7.5-kb fragments indicate the approximate location of the endpoints of the 22-kb inversion. Abbreviations for gene names follow Shinozaki et al. (1986). All fragment sizes are given in kilobases.

(Figs. 11.10–11.11). All of these mutational hotspots include regions containing long spacer sequences or unidentified open reading frames (Fig. 11.11; Jansen and Palmer, 1987a; Shinozaki et al., 1986; Hiratsuka et al., 1989). For example, the most variable segment occurs within a portion of the 10.6-kb *Sac*I region, which includes long spacer sequences and open reading frames between the genes encoding *rbc*L and *pet*A. The most conserved region of the chloroplast genome of the Asteraceae was located in the inverted repeat adjacent to the small single-copy region, which encodes the 16S, 4.5S, and 23S ribosomal RNA genes. The portion of the large single-copy region that includes the *rpo*B and *rpo*C genes shows a similarly low frequency of restriction site mutations. This is surprising in view of recent sequence comparisons of rice and tobacco (Sugiura, 1989) indicating that the RNA polymerase genes are among the most divergent genes in the chloroplast genome. Overall, our data demonstrate that cpDNA mutations in the Asteraceae are more frequent in regions containing long spacer sequences. We are currently performing more detailed mapping studies using small, gene-specific cpDNA probes from tobacco to locate precisely the position of mutations relative to coding and noncoding sequences of the chloroplast genome. These studies will provide important insights into relative rates of evolution in different portions of the chloroplast genome. Furthermore, information on the frequency of restriction site mutations in different portions of the genome will enable the selection of conserved regions of cpDNA for studies at higher taxonomic levels and rapidly changing regions for investigations at lower levels. Comparisons among the three completely sequenced chloroplast genomes (Wolfe et al., 1987; Wolfe and Sharp, 1988; Sugiura, 1989) have already improved our understanding of evolutionary rates in different genes.

Conclusions

Our efforts during the past five years have generated the largest set of data on cpDNA variation and evolution for any plant family. Phylogenetic studies have demonstrated that restriction site data can resolve systematic relationships at a wide range of taxonomic levels, including the interpopulational to intertribal levels. Furthermore, the combined use of chloroplast and nuclear rDNA data can provide exciting new insights into the origin and evolution of polyploid complexes. The availability of detailed information on rates of cpDNA evolution in different parts of the chloroplast genome makes it possible to select particular regions of this molecule for addressing systematic and evolutionary questions at most taxonomic levels. The high levels of interpopulational variation found in some species (e.g., *Krigia virginica* and *Helianthus bolanderi*), suggest that the chloroplast genome may be generally applicable at the intraspecific level in the Asteraceae.

Most studies of cpDNA variation in the Asteraceae have been conducted at the interspecific level. In most instances, sufficient variation has been documented to generate accurate estimates of phylogenetic relationships among the major lineages within the genera examined. However, in many of the interspecific studies (e.g., *Microseris,* Wallace and Jansen, 1990; *Coreopsis,* Crawford et al., 1990) it has not been possible to resolve fully relationships among closely related species within the major clades. Chloroplast DNA studies seem best suited for higher taxonomic levels in the Asteraceae, especially among genera. At this taxonomic level, there is not only sufficient variation, but the incidence of homoplasy is low enough that only one or a few equally parsimonious trees are generated in cladistic analyses.

The extensive data base on cpDNA variation has also allowed us to improve greatly our understanding of phylogenetic relationships among higher taxonomic categories in the Asteraceae. We have provided strong molecular support for the monophyly of 14 tribes and two subfamilies (Asteroideae and a new subfamily based on the Barnadesiinae). The monophyly of the third subfamily, Lactucoideae, is only moderately supported. Phylogenetic relationships among several tribes have also been resolved. Within the next few years, we will be in a position to propose a new phylogenetic classification for the Asteraceae, which we believe will be stable because it will be based on many conservatively evolving characters and rigorous cladistic methods.

One of our labs (RKJ) is continuing investigations of DNA variation in the Asteraceae in several directions. We are expanding our restriction site comparisons to include more enzymes and genera from throughout the family to test further the monophyly of the Lactucoideae, as well as to resolve relationships among the tribes. Restriction site comparisons are also underway to clarify generic and subtribal relationships in several tribes, including the Lactuceae, Tageteae, Eupatorieae, Vernonieae, Astereae, and Heliantheae (Table 11.1, asterisks). Sequence comparisons are being expanded by examining the *rbc*L gene from at least two genera from each of the tribes in the Asteraceae. Our preliminary data in this area indicate that this gene may be too conservative to provide sufficient characters to resolve phylogenetic relationships within the Asteraceae. Thus, we are also sequencing noncoding regions at both ends of the *rbc*L gene, as well as initiating sequencing of more divergent chloroplast-encoded genes. These studies, as well as those in progress in other labs, will continue to make the Asteraceae a model for molecular systematic studies in angiosperm families.

References

Anderberg, A.A. (1989) Phylogeny and reclassification of the tribe Inuleae (Asteraceae). *Can. J. Bot.* **67**, 2277–2296.

Babcock, E.B., and Stebbins, G.L., Jr. (1938) The American species of *Crepis:* their relationships and distribution as affected by polyploidy and apomixis. *Carnegie Inst. Washington* **484**, 1–200.

Bachmann, K., Chambers, K.L., and Price, H.J. (1979) Genome size and phenotypic evolution in *Microseris* (Asteraceae: Cichorieae). *Plant Syst. Evol. (Suppl.)* **2**, 41–66.

Baldwin, B.G., Kyhos, D.W., and Dvořak, J. (1990) Chloroplast DNA evolution and adaptive radiation in the Hawaiian silversword alliance (Asteraceae-Madiinae). *Ann. Missouri Bot. Gard.* **77**, 96–109.

Banks, J.A., and Birky, C.W., Jr. (1985) Chloroplast DNA diversity is low in a wild plant, *Lupinus texensis. Proc. Natl. Acad. Sci. USA* **82**, 6950–6954.

Bentham, G. (1873) Notes on the classification, history, and geographical distribution of the Compositae. *J. Linn. Soc. (Bot.)* **13**, 335–577.

Bremer, K. (1987) Tribal interrelationships of the Asteraceae. *Cladistics* **3**, 210–253.

Cabrera, A.L. (1977) Mutisieae—systematic review. In: *The Biology and Chemistry of the Compositae* (eds. V.H. Heywood, J.B. Harborne, and B.L. Turner), Academic Press, London, pp. 1039–1066.

Carlquist, S. (1976) Tribal interrelationships and phylogeny of the Asteraceae. *Aliso* **8**, 465–492.

Chambers, K.L. (1955) A biosystematic study of the annual species of *Microseris*. *Contrib. Dudley Herb.* **4**, 207–312.

Chambers, K.L. (1957) Taxonomic notes on some Compositae of the western United States. *Contr. Dudley Herb.* **5**, 57–68.

Clegg, M.T., Brown, A.H.D., and Whitfeld, P.R. (1984) Chloroplast DNA diversity in wild and cultivated barley. *Genet. Res.* **43**, 339–343.

Crawford, D.J., Palmer, J.D., and Kobayashi, M. (1990) Chloroplast DNA restriction site variation and phylogeny of *Coreopsis* section *Coreopsis* (Asteraceae). *Amer. J. Bot.* **77**, 552–558.

Cronquist, A. (1955) Phylogeny and taxonomy of the Compositae. *Amer. Midl. Natur.* **53**, 478–511.

Cronquist, A. (1977) The Compositae revisited. *Brittonia* **29**, 137–153.

Cronquist, A. (1981) *An Integrated System of Classification of Flowering Plants,* Columbia University Press, New York.

Doebley, J., Renfroe, W., and Blanton, A. (1987) Restriction site variation in the *Zea* chloroplast genome. *Genetics* **117**, 139–147.

Doyle, J.J., Doyle, J.L., Brown, A.H.D., and Grace, J.P. (1990a) Multiple origins of polyploids in the *Glycine tabacina* complex inferred from chloroplast DNA polymorphisms. *Proc. Natl. Acad. Sci. USA* **87**, 714–717.

Doyle, J.J., Doyle, J.L., and Brown, A.H.D. (1990b) A chloroplast DNA phylogeny of the wild perennial relatives of soybean (*Glycine* subgenus *Glycine*): congruence with morphological and crossing groups. *Evolution* **44**, 371–389.

Easteal, S. (1985) Generation time and the rate of molecular evolution. *Mol. Biol. Evol.* **2**, 450–453.

Felsenstein, J. (1985) Confidence limits on phylogenies: an approach using the bootstrap. *Evolution* **39**, 783–791.

Grant, V. (1981) *Plant Speciation,* Columbia University Press, New York.

Heywood, V.H., Harborne, J.B., and Turner, B.L. (1977) *The Biology and Chemistry of the Compositae,* Academic Press, London.

Heywood, V.H., and Humphries, C.J. (1977) Anthemideae—systematic review. In: *The Biology and Chemistry of the Compositae* (eds. V.H. Heywood, J.B. Harborne, and B.L. Turner), Academic Press, London, pp. 851–898.

Hiratsuka, J., Shimada, H., Whittier, R., Ishibashi, T., Sakamoto, M., Mori, M., Kondo, C., Honji, Y., Sun, C-R., Meng, B-Y., Li, Y-Q., Kanno, A., Nishizawa, Y., Hirai, A., Shinozaki, K., and Sugiura, M. (1989) The complete nucleotide sequence of the rice (*Oryza sativa*) chloroplast genome: intermolecular recombination between distinct tRNA genes accounts for a major plastid DNA inversion during the evolution of cereals. *Mol. Gen. Genet.* **217**, 185–194.

Holwerda, B.C., Jana, S., and Crosby, W.L. (1986) Chloroplast and mitochondrial DNA variation in *Hordeum vulgare* and *Hordeum spontaneum*. *Genetics* **114**, 1271–1291.

Hudson, G.S., Mahon, J.D., Anderson, P.A., Gibbs, M.J., Badger, M.R., Andrews, T.J., and Whitfeld, P.R. (1990) Comparison of rbcL genes for the large subunit of ribulose-bisphosphate carboxylase from closely related C3 and C4 plant species. *J. Biol. Chem.* **265**, 808–814.

Irmler, C., Bachmann, K.L., Chambers, K.L., Price, H.J., and Konig, A. (1982) Enzymes and quantitative morphological characters compared between the allotetraploid *Microseris decipiens* and its diploid parental species. *Beitr. Biol. Pflanzen* **57**, 269–289.

Jansen, R.K., and Palmer, J.D. (1987a) Chloroplast DNA from lettuce and *Barnadesia* (Asteraceae): structure, gene localization, and characterization of a large inversion. *Curr. Genet.* **11**, 553–564.

Jansen, R.K., and Palmer, J.D. (1987b) A chloroplast DNA inversion marks an ancient evolutionary split in the sunflower family (Asteraceae). *Proc. Natl. Acad. Sci. USA* **84**, 5818–5822.

Jansen, R.K., and Palmer, J.D. (1988) Phylogenetic implications of chloroplast DNA restriction site variation in the Mutisieae (Asteraceae). *Amer. J. Bot.* **75**, 751–764.

Jansen, R.K., Michaels, H.J., and Palmer, J.D. (1991a) Phylogeny and character evolution in the Asteraceae based on chloroplast DNA restriction site mapping. *Syst. Bot.* **16**, 98–115.

Jansen, R.K., Holsinger, K.E., Michaels, H.J. and Palmer, J.D. (1990a) Phylogenetic analysis of restriction site data at higher taxonomic levels: an example from the Asteraceae. *Evolution* **44**, 2089–2105.

Jansen, R.K., Wallace, R.S., Kim, K.-J., and Chambers, K.L. (1991b) Systematic implications of chloroplast DNA variation in the subtribe Microseridinae (Asteraceae: Lactuceae). *Amer. J. Bot.*, in press.

Jeffrey, C. (1977) Corolla forms in the Compositae—some evolutionary and taxonomic speculations. In: *The Biology and Chemistry of the Compositae* (eds. V.H. Heywood, J.B. Harborne, and B.L. Turner), Academic Press, London, pp. 111–118.

Jeffrey, C. (1978) Compositae. In: *Flowering Plants of the World* (ed. V.H. Heywood), Mayflower, New York, pp. 263–268.

Keeley, S.C., and Jansen, R.K. (1991) Evidence from chloroplast DNA for the recognition of a new tribe, the Tarchonantheae, and the tribal placement of *Pluchea* (Asteraceae). *Syst. Bot.* **16**, 173–181.

Kim, K.-J. (1989) A systematic study of *Krigia* (Asteraceae: Lactuceae), emphasizing chloroplast DNA and nuclear ribosomal DNA variations. Ph.D. Thesis, University of Texas at Austin.

Kim, K.-J., and Mabry, T.J. (1991) Evolutionary implications of intra- and interspecific variation in nuclear ribosomal DNA in *Krigia* (Asteraceae: Lactuceae). *Plant Syst. Evol.*, in press.

Kim, K.-J., Jansen, R.K., and Turner, B.L. (1991a) Evolutionary implications of intraspecific chloroplast DNA variation in dwarf dandelions (*Krigia*—Asteraceae). *Amer. J. Bot.*, submitted.

Kim, K.-J., Jansen, R.K., and Turner, B.L. (1991b) Phylogenetic and evolutionary implications of interspecific chloroplast DNA variation in dwarf dandelions (*Krigia*—Lactuceae—Asteraceae). *Syst. Bot.*, submitted.

Neale, D.B., Wheeler, N.C., and Allard, R.W. (1986) Paternal inheritance of chloroplast DNA in Douglas-fir. *Can. J. For. Res.* **16**, 1152–1154.

Ornduff, R., Saleh, N.A.M., and Bohm, B.A. (1973) The flavonoids and affinities of *Blennosperma* and *Crocidium* (Compositae). *Taxon* **22**, 407–412.

Ownbey, M. (1950) Natural hybridization and amphiploidy in the genus *Tragopogon*. *Amer. J. Bot.* **37**, 487–499.

Palmer, J.D., and Zamir, D. (1982) Chloroplast DNA evolution and phylogenetic relationships in *Lycopersicon*. *Proc. Natl. Acad. Sci. USA* **79**, 5006–5010.

Palmer, J.D., Shields, C.R., Cohen, D.S., and Orton, T.J. (1983) Chloroplast DNA evolution and the origin of amphiploid *Brassica*. *Theor. Appl. Genet.* **65**, 181–189.

Palmer, J.D., Jorgensen, R.A., and Thompson, W.F. (1985) Chloroplast DNA variation and evolution in *Pisum:* patterns of change and phylogenetic analysis. *Genetics* **109**, 195–213.

Poljakov, P.P. (1967) *Systematics and Origin of the Compositae,* Nauka, Alma-Ata (in Russian).

Price, H.J., and Bachmann, K. (1975) DNA content and evolution in the Microseridinae. *Amer. J. Bot.* **62**, 262–267.

Price, H.J., Chambers, K.L., Bachmann, K., and Riggs, J. (1983) Inheritance of nuclear 2C DNA content variation in intraspecific and interspecific hybrids of *Microseris* (Asteraceae). *Amer. J. Bot.* **70**, 1133–1138.

Raven, P.H., and Axelrod, D.I. (1974) Angiosperm biogeography and past continental movements. *Ann. Missouri Bot. Gard.* **61**, 539–673.

Rieseberg, L.H., Soltis, D.E., and Palmer, J.D. (1988) A molecular reexamination of introgression between *Helianthus annuus* and *H. bolanderi* (Compositae). *Evolution* **42**, 227–238.

Robinson, H. (1977) An analysis of the characters and relationships of the tribes Eupatorieae and Vernonieae (Asteraceae). *Syst. Bot.* **2**, 199–208.

Robinson, H. (1981) A revision of the tribal and subtribal limits of the Heliantheae (Asteraceae). *Smith. Contr. Bot.* **51**, 1–102.

Robinson, H. (1983) A generic review of the tribe Liabeae (Asteraceae). *Smith. Contr. Bot.* **54**, 1–69.

Robinson, H., and Brettell, R.D. (1973a) Tribal revisions in the Asteraceae. III. A new tribe, Liabeae. *Phytologia* **25**, 104–107.

Robinson, H., and Brettell, R.D. (1973b) Tribal revisions in the Asteraceae. VIII. A new tribe, Ursinieae. *Phytologia* **26**, 76–86.

Roose, M.L., and Gottlieb, L.D. (1976) Genetic and biochemical consequences of polyploidy in *Tragopogon*. *Evolution* **30**, 819–830.

Schilling, E.E., and Jansen, R.K. (1989) Restriction fragment analysis of chloroplast DNA and the systematics of *Viguiera* and related genera (Asteraceae:Heliantheae). *Amer. J. Bot.* **76**, 1769–1778.

Shinozaki, K., Ohme, M., Tanaka, M., Wakasugi, T., Hayashida, N., Matsubayashi, T., Zaita, N., Chunwongse, J., Obokata, J., Yamaguchi-Shinozaki, K., Ohto, C., Torazawa, K., Meng, B.Y., Sugita, M., Deno, H., Kamogashira, T., Yamada, K., Kasuda, J., Kato, A., Todoh, N., Shimada, H., and Sugiura, M. (1986) The complete nucleotide sequence of tobacco chloroplast genome: its gene organization and expression. *EMBO J.* **5,** 2043–2050.

Shinozaki, K., and Sugiura, M. (1982) The nucleotide sequence of the tobacco chloroplast gene for the large subunit of ribulose-1,5-bisphosphate carboxylase/oxygenase. *Gene* **20,** 91–102.

Small, J. (1919) The origin and development of the Compositae. XIII. General considerations. *New Phytol.* **17,** 201–234.

Soltis, D.E., and Soltis, P.S. (1989) Allopolyploid speciation in *Tragopogon:* insights from chloroplast DNA. *Amer. J. Bot.* **76,** 1119–1124.

Soltis, D.E., Soltis, P.S., Ranker, T., and Ness, B.D. (1989a) Chloroplast DNA variation in a wild plant, *Tolmiea menziesii. Genetics* **121,** 819–826.

Soltis, D.E., Soltis, P.S., and Ness, B.D. (1989b) Chloroplast DNA variation and multiple origins of autopolyploidy in *Heuchera micrantha* (Saxifragaceae). *Evolution* **43,** 650–656.

Soltis, P.S., and Soltis, D.E. (1991) Multiple origins of *Tragopogon mirus* (Compositae): evidence from rDNA. *Syst. Bot.* **16,** 407–413.

Stebbins, G.L., Jr. (1953) A new classification of the tribe Cichorieae, family Compositae. *Madroño* **12,** 33–64.

Stebbins, G.L., Jr. (1977) Developmental and comparative anatomy of the Compositae. In: *The Biology and Chemistry of the Compositae* (eds. V.H. Heywood, J.B. Harborne, and B.L. Turner), Academic Press, London, pp. 92–109.

Strother, J.L. (1977) Tageteae—systematic review. In: *The Biology and Chemistry of the Compositae* (eds. V.H. Heywood, J.B. Harborne, and B.L. Turner), Academic Press, London, pp. 769–783.

Stuessy, T.F. (1977) Heliantheae—systematic review. In: *The Biology and Chemistry of the Compositae* (eds. V.H. Heywood, J.B. Harborne, and B.L. Turner), Academic Press, London, pp. 621–671.

Sugiura, M. (1989) The chloroplast chromosomes in land plants. *Ann. Rev. Cell Biol.* **5,** 51–70.

Suh, Y., and Simpson, B.S. (1990) Phylogenetic analysis of chloroplast DNA in North American *Gutierrezia* and related genera (Asteraceae: Astereae). *Syst. Bot.* **15,** 660–670.

Sytsma, K.J., and Gottlieb, L.D. (1986) Chloroplast DNA evolution and phylogenetic relationships in *Clarkia* sect. *Peripetasma* (Onagraceae). *Evolution* **40,** 1248–1262.

Templeton, A.R. (1983) Phylogenetic inference from restriction endonuclease cleavage site maps with particular reference to the evolution of humans and apes. *Evolution* **37,** 221–244.

Thorne, R. (1983) Proposed new realignments in the angiosperms. *Nord. J. Bot.* **3**, 85–117.

Timothy, D.H., Levings, C.S. III, Pring, D.R., Conde, M.F., and Kermicle, J.L. (1979) Organelle DNA variation and systematic relationships in the genus *Zea:* teosinte. *Proc. Natl. Acad. Sci. USA* **76**, 4220–4224.

Tomb, A.S. (1977) Lactuceae—systematic review. In: *The Biology and Chemistry of the Compositae* (eds. V.H. Heywood, J.B. Harborne, and B.L. Turner), Academic Press, London, pp. 1067–1079.

Turner, B.L. (1977a) Summary of the biology and chemistry of the Compositae. In: *The Biology and Chemistry of the Compositae* (eds. V.H. Heywood, J.B. Harborne, and B.L. Turner), Academic Press, London, pp. 1105–1118.

Turner, B.L. (1977b) Fossil history and geography. In: *The Biology and Chemistry of the Compositae* (eds. V.H. Heywood, J.B. Harborne, and B.L. Turner), Academic Press, London, pp. 21–39.

Turner, B.L., and Powell, A.M. (1977) Helenieae—systematic review. In: *The Biology and Chemistry of the Compositae* (eds. V.H. Heywood, J.B. Harborne, and B.L. Turner), Academic Press, London, pp. 699–737.

Wagenitz, G. (1976) Systematics and phylogeny of the Compositae (Asteraceae). *Plant Syst. Evol.* **125**, 29–46.

Wagner, D.B., Furnier, G.R., Saghai-Maroof, M.A., Williams, S.M., Dancik, B.P., and Allard, R.W. (1987) Chloroplast DNA polymorphisms in lodgepole and jack pines and their hybrids. *Proc. Natl. Acad. Sci. USA* **84**, 2097–2100.

Wallace, R.S., and Jansen, R.K. (1990) Systematic implications of chloroplast DNA variation in *Microseris* (Asteraceae: Lactuceae). *Syst. Bot.* **15**, 606–616.

Watson, L., Jansen, R.K., and Estes, J. (1991) Tribal placement of *Marshallia* (Asteraceae) based on evidence from chloroplast DNA restriction site mapping. *Amer. J. Bot.*, in press.

Wolf, P.G., Soltis, D.E., and Soltis, P.S. (1990) Chloroplast-DNA and allozymic variation in diploid and autotetraploid *Heuchera grossulariifolia* (Saxifragaceae). *Amer. J. Bot.* **77**, 232–244.

Wolfe, K.H., and Sharp, P.M. (1988) Identification of functional open reading frames in chloroplast genomes. *Gene* **66**, 215–222.

Wolfe, K.H., Li, W.-H., and Sharp, P.M. (1987) Rates of nucleotide substitution vary greatly among plant mitochondrial, chloroplast, and nuclear DNAs. *Proc. Natl. Acad. Sci. USA* **84**, 9054–9058.

Wu, C.-I., and Li, W.-H. (1985) Evidence for higher rates of nucleotide substitution in rodents than in man. *Proc. Natl. Acad. Sci. USA* **82**, 1741–1745.

Wyatt, R.W., Odrzykoski, I.J., Stoneburner, A., Bass, H.W., and Galau, G.A. (1988) Allopolyploidy in bryophytes: multiple origins of *Plagiomnium medium*. *Proc. Natl. Acad. Sci. USA* **85**, 5601–5604.

Zurawski, G., Perrot, B., Bottomley, W., and Whitfeld, P.R. (1981) The structure of the gene for the large subunit of ribulose 1,5-bisphosphate carboxylase from spinach chloroplast DNA. *Nucleic Acids Res.* 9, 3251–3270.

12

Chloroplast DNA Restriction Site Variation and the Evolution of the Annual Habit in North American *Coreopsis* (Asteraceae)

Daniel J. Crawford, Jeffrey D. Palmer, and
Marilyn Kobayashi

Within any genus of flowering plants where both the annual and perennial habit occur, it is generally considered that the former has been derived from the latter, although each case must be decided on its own merits (Stebbins, 1958). Annual and perennial species often differ in a variety of biological attributes, including (1) degree of isolation and nature of isolating mechanisms (Grant, 1981), (2) breeding system and release of genetic variability (Stebbins, 1950, 1958), and (3) level of electrophoretically detectable variation at isozyme loci (Loveless and Hamrick, 1984). Despite the considerable differences that may exist in life history traits of annuals and perennials, little is known about the genetic bases of the different growth habits (Gottlieb, 1984; Macnair and Cumbes, 1989), although there is some evidence that few gene loci may control the difference.

Any meaningful comparative study of congeneric annuals and perennials must be based on knowledge of phylogenetic relationships among the species. Traditionally, morphology has been used to infer both the perennial ancestors of annual species, as well as whether a particular group of annual species represents a monophyletic assemblage. In the genus *Coreopsis* in North America, there is little question that the annual habit is derived (Smith, 1975; Jansen et al., 1987). Both the intuitive phylogeny of Smith (1975) and the cladistic analysis of Jansen et al. (1987) indicate that the annual habit has evolved in three different lineages of North American *Coreopsis* (Figs. 12.1 and 12.2). According to these studies, it evolved once within section *Coreopsis,* a second time in section *Calliopsis,* and a third time in the lineage that is ancestral to sections *Pugiopappus* and *Leptosyne.* In the first two sections, although annuals developed, there are also extant perennial species within each. By contrast, the hypothesis has been that species in sections *Pugiopappus* and *Leptosyne* represent the derivatives of a common ancestor, and no perennial species occur in either section. More detailed

This research was supported by NSF grant BSR-8521152 and an Ohio State University Seed Grant to DJC and by NSF grant BSR-8717600 to JDP.

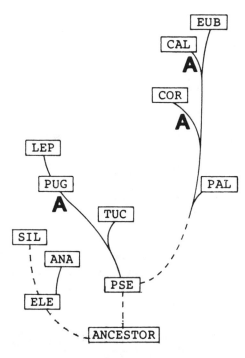

Figure 12.1. Phylogenetic hypothesis for sections of North American *Coreopsis*, redrawn and modified from Smith (1975) and Crawford and Smith (1983a). Sections toward the top are considered to be more specialized or advanced. Dashed lines denote uncertain affinities. Presence of the annual habit designated by **A.** Abbreviations for sections: *Anathysana*, ANA; *Calliopsis*, CAL; *Coreopsis*, COR; *Electra*, ELE; *Eublepharis*, EUB; *Leptosyne*, LEP; Palmatae, PAL; *Pseudoagarista*, PSE; *Pugiopappus*, PUG; *Silphidium*, SIL; *Tuckermannia*, TUC.

evidence supporting the derived nature of the annuals in *Coreopsis* will be presented below.

The value of restriction site analysis of chloroplast DNA (cpDNA) for inferring phylogenetic relationships in flowering plants has been demonstrated recently in a variety of taxa (Palmer and Zamir, 1982; Sytsma and Gottlieb, 1986a, b; Doebley et al., 1987; Jansen and Palmer, 1988; Soltis and Soltis, 1989; Soltis et al., 1989; Wendel, 1989). A number of reviews have appeared on the topic, with one of the most recent being the comprehensive overview by Palmer et al. (1988). There are a number of advantages of cpDNA data over morphologic information for constructing phylogenies (Palmer et al., 1988). Several of the most important in the context of the present discussion will be mentioned. The ability to recognize restriction enzyme cleavage site changes precisely in the cpDNAs, and to infer the direction of evolutionary change by outgroup analysis provides confidence in both the characters employed and the character site changes used to construct a cladogram. By contrast, parallel and/or convergent changes in morphologic characters associated with radiation into new habitats or changes in important life history parameters (such as becoming an annual) may obscure phylogenetic relationships. These problems are largely nonexistent for cpDNA within many genera or among closely related genera (Palmer et al., 1988).

In this chapter, we consider the use of cpDNA restriction site variation for examining the phylogenetic relationship of annuals and perennials in North Amer-

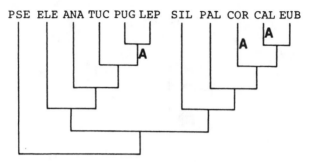

Figure 12.2. Cladogram of sectional relationships in North American *Coreopsis*, modified from Jansen et al. (1987). Presence of annuals indicated by **A**. See Jansen et al. (1987) for cladogram with all character state changes shown. Sectional abbreviations as in Fig. 12.1.

ican *Coreopsis,* and we compare our results with hypotheses formulated from other (primarily morphologic) data. The evolution of annuals within section *Coreopsis* and the two sections *Pugiopappus* and *Leptosyne* will be discussed. Section *Calliopsis,* mentioned earlier, will not be included for a variety of reasons, including the fact that a species may behave as an annual or perennial depending on the conditions under which it is grown. Thus the annual-perennial dichotomy is not clear-cut in the section. Also, extensive cpDNA data indicate that the chloroplast genomes in the three species of section *Calliopsis* are very similar, and thus the molecular data are of little phylogenetic value for inferring relationships among the species (Crawford et al., 1991).

The methods employed for the isolation of DNAs from species of *Coreopsis* (the digestion of the DNAs with restriction enzymes, separation of the resulting fragments by agarose gel electrophoresis, and blotting of the fragments onto nylon membranes) followed standard procedures (Palmer, 1986). In most cases, total DNA was isolated rather than obtaining purified cpDNA (Doyle and Doyle, 1987). Seventeen cloned restriction fragments from lettuce cpDNA were employed in filter hybridizations (Jansen and Palmer, 1987a); these fragments are known to be colinear with the chloroplast genome of *Coreopsis* (Palmer et al., 1988; Jansen and Palmer, 1987b).

The restriction site data were analyzed cladistically on an IBM microcomputer employing Wagner parsimony. This was achieved by using PAUP (Phylogenetic Analysis Using Parsimony) version 2.4.2 developed by David Swofford. Although both *Bidens coronata* and *Coreocarpus arizonicus* were used as outgroups, the direction of restriction site changes could be inferred by using only the former species.

Section *Coreopsis*

All available data for this section, which consists of five perennials (*C. auriculata, C. grandiflora, C. intermedia, C. lanceolata,* and *C. pubescens*) and four

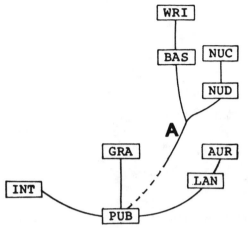

Figure 12.3. Proposed phylogeny for species of section *Coreopsis,* after Smith (1982). Species toward top of diagram are viewed as more specialized or advanced. Dashed line indicates uncertain relationships, and **A** denotes annuals. Abbreviations of species as follows: *C. auriculata,* AUR; *C. basalis,* BAS; *C. grandiflora,* GRA; *C. intermedia,* INT; *C. lanceolata,* LAN; *C. nuecensis,* NUC; *C. nuecensoides,* NUD; *C. pubescens,* PUB; *C. wrightii,* WRI.

annuals (*C. basalis, C. nuecensoides, C. nuecensis,* and *C. wrightii*), indicate that it represents a monophyletic group (Smith, 1982; Jansen et al., 1987). These include results from artificial hybridizations and cladistic analysis of morphologic characters. All species in the section are intercompatible and/or interfertile to some degree, and none of the species will cross with species from other sections (Smith, 1976, 1982). At least one morphologic feature serves as a synapomorphy for the section (Fig. 12.4).

Prior to a study of cpDNA variation in section *Coreopsis* by Crawford et al. (1990) two phylogenetic hypotheses had been produced for the section (Figs. 12.3 and 12.4). One of these hypotheses resulted from computing relative evolutionary advancement indices for each species and then drawing lines between the taxa by largely intuitive means (particularly uncertain affinities denoted by a dashed line; Fig. 12.3; Smith, 1982). The other phylogenetic hypothesis was generated by a cladistic analysis of morphologic features (Fig. 12.4; Jansen et al., 1987). The Smith (1982) study suggests that the annual habit is derived within the section and the cladistic study indicates that the annuals are a monophyletic group. In addition, Smith (1974, 1975, 1976) presented compelling cytogenetic and other biological data indicating the annual habit is the derived condition in section *Coreopsis.* The major phylogenetic questions addressed in these earlier studies were the relationships among the perennial species, which perennial species (perhaps more than one) were ancestral to the annuals, and whether the annual habit originated more than once in section *Coreopsis.* It is apparent from Fig. 12.3 that the ancestor(s) of the annuals was a matter of considerable uncertainty in the intuitive phylogeny, as denoted by the dashed line (Smith, 1982), and the cladistic analysis of morphologic features did little to resolve this problem because the annuals and perennials form an unresolved trichotomy (Fig. 12.4). The cladogram for section *Coreopsis* was based on a total of 15 character state changes, with nine of them representing parallelisms or reversals (Fig. 12.4).

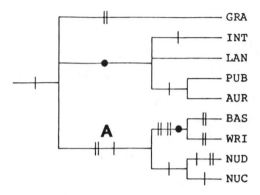

Figure 12.4. Cladistic relationships among species of section *Coreopsis*, from Jansen et al. (1987). Single lines represent unique character state changes, double bars are parallelisms, and dots designate reversals. Annual habit designated by **A.** Species designations same as in Fig. 12.3.

More recent analyses of the data given by Jansen et al. (1987) produced a strict consensus tree (i.e., placed together only those groups that appeared in the first 50 trees), as shown in Fig. 12.5. This tree likewise did little to elucidate relationships between annual and perennial species.

The purpose of the study of cpDNA variation in section *Coreopsis* was to elucidate further phylogenetic relationships among the species (Crawford et al., 1990). A total of 26 restriction enzymes was employed, and these surveyed approximately 1,060 cleavage sites, or 3.5% of the 151-kb genome. Fourteen restriction site mutations were detected among the 17 DNAs of the nine species, and one small insertion was found. The cladogram resulting from analysis of the restriction site data is shown in Fig. 12.6. Only the minimal 14 mutations must be postulated to account for the observed differences in the cpDNAs of section *Coreopsis,* that is, there is no homoplasy in the cladogram. Seven species group together as a result of sharing five derived restriction site mutations and the one length mutation. By contrast, two of the perennial species, *C. auriculata* and *C. pubescens,* exhibit no mutations (Fig. 12.6). This indicates that, with regard to cpDNA, the three remaining perennial species, *C. intermedia, C. grandiflora,* and *C. lanceolata,* are closer phylogenetically to the annual species than they are to the two aforementioned perennial species. These data establish that the perennial habit is primitive in section *Coreopsis,* and they effectively eliminate *C. auriculata* and *C. pubescens* as possible ancestors to the annual species, something that was not accomplished with either the intuitive phylogeny or the cladistic analysis of morphologic characters (Figs. 12.3–12.5). It is not possible, however, to infer from the cpDNA data which of the three perennial species, *C. grandiflora, C. intermedia,* or *C. lanceolata,* is ancestral to the annual species. In fact, the data suggest that none of the three extant species is the ancestor because the three perennials share six derived restriction site mutations not found in the annual species (Fig. 12.6).

Relationships among the annuals in section *Coreopsis,* as inferred from cpDNA restriction site changes, are different from the two previous phylogenies (Figs.

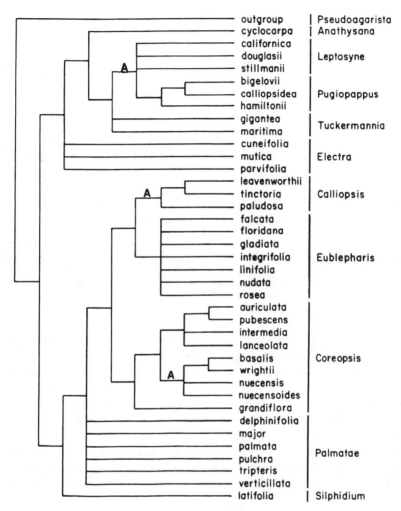

Figure 12.5. Strict consensus tree for species and sections of North American *Coreopsis* based on morphologic data given in Jansen et al. (1987). The tree shows those groups occurring in all 50 trees generated by Wagner parsimony using the PAUP program. **A** indicates the annual habit.

12.3–12.6). The prevailing view has been that *C. nuecensoides* and *C. nuecensis* represent one lineage, whereas the pair *C. basalis* and *C. wrightii* constitute the other (Smith, 1974, 1975, 1976, 1982; Crawford and Smith, 1982a, b). In addition to morphologic considerations, allozymes have been used to argue for these relationships (Crawford and Smith, 1982a, b), but a recent study comparing many more loci reveals that all four species are almost equally divergent from one another (Cosner and Crawford, 1990). The morphologic features uniting *C.*

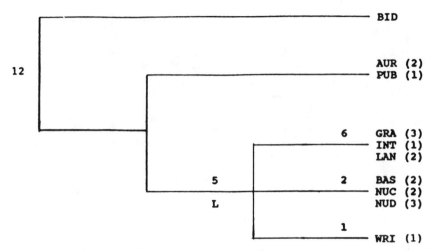

Figure 12.6. Cladistic relationships in section *Coreopsis* based on restriction site mutations of cpDNA, from Crawford et al. (1990). L denotes one length mutation and numbers refer to mutations along a branch. Numbers in parentheses denote numbers of populations examined for each species. Abbreviations for species as in Fig. 12.3.

basalis and *C. wrightii* consist of two parallelisms and one reversal; with regard to the other species pair, only a reduced chromosome number is unique to them (Fig. 12.4). It may be seen in the consensus tree (where chromosome number was not used) that *C. nuecensoides* and *C. nuecensis* did not form a group in all 50 trees (Fig. 12.5). However, it is important to emphasize that elegant cytogenetic studies by Smith (1974) argue strongly for *C. nuecensis* and *C. nuecensoides* being a monophyletic group, with the aneuploid reduction series seen in these species clearly the result of reciprocal translocations. Even though *C. wrightii* has been viewed as a derivative of *C. basalis* (Smith, 1982), the data upon which this inference was made are not as strong as the information for the other species pair.

The cpDNA data, by contrast, suggest that *C. basalis* belongs to the same lineage as *C. nuecensoides* and *C. nuecensis,* and *C. wrightii* constitutes another line. Thus, the cpDNA phylogeny does not view *C. basalis* and *C. wrightii* as a strictly monophyletic group, with the latter derived relative to the former. Rather, *C. basalis* is allied with the other two annual species. This represents a basic difference from previous phylogenies, and one might argue that it is based on only two restriction site mutations for the three species and only one autapomorphy in the DNA of *C. wrightii.* It would be desirable to have more characters; however, the lack of homoplasy in the cladogram (i.e., the species show very little divergence in their cpDNAs) argues against the shared site changes being the result of independent mutations. It is fair to say, therefore, that cpDNA restriction site data support the hypothesis of relationships among the two species of annuals for which the best biological data were available (i.e., *C. nuecensoides* and *C. nuecensis*), but they do not support the hypothesis for the two species (*C. basalis* and *C. wrightii*) for which previous data were much less compelling.

Hybridization cannot be discounted completely as an explanation for *C. basalis* having the same cpDNA genome as *C. nuecensoides* and *C. nuecensis;* however, this seems highly unlikely because of the high sterility of synthetic interspecific hybrids (Smith, 1976, 1982). Also there is no evidence of naturally occurring hybrids between *C. basalis* and the other two species.

In summary, in section *Coreopsis,* cpDNA data show two basic evolutionary lineages, one defined by five restriction site changes and containing three perennial species and the four annuals, and another consisting of the two perennials *C. auriculata* and *C. pubescens* which have no mutations, either individually or collectively. This indicates that, with respect to cpDNA, the former group of perennials is more closely related to the annual species than it is to the other two perennials. Also, the cpDNA data suggest that the annuals diverged from the perennials prior to the evolution of the three perennial species *C. grandiflora, C. intermedia,* and *C. lanceolata.* If this is true, it has important implications for attempting to understand the origin of the annuals from perennials, and perhaps it would be more meaningful to compare the annuals to the three perennial species collectively. Information from cpDNA has not been useful for resolving the question of whether the annuals originated more than once in section *Coreopsis.*

These results for section *Coreopsis* demonstrate both the strengths and limitations of using cpDNA for studying phylogenetic relationships among congeneric species. The low level of divergence at this taxonomic level effectively eliminates homoplasy as a problem. On the other hand, lack of divergence does not produce adequate variation to resolve relationships among all species.

Sections *Pugiopappus* and *Leptosyne*

The six species comprising sections *Leptosyne* and *Pugiopappus* occur almost exclusively in California, where they grow primarily in cismontane and desert habitats in southern California. These species were the subject of a detailed taxonomic study by Sharsmith (1938), and more recently Smith (1984) updated taxonomic information and provided data from artificial interspecific hybridizations. Both Sharsmith and Smith assigned the three species *C. bigelovii, C. calliopsidea,* and *C. hamiltonii* to section *Pugiopappus,* and *C. californica, C. douglasii,* and *C. stillmanii* were placed in section *Leptosyne.* These two workers viewed the sections as distinct, and the cladistic study of morphologic features by Jansen et al. (1987) suggested that each section is monophyletic, although *Pugiopappus* is supported by only one synapomorphy and *Leptosyne* is defined by two parallelisms (Fig. 12.7). However, the two sections collectively are defined by six synapomorphies, half of which are parallelisms. On the consensus tree, section *Pugiopappus* is monophyletic but species assigned to section *Lepto-*

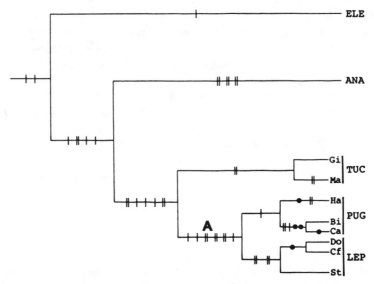

Figure 12.7. Cladistic relationships for western United States and Mexican sections of *Coreopsis,* from Jansen et al. (1987). For the complete cladogram with character state changes, the reader should see the original publication, or refer to Fig. 12.2 for an overview of sectional relationships. Sectional abbreviations as in Fig. 12.1, and **A** denotes the annual habit. Species abbreviations as follows: *C. bigelovii,* Bi; *C. calliopsidea,* Ca; *C. californica,* Cf; *C. douglasii,* Do; *C. gigantea,* Gi; *C. hamiltonii,* Ha; *C. maritima,* Ma; *C. stillmanii,* St.

syne do not always group together (Fig. 12.5). Although Smith (1975) initially considered *Leptosyne* to be ancestral and *Pugiopappus* derived from it, later versions of the phylogeny reversed this order (Crawford and Smith 1983a, b; Fig. 12.1).

Results of an extensive program of artificial interspecific hybridizations carried out by Smith (1975, 1976) have been most useful (together with other data) for delimiting most sections in North American *Coreopsis.* In nearly all cases, species assigned to a given section exhibit some degree of intercompatibility and/or interfertility, and crosses between sections fail in nearly every instance. The six species of annuals being considered here represent exceptions to the usual situation because all are totally cross-incompatible (Smith, 1984). Thus, the delimitation of sections has been, by necessity, almost totally on the basis of morphologic features. Several biological attributes of the species in sections *Pugiopappus* and *Leptosyne* suggest that they (and thus the annual habit) represent derived or specialized plants relative to the common perennial situation in North American *Coreopsis.* The plants are very reduced in size, occupy relatively young habitats in California, and their achenes (with the exception of *C. calliopsidea*) contain inhibitors that must be leached out by rain before germination will occur. The latter feature represents an adaptation to the habitats in which these species grow.

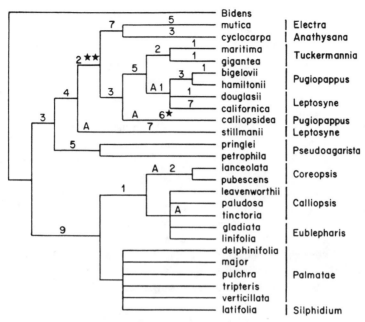

Figure 12.8. Cladogram for species and sections of North American *Coreopsis* based on cpDNA restriction site mutations (from Crawford et al., 1991). *Bidens coronata* and *Coreocarpus arizonicus* were used as outgroups. Numbers refer to restriction site changes along a branch. Stars refer to parallelisms or reversals. **A** indicates annual habit.

A study of variation in restriction enzyme cleavage sites has been carried out to address the questions of whether the annual habit is indeed derived, whether it evolved once or multiple times in California *Coreopsis* (i.e., do the six species constitute a monophyletic group?), and whether each of the two sections as now delimited is monophyletic. This is part of a larger study of phylogenetic relationships among the sections of North American *Coreopsis* (Crawford et al., 1991) and only those sections pertinent to the two portions of annuals will be considered here in any detail. The methods employed were the same as described in Crawford et al. (1990).

One most parsimonious cladogram resulted from the analysis of restriction site variation (Fig. 12.8). The "branch and bound" function of PAUP failed to produce any other equally parsimonious trees. The cladogram has a consistency index of 0.94, that is, there is only 6% homoplasy. The distribution of the perennial and annual habits in this cladogram allows one to conclude that the annuals represent the derived condition in North American *Coreopsis* as a whole. Phylogenetic relationships among the six California annual species depicted in this cladogram differ in several fundamental respects from all previous hypotheses presented for them. The cpDNA restriction site data not only suggest that the six annuals do not collectively form a monophyletic group but also indicate that each section as now recognized is not monophyletic (Fig. 12.8).

One interpretation of the cpDNA data is that the annual habit has originated *at least* three times in the six species of California *Coreopsis* rather than once, as was hypothesized in all previous phylogenies. One origin is in *Coreopsis stillmanii*, a second is in the common ancestor of *C. bigelovii*, *C. hamiltonii*, *C. douglasii*, and *C. californica*, and a third origin is in *C. calliopsidea* (Fig. 12.8). A second equally parsimonious explanation based solely on the cpDNA cladogram is that annuals originated initially only in the common ancestor of *C. stillmanii* and its sister clades composed of both annuals and perennials. This was followed by two cases of the derivation of perennials from annuals, once in the common ancestor of sections *Electra* and *Anathysana*, and once in section *Tuckermannia*. Although the cpDNA cladogram does not differentiate *per se* between these two alternative hypotheses, the weight of biological data argues strongly for the first hypothesis; in other words, the annuals represent the derived condition. The first hypothesis requires only that the annuals originated independently as they radiated into and adapted to the habitats they now occupy. The alternative hypothesis requires that the annual habit evolved early in the lineages leading to the Californian and Mexican plants, and this was followed by two independent reversals to the perennial habit. Evidence was presented earlier that the annual species in California represent the derived condition, that is, they are reduced in stature and have specialized germination requirements that represent an adaptation to the very young habitats where they now occur (Raven and Axelrod, 1978). It is therefore unlikely that the woody plants of sections *Electra* and *Anathysana* growing in the mountains of Mexico represent the derived condition.

Sharsmith (1938) and Smith (1984) both commented on relationships among species within each of the sections *Pugiopappus* and *Leptosyne*, and Jansen et al. (1987) produced a cladogram depicting relationships within each section (Fig. 12.7). Within section *Pugiopappus*, Sharsmith (1938) viewed *C. bigelovii* as the ancestral element from which *C. calliopsidea* and *C. hamiltonii* diverged independently, whereas Smith (1984) treated *C. calliopsidea* as ancestral with *C. bigelovii* and *C. hamiltonii* as the derived elements. The cladogram based on morphologic features places *C. bigelovii* and *C. calliopsidea* as a morphologic group with *C. hamiltonii* constituting another line (Fig. 12.7), and the same is true of the consensus tree (Fig. 12.5). It is difficult to compare the narrative phylogenies of Sharsmith (1938) and Smith (1984) with cladograms; nevertheless, it is clear that the view of Smith (1984) that *C. bigelovii* and *C. hamiltonii* are treated as derived is more similar to the cpDNA phylogeny (Fig. 12.8) than is Sharsmith's (1938) hypothesis that *C. calliopsidea* and *C. hamiltonii* are derived from *C. bigelovii*. It is also clear that cladistic relationships for species of section *Pugiopappus* are quite different when using morphology as contrasted with cpDNA (Figs. 12.5, 12.7, 12.8).

With regard to section *Leptosyne*, Sharsmith (1938) stated only that she considered *C. californica* and *C. douglasii* as more closely related to each other than either is to *C. stillmanii,* and Smith (1984) was in basic agreement with this view because he treated *C. stillmanii* as the primitive element with the other two species derived from it. The cladogram of Jansen et al. (1987) generated from morphology (Fig. 12.7) and cpDNA are concordant with the Sharsmith and Smith hypotheses, in the sense that both *C. californica* and *C. douglasii* are treated as more closely related to each other than either is to *C. stillmanii* (Fig. 12.8). In contrast, the consensus tree based on morphologic features fails to resolve relationships among the three species (Fig. 12.5).

The questions as to why the positions of *C. calliopsidea* and *C. stillmanii* are different in the cpDNA phylogeny as compared to other phylogenies and why the sections as traditionally recognized do not form monophyletic groups remain unanswered. Nevertheless, several comments are in order. First, the sections are delimited almost entirely on the basis of achene morphology and the nature of the pappus (Sharsmith, 1938; Smith, 1984). Features of the pappus and achenes in the Asteraceae have been shown to be under simple genetic control (Gottlieb, 1984), and this is true for *Coreopsis* (Smith, 1973) and the closely related genus *Coreocarpus* (Smith, 1989). It seems likely, therefore, that one could be misled if too much emphasis were placed on these features when inferring phylogenetic relationships. Sharsmith (1938) commented on problems of ascertaining relationships in section *Pugiopappus* by indicating that different inferences would be made if one emphasized vegetative rather than achenial characters. She also observed that *C. calliopsidea* and *C. stillmanii,* although from different sections, both have cauline rather than basal leaves as in the other four species, and that this character is probably primitive. Thus, even though achenial characters might appear to represent sound diagnostic characters, their validity for phylogenetic purposes is open to question. Also, because all species are totally cross-incompatible (Smith, 1984), it is not possible to assess relationships by determining the level of cross-compatibility and/or interfertility as has been done for other sections of North American *Coreopsis* (Smith, 1975, 1976).

In comparing the cpDNA cladogram with other phylogenetic hypotheses for the California annuals of *Coreopsis,* one point deserves emphasis. The number of character state changes supporting many of the groups is quite high, and there is nearly no homoplasy in the cladogram (Fig. 12.8). Also, it is possible to ascertain with considerable reliability the evolutionary direction of the site changes by outgroup analysis. By contrast, the majority of morphologic features in the cladogram represent parallelisms and reversals (Fig. 12.7). Whereas no one data set should be employed to the exclusion of all others for phylogenetic reconstruction, the cpDNA data do cast doubt on previous phylogenetic concepts in this group, particularly the sections as now delimited. Additional data sets, such as allozymes and nuclear DNA (rDNA restriction site or length mutations and sequencing), would be highly desirable for elucidating relationships among the annual species from California.

The following points may be made relative to cpDNA and the origin of the annual habit in North American *Coreopsis*. Within the monophyletic section *Coreopsis*, it appears that certain perennial species have diverged from the lineage of perennials that gave rise to the annuals subsequent to the origin of the annual lineage itself. It is also evident that morphologic features, particularly if too much emphasis is placed on certain characters without additional cytogenetic or other biological data, may be inadequate for resolving phylogenetic relationships. This appears to be particularly true for the California species. Lastly, including section *Calliopsis*, the annual habit may have originated five or more separate times in North American *Coreopsis* (Fig. 12.5).

References

Cosner, M.B., and Crawford, D.J. (1990) Allozyme variation in *Coreopsis* sect. *Coreopsis* (Compositae). *Syst. Bot.* **15**, 256–265.

Crawford, D.J., and Smith, E.B. (1982a) Allozyme variation in *Coreopsis nuecensoides* and *C. nuecensis* (Compositae), a progenitor-derivative species pair. *Evolution* **36**, 379–386.

Crawford, D.J., and Smith, E.B. (1982b) Allozyme divergence between *Coreopsis basalis* and *C. wrightii* (Compositae). *Syst. Bot.* **7**, 359–364.

Crawford, D.J., and Smith, E.B. (1983a) The distribution of floral anthochlor pigments in North American *Coreopsis* (Compositae): taxonomic and phyletic interpretations. *Amer. J. Bot.* **70**, 355–362.

Crawford, D.J., and Smith, E.B. (1983b) Leaf flavonoid chemistry of North American *Coreopsis* (Compositae): intra- and intersectional variation. *Bot. Gaz.* **144**, 577–583.

Crawford, D.J., Palmer, J.D., and Kobayashi, M. (1990) Chloroplast DNA restriction site variation and the phylogeny of *Coreopsis* section *Coreopsis* (Asteraceae). *Amer. J. Bot.* **77**, 552–558.

Crawford, D.J., Palmer, J.D., and Kobayashi, M. (1991) Chloroplast DNA restriction site variation, phylogenetic relationships and character evolution among sections of North American *Coreopsis* (Asteraceae). *Syst. Bot.* **16**, 211–224.

Doebley, J., Renfroe, W., and Blanton, A. (1987) Restriction site variation in the *Zea* chloroplast genome. *Genetics* **117**, 139–147.

Doyle, J.J., and Doyle, J.L. (1987) A rapid DNA isolation procedure for small quantities of fresh leaf tissue. *Phytochem. Bull.* **19**, 11–15.

Gottlieb, L.D. (1984) Genetics and morphological evolution in plants. *Amer. Natur.* **123**, 681–709.

Grant, V. (1981) *Plant Speciation,* Columbia University Press, New York.

Jansen, R.K., and Palmer, J.D. (1987a) Chloroplast DNA from lettuce and *Barnadesia* (Asteraceae): structure, gene localization, and characterization of a large inversion. *Curr. Genet.* **11**, 553–564.

Jansen, R.K., and Palmer, J.D. (1987b) A chloroplast DNA inversion marks an ancient evolutionary split in the sunflower family (Asteraceae). *Proc. Natl. Acad. Sci. USA* **84**, 5818–5822.

Jansen, R.K., and Palmer, J.D. (1988) Phylogenetic implications of chloroplast DNA restriction site variation in the Mutisieae (Asteraceae). *Amer. J. Bot.* **75**, 753–766.

Jansen, R.K., Smith, E.B., and Crawford, D.J. (1987) A cladistic study of North American *Coreopsis* (Asteraceae: Heliantheae). *Plant Syst. Evol.* **157**, 73–84.

Loveless, M.D., and Hamrick, J.L. (1984) Ecological determinants of genetic structure in plant populations. *Ann. Rev. Ecol. Syst.* **15**, 65–95.

Macnair, M.R., and Cumbes, Q.J. (1989) The genetic architecture of interspecific variation in *Mimulus*. *Genetics* **122**, 211–222.

Palmer, J.D. (1986) Isolation and structural analysis of chloroplast DNA. *Meth. Enzymol.* **118**, 167–186.

Palmer, J.D., and Zamir, D. (1982) Chloroplast DNA evolution and phylogenetic relationships in *Lycopersicon*. *Proc. Natl. Acad. Sci. USA* **79**, 5006–5010.

Palmer, J.D., Jansen, R.K., Michaels, H.J., Chase, M.W., and Manhart, J.R. (1988) Chloroplast DNA variation and plant phylogeny. *Ann. Missouri Bot. Gard.* **75**, 1180–1206.

Raven, P.H., and Axelrod, D.I. (1978) Origin and relationships of the California flora. *Univ. Calif. Publ. Bot.* **72**, 1–134.

Sharsmith, H.K. (1938) The native Californian species of the genus *Coreopsis* L. *Madroño* **4**, 209–231.

Smith, E.B. (1973) A biosystematic study of *Coreopsis saxicola* (Compositae). *Brittonia* **25**, 200–208.

Smith, E.B. (1974) *Coreopsis nuecensis* (Compositae) and a related new species from southern Texas. *Brittonia* **26**, 161–171.

Smith, E.B. (1975) The chromosome numbers of North American *Coreopsis*, with phyletic interpretations. *Bot. Gaz.* **136**, 78–86.

Smith, E.B. (1976) A biosystematic survey of *Coreopsis* in eastern United States and Canada. *Sida* **6**, 123–215.

Smith, E.B. (1982) Phyletic trends in section *Coreopsis* of the genus *Coreopsis* (Compositae). *Bot. Gaz.* **143**, 121–124.

Smith, E.B. (1984) Biosystematic study and typification of the Californian *Coreopsis* (Compositae) sections *Tuckermannia, Pugiopappus,* and *Euleptosyne*. *Sida* **10**, 276–289.

Smith, E.B. (1989) A biosystematic study and revision of the genus *Coreocarpus* (Compositae). *Syst. Bot.* **14**, 448–472.

Soltis, D.E., and Soltis, P.S. (1989) Allopolyploid speciation in *Tragopogon:* insights from chloroplast DNA. *Amer. J. Bot.* **76**, 1119–1124.

Soltis D.E., Soltis, P.S., and Ness, B.D. (1989) Chloroplast-DNA variation and multiple origins of autopolyploidy in *Heuchera micrantha* (Saxifragaceae). *Evolution* **43**, 650–656.

Stebbins, G.L. (1950) *Variation and Evolution in Plants,* Columbia University Press, New York.

Stebbins, G.L. (1958) Longevity, habit, and release of genetic variability in higher plants. *Cold Spring Harbor Symp. Quant. Biol.* **23,** 365–378.

Sytsma, K.J., and Gottlieb, L.D. (1986a) Chloroplast DNA evidence for the derivation of the genus *Heterogaura* from *Clarkia* (Onagraceae). *Proc. Natl. Acad. Sci. USA* **83,** 5554–5557.

Sytsma, K.J., and Gottlieb, L.D. (1986b) Chloroplast DNA evolution and phylogenetic relationships in *Clarkia* sect. *Peripetasma* (Onagraceae). *Evolution* **40,** 1248–1262.

Wendel, J.F. (1989) New World tetraploid cottons contain Old World cytoplasm. *Proc. Natl. Acad. Sci. USA* **86,** 4132–4136.

13

Molecular Systematics of Onagraceae: Examples from *Clarkia* and *Fuchsia*

Kenneth J. Sytsma and *James F. Smith*

Phylogenetic information is useful for experimentalists because it directs specific research into many worthwhile questions that may have seemed unapproachable

Gottlieb (1988a)

This insight is perhaps nowhere more appropriate than for a number of plant families, several of which are discussed in this book (see Chapters 10 and 11), that are the focus of extraordinarily detailed, multidisciplinary biosystematic studies. The Onagraceae, comprising seven tribes, 16 genera, and approximately 650 species (Raven, 1988), is undoubtedly one of the most thoroughly examined angiosperm families of moderate size (Raven, 1979, 1988; Sytsma and Smith, 1988). Abundant and detailed biosystematic information exists for cytology, breeding systems, morphology, vegetative and floral anatomy, and flavonoid chemistry (see the review in Raven, 1988). More recent systematic studies have used proteins and formal cladistic analyses (Gottlieb, 1986, 1988; Crisci and Berry, 1990). Given the wealth of information from traditional and more recent systematic approaches, molecular phylogenetic analyses in the Onagraceae are especially applicable and justified in order to place other data bases in a more accurate or detailed phylogenetic context.

The Onagraceae has served as an exemplary model for evolutionary studies dealing with, among other topics, catastrophic speciation (Lewis and Raven, 1958; Lewis, 1962), chromosomal variation (Lewis, 1953a, b, 1980; Kurabayashi et al., 1962; Small 1972a, b), breeding systems (Godley, 1955, 1963; Vasek, 1958; MacSwain et al., 1973; Arroyo and Raven, 1975; Raven, 1979), hybridization (Raven and Raven, 1976; Seavey, 1977); progenitor-derivative species relationships (Lewis and Roberts, 1956; Gottlieb, 1973, 1974a; Gottlieb and Pilz, 1976), and gene duplications (Gottlieb, 1974b, 1977, 1982, 1983, 1988b; Gottlieb and Weeden, 1979; Pichersky and Gottlieb, 1983; Odrzykoski and Gottlieb, 1984; Soltis et al., 1987). The application of DNA variation to evolutionary and systematic problems in the Onagraceae thus makes possible potentially informative comparisons among a diversity of biosystematic data sets.

In this chapter, we summarize the current information gathered with DNA methods as they relate to evolutionary and systematic questions in the Onagraceae, specifically in two of the most thoroughly studied genera, *Clarkia* and *Fuchsia*. The current DNA-based phylogenetic models for both *Clarkia* and *Fuchsia* are presented. An assessment of congruency between the different kinds of traditional and other molecular data and DNA data follows. Lastly, we examine the general problem of using both DNA and morphology in systematic and evolutionary studies and illustrate how one robust data set can provide additional information about character state changes in a less robust set.

Current Molecular Phylogenetic Knowledge

Molecular Phylogenetics of Clarkia *Sections*

Clarkia is a distinctive genus in the tribe Onagreae and comprises approximately 44 annual species mostly confined to western, temperate North America. Lewis and Lewis (1955) combined the then-recognized genera *Clarkia, Eucharidium, Godetia,* and *Phaeostoma* on the bases of floral and vegetative morphology, chromosome number, and crossing relationships. Diploid species were placed in seven sections: *Sympherica* (formerly *Peripetasma*), *Fibula, Phaeostoma, Eucharidium, Godetia, Rhodanthos* (formerly *Primigenia*), and *Myxocarpa*. Polyploid species thought or known to be of intersectional hybrid origin were placed in the strictly polyploid sections *Clarkia, Biortis,* and *Connubium,* whereas the polyploid species of intrasectional hybrid origin were placed in the sections of their diploid progenitors. Morphological and cytological data (Raven, 1979, 1988; Tobe and Raven, 1985, 1986) and genetic evidence (Gottlieb and Weeden, 1979; Pichersky and Gottlieb, 1983) strongly suggest that *Clarkia* is monophyletic.

Relationships among sections of *Clarkia* based on morphology, karyology, and crossing relationships as first postulated (Lewis and Lewis, 1955) and later modified (Lewis, 1980) are presented in Fig. 13.1. The large systematic and evolutionary information base available in *Clarkia* and the specific progenitor-derivative species relationships identified in the genus have made *Clarkia* a model system for many subsequent evolutionary studies spanning 35 years (Lewis, 1953a, b, 1962, 1980; Lewis and Roberts, 1956; Lewis and Raven, 1958; Small et al., 1971; Small, 1972a, b; Gottlieb, 1973, 1974a, 1986, 1988b; Gottlieb and Weeden, 1979; Averett and Raven, 1984; Averett et al., 1986; Soltis, 1986; Smith-Huerta, 1986; Sytsma and Gottlieb, 1986a, b; Soltis et al., 1987; Holsinger and Gottlieb, 1988; Raven, 1988; Narayan, 1988; Sytsma and Smith, 1988; Sytsma et al., 1990).

The cpDNAs of 23 *Clarkia* species have been examined in a phylogenetic manner (Sytsma and Gottlieb, 1986a, b; Sytsma and Smith, 1988; Sytsma et al., 1990, unpublished data). The best current estimate of relationships in *Clarkia* based on parsimony (Wagner and Dollo) analysis of cpDNA restriction site data is illustrated in Fig. 13.2. Because *Clarkia* cpDNA is maternally inherited (Syt-

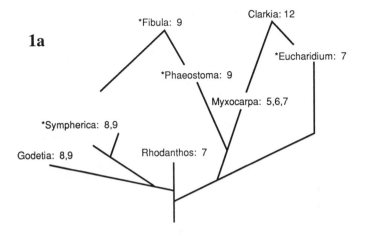

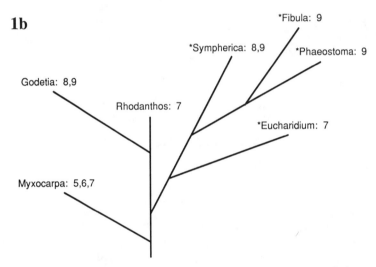

Figure 13.1. A—Phyletic relationships among *Clarkia* sections based on morphology, cytology, and crossing relationships (after Lewis and Lewis, 1955). B—Phyletic relationships among *Clarkia* sections based on Lewis and Lewis (1955) but also incorporating the distribution of the phosphoglucose isomerase gene duplication presented by Gottlieb and Weeden (1979) (after Lewis, 1980).

sma et al., 1990), the cladogram represents maternal relationships. This clado-gram depicts the early divergence of sect. *Myxocarpa,* with a later trichotomy of sect. *Rhodanthos,* sect. *Eucharidium,* and the lineage including sects. *Sympherica, Phaeostoma, Fibula,* and *Godetia.* It should be stressed, however, that when trees only one step longer are considered, the resulting consensus cladogram (see Sytsma et al., 1990) provides considerably less information about relationships among sections. Sections *Sympherica, Phaeostoma, Fibula,* and *Godetia* form a

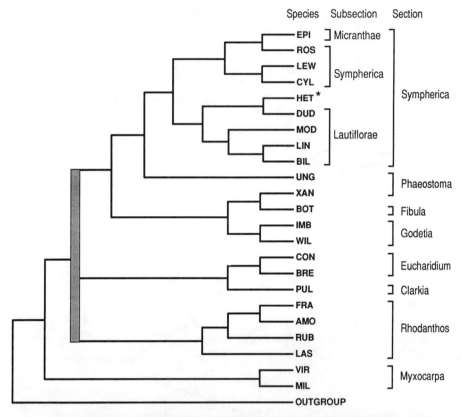

Figure 13.2. Chloroplast DNA cladogram of *Clarkia* (after Sytsma and Gottlieb, 1986b; Sytsma et al., 1990, unpublished data). The consensus cladogram depicts relationships among sections using 18 restriction enzymes and within sect. *Sympherica* (=*Peripetasma*) using 29 enzymes. Species are abbreviated following Sytsma and Gottlieb (1986b) and Sytsma et al. (1990). The asterisk indicates the position of *Heterogaura heterandra.*

monophyletic lineage in all most parsimonious trees, although the bootstrap support for this grouping (27%) is weak. *Clarkia bottae* of sect. *Fibula* is closely related to *C. xantiana* of sect. *Phaeostoma*. In fact, these two species share a more recent common ancestor than does *C. xantiana* with *C. unguiculata* of the same section. These results support, but do not prove, the suggestion that sect. *Fibula* has a diploid hybrid origin between species of sect. *Phaeostoma* and *Sympherica*. The stems, leaves, and position of buds in sect. *Fibula* are similar to those in sect. *Phaeostoma,* and the flowers are similar to those in sect. *Sympherica* (Lewis and Lewis, 1955). However, the relationship is with *C. xantiana* rather than with *C. unguiculata* as originally proposed. Further analysis of nuclear rDNA may provide the requisite data to demonstrate the hybrid origin of *C. bottae*. The placement of *C. unguiculata* at the base of sect. *Sympherica* is somewhat supported, suggestive that sect. *Phaeostoma* might be paraphyletic.

The grouping of sect. *Godetia* with sects. *Sympherica, Phaeostoma,* and *Fibula* places all *Clarkia* species with $n = 8$ or 9 together and excludes all those with $n = 7$. Because $n = 7$ has been argued to be the plesiomorphic condition in *Clarkia* (Lewis, 1953a, b, 1980; Small, 1972b), the aneuploid increase to $n = 8$ and 9 would have had to have occurred only once in *Clarkia*. Lastly, the diploid sect. *Eucharidium* and the allotetraploid sect. *Clarkia* form a monophyletic lineage.

Only two diploid sections (*Myxocarpa* and *Eucharidium*) are ever placed at the base of *Clarkia*. Section *Myxocarpa* is the sister group to the rest of the genus in the majority of the Wagner cladograms and in all Dollo cladograms. In the remaining cladograms, sect. *Myxocarpa* is the second lineage to diverge following the early divergence of sect. *Eucharidium*. These two diploid sections possess a number of derived morphological and cytological character states. Section *Myxocarpa* shows a strong aneuploid reduction from $n = 7$ to $n = 5$, and is the only section with $n = 5$. Petals in the section are constricted into a claw that expands into a pair of lateral teeth or lobes near the base; flaps of sterile tissue are present at the base of stamens (Lewis and Lewis, 1955; Lewis, 1980). The two species of sect. *Eucharidium,* undoubtedly the most divergent species morphologically in the genus, retain the plesiomorphic chromosome number of $n = 7$ but differ from all other diploid clarkias in possessing (1) distinctive 3-limbed petals, (2) an elongated and narrow floral tube adapted to pollination by butterflies, moths, and long-tongued flies rather than bees, (3) a reduction of stamens from eight to four, and (4) distinctive pollen exine (Lewis and Lewis, 1955; Small et al., 1971; MacSwain et al., 1973; Lewis, 1980). Despite being viewed as the extant section with the largest suite of primitive morphological character states (Lewis and Lewis, 1955; Lewis, 1980), sect. *Rhodanthos* diverges later in the cladogram.

In summary, the *Clarkia* cpDNA cladogram indicates that species within sections generally form well-demarcated lineages similar to those proposed on more traditional information (Lewis and Lewis, 1955). Resolution of relationships among sections is not yet possible due to the low number of synapomorphic characters shared by two or more sections. Further analysis of cpDNA with more restriction enzymes or analysis of the nuclear ribosomal DNA (or other nuclear genes) may or may not resolve these relationships. Rapid and early radiation in *Clarkia* may be responsible for the lack of synapomorphic characters (Sytsma et al., 1990).

Molecular Phylogenetics of
Clarkia *Sect.* Sympherica

The phylogenetic relationships within sect. *Sympherica* (= *Peripetasma*) have been of special interest because of evolutionary and systematic questions directed at species within the section. Subsection *Micrantheae* consists of one strictly self-pollinating species with small, inconspicuous, white flowers. Subsections

Sympherica and *Lautiflorae* consist of primarily outcrossing species with large, showy, colorful flowers. The latter two outcrossing subsections also differ from each other in coloration and flecking of petals and in the shape and ribbing of capsules.

All species of sect. *Sympherica* have been examined for cpDNA restriction site variation (Sytsma and Gottlieb, 1986a, b), and their placement within the genus is depicted in the *Clarkia* consensus cladogram (Fig. 13.2). The species of subsect. *Lautiflorae* form a significant monophyletic group (98% confidence by bootstrap analysis). Subsection *Sympherica*, however, is paraphyletic with a close relationship (100% confidence level) between *Clarkia lewisii* and *C. cylindrica*, but with *C. rostrata* forming a tight relationship (100% confidence level) with *C. epilobioides*, the single species of subsect. *Micrantheae*. These relationships and those with the genus *Heterogaura* are discussed in detail later.

Molecular Phylogenetics of Fuchsia *Sections*

Fuchsia is a distinctive genus in the Onagraceae with its fleshy fruits, corolloid calyx and unspecialized features in floral and vegetative anatomy (Raven, 1979). It consists of 105 species in 10 sections (Berry, 1982; Raven, 1988; Berry et al., 1988). This genus of outcrossing shrubs and some trees is centered in tropical Andean cloud forests and temperate forests in the southernmost Andes and mountains of southeastern Brazil. In this region are found 60 species of sect. *Fuchsia*, 14 species of sect. *Hemsleyella*, eight species of sect. *Quelusia*, and the monotypic sects. *Kierschlegeria* and *Pachyrrhiza* occur in this region. The 12 species of sects. *Ellobium*, *Encliandra*, *Jimenezia*, and *Schufia* occur in Mexico and Central America. The remaining four species are found only in New Zealand and Tahiti and form the distinctive Old World sect. *Skinnera* to be discussed in detail below.

Most lines of evidence point to an origin of *Fuchsia* in austral temperate forests of South America (or possibly Antarctica and Australia) in the Oligocene or Miocene (Raven and Axelrod, 1974; Berry, 1982; Raven, 1988; Berry et al., 1990). Species of the South American sect. *Quelusia*, restricted to temperate regions of South America, possess the largest suite of generalized characters in the genus and may represent the extant section most similar to ancestral fuchsias. These characters include shrubby habit, bisexual flowers, well-developed petals, bird pollination, numerous seeds, and segmented-beaded viscin pollen threads (Skvarla et al., 1978; Berry, 1982; Nowicke et al., 1984; Averett et al., 1986; Raven, 1988). The first offshoot in *Fuchsia* probably involved the lineage that dispersed to New Zealand and subsequently Tahiti (Fig. 13.3). This lineage, now comprising four species in sect. *Skinnera*, diverged at least 25–30 million years ago based on pollen fossils (Pocknall and Mildenhall, 1984; Daghlian et al., 1985; Berry et al., 1990). *Skinnera* is the most distinctive section in the genus with the advanced conditions of male sterility (Godley, 1955), reduced petals,

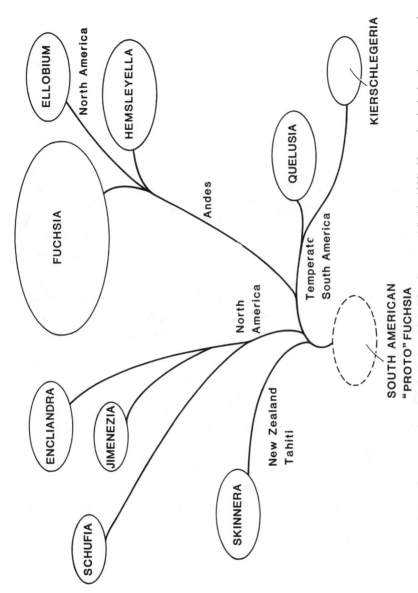

Figure 13.3. Schematic representation of evolution in *Fuchsia* (after Sytsma and Smith, 1988). No section is depicted as ancestral to the genus. Section *Quelusia*, however, is depicted as retaining the greatest number of plesiomorphic character states and sect. *Skinnera* as the basal lineage within the genus (after Sytsma and Smith, 1988).

and varying life forms that include a tree, a scandent shrub, and an almost herbaceous creeper.

Other early dispersal events probably included the ancestor to the morphologically similar sections *Encliandra, Jimenezia,* and *Schufia* of Mexico and Central America (Fig. 13.3). The two most species-rich sections, *Fuchsia* and *Hemsleyella,* primarily inhabit the moist slopes of the tropical Andes. These two sections most certainly evolved rapidly as the Andes uplifted to their present height over the past few million years (Berry, 1982; Raven, 1988). The fourth Mexican and Central American section, *Ellobium,* is related to these two Andean sections and represents an additional, and probably Neogene, dispersal northward (Fig. 13.3).

The cpDNAs of 17 species of *Fuchsia* (plus two hybrids), representative of all 10 sections, have been examined for restriction site variation, some in preliminary fashion (Sytsma and Smith, 1988) and others in considerable detail (Sytsma et al., 1991). A consensus cpDNA cladogram (Fig. 13.4) depicts the best estimate of maternal relationships (Sytsma et al., 1991) of the genus *Fuchsia.* The Old World sect. *Skinnera* and the monotypic Central American sect. *Jimenezia* are the first lineages to split off from the presumed ancestral *Fuchsia* stock in temperate austral forests. The consensus tree places sect. *Skinnera* as the sister group to all other *Fuchsia* sections including *Jimenezia.* The ancient split of sect. *Skinnera* thus mirrors the fossil pollen record and the spectacular morphological and flavonoid divergence seen within the section (see below).

The nearly basal position of sect. *Jimenezia* is unexpected and contrary to relationships judged from floral characters. *Fuchsia jimenezii* has been suggested to occupy an intermediate position between the Central American sections *Encliandra* and *Schufia* (Berry, 1982) based on shared derived characters of small flowers, smooth viscin threads, special lobed nectaries, and one set of stamens reflexed with the floral tube (Breedlove, 1969; Breedlove et al., 1982; Berry, 1982).

The remaining eight sections of *Fuchsia* are a monophyletic lineage (Sytsma and Smith, 1988), but relationships among these sections are not yet resolved. Only three lineages are apparent. The first lineage comprises sects. *Kierschlegeria* and *Quelusia,* previously considered related based on their temperate South American distribution and polyploid nature (Raven, 1988). The second lineage consists solely of sect. *Hemsleyella.* The last lineage includes an unresolved polychotomy comprising the five sects. *Schufia, Encliandra, Fuchsia, Ellobium,* and *Pachyrrhiza.* The lack of substantial numbers of cpDNA site synapomorphies linking any of these sections (except for *Kierschlegeria* and *Quelusia*) is reminiscent of the situation among sections of *Clarkia* and suggestive that *Fuchsia* diverged rapidly and profusely following the early separation of sect. *Skinnera* and sect. *Jimenezia.*

Molecular Phylogenetics of Fuchsia *Sect.* Skinnera

Understanding the relationships of the species comprising *Fuchsia* sect. *Skinnera* in the Old World has been of critical importance for addressing questions

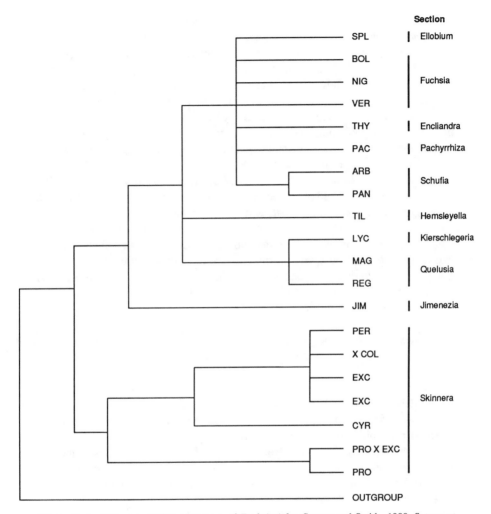

Figure 13.4. Chloroplast DNA cladogram of *Fuchsia* (after Sytsma and Smith, 1988; Sytsma et al., 1991). The consensus cladogram depicts relationships among sections using 11 restriction enzymes and within sect. *Skinnera* using 30 enzymes. Four deletions and one insertion are additionally shared by all members of sect. *Skinnera* exclusive of *F. procumbens*. Species are abbreviated following Sytsma and Smith (1988) and Sytsma et al. (1991).

dealing with evolution of morphological, chemical, and breeding-system characters and in understanding biogeographic distributions (see "DNA and Morphology," in this chapter). Four species are recognized in the section and have been the subject of a morphologically based cladistic analysis (Crisci and Berry, 1990). Three occur on New Zealand: *Fuchsia excorticata,* the widespread tree fuchsia; the shrubby or lianoid *F. perscandens;* and *F. procumbens,* a rare, almost herbaceous, creeper endemic to the northern third of North Island. One frequent and naturally occurring hybrid, *Fuchsia × colensoi,* is a cross between *F. perscandens* and *F. excorticata* (Allan, 1927, 1928). *Fuchsia cyrtandroides,* a small tree, is restricted to high peaks in Tahiti, 3,800 km northeast of New Zealand. All the New Zealand taxa are gynodioeceous, a condition governed by a dominant gene (Godley, 1955, 1963). In addition, a linked dominant gene for female sterility present in *F. procumbens* causes subdioecy. *Fuchsia cyrtandroides* from Tahiti lacks the dominant alleles and has only hermaphroditic flowers. All New Zealand taxa also possess sulphated flavones, compounds lacking in *F. cyrtandroides* and other Onagraceae (Averett et al., 1986; Williams and Garnock-Jones, 1986).

Restriction site variation in the cpDNAs of all species and the interspecific hybrid have been analyzed phylogenetically (Fig. 13.4; Sytsma et al., 1991). The creeping *Fuchsia procumbens* is situated basally within the section and represents the sister group to all other members of the section. Eleven restriction site mutations, four deletions, and one insertion define as a monophyletic lineage the two other New Zealand species (plus their natural interspecific hybrid) and the Tahitian *F. cyrtandroides*. The shrubby, sometimes lianoid *F. perscandens* and the large tree *F. excorticata* are indistinguishable based on DNA characters and represent a very close maternal lineage.

Molecules and Morphology in
Clarkia and *Fuchsia*

The detailed phylogenetic relationships within *Clarkia* and *Fuchsia* obtained from cpDNA analysis permit informative comparisons of morphology, proteins, and DNA. The general issue of molecules and morphology as sets of data for inferring plant phylogeny has been examined elsewhere in some detail (Sytsma, 1990). Here we wish to describe, from *Clarkia* and *Fuchsia,* examples of lineages in which morphological divergence and DNA divergence are both correlated and not correlated. Four possible situations exist in reference to relative amounts of divergence in morphology and DNA. Two involve either low or high amounts of divergence, but the amount of divergence is similar for both morphology and DNA. The other two involve discordant amounts of divergence, with high amounts of divergence in morphology but low amounts of divergence in DNA, and vice versa.

Low Morphological and DNA Divergence

Clarkia biloba *and* Clarkia lingulata

A large amount of systematic information indicates that *C. biloba* and *C. lingulata* are one of the closest pairs of species within the genus. *Clarkia lingulata* is known from only one region in the southern margin of the range of *C. biloba*.

The two species differ morphologically only in the absence of bilobed petals in
C. lingulata. In fact, the two species cannot be separated based on morphological
characters prior to flowering. Despite their morphological similarity, however,
the two are maintained as separate species, because a strong reproductive isolating
barrier exists in the form of an additional chromosome in *C. lingulata* ($n = 9$)
relative to *C. biloba* ($n = 8$). The low amount of morphological divergence
between the two species is mirrored in cpDNA divergence. The two species form
one of the closest species pairs known in *Clarkia* with an average nucleotide
genetic distance (*p* value) of 0.17% based on a screening with 29 restriction
enzymes (Fig. 13.2).

Fuchsia excorticata *and* F. perscandens

The two woody species of *Fuchsia* in New Zealand, *F. excorticata* and *F.
perscandens,* have been demonstrated to be sister species based on vegetative,
floral, and flavonoid characters (Crisci and Berry, 1990). Although *F. excorticata*
is a tree and *F. perscandens* is a shrub or shrubby liana, the two species are
nearly identical in floral morphology with a constricted floral tube and male
sterility. Moreover, unlike with *Clarkia biloba* and *C. lingulata,* a naturally
occurring and fully fertile hybrid in areas of sympatry indicates that the two
fuchsias are similar in chromosomal and genetic features. An analysis of the
cpDNA of the two species and their hybrid with 30 restriction enzymes showed
virtually no nucleotide divergence. One population of *F. excorticata* exhibited a
single autapomorphic site mutation relative to another conspecific population and
to *F. perscandens* (Sytsma et al., 1991; Fig. 13.4).

High Morphological and DNA Divergence

Clarkia *sects*. Myxocarpa *and* Eucharidium

Sections *Myxocarpa* and *Eucharidium* are two of the most distinctive sections
in *Clarkia*. As described above, these two sections exhibit numerous changes in
floral, vegetative, and chromosomal features relative to other species of *Clarkia*.
The morphological divergence of sect. *Eucharidium* is sufficiently pronounced
that it had been segregated out of *Clarkia* prior to the monographic work of Lewis
and Lewis (1955). In the latest sectional alignment of *Clarkia,* Lewis (1980)
places sect. *Myxocarpa* as the basal lineage in the genus (Fig. 13.1B). Restriction
site mapping and subsequent phylogenetic analysis of cpDNA indicate that these
are the only two sections implicated as basal lineages within the genus (Sytsma
et al., 1990; Fig. 13.2). The morphological and molecular divergence of the two
sections is even more striking when it is remembered that these two sections,
now completely allopatric, were involved in the polyploid origin of *C. pulchella*,

a wide-ranging species not overlapping with the ranges of the two parental species (see "DNA and Origin of Polyploids," in this chapter).

Fuchsia *sect.* Skinnera

The Old World sect. *Skinnera* is distinctive from all other sections of *Fuchsia* by the presence of blue pollen in all species, petals strongly reduced or lacking, male sterility governed by a dominant gene, and a diversity of growth forms from a small, almost herbaceous, creeper to a large tree (Crisci and Berry, 1990). The section is so distinctive morphologically, as well as in the ubiquitous presence of flavones and in its peculiar Old World distribution, that it has been considered one of the basal lineages within the genus (Berry, 1982; Fig. 13.3). Moreover, fossil pollen of *Fuchsia* has been found in 30-million-year-old rocks in New Zealand (Pocknall and Mildenhall, 1984) and Australia (Berry et al., 1990), attesting to the probable antiquity of the Old World section. The amount of cpDNA divergence between sect. *Skinnera* and all other New World sections reflects this great morphological divergence (Sytsma and Smith, 1988; Sytsma et al., 1991; Fig. 13.4). For example, between sect. *Fuchsia* (Andean) and sect. *Skinnera*, 116 restriction site mutations have been recorded with 30 restriction enzymes for a maximum nucleotide divergence of 2.7% (Sytsma et al., 1991). This value approaches the greatest p value recorded in *Clarkia* (3.5%; Sytsma et al., 1990) and half the largest p value found within the entire Asteraceae (5.4%; Jansen and Palmer, 1988).

Low Morphological Divergence and High DNA Divergence

Clarkia rostrata *and* Clarkia lewisii/Clarkia cylindrica

Three species have been traditionally placed within subsect. *Sympherica: C. rostrata, C. lewisii,* and *C. cylindrica* (Lewis and Lewis, 1955; Davis, 1970; Fig. 13.2). The close phylogenetic relationships of the three species relative to other species in sect. *Sympherica* had not been questioned until recently. Lewis and Lewis (1955) first considered the populations now composing *C. rostrata* as unusual northern members of *C. cylindrica,* but Davis (1970) later segregated *C. rostrata* and indicated that it was more similar to *C. lewisii* than to *C. cylindrica.* Moreover, *C. rostrata* can be crossed successfully with the former but not the latter (Davis, 1970). Davis (1970) suggested that *C. rostrata* and *C. lewisii* may have become restricted in their distribution and perhaps even been involved in the origin of *C. cylindrica.* This suggestion was based on the northern distribution and wide separation of *C. rostrata* and *C. lewisii,* the continuous and more southern distribution of the two subspecies of *C. cylindrica,* and the directional trend of speciation in *Clarkia* from north to south and from mesic to more xeric habitats (Lewis, 1962; Lewis and Roberts, 1956; Lewis and Raven, 1958; Vasek,

1958; Gottlieb, 1974a). *Clarkia rostrata* is difficult to separate from the other two species when locality information and mature fruits are not available. This difficulty disappears when either nuclear ribosomal DNA (rDNA) or cpDNA are examined (Sytsma and Gottlieb, 1986b; Sytsma and Smith, 1988). Although *C. lewisii* and *C. cylindrica* are a close sister-species pair, *C. rostrata* differs from the two by 47–52 restriction site mutations. In addition, *C. rostrata* is more closely related to *C. epilobioides* of subsect. *Micranthae* than it is to the species that it so closely resembles in morphology.

This example strongly suggests that uncoupling of morphological divergence and molecular divergence has occurred in this group. Most likely, *C. rostrata* has such a remarkable overall similarity to *C. lewisii* and *C. cylindrica* not because of convergence, but rather because of an extreme slo·/down of morphological evolution in the three species since their common ancestor (Sytsma and Smith, 1988). The overall morphological similarity of *C. rostrata* to the other two is thus due to almost complete retention of symplesiomorphic features in the three species. This scenario is supported by (nuclear-encoded) isozyme number (as described later in the section entitled "Comparison of Isozymes and Chloroplast DNA").

High Morphological Divergence and Low DNA Divergence

Clarkia epilobioides *and* Clarkia rostrata

Just as a low amount of morphological change but fairly substantial DNA divergence has occurred between *C. rostrata* and *C. lewisii* and *C. cylindrica*, the opposite kind of divergence is seen between *C. rostrata* and *C. epilobioides*. The latter two species are a very close species pair within *Clarkia*. Their *p* value of 0.0065 is one of the lowest known in the genus. Gene number differences with isozymes also support this relationship (see the next section, "Comparison of Isozymes and Chloroplast DNA"). This close genetic relationship, however, is not evident with morphological comparisons. As noted earlier, *C. epilobioides* is an obligate inbreeder with inconspicuous, white flowers whereas *C. rostrata* is an outcrosser with large, colorful flowers. *Clarkia epilobioides*, being closely related to *C. rostrata*, exemplifies a lineage that has undergone tremendous morphological divergence relative to its sister lineage. Because it is exclusively inbreeding, *C. epilobioides* was certainly derived from an outcrossing taxon constituting the direct ancestor of *C. epilobioides* and *C. rostrata*.

Heterogaura heterandra *and* Clarkia dudleyana

The most dramatic example of rapid morphological divergence relative to DNA change known in the Onagraceae involves *Heterogaura*. *Heterogaura* is a monotypic genus most similar morphologically to *Clarkia* and occurs on the

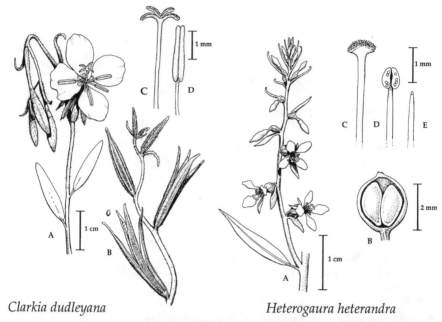

Clarkia dudleyana   Heterogaura heterandra

Figure 13.5. Flowers and fruits of *Clarkia dudleyana* and *Heterogaura heterandra* of the Onagra-ceae: A—inflorescence; B—fruit; C—style; D—stamen; E—staminode (after Sytsma, 1990).

slopes of the Sierra Nevada in California and Oregon. *Heterogaura heterandra* differs markedly from *Clarkia* in having only four fertile anthers (four are sterile), an unlobed stigma, and a round, nutlike, indehiscent fruit with one or two seeds (Fig. 13.5). In contrast, species of *Clarkia* generally have eight fertile anthers, four-lobed stigmas (although self-pollinating species have short lobes), and elon-gated and many-seeded capsules that dehisce along four septa. The floral and fruit differences between the two genera are so distinctive that they have been maintained as separate genera since 1864 (Rothrock, 1864).

Restriction site analysis of cpDNA and rDNA surprisingly placed *Heterogaura heterandra* firmly within *Clarkia* subsect. *Lautiflorae* with *C. dudleyana* as its sister species (Fig. 13.2; Sytsma and Gottlieb, 1986a, b; Sytsma and Smith, 1988). The two species share nine cpDNA synapomorphies despite the extensive morphological divergence that separates the two. Synapomorphic morphological characters uniting *Heterogaura* with subsect. *Lautiflorae* and specifically *C. dudleyana* undoubtedly exist (although none has yet been found), but they may be found *a posteriori* and as a direct result of the compelling cpDNA and rDNA data. It is of interest that both examples presented here, illustrating rapid morphological divergence within specific *Clarkia* lineages, involve two highly self-pollinating species.

Surely as work continues on the molecular systematics of the Onagraceae, and other plant families, more examples of correlated and noncorrelated rates of morphological and DNA change will be found. Interpretations of discrepancies in rates of change have been relatively easy with the Onagraceae due to the tremendous body of knowledge that exists for the biology and natural history of the family. Future work that focuses on discrepancies in rates of change in other lesser known families will depend on obtaining such data.

Comparison of Isozymes and Chloroplast DNA

Nine different duplications of nuclear genes encoding isozymes have been identified within diploid *Clarkia* species (Gottlieb, 1988b), making it probably the most thoroughly studied genus of plants or animals for this type of mutation. The unlikely series of events that must occur to facilitate the presence of duplicate and, at the same time, independently assorting nuclear genes (Gottlieb, 1981, 1982, 1983) is compelling evidence for the rarity of such gene duplications and thus for their intrinsic phylogenetic significance in defining monophyletic lineages. Despite the rarity of these duplications, three possibilities should be considered when using such characters for phylogenetic purposes: (1) convergent origin of the duplication within two distinct lineages; (2) subsequent silencing of either gene in one or several lineages; and (3) the duplication being the plesiomorphic condition at the hierarchical level examined. The evidence to date in *Clarkia* indicates that the phylogenetic application of such gene duplications is indeed not always straightforward because of the latter two possibilities; the convergent origin of a duplication has not yet been documented within *Clarkia*. These problems indicate that questions dealing with homology of genes must be addressed, and that these questions might ultimately only be answered with nucleotide sequence information (Gottlieb, 1988b). Duplications of four isozyme-coding genes within *Clarkia* demonstrate the range of problems and systematic utility inherent in this class of molecular characters.

Triose Phosphate Isomerase

Duplications of genes encoding both the plastid and cytosolic triose phosphate isomerase (TPI) are present in all diploid species of *Clarkia* examined (Pichersky and Gottlieb, 1983). The duplication for cytosolic TPI occurs in nine other genera of the Onagraceae, suggesting its great antiquity at least at the family level. In contrast, the duplication for plastid TPI has not been found outside *Clarkia* (except in *Heterogaura*), indicating that this duplication unites all species of *Clarkia* but has no phylogenetic utility within the genus.

6-Phosphogluconate Dehydrogenase

Duplicate genes of both cytosolic and plastid 6-phosphogluconate dehydrogenase (6PGDH) occur in all diploid sections of *Clarkia,* indicating that the duplications are at least as old as the genus (Odrzykoski and Gottlieb, 1984). However,
unlike those for TPI, the duplications are not found in all species, allowing them
to be potentially useful for phylogenetic inference within *Clarkia.* Four species
within sect. *Sympherica* lack one or both 6PGDH duplications. *Clarkia rostrata*
and *C. epilobioides* have lost both duplications, whereas *C. lewisii* and *C.
cylindrica* have a duplicated gene for plastid 6PGDH but only a single gene for
cytosolic 6PGDH. Odrzykoski and Gottlieb (1984) interpreted the loss of one
cytosolic 6PGDH gene as support for the monophyletic nature of the four species,
and the loss of a plastid 6PGDH gene as evidence uniting *C. rostrata* and *C.
epilobioides.* The cpDNA cladogram (Fig. 13.2; Sytsma and Gottlieb, 1986b)
substantiates this suggestion that *C. rostrata* is indeed phylogenetically closer to
C. epilobioides of the monotypic subsect. *Micranthae* than it is to the two
morphologically similar species, *C. cylindrica* and *C. lewisii.* These cpDNA
results were the first alternative genetic confirmation of phylogenetic relationships
based on gene duplication data and greatly strengthen the utility of the latter
approach in systematics.

Phosphoglucose Isomerase

Duplication of the gene encoding the cytosolic phosphoglucose isomerase
(PGI), unlike those for TPI or 6PGDH, is known from only four of the diploid
sections in *Clarkia* (Gottlieb and Weeden, 1979). The four sections (*Sympherica,
Phaeostoma, Fibula,* and *Eucharidium*) have been proposed to be monophyletic,
and *C. rostrata* of sect. *Sympherica,* the only species of these four sections
lacking the duplication, was hypothesized to have silenced one PGI-encoding
locus (Gottlieb and Weeden, 1979). Lewis (1980) accepted the weight of this
argument and placed the four sections, not all previously considered related
(Lewis and Lewis, 1955), in one lineage. The cpDNA cladogram (Fig. 13.2;
Sytsma et al., 1990) supports the monophyletic nature of the first three sections,
as does nuclear rDNA data (Sytsma and Smith, 1988), and hence the hypothesized
silencing of one PGI locus in *Clarkia rostrata.* If the sectional grouping in Fig.
13.2 is correct, then the distribution of duplicated PGI genes must be accounted
for by a convergent gain in *Eucharidium* or a loss in an ancestor of sect. *Godetia,*
and perhaps also in sect. *Rhodanthos.* However, in view of the limited support
given to some sectional groupings by the presently available cpDNA data, the
original prediction based on the distribution of the duplication of the gene encoding cytosolic PGI cannot be conclusively refuted.

Definitive resolution of the relationships of the four sections with duplicated
PGI genes, must await either significantly more informative cpDNA restriction

site mutations or nucleotide sequencing of the PGI genes from sections with and without the duplication (Gottlieb, 1988b). The nucleotide sequence analysis of the various genes encoding PGI should be able to differentiate among several possible scenarios involving the duplication: (1) that the four sections with the duplication are strictly monophyletic and the cpDNA results do not reflect the phylogeny of the nuclear genome; (2) that sect. *Eucharidium* has independently gained the duplication relative to the other three sections; and (3) that the duplication occurred *ancestrally* to the genus *Clarkia* and that differential gene silencing (as with genes for PGI in *C. rostrata* and 6PGDH in sect. *Sympherica*) has occurred two or three times in lineages comprising the sections *Rhodanthos*, *Godetia*, and *Myxocarpa*.

Phosphoglucomutase

In contrast to the situation in 6PGDH or perhaps even PGI, the distribution of duplications encoding PGM in *Clarkia* is almost certainly indicative of independent losses (or gains) in unrelated lineages. The gene for plastid PGM is duplicated in all diploid species except *C. concinna* and *C. lassenensis* (Soltis et al., 1987). These two species are unequivocally assigned to two divergent sections as seen with cpDNA (Fig. 13.2) or with morphology (Lewis and Lewis, 1955). Because the placement of the two species in different sections is in agreement with both DNA and morphology, this is probably the best example known of independent loss of a duplication.

The presence of the duplication for cytosolic PGM in *C. arcuata* (sect. *Rhodanthos*) and in all species of sects. *Godetia* and *Myxocarpa* has been argued to be consistent with a previous (Lewis and Lewis, 1955) model of sectional relationships (Soltis et al., 1987; Gottlieb, 1988b). This model would place sect. *Rhodanthos* as "probably directly ancestral" to sect. *Godetia* and "perhaps" to sect. *Myxocarpa* as well. This would imply that sect. *Rhodanthos* is basal with regard to these two other sections and also paraphyletic with *C. arcuata* as the sister group to sects. *Godetia* and *Myxocarpa*. The cpDNA cladogram (Fig. 13.2) suggests that sect. *Myxocarpa*, not sect. *Rhodanthos*, is basal in the genus. The alternative proposal, that the duplication for cytosolic PGM may have had independent origins in *C. arcuata* and sections *Godetia* and *Myxocarpa* (Soltis et al., 1987), is more compatible with the cpDNA results. The phylogenetic placement of *C. arcuata*, of critical importance in understanding the nature of the PGM gene duplication, is now being determined with cpDNA restriction site mapping.

DNA and Catastrophic Speciation

Evidence for saltational, or catastrophic, speciation in plants has come from several examples in the Onagraceae (Lewis, 1953a, 1962, 1980; Lewis and

Roberts, 1956; Lewis and Raven, 1958). The catastrophic speciation model applied to *Clarkia* (see Gottlieb, 1988b) includes:

1. Species are regarded as progenitor and derivative rather than as sister species
2. The new species differed from its parent by chromosomal rearrangements, presumably fixed by a series of population crashes
3. The process of reproductive isolation (speciation) was rapid and abrupt
4. Speciation was independent of the evolution of new adaptations and thus largely fortuitous
5. Speciation occurred at the xeric, southern borders of the range of the parent species

Gottlieb (1973, 1974a, 1981) further stated that the derivative species must share high genetic similarity to its progenitor and also possess a subset of alleles of the progenitor species. Two contrasting putative examples of such catastrophic speciation have been postulated in *Clarkia* and examined with morphology, cytology, isozymes, and cpDNA.

The first example involves the derivation of *Clarkia lingulata* from *C. biloba* within sect. *Sympherica* subsect. *Lautiflorae*. The putative progenitor species is adapted to mesic woodlands from northern to central California, possesses eight haploid chromosomes, and is distinguished by bilobed petals (Fig. 13.6a). *Clarkia lingulata,* the derivative species, is restricted to the Merced River drainage at the southern extent of *C. biloba,* possesses nine haploid chromosomes, and is morphologically indistinguishable from *C. biloba* except for its lingulate rather than lobed petals (Fig. 13.6a). Lewis and Roberts (1956) proposed that *C. lingulata* was derived from the southern populations of *C. biloba* via a number of chromosomal translocations resulting in nine rather than eight chromosomes. This instantaneous reproductive barrier was thought to have been abetted by the dramatic population crashes evident in the southern populations of *C. biloba* and other mesic species (Lewis, 1953b). Gottlieb (1974a) demonstrated that the two species have a high genetic identity ($I = 0.98$) and that all alleles of *C. lingulata* can be found within the progenitor species. Chloroplast DNA restriction site analysis confirms the close phylogenetic relationship of the two species (Fig. 13.2; Sytsma and Gottlieb, 1986a, b; Sytsma and Smith, 1988). The two species have a p value of 0.0017, one of the two lowest interspecific values seen in *Clarkia.*

The second example concerns three species of sect. *Rhodanthos: Clarkia amoena, C. rubicunda,* and *C. franciscana.* The first species is restricted to the more mesic coastal mountains north of San Francisco, the second species to the more xeric coastal mountains south of San Francisco, and the third endemic to the Presidio within San Francisco (Fig. 13.6b). Lewis and Raven (1958)

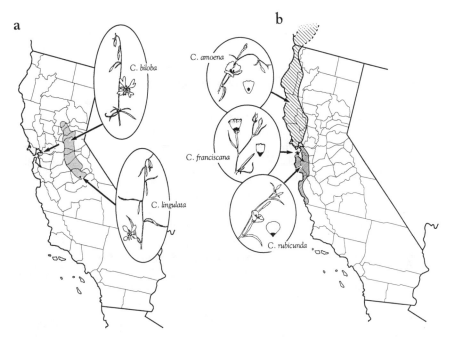

Figure 13.6. A—Distribution and inflorescence of the putative progenitor-derivative species pair *Clarkia biloba* and *C. lingulata* (after Lewis and Lewis, 1955; Lewis and Roberts, 1956). B— Distribution and inflorescence of *Clarkia amoena* and the putative progenitor-derivative species pair *C. rubicunda* and *C. franciscana* (after Lewis and Lewis, 1955; Lewis and Raven, 1958).

demonstrated that the three species differ from each other by similar numbers of chromosomal rearrangements. However, F_1 offspring of *C. franciscana* × *C. rubicunda* showed fewer univalents in meiosis than did F_1 offspring of the other crosses (Lewis and Raven, 1958). Moreover, the pigment band is basal in the petals of *C. franciscana* and *C. rubicunda* but is central in *C. amoena* (Lewis and Raven, 1958). These observations, as well as the geographic and habitat distributions, led Lewis and Raven (1958) to postulate specific hypotheses concerning the evolution of the three species. They concluded that the southern and more xerically tolerated *C. rubicunda* was derived from a subspecies of the more northern and mesophytic *C. amoena,* but that *C. franciscana* was derived in a catastrophic manner from *C. rubicunda*. These hypotheses, however, were not verified by isozymic analysis. *Clarkia franciscana* has very low genetic similarity with both *C. rubicunda* and *C. amoena* (approximately $I = 0.28$) and possesses alleles in six of eight enzyme systems not found in either of the putative progenitor species (Gottlieb, 1973, 1981). Additionally, a unique gene duplication of alcohol dehydrogenase is present in *C. franciscana* (Gottlieb, 1974b).

Preliminary cpDNA restriction site analysis of the three species provides an alternative set of relationships (Fig. 13.2; Sytsma et al., unpublished data).

Clarkia amoena (ssp. *huntiana*) from Marin County immediately north of San Francisco is identified as the closest species to *C. franciscana*. The two species have a *p* value of 0.0016, the lowest interspecific divergence seen in *Clarkia*. This value is almost identical to that found between *C. biloba* and *C. lingulata*, the well-corroborated progenitor-derivative species pair. *Clarkia rubicunda* exhibits 3–4 times the cpDNA divergence with *C. franciscana* as does *C. amoena*. These data suggest that *C. franciscana* was recently derived from the southern populations of *C. amoena*. The co-occurence of the basal pigment band in petals of *C. rubicunda* and *C. franciscana* could be interpreted as a symplesiomorph, not necessarily indicative of a close phylogenetic relationship. Additionally, the position of the pigment band or spot in petals of *C. gracilis* (also sect. *Rhodanthos*) is governed by alleles of a single locus (Gottlieb and Ford, 1988). It could be argued that an easily changed genetic difference, such as petal spot position, does not provide strong evidence for phylogenetic relationships (Gottlieb, 1989). The incongruity of the cpDNA and isozyme data seen in this example, but not in the *C. biloba/C. lingulata* example, is more difficult to explain. If the cpDNA results accurately reflect the evolutionary relationships of *C. franciscana,* the highly inbred nature of the species, as well as its documented severe population crashes (Lewis and Raven, 1958), might have contributed to fixation of novel or rare alleles and a gene duplication subsequent to its derivation from *C. amoena.*

DNA and Origin of Polyploids

Maternally inherited cpDNA, often in conjunction with isozymes or nuclear rDNA, has been useful in documenting parentage of allopolyploids (see Chapter 8). A number of polyploids, several of intersectional origin, are known in *Clarkia*. The most unusual polyploid in the genus is *C. pulchella* ($n = 12$), the lone member of sect. *Clarkia,* believed to be an allotetraploid derivative of an intersectional cross between *C. concinna* ($n = 7$, sect. *Eucharidium*) and *C. virgata* ($n = 5$, sect. *Myxocarpa*) (Lewis and Lewis, 1955). This polyploid is of considerable systematic significance as the putative relationship of sect. *Eucharidium* with *C. pulchella* was used to maintain the section within *Clarkia*. The origin of *C. pulchella* via hybridization between sects. *Eucharidium* and *Myxocarpa* is supported by its additivity for chromosome number and for characters of the anthers, petals, and pollen (Lewis and Lewis, 1955; Sytsma et al., 1990).

The cpDNA analysis indicates that a species in sect. *Eucharidium* acted as the maternal parent in the hybridization event leading to *C. pulchella* (Fig. 13.2; Sytsma et al., 1990). The cpDNA cladogram (Fig. 13.2) reveals that *C. pulchella* diverged from its diploid progenitor in sect. *Eucharidium* prior to the split of the two extant diploid species, *C. concinna* and *C. breweri.*

Morphological characters, such as lateral projections on the claw of petals and sterile flaps at the base of stamens, both strictly confined to *C. pulchella* and sect. *Myxocarpa,* the presence of species with $n = 5$ in sect. *Myxocarpa,* and possibly the identical 2C nuclear DNA content of *C. pulchella* and *C. virgata* (sect. *Myxocarpa*) (Narayan, 1988) strongly implicate this section as the paternal diploid parent. Of the two extant species of sect. *Myxocarpa* with the prerequisite five chromosomes, *C. virgata,* rather than the recently derived *C. australis* (Small, 1971, 1972b), is the most likely paternal candidate (Sytsma et al., 1990). The two putative diploid parents are, however, now completely allopatric and do not overlap with *C. pulchella.* The hybridization and chromosome doubling events must not have occurred recently judging from cpDNA divergence, the basal nature of sect. *Myxocarpa* and possibly sect. *Eucharidium,* and the present-day allopatric nature of *C. pulchella* and the two parental diploid sections.

DNA and Morphology

How Both Sets of Characters Can Be Used in the Same Study: An Example from Fuchsia *section* Skinnera

Various methods of dealing with both molecular and morphological characters in the inference of phylogeny have been suggested (Hillis, 1987; Sytsma, 1990; see also Donoghue and Sanderson, Chapter 15, this volume). Four different ways to treat these data sets are possible (Sytsma, 1990):

1. Combine the two sets of characters with equal weighting for each character. In this way the morphological characters might provide strength to branches that are weak based on molecular characters, and vice versa. A variant on this approach, but perhaps difficult to justify and implement, is to provide equal weighting for each set of characters, thus not discriminating against the (usually) fewer morphological characters.

2. Analyze the two sets of characters independently and construct a consensus cladogram that depicts only relationships recognized by both approaches.

3. Use only the molecular characters to generate a cladogram and secondarily overlay the morphological characters onto the molecular cladogram.

4. Use only the morphological characters to generate a cladogram and secondarily overlay the molecular characters onto the morphological cladogram.

To illustrate this problem of using both molecular and morphological data sets, these four methods are used to examine relationships within *Fuchsia* sect. *Skinnera.* This section has been the focus of several competing hypotheses concerning the evolution and relationships of component taxa and the nature of changes in habit, morphology, floral biology, flavonoids, and distribution. In addition, cladistic analyses of morphological characters (with flavonoid data) and of cpDNA restriction site mutations have been performed separately (Williams and Garnock-Jones, 1986; Crisci and Berry, 1990; Sytsma et al., 1991). The morphological data set of Crisci and Berry (1990) is used rather than that of

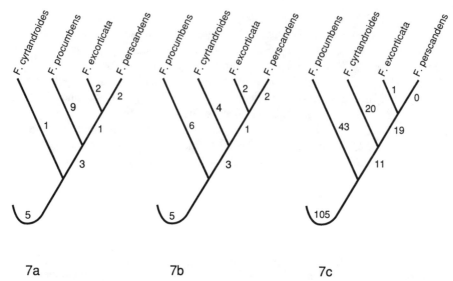

7a   **7b**   **7c**

Figure 13.7. A,B—The two equally parsimonious morphological/chemical cladograms of *Fuchsia* sect. *Skinnera* (after Crisci and Berry, 1990). C—Most parsimonious cpDNA cladogram of *Fuchsia* sect. *Skinnera* (after Sytsma et al., 1991; see also Fig. 13.4).

Williams and Garnock-Jones (1986) because the former is a more exhaustive analysis and corrects a number of errors in the latter.

The morphological data set has 17 characters and generates two equally parsimonious cladograms of 23 steps (Fig. 13.7A and B). The cpDNA data set has 194 characters (excluding length mutations) and generates one most parsimonious cladogram of 199 steps (Fig. 13.7C, see also Fig. 13.4), identical to one of the morphology-based cladograms (Fig. 13.7B). The discrepancy among the possible cladograms includes the identification of the basal species within the section, either the New Zealand *Fuchsia procumbens* or the Tahitian *F. cyrtandroides*. The discrepancy has considerable ramifications on how biogeographic patterns and character-state changes are to be interpreted within sect. *Skinnera* (Sytsma et al., 1991).

Combining the two data sets generates a single cladogram identical to the cpDNA cladogram and one of the two derived with morphological characters (Fig. 13.8A). Combining the data sets thus resolves a basal lineage in the section (*Fuchsia procumbens*), a branching pattern that was ambiguous with morphology but well supported with cpDNA. This result is arguably not unexpected because of the larger number of characters that support a basal clade in the cpDNA cladogram relative to that in the morphology cladogram. Moreover, the cpDNA cladogram exhibits considerably less homoplasy than does the morphology cladogram (3% versus 23%, autapomorphs included).

A consensus cladogram (Adams or Strict) of the cladograms derived indepen-

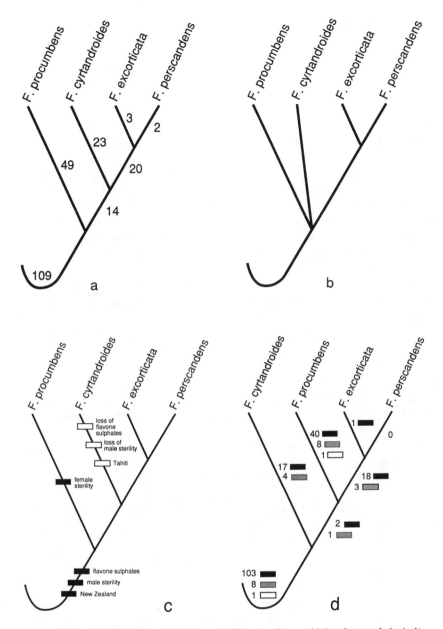

Figure 13.8. A—Cladogram of *Fuchsia* sect. *Skinnera* after combining the morphological/
chemical data set (Crisci and Berry, 1990) and cpDNA data set (Sytsma et al., 1991). B—Strict
(and Adams) consensus cladogram of *Fuchsia* sect. *Skinnera* based on the two morphological/
chemical cladograms and the one cpDNA cladogram. C—The cpDNA cladogram (Fig. 13.7C)
overlain with selected morphological, chemical, and biogeographical characters. Open bars
indicate reversals. D—The first morphological/chemical cladogram (see Fig. 13.7A) overlain
with cpDNA restriction site mutations. Numbers refer to restriction site mutations along each
branch. This cladogram is eight steps longer than the cpDNA cladogram in Fig. 13.7C. Solid
bars represent unique character state changes, grey bars indicate "likely" convergent mutations,
and open bars indicate "unlikely" convergent mutations.

dently with the two approaches is shown in Fig. 13.8B. The basal node is an unresolved trichotomy reflecting the ambiguity generated by the morphological analysis (Fig. 13.7A, B). It could be argued that consensus analysis might be too conservative in this example by obscuring the basal nature of *Fuchsia procumbens*.

Overlaying the morphological character states onto the cpDNA cladogram permits some insights into biogeography and morphological evolution within the section (Fig. 13.8C). First, the basal nature of *Fuchsia procumbens* within the section is in agreement with the presence of *Fuchsia* on New Zealand (known fossils at 30 myr old) (Pocknall and Mildenhall, 1984; Daghlian et al., 1985) well before its appearance on Tahiti (less than 2 myr old) (Dymond, 1975). Second, gynodioecy should be viewed as the plesiomorphic condition in sect. *Skinnera* and the presence of only hermaphroditic flowers in *F. cyrtandroides* as the apomorphic condition. The independent acquisition of gynodioecy in two New Zealand lineages is equally parsimonious (assuming Wagner parsimony) but less likely based on Dollo parsimony. Third, the presence of sulphated flavones is the plesiomorphic condition within sect. *Skinnera,* despite their absence from other sections of *Fuchsia,* and the secondary loss of these compounds in *F. cyrtandroides* is the apomorphic state.

Overlaying the molecular data onto the morphology cladogram that is not congruent with cpDNA (Fig. 13.7A) generates a cladogram that is eight steps longer than when overlayed on the morphology cladogram congruent with that based on cpDNA (Fig. 13.8D). These additional steps include one unlikely convergent gain or loss followed by a gain. In addition, the four deletions and one insertion shared by all species of sect. *Skinnera* except *F. procumbens* become homoplastic characters (i.e., convergences between *F. cyrtandroides* and *F. excorticata*/*F. perscandens* or secondary reversals in *F. procumbens*). The cpDNA data provide little character support for the monophyly of all New Zealand taxa; only three synapomorphies (one homoplastic) define this clade. The increased numbers of restriction site mutations, the homoplastic nature of five insertion/deletion events, and the extremely weak support for a New Zealand clade seen when cpDNA characters are overlain on the first morphological cladogram, indicate that the discrepancy between these two cladograms is indeed serious.

This example of *Fuchsia* sect. *Skinnera* indicates the types of secondary analyses that are possible when confronted with discordant data sets. In this case, there is strong support for relationships suggested by cpDNA even though morphology provides ambiguous results. These types of secondary analysis will have perhaps more impact in other kinds of studies. Combining data sets might be informative in studies in which morphology and DNA are each weak in clarifying different clades. Donoghue and Sanderson (Chapter 15, this volume) describe an example in which morphology aids in resolving relationships that are ambiguous with DNA alone and vice versa. In contrast, combining data sets would be misleading, and assessing each data set separately would be more appropriate in groups where hybridization and/or introgression occur (see Chapter 7 by Rieseberg and Brunsfeld, this volume; Sytsma, 1990; Smith and Sytsma, 1990). In these cases morphology and/or different DNA data sets (e.g., maternally inherited versus biparentally inherited DNAs) are necessarily contradictory. Although it is not yet entirely clear how to utilize both morphological and molecular character sets in inferring plant phylogeny, these examples underscore the need to examine as many data sets as possible.

Conclusions

This on-going multidisciplinary approach in the Onagraceae has provided detailed and interesting, yet sometimes controversial, evolutionary results. But, in the words of the foremost researcher on the Onagraceae,

> "Such results, however, have meaning only in a context in which the relation-ships of the plant groups being considered are well understood; therefore, data sets such as those available for Onagraceae are particularly valuable. With the current availability of cladistic methods of analysis that exceed those employed in the past in their rigor and ability to compare different subsets of information, the family should become an even more instructive example in the future" Raven (1988, pp. 85, 87).

References

Allan, H.H. (1927) Illustrations of wild hybrids in the New Zealand flora V. *Genetica* **9**, 499–514.

Allan, H.H. (1928) *New Zealand Trees and Shrubs and How to Identify Them,* Whitcombe and Tombs, Auckland.

Arroyo, M.T., and Raven, P.H. (1975) The evolution of subdioecy in morphologically gynodioecious species of *Fuchsia* sect. *Encliandra* (Onagraceae). *Evolution* **29**, 500–511.

Averett, J.E., Hahn, W.J., Berry, P.E., and Raven, P.H. (1986) Flavonoids and flavonoid evolution in *Fuchsia* (Onagraceae). *Amer. J. Bot.* **73**, 1525–1534.

Averett, J.E., and Raven, P.H. (1984) Flavonoids of Onagraceae. *Ann. Missouri Bot. Gard.* **71**, 30–34.

Berry, P.E. (1982) The systematics and evolution of *Fuchsia* sect. *Fuchsia* (Onagraceae). *Ann. Missouri Bot. Gard.* **69**, 1–198.

Berry, P.E., Stein, B.A., Carlquist, S., and Nowicke, J.W. (1988) *Fuchsia pachyrrhiza* (Onagraceae), a tuberous new species and section of *Fuchsia* from western Peru. *Syst. Bot.* **13**, 483–492.

Berry, P.E., Skvarla, J.J., Partridge, A.D., and MacPhail, M.K. (1990) *Fuchsia* pollen from the Tertiary of Australia. *Austr. Syst. Bot.,* submitted.

Breedlove, D.E. (1969) The systematics of *Fuchsia* sect. *Encliandra* (Onagraceae). *Univ. Calif. Publ. Bot.* **53**, 1–69.

Breedlove, D.E., Berry, P.E., and Raven, P.H. (1982) The Mexican and Central American species of *Fuchsia* (Onagraceae) except for sect. *Encliandra*. *Ann. Missouri Bot. Gard.* **69**, 209–234.

Crisci, J.V., and Berry, P.E. (1990) A cladistic analysis of the Australasian and Pacific species of *Fuchsia* (Onagraceae). *Ann. Missouri Bot. Gard.* **77**, 517–522.

Daghlian, C.P., Skvarla, J.J., Pocknall, D., and Raven, P.H. (1985) *Fuchsia* pollen from the early Miocene of New Zealand. *Amer. J. Bot.* **72**, 1039–1047.

Davis, W.S. (1970) The systematics of *Clarkia bottae, C. cylindrica,* and a related species, *C. rostrata. Brittonia* **22**, 270–284.

Dymond, J. (1975) K-Ar ages of Tahiti and Moorea, Society Islands, and implications for the hot-spot model. *Geology* **3**, 236–240.

Godley, E.J. (1955) Breeding systems in New Zealand plants. 1. *Fuchsia. Ann. Bott.* **19**, 549–559.

Godley, E.J. (1963) Breeding systems in New Zealand plants. 2. Genetics of the sex forms in *Fuchsia procumbens. NZ J. Bot.* **1**, 48–52.

Gottlieb, L.D. (1973) Enzyme differentiation and phylogeny in *Clarkia franciscana, C. rubicunda* and *C. amoena. Evolution* **27**, 205–214.

Gottlieb, L.D. (1974a) Genetic confirmation of the origin of *Clarkia lingulata. Evolution* **28**, 244–250.

Gottlieb, L.D. (1974b) Gene duplication and fixed heterozygosity for alcohol dehydrogenase in the diploid plant *Clarkia franciscana. Proc. Natl. Acad. Sci. USA* **71**, 1816–1818.

Gottlieb, L.D. (1977) Evidence for duplication and divergence of the structural gene for phosphoglucose isomerase in diploid species of *Clarkia. Genetics* **86**, 289–307.

Gottlieb, L.D. (1981) Electrophoretic evidence and plant populations. *Progress in Phytochem.* **7**, 1–46.

Gottlieb, L.D. (1982) Conservation and duplication of isozymes in plants. *Science* **216**, 373–380.

Gottlieb, L.D. (1983) Isozyme number and phylogeny. In: *Proteins and Nucleic Acids in Plant Systematics* (eds. U. Jensen and D.E. Fairbrothers), Springer-Verlag, Berlin, pp. 209–221.

Gottlieb, L.D. (1986) Genetic differentiation, speciation, and phylogeny in *Clarkia* (Onagraceae). In: *Modern Aspects of Species* (eds. K. Iwatsuki, P.H. Raven, and W.J. Bock), University of Tokyo Press, Tokyo, pp. 145–160.

Gottlieb, L.D. (1988a) In: *Plant Evolutionary Biology* (eds. L.D. Gottlieb and S.K. Jain), Chapman and Hall, London, p. 153.

Gottlieb, L.D. (1988b) Towards molecular genetics in *Clarkia:* gene duplications and molecular characterization of PGI genes. *Ann. Missouri Bot. Gard.* **75**, 1169–1179.

Gottlieb, L.D. (1989) Floral pigmentation patterns in *Clarkia* (Onagraceae). *Madroño* **36**, 1–8.

Gottlieb, L.D., and Ford, V.S. (1988) Genetic studies of the pattern of floral pigmentation in *Clarkia gracilis. Heredity* **60**, 237–246.

Gottlieb, L.D., and Jain, S.K. (eds.) (1988) *Plant Evolutionary Biology,* Chapman and Hall, London.

Gottlieb, L.D., and Pilz, G. (1976) Genetic similarity between *Gaura longifolia* and its obligately outcrossing derivative *G. demareei. Syst. Bot.* **1**, 181–187.

Gottlieb, L.D., and Weeden, N.F. (1979) Gene duplication and phylogeny in *Clarkia. Evolution* **33**, 1024–1039.

Hillis, D.M. (1987) Molecular versus morphological approaches to systematics. *Ann. Rev. Ecol. Syst.* **18**, 23–42.

Holsinger, K.E., and Gottlieb, L.D. (1988) Isozyme variability in the tetraploid *Clarkia gracilis* (Onagraceae) and its diploid relatives. *Syst. Bot.* **13**, 1–6.

Jansen, R.K., and Palmer, J.D. (1988) Phylogenetic implications of chloroplast DNA restriction site variation in the Mutisieae (Asteraceae). *Amer. J. Bot.* **75**, 753–766.

Kurabayashi, M., Lewis, H., and Raven, P.H. (1962) A comparative study of mitosis in Onagraceae. *Amer. J. Bot.* **49**, 1003–1026.

Lewis, H. (1953a) The mechanism of evolution in the genus *Clarkia. Evolution* **7**, 1–20.

Lewis, H. (1953b) Chromosome phylogeny and habitat preference in *Clarkia. Evolution* **7**, 102–109.

Lewis, H. (1962) Catastrophic selection as a factor in speciation. *Evolution* **16**, 257–271.

Lewis, H. (1980) The mode of evolution in *Clarkia.* Symposium paper presented at International Congress Systematics and Evolutionary Biol. II, Vancouver, BC.

Lewis, H., and Lewis, M.E. (1955) The genus *Clarkia. Univ. Calif. Publ. Bot.* **20**, 241–392.

Lewis, H., and Raven, P.H. (1958) Rapid evolution in *Clarkia. Evolution* **12**, 319–336.

Lewis, H., and Roberts, M.R. (1956) The origin of *Clarkia lingulata. Evolution* **10**, 126–138.

MacSwain, J.W., Raven, P.H., and Thorp, R.W. (1973) Comparative behavior of bees and Onagraceae. IV. *Clarkia* bees of the Western United States. *Univ. Calif. Publ. Entomol.* **70**, 1–80.

Narayan, R.K.J. (1988) Evolutionary significance of DNA variation in plants. *Evol. Trends Plants* **2**, 121–130.

Nowicke, J.W., Skvarla, J.J., Raven, P.H., and Berry, P.E. (1984) A palynological study of the genus *Fuchsia* (Onagraceae). *Ann. Missouri Bot. Gard.* **71**, 35–91.

Odrzykoski, I.J., and Gottlieb, L.D. (1984) Duplications of genes coding 6-phosphogluconate dehydrogenase in *Clarkia* (Onagraceae) and their phylogenetic implications. *Syst. Bot.* **9**, 479–489.

Pichersky, E., and Gottlieb, L.D. (1983) Evidence for duplication of the structural genes coding plastid and cytosolic isozymes of triose phosphate isomerase in diploid species of *Clarkia. Genetics* **105**, 421–436.

Pocknall, D.T., and Mildenhall, D.C. (1984) Late Oligocene—early Miocene spores and pollen from Southland, New Zealand. *NZ Geol. Surv. Paleontol. Bull.* **51**, 1–66.

Raven, P.H. (1979) A survey of reproductive biology in Onagraceae. *NZ J. Bot.* **17**, 575–593.

Raven, P.H. (1988) Onagraceae as a model of plant evolution. In: *Plant Evolutionary Biology* (eds L.D. Gottlieb and S.K. Jain), Chapman and Hall, London, pp. 85–107.

Raven, P.H., and Axelrod, D.I. (1974) Angiosperm biogeography and past continental movements. *Ann. Missouri Bot. Gard.* **61**, 539–673.

Raven, P.H., and Raven, T.E. (1976) The genus *Epilobium* in Australasia: a systematic and evolutionary study. *NZ Dept. Sci. Ind. Res. Bull.* **216**, 1–321.

Rothrock, J.T. (1864) Synopsis of North American Gaurineae. *Proc. Amer. Acad.* **6**, 347–354.

Seavey, S.R. (1977) Segregation of translocated chromosomes in *Epilobium* and *Boisduvalia* hybrids (Onagraceae). *Syst. Bot.* **2**, 109–121.

Skvarla, J.J., Raven, P.H., Chissoe, W.F., and Sharp, M. (1978) An ultrastructural study of viscin threads in Onagraceae pollen. *Pollen et Spores* **20**, 5–143.

Small, E. (1971) The systematics of *Clarkia*, section *Myxocarpa*. *Canad. J. Bot.* **49**, 1211–1217.

Small, E. (1972a) Adaptation in *Clarkia*, section *Myxocarpa*. *Ecology* **53**, 808–818.

Small, E. (1972b) Tempo of adaptive change during the rapid evolution of chromosomal isolates. *Taxon* **21**, 559–565.

Small, E., Bassett, I.J., Crompton, C.W., and Lewis, H. (1971) Pollen phylogeny in *Clarkia*. *Taxon* **20**, 739–746.

Smith, R.L., and Sytsma, K.J. (1990) Evolution of *Populus nigra* (sect. *Aigeiros*): introgressive hybridization and the chloroplast contribution of *Populus alba* (sect. *Populus*). *Amer. J. Bot.* **77**, 1176–1187.

Smith-Huerta, N.L. (1986) Isozymic diversity in three allotetraploid *Clarkia* species and their putative diploid progenitors. *J. Hered.* **77**, 349–354.

Soltis, P.S. (1986) Anthocyanidin variation in *Clarkia*. *Biochem. Syst. Ecol.* **14**, 487–489.

Soltis, P.S., Soltis, D.E., and Gottlieb, L.D. (1987) Phosphoglucomutase gene duplications in *Clarkia* (Onagraceae) and their phylogenetic implications. *Evolution* **41**, 667–671.

Sytsma, K.J. (1990) DNA and morphology: inference of plant phylogeny. *Trends Ecol. Evol.* **5**, 104–110.

Sytsma, K.J., and Gottlieb, L.D. (1986a) Chloroplast DNA evidence for the origin of the genus *Heterogaura* from a species of *Clarkia* (Onagraceae). *Proc. Natl. Acad. Sci. USA* **83**, 5554–5557.

Sytsma, K.J., and Gottlieb, L.D. (1986b) Chloroplast DNA evolution and phylogenetic relationships in *Clarkia* sect. *Peripetasma* (Onagraceae). *Evolution* **40**, 1248–1261.

Sytsma, K.J., and Smith, J.F. (1988) DNA and morphology: comparisons in the Onagraceae. *Ann. Missouri Bot. Gard.* **75**, 1217–1237.

Sytsma, K.J., Smith, J.F., and Berry, P.E. (1991) The use of chloroplast DNA to assess biogeography and evolution of morphology, breeding systems, flavonoids, and chloroplast DNA in *Fuchsia* sect. *Skinnera* (Onagraceae). *Syst. Bot.* **16**, 257–269.

Sytsma, K.J., Smith, J.F., and Gottlieb, L.D. (1990) Phylogenetics in *Clarkia* (Onagraceae): restriction site mapping of chloroplast DNA. *Syst. Bot.* **15**, 280–295.

Tobe, H., and Raven, P.H. (1985) The histogenesis and evolution of integuments in Onagraceae. *Ann. Missouri Bot. Gard.* **72**, 451–468.

Tobe, H., and Raven, P.H. (1986) Evolution of polysporangiate anthers in Onagraceae. *Amer. J. Bot.* **73**, 475-488.

Vasek, F.C. (1958) The relationships of *Clarkia exilis* to *Clarkia unguiculata*. *Amer. J. Bot.* **45**, 150–162.

Williams, C.A., and Garnock-Jones, P.J. (1986) Leaf flavonoids and other phenolic glycosides and the taxonomy and phylogeny of *Fuchsia* sect. *Skinnera* (Onagraceae). *Phytochem.* **25**, 2547–2549.

14

Floral Morphology and Chromosome Number in Subtribe Oncidiinae (Orchidaceae): Evolutionary Insights From a Phylogenetic Analysis of Chloroplast DNA Restriction Site Variation

Mark W. Chase and *Jeffrey D. Palmer*

Although the Orchidaceae are one of the two largest families of angiosperms, they have been neglected as subjects for evolutionary studies by most plant systematists. This omission is surprising for two reasons: the family exhibits considerable heterogeneity and available evidence supports the hypothesis that it has been evolutionarily active (Dressler, 1981). To a large extent, the tropical distribution of orchids and their complex floral morphology have been historically responsible for dulling the interest of potential students (for example, considerably greater numbers of systematists have focused on the Asteraceae and Fabaceae). Nevertheless, interest in evolutionary issues concerning orchids has recently increased (Ackerman, 1986, 1989; Benzing, 1987; Dafni, 1987). Molecular systematics is likely to have a major impact on such studies in the future, and it is in this light that the following study on the orchid subtribe Oncidiinae is presented.

The Oncidiinae are one of the largest (70 genera; 1,500 species) and floristically most important orchid subtribes in the Neotropics. It is a member of one of the most highly evolved (in terms of floral and vegetative features) tribes in the Orchidaceae, the Maxillarieae (*sensu* Dressler, 1986, 1989), which are also strictly New World in distribution. Florally and vegetatively, this tribe is one of the most diverse, but it is held together by a synapomorphic seed condition, termed the *Maxillaria*-type (Ziegler, 1981; Chase and Pippen, 1988). The relationship of the Maxillarieae to other tribes is uncertain (Dressler, 1986). In this tribe, as well as most others in the Orchidaceae, the subtribal boundaries are relatively clear, even though interrelations of subtribes are unclear. Floral features have been emphasized traditionally in orchid systematics; only recently studies on other potential sources of information have been initiated (reviews in Dressler, 1986, 1989). Within subtribes, a similar lack of information exists. Distinct gaps clearly mark groups of genera, but these groups are typically distinguished by unique morphologies that reveal little about phylogenetic relationships. An additional source of data that could span the discontinuities in morphology has long been needed; variation in chloroplast DNA (cpDNA) is potentially one such type of information.

In addition to the phylogenetic problems posed by unique characteristics, a secondary problem in the Oncidiinae has been the general lack of taxonomic consistency between floral features, habit, and chromosome number (i.e., the traditionally recognized genera, founded on floral features, are polymorphic in terms of habit and chromosome number). Present in this single subtribe are plants with a six-fold range in chromosome number ($n = 5$–30, which is greater than the range present in the rest of the family). In some cases, species that differ by this six-fold range in chromosome number have floral features that traditionally placed them in the same genus, whereas other species with similar chromosome numbers have radically different flowers or unique vegetative formats (Chase, 1986a). Similar paradoxes occur with respect to each pairwise combination of floral morphology, habit, and chromosome number. Chase (1986a) attempted to place this inconsistency in traditional systematic information within a general evolutionary perspective. An analysis (Chase, 1987a) of the complex pollinaria of *Oncidium* (ca. 500 species in its traditional circumscription; Garay and Stacy, 1974) and several genera closely related to it supported hypotheses that gross floral morphology is highly plastic and that current generic and infrageneric limits are artificial. That study did not indicate any simple method by which the taxonomic problems could be solved.

Studies of cpDNA variation in the Oncidiinae (Chase and Palmer 1989) have begun to demonstrate that each of these sources of information—chromosome number, habit, and floral morphology—can yield answers about systematic and evolutionary questions if they are properly interpreted. Because the Oncidiinae are so diverse in terms of floral, vegetative, and chromosomal features, it was chosen as a model for studies that should prove useful in studying many orchid groups, as well as other lilioid monocots. The Liliaceae presents perhaps an analogous problem in which a lack of correspondence between floral and vegetative traits has been the source of long-standing disagreements about familial limits. This chapter focuses on two particular aspects of the Oncidiinae, morphology and chromosome number, to illustrate more clearly the nature of the problems faced. Taxonomic changes will not be considered in great detail until a more comprehensive, ongoing survey of species in the Oncidiinae is completed.

Two principal methods of analysis are employed: comparative mapping of restriction sites in cpDNA and examining isozymes (i.e., numbers of loci coding for various proteins). The former type of study is emphasized here. It is employed in constructing phylogenies to test hypotheses about which chromosome numbers are ancestral, as well as to test opposing ideas about vegetative and floral evolution. The latter type of inquiry establishes whether or not allopolyploid changes were involved in the production of the observed variation in chromosome number. Isozyme data were reported previously (Chase and Olmstead, 1988), and methods and specific results are not presented here, although the conclusions are discussed because they have a bearing on the evidence from cpDNA.

Methods of Phylogenetic Analysis

Variation in cpDNA from 51 species in the Oncidiinae and one species from the closely related subtribe Maxillariinae was analyzed by filter hybridization and autoradiography, as described in Palmer et al. (1988) and Chase and Palmer (1989). Subtribe Maxillariinae was chosen as an outgroup because it shares a derived habit condition with the Oncidiinae, and a preliminary tribal analysis of cpDNA of the Epidendroideae *sensu lato* demonstrated that the Maxillariinae were, at some level, sister group to a monophyletic Oncidiinae. Ten restriction endonucleases were used, and complete restriction site maps were constructed for each species. The survey reported here represents a part of a larger study of species in the Oncidiinae (the complete study encompasses 160 species). Dollo parsimony, which completely prohibits parallel site gains but allows parallel losses, and Wagner parsimony, which weighs gains and losses equally, were also used to construct trees. The Dollo and Wagner parsimony consensus trees are presented in Figs. 14.1 and 14.2, respectively (the data matrix for this analysis is in the Appendix and clade descriptions are given in Table 14.1). Wagner parsimony is not rigorous enough in its equal weighting of site gains and losses; however, we believe that Dollo parsimony is far too harsh.

All restriction site data were analyzed by PAUP (Phylogenetic Analysis Using Parsimony, version 3.0 by D. Swofford), principally using a method of character-state weighing, such that homoplasious site gains are minimized (Fig. 14.3; this procedure is described in detail in Albert et al., Chapter 16, this volume). The inclusion of step matrices and variable weighting of gains and losses within a character in PAUP 3.0 makes feasible an alternative to Wagner and Dollo.

We would like to point out several features of the three trees presented (Figs. 14.1–14.3). The methods for determining the proper weights to be applied are described in Albert et al., Chapter 16, this volume. In the case of the analysis that produced Fig. 14.3, site gains were weighted 1.3:1 relative to site losses. The Dollo analysis produced 172 equally parsimonious trees, and, from the standpoint of having studied the floral and vegetative morphology of these plants, we believe that, in spite of the fair degree of resolution found, the monophyletic lineages identified are unreasonable. In contrast, the Wagner analysis found 151 equally parsimonious trees, but the major feature of the consensus tree is a basal polychotomy. With character-state weighting, 12 equally parsimonious trees were identified; these topologies were found using Wagner parsimony. By minimizing the site gains without requiring them to be unique, a great deal more resolution has been achieved. Wagner rather than Dollo parsimony found these same topologies but also 139 others that did not minimize parallel site gains.

Table 14.1. Comparison of some aspects of the sixteen clades identified in Fig. 14.3.

Clade	Distribution	Altitude Range	Habit	Floral Traits—Pollination	Host Site Preferences
Altissimum	Neotropics	0–2,000 m	standard†	mostly *Oncidium*-type—deceit	trunk-limb
Rodriguezia	Neotropics	0–1,500 m	variously modified terete leaves, fan-shaped	nectaries of various types	twigs
Cischweinfia	Central America South America*	500–1,500 m	standard†	lip-column fusion—deceit	trunk-limb
Psygmorchis	Neotropics	0–500 m	fan-shaped monopodial	*Oncidium*-type—deceit	twigs
Crispum	southern Brazil	500–1,500 m	standard†	mostly *Oncidium* type—deceit	trunk-limb
Palumbina	Central America	1,000–2,000 m	standard†	*Oncidium*-type—deceit	trunk-limb
Rossioglossum	Mexico-Central America	1,000–2,500 m	standard†	*Oncidium*-type—deceit	trunk-limb
Lophiaris	Neotropics	0–2,000 m	modified: mule-ear and rat-tail types	*Oncidium*-type—deceit (*Trichocentrum*-euglossine)	trunk-limb
Ticoglossum	Central America	1,000–2,500 m	standard†	*Oncidium*-type—deceit	trunk-limb
Lembglossum	Mexico, Central and South America	1,000–2,500 m	standard†	*Oncidium*-type—deceit	trunk-limb
Ada	Central and South America	1,000–2,500 m	standard†	mostly *Oncidium*-type—deceit	trunk-limb
Brassia	Neotropics	200–2,000 m	standard†	*Oncidium*-type—deceit	trunk-limb
Symphyglossum	Andes	1,500–3,000 m	standard†	lip-column fusion—bird	trunk-limb
Macranthum	Andes	2,000–páramo	standard†	*Oncidium*-type—deceit	trunk-limb or terrestrial
Otoglossum	Central and South America	1,000–2,500 m	standard†	*Oncidium*-type—deceit	low shrubs or terrestrial
Trichopilia	Neotropics§	1,000–2,000 m	standard† to terete leaves	lip-column fusion—deceit	trunk-limb

*South America excluding the Amazon basin and southern Brazil.

†Standard habit: sympodial growth, pseudobulbs with conduplicate leaves.

§Neotropics, except not present in the Caribbean.

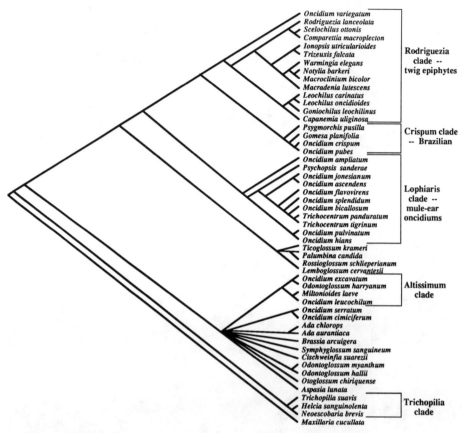

Figure 14.1. Semi-strict consensus tree from all equally parsimonious trees (172) using Dollo parsimony analysis.

No bootstrap analyses (Felsenstein, 1985) were performed. From a practical standpoint, running a bootstrap analysis would result in prohibitively long computer runs when combined with the already lengthy running times of step matrices. More importantly, we maintain that trees found by optimizing site gains benefit not at all from bootstrap analysis. Furthermore, one site gain that marks a basal node of a tree will always have a low bootstrap value if it is part of a large data set (100 or more characters), but, if there is no homoplasy or only homoplasy due to site losses at that node, then we conclude that the bifurcation should not be rejected.

Whereas analyses of equally weighted characters (in which no *a priori* evolutionary model exists) could consider trees several steps longer as nearly as good explanations of the data, this standard should not be applied to weighted characters. Occurrence and probabilities of restriction site gains and losses in cpDNA can be modeled (Albert et al., Chapter 16, this volume), and this knowl-

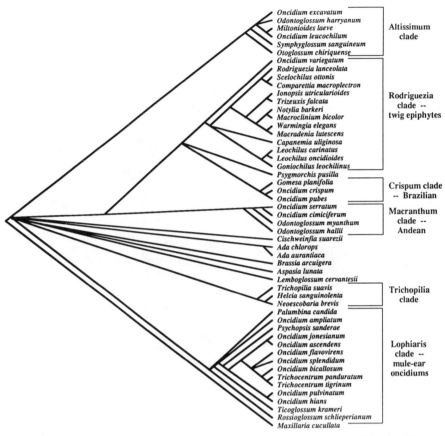

Figure 14.2. Semi-strict consensus tree from all equally parsimonious trees (151) using Wagner parsimony analysis. Note lack of resolution of most clades.

edge of molecular events should lead us to reject any hypothesis that does not minimize parallel site gains.

Relative to the point discussed above, we shall give no further consideration to the consensus trees found by Dollo (Fig. 14.1) or Wagner (Fig. 14.2) parsimony. From the perspective of minimizing homoplasious site gains while not restricting parallel site losses, the consensus tree of the 12 equally parsimonious trees found using the character-state weighting method (Fig. 14.3) will be used to consider chromosomal and morphological evolution. We regard this consensus tree to be the best explanation of the molecular data.

Evidence for Aneuploid Reduction in Chromosome Number

The invariant numbers of coding loci found by Chase and Olmstead (1988) in taxa that span the range of chromosome number are not compatible with an

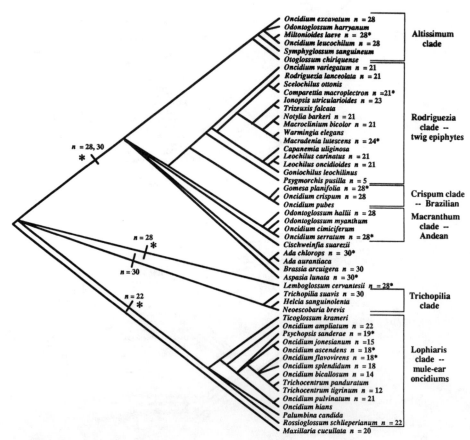

Figure 14.3. Semi-strict consensus tree from all equally parsimonious trees (12) using the character-state weighting method of PAUP 3.0. Chromosome numbers have been added (all from Tanaka and Kamemoto, 1984, except for the two species of *Leochilus,* which are from Chase, 1986b). In some cases the exact species in this analysis has not been counted, but when a closely related species has been determined this count was included (and marked with an asterisk). Chromosome numbers of the most basal member of each clade have been added to the four basal lineages within the subtribe; an asterisk on these basal clades denotes occurrence of predominantly *Oncidium*-type floral features.

hypothesis of hybridization accompanied by doubled ploidy. Instead, these results would be expected if changes in chromosome numbers were accomplished by aneuploidy. Autopolyploidy cannot be eliminated by analyses of coding loci, because undiverged, duplicated loci produce identical proteins. We have begun to quantitate DNA amounts (by fluorescence microscopy; Coleman et al., 1981; Coleman and Goff, 1985) of species in the Oncidiinae in order to look for evidence of autopolyploidy. Our results to date are not consistent with hypotheses that account for differences in chromosome number by changes in ploidy (Albert et al., unpublished data).

To illustrate the distribution of variation in chromosome number, published counts have been placed to the right of the species in Fig. 14.3. Chromosome numbers of $n = 28$, 30 were associated with vegetative features found in outgroups, such as the Maxillariinae, suggesting that this number is the ancestral condition for the Oncidiinae (Chase, 1986a). In contrast, lower numbers are found in derived groups, such as the Rodriguezia clade ($n = 5$–24) and the Lophiaris clade ($n = 14$–22), and are associated with modified vegetative features and atypical habitats (often more xeric conditions). Relative to the cpDNA phylogenies, no consistent downward trend is obvious in the Rodriguezia group, but in the Lophiaris clade a descending series occurs. *Trichocentrum* represents the lowest number ($n = 12$) and occupies the most derived position within the Lophiaris group. Figure 14.3 offers support for the hypothesis that $n = 28$, 30 is ancestral for the Oncidiinae, as does the frequent occurrence of these two numbers in most clades. One of the three equally parsimonious positions *Lemboglossum* occupies is the most basal member of the Lophiaris clade. From the standpoint of its distribution and floral and vegetative features, we argue that it is in fact a member of the Lophiaris clade, and, if it is so placed, then all major clades contain at least basal members with chromosome numbers of $n = 28$ or 30.

Explaining how the chromosome numbers within the Oncidiinae relate to that of the outgroup, represented in this case by *Maxillaria cucullata* with $n = 20$ ($n = 19$ or 20 are the most common numbers in the Orchidaceae), presents a more complex problem. An alternative explanation is that $n = 20$ is ancestral in the Oncidiinae and different groups within the subtribe have independently undergone aneuploid increases as well as decreases. This hypothesis is not nearly as parsimonious with respect to the cpDNA phylogeny (it requires more lineages to undergo parallel changes). Regardless of the direction of aneuploid change, the Oncidiinae are extraordinarily dynamic chromosomally, and few other groups of flowering plants are known to approach this extreme degree of aneuploid alteration and heterogeneity.

Morphological Evolution:
Evidence for Plesiomorphy of *Oncidium*-type Floral Traits

Obtaining a clear understanding of morphological variation has always been difficult to achieve in this group of orchids. The taxonomy of the Oncidiinae, as in most other Orchidaceae, has been based almost exclusively on floral traits. Geography and vegetative features have rarely played a major role in generic considerations. The major problem has been the separation of the largest genus of the subtribe, *Oncidium*, from apparently closely related genera, such as *Brassia*, *Miltonia*, and *Odontoglossum* (Lindley, 1837; Garay, 1963; Dressler and Williams, 1975). The tactic several authors have chosen is to recognize ever more finely delimited generic concepts, and, in recent years, there has been a proliferation of genera. The cpDNA study presented here was initiated with the hope that it would assist in the process of redefining genera and illuminating morphological evolution, both vegetative and floral. The cladogram presented in Fig. 14.3 clearly supports the contention (Dressler and Williams, 1975, Chase 1986a, 1987a) that the present generic scheme is decidedly artificial. These phylogenies argue for an extremely complex evolutionary history for the Oncidiinae. An alternative view is that the monophyletic groups identified, each of which appears to cluster species almost randomly regardless of genus, are an artifact of some as yet undiscovered peculiarity of cpDNA evolution, a clear case in which morphology and molecules do not agree. To the contrary, the morphological evidence has argued for a "new vision" since the time of Lindley, and no worker has attempted to construct an overall systematic perspective on the old generic scheme before erecting new genera.

Orchids are notorious for permitting artificial crosses between distantly related genera. Isolating mechanisms within the Orchidaceae appear to be based principally on pollinator-specific relationships (van der Pijl and Dodson, 1966; Dressler, 1981), and although frequencies of natural interspecific hybridization within flowering plant families have never been rigorously surveyed, we suspect that natural populations of orchids are probably less profligate than other groups. If the genealogical relationships expressed in Fig. 14.3 represent reality, then many of the "intergeneric hybrids" produced artificially over the past 150 years should be considered intrageneric.

Furthermore, naturally occurring intergeneric hybrids within the Oncidiinae are rare (Chase, personal observation based on field experience, herbarium study, and paucity of reports in the literature), and all of those reported (*Miltonia–Aspasia, Miltonia–Oncidium*, and *Oncidium–Ornithophora;* Dressler, 1981) could also be considered intrageneric relative to the phylogeny of Fig. 14.3. If one were to hypothesize natural hybridization as the cause of generic inconsistencies within cpDNA phylogenies, for instance the placement of *Trichocentrum* within the Lophiaris clade rather than as a sister group to it, then one should be able to produce artificially such an admixture of floral and vegetative features or at least to suggest parents that could produce an intermediate with the features of *Trichocentrum* (at least one parent would have to be a "mule-ear" type, since this is the type of habit and cpDNA that *Trichocentrum* possesses). The unique floral features and low chromosome number found in *Trichocentrum* cannot have been produced by hybridization between any extant members of the Oncidiinae. Similar arguments can be advanced against "hybridization hypotheses" for most of the pairs of radically different sister species (in terms of floral features) in the cpDNA phylogeny (Fig. 14.3). Unless one wishes to begin creating hypothetical, extinct male parents (the female parents being members of the cpDNA clades with which these "hybrids" are aligned) to account for all the generic discrepancies, then we are left with the impression that floral features must be easily and drastically modified without major changes in genetic makeup. Although we cannot eliminate a role for hybridization in the Oncidiinae, neither are we aware of any evidence that would support a major role. Furthermore, this cpDNA phylogeny for the subtribe does suggest reasons why, without resorting to hypotheses of hybridization, delimiting a "natural" *Oncidium* has been difficult.

The cladogram (Fig. 14.3) presented here has, as its major feature, a basal tetrachotomy. The basal node within the subtribe is unresolved due to three site gains in *Lemboglossum*, any two of which are homoplasious from the perspective of the third. If *Lemboglossum* were removed from the data set, then only two equally parsimonious trees would have been found, these due to two homoplasious site gains found in *Psychopsis* (Chase and Palmer, 1989). At least three hypotheses could explain the basal tetrachotomy, two molecular and one morphological:

1. Two of the three site gains in *Lemboglossum* may in fact not be homologous with the site gains* that produce basal nodes within the Oncidiinae, and this would resolve the position of *Lemboglossum* with respect to the basal nodes.

2. Past episodes existed in which the rate of molecular evolution has increased or decreased, and one period of slow rates occurred at the time during which the ancestral lineages of the Oncidiinae were diverging, thus leaving little molecular evidence of this early diversification.

3. The rate of molecular change remained more or less constant, but morphological evolution was punctuated and sometimes rapid. A number of eventually distinctive lineages diverged more or less simultaneously from an ancestral and perhaps molecularly polymorphic stock, and either little or only conflicting evidence remains in the cpDNA of these plants from this evolutionarily active, but brief, episode. This phyletic, and perhaps also adaptive, radiation may in fact not have been instantaneous, but from the present point in time the process appears nearly so.

We favor a combination of the first and third scenarios, but until we understand molecular evolution better, we cannot rule out the second. In fact, these three hypotheses are not mutually exclusive and could be collectively responsible for the tetrachotomy.

The morphology of the various oncidioid lineages supports the third hypothesis. The traditional genera are admixtures of character states for various floral features. In 1837, Lindley made the following distinctions: "The genera Brassia, Miltonia, Cyrtochilum [*Oncidium macranthum* and its relatives from Fig. 14.3], Odontoglossum and Oncidium are closely related. . . . Oncidium has a column with two

---

*Inability to distinguish between similarly sized fragments produced by different site mutations may sometimes lead to the conclusion that truly nonhomologous site gains are homologous; sequencing the region in which the homoplasious site gains occur would test their homology, as would doing double-digests and running the fragments out farther on standard agarose gels so that smaller differences in size could be detected.

ears, and a distinct lobed lip; Miltonia a column with two ears, and an entire lip partially united to it at the base; Odontoglossum a winged column, and an entire lip partially united to it at the base; Cyrtochilum a winged column and a distinct entire lip; and finally Brassia has a column that is neither winged nor eared, and a distinct lip. . . . Moreover, Oncidium and Cyrtochilum should have unguiculate sepals and petals, while the other genera have them sessile."

These distinctions were insufficient even in the time of Lindley to separate the oncidioid genera (intermediates and other combinations also exist), but they are still in use today for lack of a better system. These genera appear either to have mosaically retained or developed in parallel many of the same traits. If one emphasizes lip shape, then *Brassia* and *Cyrtochilum* form a closely related pair; if a winged column, then *Cyrtochilum* and *Odontoglossum;* if lip-column fusion, then *Miltonia* and *Odontoglossum*. In this sense, morphological relationships in the Oncidiinae are unresolvable. Adding to the problem is the fact that all these traits are unpolarizable by outgroup comparison; they exist in this format in no other group of orchids. If morphological cladistics are to be feasible in the Oncidiinae, gross floral traits must be excluded and new micromorphological characters sought.

From the standpoint of cladistic classification, two types of artificial genera are identified by the molecular analyses. The first type is derivative; these sorts of progenitor-derivative relationships have been identified in other molecular studies (Sytsma and Gottlieb, 1986). *Trichocentrum* (the most derivative member of the Lophiaris clade; Fig. 14.3) shares a unique habit with other species in this lineage but has a radically different floral morphology. Furthermore, it is pollinated by fragrance-collecting euglossine bees, whereas its close relatives operate by deceit pollination (van der Pijl and Dodson, 1966). *Trichocentrum* should be reduced to synonymy with whatever genus the Lophiaris clade becomes.

The second type of unnatural genus is represented by *Odontoglossum* and *Oncidium*. These two taxa are polyphyletic; the floral traits that define them appear to have evolved independently several times. Chase (1987a) concluded that several of the groups making up *Odontoglossum* appeared to be more closely related to various groups of *Oncidium* than they were to each other, and the molecular data bear this out for two groups of species in *Odontoglossum*. All molecularly distinct lineages, except for the Trichopilia clade, contain at least some members that have traditionally been considered members of *Oncidium*. From a morphological standpoint, the identification of an orchid as a member of the Oncidiinae is easy. A phenetic gap demarcating oncidioid orchids from all others is abundantly clear, but this gap renders outgroup analysis of floral traits useless in an attempt to resolve differences between competing hypotheses concerning floral evolution within the subtribe. Either the widespread *Oncidium*-type floral features (open flowers offering no reward and pollination by deceit) have arisen independently in each of these lineages or this set of floral features is the ancestral (plesiomorphic) one for the subtribe and has been retained by at least some members of all clades (except in *Trichopilia* and its relatives; Fig. 14.3). We favor the latter case, in which switches either to other types of deceit pollination or to offering some type of reward have resulted in the divergent floral morphologies that have been segregated generically.

Analysis of cpDNA has not supported concepts of relationships based solely on floral traits. Not surprisingly (especially in light of the triviality of the features cited by Lindley above), gross floral morphology appears particularly unreliable, but at least some of the 16 clades identified in this analysis, for example the Trichopilia clade (*Helcia, Neoescobaria,* and *Trichopilia*), nevertheless may be identified reliably by a unique set of shared floral features. Other groups can be identified best by life-history traits (Rodriguezia clade), geography (Brazilian Crispum and Andean Macranthum clades), altitude (the high Andean Macranthum and largely low- to middle-elevation Altissimum groups), or habit (the Lophiaris clade). The Rodriguezia and Lophiaris groups can also be identified by seed features (Chase, 1987b; Chase and Pippen, 1988). Those groups that have specialized altitudinally and geographically are particularly hard to define in terms of floral morphology. They may in fact have some diagnostic, but as yet undiscovered, vegetative traits. The answer to the question of which features are the most reliable phylogenetically and systematically is that such matters must be examined on a case-by-case basis. Generally speaking, nonfloral traits are more reliable, but it cannot be stated *a priori* that any single aspect—floral features, habit, geographical distribution, life-history traits, or cpDNA variation—may always be emphasized at all levels to the exclusion of the others.

The analysis of cpDNA presented here leads us to hypothesize an evolutionary scenario that is complex and novel (genus *Oncidium* defined by plesiomorphic floral traits; closely related species with radically different features). This interpretation is compatible with ideas that morphological evolution may be episodic and that major morphological shifts may be accomplished with little genetic modification. Many of the morphologically different pairs of species in this analysis have few cpDNA mutations to separate them. Additionally, these molecular phylogenies reveal, not surprisingly, that species are mosaics of ancestral and derived traits (both morphological and molecular). Species at any level of phylogenetic advancement may exhibit the *Oncidium*-type of floral morphology, and this floral format may be combined with extremely derived vegetative traits, highly modified life-history traits, or low chromosome number, as in *Psygmorchis*. The converse situation also occurs: *Symphyglossum* has a highly modified floral morphology adapted for hummingbird pollination but has no apparent nonreproductive specializations. Some taxa, such as *Rossioglossum, Lemboglossum,* and many species of *Oncidium,* appear to be similar to what we would imagine to be the progenitor of the Oncidiinae. Simplistic reliance on gross floral morphology has created a suite of artificial genera in this subtribe, and an apparently complex evolutionary history has prevented workers from accurately assessing phylogenetic relationships and creating a more natural taxonomy.

Molecular studies have not presented us with any simple solutions for the taxonomic problems in the Oncidiinae, but they have led us to a new hypotheses of morphological and chromosomal evolution. This new perspective brings with it new challenges. If evolution is episodic and major morphological change can be accomplished by minor genetic restructuring, then we should be able to detect evidence of how such modifications are accomplished ontogenetically. Analyses of variation in cpDNA not only give us an independent data set to test contrasting explanations of morphological characters but also assist in the formulation of new hypotheses and direct our focus to specific taxa that are likely candidates for further investigation. In this sense, molecular phylogenetics are only mere entry points into more extensive studies of the evolutionary mechanisms that control morphological and chromosomal change.

References

Ackerman, J.D. (1986) Mechanisms and evolution of food-deceptive pollination systems in orchids. *Lindleyana* **1**, 108–113.

Ackerman, J.D. (1989) Limitations to sexual reproduction in *Encyclia krugii* (Orchidaceae). *Syst. Bot.* **14**, 101–109.

Benzing, D.H. (1987) Major patterns and processes in orchid evolution: a critical synthesis. In: *Orchid Biology, Reviews and Perspectives* (ed. J. Arditti), Cornell University Press, Ithaca, New York, pp. 35–77.

Chase, M.W. (1986a) A reappraisal of the oncidioid orchids. *Syst. Bot.* **11**, 477–491.

Chase, M.W. (1986b) A monograph of *Leochilus* (Orchidaceae). *Syst. Bot. Monog.* **14**, 1–97.

Chase, M.W. (1987a) Systematic implications of pollinarium morphology in *Oncidium* Sw., *Odontoglossum* Kunth, and allied genera (Orchidaceae). *Lindleyana* **2**, 8–28.

Chase, M.W. (1987b) Obligate twig epiphytism in the Oncidiinae and other Neotropical orchids. *Selbyana* **10**, 24–30.

Chase, M.W., and Olmstead, R.G. (1988) Isozyme number in subtribe Oncidiinae (Orchidaceae): an evaluation of polyploidy. *Amer. J. Bot.* **75**, 1080–1085.

Chase, M.W., and Pippen, J.S. (1988) Seed morphology in the Oncidiinae and related subtribes (Orchidaceae). *Syst. Bot.* **13**, 313–323.

Chase, M.W., and Palmer, J.D. (1989) Chloroplast DNA systematics of lilioid monocots: resources, feasibility, and an example from the Orchidaceae. *Amer. J. Bot.* **76**, 1720–1730.

Coleman, A.W., and Goff, L.J. (1985) Applications of fluorochromes to pollen biology. I. Mithramycin and 4′,6-diamidino-2-phenylindole (DAPI) as vital stains and for quantitation of nuclear DNA. *Stain Technol.* **60**, 145–154.

Coleman, A.W., Maguire, M.J., and Coleman, J.R. (1981) Mithramycin- and 4′,6-diamidino-2-phenylindole (DAPI)-DNA staining for fluorescence microspectrophotometric measurement of DNA in nuclei, plastids, and virus particles. *J. Histochem. Cytochem.* **19**, 959–968.

Dafni, A. (1987) Pollination in *Orchis* and related genera: evolution from reward to deception. In: *Orchid Biology, Reviews and Perspectives* (ed. J. Arditti), Cornell University Press, Ithaca, New York, pp. 79–104.

Dressler, R.L. (1981) *The Orchids: Natural History and Classification*, Harvard University Press, Cambridge, MA.

Dressler, R.L. (1986) Recent advances in orchid phylogeny. *Lindleyana* **1**, 5–20.

Dressler, R.L. (1989) The vandoid orchids: a polyphyletic grade? *Lindleyana* **4**, 89–93.

Dressler, R.L., and Williams, N.H. (1975) El complejo Oncidoglossum confusum. *Orquídea* **4**, 322–340.

Felsenstein, J. (1985) Confidence limits on phylogenies: an approach using the bootstrap. *Evolution* **39**, 783–791.

Garay, L.A. (1963) *Oliveriana* and its position in the Oncidieae. *Amer. Orchid Soc. Bull.* **32**, 18–24.

Garay, L.A., and Stacy, J.E. (1974) Synopsis of the genus *Oncidium*. *Bradea* **1**, 393–429.

Lindley, J. (1837) *Miltonia spectabilis*. *Edwards Bot. Reg.* **23**, t1992.

Palmer, J.D., Jansen, R.K., Michaels, H.J., Chase, M.W., and Manhart, J.R. (1988) Chloroplast DNA variation and plant phylogeny. *Ann. Missouri Bot. Gard.* **75**, 1180–1206.

Sytsma, K.J., and Gottlieb, L.D. (1986) Chloroplast DNA evidence for the derivation of the genus *Heterogaura* from *Clarkia* (Onagraceae). *Proc. Natl. Acad. Sci. USA* **83**, 5554–5557.

Tanaka, R., and Kamemoto, H. (1984) Chromosomes in the orchids: counting and numbers. In: *Orchid Biology: Reviews and Perspectives* (ed. J. Arditti), Cornell University Press, Ithaca, New York, pp. 323–410.

van der Pijl, L., and Dodson, C.H. (1966) *Orchid Flowers: Their Pollination and Evolution*, University of Miami Press, Coral Gables, FL.

Ziegler, B. (1981) Mikromorphologie der Orchideen-samen unter Berücksichtigung taxonomischer Aspekte, Ph.D. Dissertation, Ruprecht Karls-Universität, Heidelberg, West Germany.

Appendix. Data matrix used in this analysis of restriction site variation in cpDNA (PAUP 3.0); "0" denotes site absence, and "1" denotes site presence

Ada aurantiaca	0011110001011001001110001011111101100110011001110001100010000001011000010110101010100011110101010001111110
Ada chlorops	0011110001011001001110001011111101100110011001110001100010000001011000010110101010100011110101010001111110
Aspasia lunata	0111110001011001001110001011111101100110011001110001100010000001011000010110101010100011110101010001111110
Brassia arcuigera	0111110001011001001110001011111101100110011001110001100010000001011000010110101010100011110101010001111110
Capanemia superflua	0010110101011001001110010011111110011001100100000011000100000001011000010110101010100011110101010001111110
Cischweinfia suarezii	0111110001011001001110001011111101100110011001110001100010000001011000010110101010100011110101010001111110
Comparettia macroplectron	0010100101001001110110100111110011001100100000001100010000000110110101010100011110101010001111110
Gomesa planifolia	0111110101001010011110001011111011001100110011100011000100000010110001011010101010100011110101010001111110
Goniochilus leochilinus	0010110101001001001110001011111011001100110011000011000100000001011000101010101010100011001010001001110010
Helcia sanguinolenta	0111110001011001001111111101101110101100110011000100010111100010000001011000010110101010100011110101010001111110
Ionopsis utricularioides	0010010110010011001001011111101101011001100110000110000101010001001110000100000011
Lemboglossum cervantesii	0111110001011001001110101011111011001100110011000100010000011010110001011010101010100011110101010001111110
Leochilus carinatus	0010100101000010010010011111011001100110011100001100010000100010101010101010101010100011101011000010011000
Leochilus oncidioides	0010010101001001001110101011111011001100110011000011000100001010110101010101010101010001110101100010011000
Macroclinium bicolor	0010100101001001001101011111011001100110011000011000100000001011010101010101010101010001110001111100011
Macradenia lutescens	0010100101000110011100101011110110011001100110000110000101000011000010101010101010100011110011000110100011
Miltonioides laeve	1111010001100001110001011111111111110100110011000100010000011000010100000100011000010110101010001011000011
Neoescobaria brevis	0111110001100010000111111101110011001100110011000100010000011101101100010110101010100011110101010001111110
Notylia barkeri	0010110110010011001000111101011001100110011000011000100110110001110101010100011110001111110001
Odontoglossum hallii	0111110000100100111011101010111110110011001100110000100010000011010110001011010101010100011110101010001111110
Odontoglossum harryanum	0111010001000001110001011111010110111001100100001100010000011110110001011010101010100011110101010001111110
Odontoglossum myanthum	0111110000110010001110110011101100110011001100110000010000110110001011010101010100011110101010001111110
Oncidium ampliatum	0111110011000111001111111011010111001100110011010000100010000011010110001011010101010100011110101010001111110
Oncidium ascendens	0111110101100011110101011101011001100110011101000100010000000100010100110100011101010001011010101010001111110
Oncidium bicallosum	0111110001100011100101110110101011001100111001100110010100000011010100000100100011010101010001111110
Oncidium cimiciferum	0111110001100010000110010110110101100110011101000110010000000011100000100000111101011010101010001111110

Oncidium crispum

Oncidium excavatum

Oncidium flavovirens

Oncidium hians

Oncidium jonesianum

Oncidium leucochilum

Oncidium pubes

Oncidium pulvinatum

Oncidium serratum

Oncidium splendidum

Oncidium variegatum

Otoglossum chiriquense

Palumbina candida

Psychopsis sanderae

Psygmorchis pusilla

Rodriguezia lanceolata

Rossioglossum schlieperianum

Scelochilus ottonis

Symphyglossum sanguineum

Ticoglosssum krameri

Trichocentrum panduriforme

Trichocentrum tigrinum

Tricopilia suavis

Trizeuxis falcata

Warmingia elegans

Maxillaria cucullata

15

The Suitability of Molecular and Morphological Evidence in Reconstructing Plant Phylogeny

Michael J. Donoghue and *Michael J. Sanderson*

Renewed interest in phylogenies over the last few decades coincides with a growing sense that it will actually be possible to obtain an accurate picture of evolutionary history. Indeed, the prospects of retrieving phylogeny now seem better than ever, owing to basic theoretical advance (due mainly to Hennig, 1966), the availability of computer programs that can handle large data sets, and the accessibility of new sources of evidence, especially molecular characters.

It is our impression that methods of phylogenetic inference—their assumption and reliability—have received more attention than the data upon which phylogenies are based. However, the rapidly increasing use of molecular techniques has focused attention on the pros and cons of molecular versus morphological evidence. Hillis (1987) and Patterson (1987) have reviewed the main arguments, and a summary of results for a number of major groups is available in the proceedings of the recent Nobel symposium (Fernholm et al., 1989). Regarding plant phylogeny in particular, only Sytsma (1990) has attempted a general survey and comparison of molecular and morphological studies. However, reviews of the use of particular molecules (e.g., Palmer et al., 1988, on chloroplast DNA) include useful discussion of molecular versus morphological results, and comparisons have been made within several angiosperm families (e.g., see in this volume, Chapter 10 by Doyle et al. on Fabaceae, Chapter 11 by Jansen et al. on Asteraceae, and Chapter 13 by Sytsma and Smith on Onagraceae).

This chapter is not a general review. Few generalizations seem possible at present, because there are too few careful morphological and molecular cladistic

We are grateful to all those who generously shared their unpublished research and/or commented on the manuscript: Bil Alverson, Steve Downie, Jeff Doyle, Bob Jansen, Toby Kellogg, Brent Mishler, Dick Olmstead, Jeff Palmer, Ward Wheeler, and Liz Zimmer. MJD is indebted to Ken Sytsma and his students for facilitating his molecular education and also gratefully acknowledges the support of the NSF (BSR-8822658). MJS thanks Jeff Doyle for his assistance and the Sloan Foundation for financial support. Finally, we thank the editors and reviewers for their patience and constructive criticisms.

studies of the same groups of plants. We anticipate that much more detailed comparisons will be possible within a few years, although (surprisingly) the limiting factor is likely to be the number of solid morphological analyses. In the meantime, our aim is to examine the view that an accurate picture of phylogeny can be obtained solely on the basis of molecular data—that morphological data can be set aside safely at the outset of an analysis and mapped onto the molecular phylogeny later.

This view seems to be popular. Sibley and Ahlquist (1987, p. 118), for example, assert that "the molecules can reconstruct the phylogeny with a high degree of accuracy. Given the phylogeny, the morphologist will be able to interpret structure and to separate similarities due to common ancestry from those resulting from convergence." Much the same view has been expressed by Gould (1985), and it appears to be widespread in botanical circles. For example, in reference to cases of adaptive radiation, Sytsma et al. (1991) specifically recommend a "two-step process" in which trees based only on molecular data are used to interpret the evolution of morphological characters. Morphological evidence, they say, should be avoided, because it "can often be phylogenetically uninformative or even misleading because of the operation of strong selection resulting in homoplasy, difficulty in ordering or even polarizing character states, the high number of autapomorphies, and the lack of well defined synapomorphies." Likewise, Gottlieb (1988, p. 1170) contends that "the molecular data are self-sufficient in that their usefulness does not depend on concordance with other lines of phenotypic evidence." Rather than being relevant in assessing relatedness, "the data of morphology, the traditional source of information about phylogeny, should be viewed as relevant to studies of plant development." Much the same view has been expressed by other botanists and is, we believe, widely held.

Here we consider theoretical arguments and selected empirical studies of plant phylogeny that bear on the view that phylogenetic hypotheses based on molecular data alone are more reliable than those based on morphology or on a combination of evidence. We conclude that, at best, this outlook is premature, and, at worst, it will stand in the way of achieving an accurate picture of phylogeny. We will deal only in passing with some kinds of difficulties that might arise in interpreting molecular results (e.g., different modes of inheritance of different genomes, or inadvertent analysis of paralogous genes; Doyle, 1987; Kawata, 1987; Patterson, 1987; Avise, 1989), or arguments in favor of morphological data (e.g., the likelihood of more thorough sampling of organisms/taxa; Hillis, 1987; Donoghue et al., 1989). Almost all of our discussion concerns broad (or "higher level") phylogenetic questions, both for practical purposes (to narrow the scope), and because these are the sorts of problems we have personally pursued. However, we believe that our general conclusions also apply to population and "species-level" problems, which are the focus of several chapters in this book.

We will be disappointed if our analysis is interpreted as a reaction against molecular approaches to phylogeny reconstruction. Nothing could be further from the truth, as should be evident from the fact that both of us are pursuing molecular studies (MJD on *Viburnum,* MJS on *Astragalus*). Our argument is not against the use of molecular data; rather, it is against ignoring relevant morphological evidence. On the positive side, we hope to focus attention on a set of issues that must be confronted in adopting the view that *both* morphological and molecular data should be used in reconstructing phylogeny.

Homoplasy

Here we consider a series of issues related to homoplasy (convergence, parallelism, reversal) and its impact on phylogeny reconstruction. The first four sections deal mainly with theoretical arguments on homoplasy in relation to reliability, selection, character complexity, and environmental variation. We have deferred a discussion of levels of homoplasy seen in real data sets to the last section, because the homoplasy reported in cladistic studies may have a variety of causes.

Homoplasy and Reliability

Many arguments for the superiority of molecular data rest on the assumption that homoplasy is directly related to reliability. If more reliable results are obtained when there is less homoplasy and if molecular data are less homoplastic than morphological data, then it follows that molecular data are superior. We consider the first part of this equation to be questionable; that is, the relationship between level of homoplasy and reliability or confidence is weak at best.

The standard intuition about the effect of homoplasy stems in part from the view that it is merely a "mistake" (Mickevich and Weller, 1990), which implies that homoplastic characters cannot be useful in reconstructing phylogeny. However, this is clearly false, because independent gains (or losses) can certainly function as synapomorphies of the two or more clades in which they evolved (e.g., the independent evolution of an inferior ovary, or of a particular structural rearrangement of the genome; see below). A homoplastic character might be *misleading,* but this depends in a complicated way on how it interacts with other characters in the data set. Homoplasy—even a large amount of it—does not by itself guarantee an inaccurate tree, particularly in large studies, which tend to have high levels of homoplasy simply by virtue of the number of taxa involved (Sanderson and Donoghue, 1989).

The overall amount of homoplasy is not as critical as its distribution (Jansen et al., 1990). Thus, it is not difficult to construct data sets in which there is a high level of homoplasy *and* a high level of confidence (Sanderson and Donoghue, 1989). This is true, at least, if confidence is estimated by resampling techniques such as the bootstrap (Felsenstein, 1985; Sanderson, 1989). Conversely, because reliability is a function of the weight of multiple, independent synapomorphies

(Hennig, 1966; Sanderson, 1989), confidence can be low even if homoplasy is low or nonexistent, such as when only one synapomorphy supports each clade.

Homoplasy and Selection

The second assumption of the argument presented above is that molecular characters are less homoplastic than morphological characters. This view is typically linked to two other assumptions: (1) selectively neutral characters are less likely to show homoplasy than those subject to selection (a view advanced initially by morphological systematists to aid in character selection/weighting; e.g., Mayr, 1969), and (2) molecular data as a whole are less likely to be subject to selection than morphological data. The second assumption implies that most morphological changes are adaptations (but see Gould and Lewontin, 1979) and that most molecular changes are neutral (but see Kreitman and Aguade, 1986). Nevertheless, instances of apparent sequence conservation are generally interpreted as reflecting selective constraints (e.g., Kimura, 1983; Patterson, 1988).

But even the first assumption is misguided. That is, the selective value of a character does not necessarily bear any particular relationship to the amount of homoplasy it exhibits. Features that are or were subject to selection need not have evolved more than once and may be highly conserved. It is not difficult, for example, to think of morphological traits that are presumed to be adaptations and are also thought to have evolved only once (e.g., the closed carpel of angiosperms). Some morphological characters might actually become less likely to undergo change (including homoplasy) owing to increased "burden," that is, by virtue of the evolution of dependent traits that constrain further evolution (Riedl, 1978; Donoghue, 1989).

Just as selection need not result in a high level of homoplasy, neutrality does not insure a low level. Mutations can occur at the same nucleotide site during the evolution of a lineage and, all things being equal, the probability of such multiple hits increases with time. When the number of possible states is highly constrained, as it is in the case of nucleotides, the chance that mutations at a particular site will result in homoplasy is quite high (Mishler et al., 1988). One might therefore expect high levels of homoplasy in neutral molecular characters given a sufficient amount of time. Archie (1989b) has shown that the consistency indices of two plant nucleotide data sets (derived by Bremer, 1988, from the amino acid studies of Martin et al., 1983, 1985) do not differ from those expected in randomly generated data, perhaps indicating near saturation with homoplastic multiple hits.

Homoplasy and Character Complexity

In the case of complex morphological features, it has long been argued that homology can be reliably determined at the outset through a detailed comparison of position, structure, and development (i.e., Remane's primary criteria of homology; Remane, 1952; Kaplan, 1984); any characters that have passed such a rigorous examination are unlikely to be homoplastic. The key to this argument is "complexity" (Donoghue, 1991). In the case of "simple" characters (e.g., effectively lacking development) the determination of homology rests more or less completely on congruence with other characters (i.e., Remane's auxiliary criteria; Remane, 1952). Patterson (1988) used this reasoning to reach the unexpected conclusion that similarity is a better guide to homology than congruence in the case of molecular data. But it is important to recognize that his assessment applies only to the comparison of whole sequences, where the level of complexity is sufficient (but not too great) to establish probabilities of convergence (Donoghue, 1991). It does not apply to individual nucleotide sites, where homology is established in the act of aligning sequences.

Based on similar logic, structural modifications of the genome (e.g., inversions, large insertions and deletions/transfers) have been touted as especially reliable indicators of phylogenetic relationship (e.g., Palmer et al., 1988; but see Doyle, 1987). Given the very large number of possible rearrangements, and the possible functional consequences of such modifications, it is highly unlikely that the same one would arise more than once. Furthermore, any doubt regarding homology could be resolved by sequencing through the critical regions to determine if the similarity extends to the individual nucleotide level. It is now clear, however, that even major structural rearrangements are not infallible guides to phylogeny—they too can arise independently. Perhaps the most obvious example is the (presumably) independent loss of one copy of the chloroplast DNA (cpDNA) inverted repeat in conifers (Strauss et al., 1988), within legumes (Lavin et al., 1990), and within Geraniaceae (Downie and Palmer, Chapter 2, this volume).

A second example demonstrates that molecular and morphological systematists employ the same logic in such cases. Downie et al. (1991; Downie and Palmer, Chapter 2, this volume) have shown that the loss of the cpDNA *rpl2* intron, which was originally thought to be unique to caryophillids (Zurawski et al., 1984; Palmer et al., 1988), has also occurred in several other taxa (Convolvulaceae, *Cuscuta, Drosera,* two genera of Geraniaceae, *Menyanthes,* and Saxifragaceae *sensu stricto*). Although this distribution suggests a number of independent origins, Downie and coauthors argue that the loss of this intron can still be a powerful indicator of relationships *within* the different lines in which it has occurred. This parallels an argument made by Donoghue (1983) concerning the evidential significance of a morphological feature, *Adoxa*-type embryo sac development. Although its presence in *Adoxa, Sambucus,* and some species of *Tulipa, Erythronium,* and *Ulmus* strongly suggests that it has evolved a number of times, it still might provide evidence of a direct connection between *Sambucus* and *Adoxa,* in view of the other characters that also suggest a close relationship. Nevertheless, even at this level, homology is ultimately judged by congruence with other data (Patterson, 1982). It could still turn out that the *Adoxa*-type embryo sac evolved independently in *Adoxa* and *Sambucus,* even if they are closely related, or that the *rpl2* intron was lost several times within Geraniaceae, for example.

Homoplasy, Environmental Variation, and Subjectivity

Morphological traits may be subject to considerable variation solely as a function of environment. This seems especially true in plants, which are notorious for plasticity in such features as body and leaf size (Stebbins, 1950; Schlichting, 1986; Sultan, 1987). Although it is conceivable that this kind of variation would lead a phylogenetic analysis astray, we are unaware of any example where this has actually occurred. There appear to be several reasons why. First, systematists working on higher level phylogenetic problems generally deal with characters that are not prone to environmental variation. Plants of *Astragalus,* for example, always produce zygomorphic flowers; subjecting them to more water or light will not induce actinomorphy. Second, occasional variation is recognized for what it is in most cases; for example, the presence of a four-merous flower in a lineage characterized by parts in fives. Plant systematists pursuing morphological studies at the generic level or below typically examine hundreds or thousands of individual specimens and will often see plants in a variety of habitats in the field, prior to deciding on appropriate characters for analysis. The availability of such extensive information about variation and plasticity seems to compensate to some extent for lack of information about the genetic and developmental basis of morphological characters. In particularly difficult cases, the extent of plasticity can be tested (e.g., Davis, 1983, 1987).

A concrete indication that the problem of plasticity has not had a major impact on phylogeny reconstruction is seen in comparing studies of plants and animals. If plants tend to be more plastic than animals, as is commonly believed, and if plasticity leads to problems in phylogenetic analysis, one might expect to see more homoplasy in plant studies than in animal studies. In fact, Fig. 15.1 indicates that the level of homoplasy in plant and animal studies does not differ significantly (Sanderson and Donoghue, 1989; but see Syvanen et al., 1989, who we believe were misled by their comparison of trees based only on cytochrome c). This suggests that plant systematists are not really being fooled by plasticity; rather, they have become adept at delimiting characters even in the face of considerable environmental variation.

Environmental variation is related to the issue of subjectivity in delimiting states, since it might blur discrete differences into a continuum. Alternatively, apparently discrete states may represent an underlying continuous variable subject to a threshold effect. This observation highlights a more general problem, namely the appropriate way in which to subdivide a system into characters and states (or the necessity of doing so at all; Felsenstein, 1988). In morphology there are certainly constraints on subdividing characters into states. Thus, as Wagner

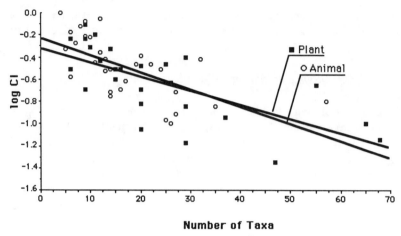

Figure 15.1. Log of consistency index versus number of taxa for 26 plant and 30 animal cladistic studies; redrawn from Sanderson and Donoghue (1989).

(1989) has emphasized, some degree of developmental individuation is critical. As an example, he discussed the morphology of the androecium in *Convolvulus,* in which stamens of varying length within the corolla are arranged in just two patterns that are mirror images of one another. Under these circumstances it does not make sense to treat the length of particular stamens as character states, because this attribute lacks its own genetic or developmental basis.

In the case of molecular data the appropriate characters and states have seemed more obvious, but it is not completely clearcut. Thus, in sequence data it is generally assumed that each nucleotide position (where homology is assessed in the alignment phase) is a character and the four possible bases are the appropriate states. But there are other ways to delimit the states at each site; for example, bases might be partitioned into purines and pyrimidines.

Even if there were some one "correct" or "natural" atomization, this does not mean that the wrong tree will be obtained by subdividing in another way. If this were not so, it is hard to imagine how any progress could have been made in reconstructing phylogeny, since idiosyncracies of individual investigators would have muddled the whole picture. Whether, and under what circumstances, different atomizations will yield different results has not been explored directly, but could easily be tested using sequence or morphometric data. It may turn out to be rather difficult to concoct circumstances under which different subdivisions give very different results.

Levels of Homoplasy in Real Data Sets

Leaving aside theoretical expectations, we can now ask whether there actually appears to be more homoplasy in cladistic studies based on morphological versus molecular data. A comparison of 42 morphological and 18 molecular cladistic stud-

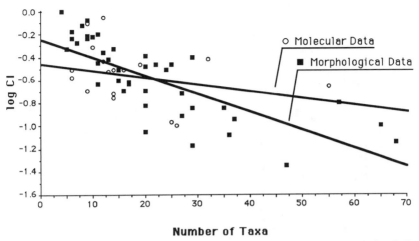

Figure 15.2. Log of consistency index versus number of taxa for 42 morphological and 18 molecular cladistic studies; redrawn from Sanderson and Donoghue (1989).

ies (Sanderson and Donoghue, 1989) indicated no significant difference in homoplasy as measured by the consistency index (Kluge and Farris, 1969). As shown in Fig. 15.2, when the number of taxa is taken into consideration it is not possible to conclude that one type of data consistently shows more homoplasy than the other.

It is important to note, however, that our analysis is very preliminary and that there were too few molecular studies to draw a statistically robust conclusion. More molecular studies are needed, especially analyses that include many terminal taxa. Furthermore, as more cladistic analyses become available, it would be desirable to subdivide the molecular studies (as well as the morphological studies). It is possible, for example, that a different pattern would emerge in comparing only analyses based on cpDNA, or those based on protein electrophoresis. Thus, we note that for small numbers of taxa the lowest consistency indices are found in electrophoretic studies, whereas the highest consistency index for a large number of taxa is based on cpDNA restriction site data (Jansen et al., 1990). It may also turn out that restriction fragment, nucleotide sequence, and amino acid data sets will differ significantly in consistency, since the processes governing the tempo and mode of evolution may be different in each. Additional parameters will also need to be considered, such as the relationship between the amount of homoplasy and the degree of resolution of relationships.

Neutrality and Rates of Evolution

Besides its supposed connection to lower levels of homoplasy, neutrality has also been regarded as an advantage of molecular data because neutral characters are more likely to evolve in a clocklike fashion. From the outset of molecular studies of phylogeny, clocklike evolution has been considered among the most

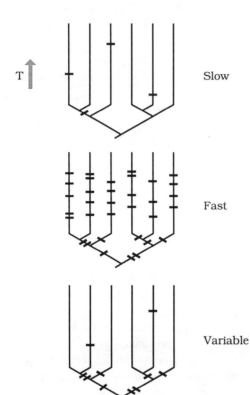

Slow

Fast

Variable

Figure 15.3. Clocklike evolution (whether slow or fast) is not desirable when reconstructing a rapid, ancient radiation. Character changes are shown as bars on branches.

important benefits of molecular evidence (see references in Clegg, 1990; Melnick, 1990). One reason for this is that variable rates of evolution could mislead some methods of inference, particularly those based on overall similarity. This argument is no longer very compelling, because methods are now available that are far more robust to variation in rates of evolution—cladistic parsimony methods with outgroup rooting, for example (e.g., Hillis, 1987; Sober, 1988). In any case, the molecular clock has since been called into question (e.g., Britten, 1986; Gillespie, 1986; Doyle et al., 1990; Melnick, 1990; Scherer, 1990).

However, even if we knew at the outset that a particular molecule evolved in a clocklike manner, this might not be desirable from the standpoint of phylogeny reconstruction (as opposed, for example, to age estimation). To appreciate why, consider Fig. 15.3, wherein we depict a group that radiated over a short period of time in the distant past, with little subsequent cladogenesis. Under these circumstances, a molecule evolving at a slow but constant rate may show too few changes during the critical period of diversification (Fig. 15.3, top). The result, commonly observed, will be an unresolved tree. On the other hand, in the case of a rapidly evolving molecule (Fig. 15.3, center), changes might mark each of the branches of the radiation, but continued rapid evolution would yield an increasingly "noisy" picture (Lanyon, 1988; Donoghue et al., 1989). That is, homoplasy is likely to increase and be distributed so as to eventually overwhelm the signal in the data (Felsenstein, 1978). This might account, for example, for difficulties in assessing cyanophyte phylogeny (Bremer and Bremer, 1989), relationships among major lines of land plants (Mishler et al., 1990), and even among major seed plant groups (Zimmer et al., 1989).

Ironically, changes of rate might actually be desirable—even necessary—to achieve a resolved and accurate assessment of relationships (Lanyon, 1988; Donoghue et al., 1989). Ideally, we would like to focus on characters that evolved like those shown in Fig. 15.3 (bottom); that is, those that underwent sufficient change during the radiation, but changed very little after that time. It might be argued that this pattern of evolution is more likely in the case of morphological characters, where there may be selection associated with cladogenetic events (Olmstead, 1989) and/or subsequent constraints on change as a function of selection (Lanyon, 1988) or increasing burden (Donoghue et al., 1989). Thus, morphological characters could more faithfully retain information on the sequence of events in ancient, rapid radiations, especially when it is possible to include fossils, since these can provide relatively unmodified branches near the base of the radiation of interest (Donoghue et al., 1989).

It is now widely appreciated that different molecules and different parts of the same molecule can evolve at different rates, which means that choices must be made regarding which data are appropriate for which phylogenetic problems (e.g., Palmer et al., 1988; Sytsma, 1990). It is not entirely obvious, however, how to evaluate the limits of particular molecules (especially in view of the usually limited sampling within terminal taxa), or how to identify and treat rate variation within a molecule (Mishler et al., 1988). In some cases, limits are imposed by the method of analysis. Thus, in the case of restriction fragment comparisons, reliable estimates of homology become virtually impossible in considering very distantly related species (e.g., Palmer et al., 1988). In other cases, an initial analysis of the data might serve as a warning of unreliable results; for example, if the level of homoplasy is similar to that found in random data sets (Archie, 1989a) or if incongruent results are obtained when different subsets of the data are analyzed. Smith (1989) used well-supported morphological clado-grams and divergence dates for echinoderms to test the limits of RNA sequence data, and concluded that 18S rRNA data yield reliable results for divergences within the last 100 m yrs, whereas beyond this point estimations are too prone to error due to saturation effects (regardless of whether one analyzes paired sites, unpaired sites, or transversions).

Independence

Character Independence and Weighting

The independence of morphological features from one another may be doubtful in some cases. If two or more "characters" are linked in the sense that they always undergo evolutionary change together, then they should be counted as only one bit of evidence to prevent them from outweighing a smaller number of truly independent characters. The argument is sometimes made that the same difficulties do not plague molecular data or that the problem is much reduced. Thus, it is commonly assumed that changes at different nucleotide sites are independent of one another and that each is to be counted as a single bit of evidence.

In response to this argument, it is important to recognize that, although the situation may be difficult in the case of morphological data, it is by no means impossible to establish independence. Sepal size and petal size may be strictly correlated in a particular taxon, but petal size, type of stomate development, and pollen exine structure are quite likely to be evolutionarily uncoupled, and each of these traits can legitimately provide evidence on phylogeny. Moreover, it is possible to evaluate hypotheses of morphological independence by gathering appropriate genetic and developmental data.

Although some molecular characters are very probably independent of one another (e.g., third sites in different codons), there are also well-known exceptions. The application of some techniques is very likely to result in correlations; for example, the use of restriction fragment patterns derived from random probes of the nuclear genome (e.g., Song et al., 1988). Of much more general importance are the effects of secondary structure. Thus, there may be compensatory base changes where nucleotides are paired, as in the stem regions of rRNA molecules (Hixson and Brown, 1986; Wolters and Erdmann, 1986; Steele et al., 1988; Wheeler and Honeycutt, 1988), or gene conversion activity promoting homogenization of the inverted repeats of cpDNA (Palmer, 1985, 1987). In the absence of information on secondary structure, paired sites may inadvertently be treated as independent, thereby overweighting what is basically a single underlying change (although mismatches are sometimes maintained). Thus, Wheeler and Honeycutt (1988) suggested that nucleotide positions in paired regions of rRNA "should be downweighted, perhaps by one-half, or even excluded." In contrast, Patterson (1989) and Smith (1989) report instances in which paired sites actually appear to perform better than unpaired sites and suggest that this may be related to divergence time.

More complicated problems arise through interactions among sites that may be separated by some distance in the sequence (cf., Appels and Honeycutt, 1988). For example, changes in one site can initiate selection for a compensatory change at a distant site, perhaps even in a gene coding for a separate but functionally interrelated protein. Such effects might be especially important in enzymes such as RUBISCO, which are constructed from separately encoded subunits. Indeed, in this case a change in the cpDNA might influence selection in nuclear DNA or vice versa. In most cases we are still blissfully unaware of such functional constraints on evolution.

It should be clear from the foregoing discussion that the issue of character independence can be rephrased in terms of character weighting. The argument is sometimes made that it is possible to establish objective, *a priori* weighting schemes for molecular data based on the likelihood of character change, and that this allows the use of methods such as maximum likelihood. From time to time, the same has been said of morphological data, and quantitative methods have even been devised (e.g., Riedl, 1978). However, whereas most morphologists have been suspicious of such procedures (tending to reject weighting altogether or make use of *a posteriori* weighting, such as the successive approximations method; Farris, 1969; Carpenter, 1988), molecular systematists have been much more confident that the relative likelihood of character state changes can be derived from first principles or prior empirical data. It is widely accepted, for example, that third sites within a codon are less constrained (Kimura, 1983), that transitions are more likely than transversions (Lake, 1987), and that an independent loss of a restriction site is more likely than an independent gain (DeBry and Slade, 1985; Templeton, 1983).

Before proceeding, it is important to consider that even if these observations about the relative likelihood of change are correct, it is not clear when and how differential weighting will affect the outcome of phylogeny reconstruction. Nor is it clear that not weighting characters differently will render the outcome especially sensitive to differences in likelihood of character change. Despite earnest efforts to comprehend the relationship between likelihood of character change and parsimony, this connection remains poorly understood (Sober, 1988).

In any case, hypotheses about likelihood of change are not as easy to translate into a weighting scheme as they may appear. Thus, the rate of evolution of third-position sites can be influenced by a number of factors. For example, such changes might be constrained by the presence of a functional open reading frame encoded on the opposite strand (c.f., Zurawski and Clegg, 1987). It has also been noted that transition-transversion ratios differ in different genomes (e.g., from 30:1 in animal mitochondrial DNA to 1.5–2:1 in cpDNA; Palmer, 1987; Zurawski and Clegg, 1987), and even in different genes within a genome (Zurawski and Clegg, 1987; Wheeler, 1990). However, such conclusions have only rarely been tested empirically (using cladistic analysis), and the results of such studies have not always fit the preconceptions. For example, Doyle (1991) constructed phylogenies based on 16 glutamine synthetase sequences. He found few differences in the level of homoplasy or confidence in comparing subsets of the data consisting of transitions versus transversions or third versus first and second sites within codons. Furthermore, methods designed to deal with differences in likelihood of change make their own assumptions. Thus, Lake's method of invariants ("evolutionary parsimony"; Lake, 1987; Holmquist et al., 1988) assumes that transversions of the two types are balanced, and it may be sensitive to deviations from this assumption resulting, for example, from differences in $G+C$ content (Gouy and Li, 1989; Sidow and Wilson, 1990).

In the case of restriction sites, it is quite clear that independent gains can occur. It is on this basis, in fact, that Albert et al. (Chapter 16, this volume) argue against the use of Dollo parsimony. Their derivation of a likelihood weighting scheme for restriction site data demonstrates the assumptions inherent in such calculations. Some of these are clearly unrealistic, for example, a constant rate of substitution across lineages (e.g., Wu and Li, 1985). It also emerges that such weighting schemes are not generalizable (to other genomes, for example, or possibly even to different genes in the same genome) since they are tied to particular estimates of substitution rate, which themselves are dependent on assumptions about cladistic relationships and divergence times (Brown et al., 1979; Wolfe et al., 1987). The most that one can hope for under such circumstances is what Albert et al. actually report, namely that weights fall within relatively narrow limits over the entire range of phylogenetic problems to which the data in question are applicable.

Another weighting problem concerns the relative value of structural mutations versus site mutations. Palmer et al. (1988, p. 1188) argued that "the extreme rarity and lack of homoplasy of major rearrangements makes each one a single character in a way that nucleotide substitutions, which inevitably will be afflicted with certain levels of homoplasy, can never be," and that such rearrangements "should be weighted much more heavily than a single nucleotide substitution or restriction site mutation." However, it is unclear exactly which structural rearrangements warrant such treatment and how much more they should be weighted (Olmstead et al., 1990). It is clear that the outcome of an analysis can be substantially affected by such decisions. This sensitivity is illustrated by the analyses of prochlorophyte relationships conducted by Morden and Golden (1989a, b) based on *psb*A sequence data. When the presence or absence of a seven-amino-acid domain is treated as the equivalent of one or two amino acid substitutions, *Prochlorothrix* may be nested among blue-green algal groups (consistent with the 16S rRNA result of Turner et al., 1989). However, when it is weighted any more heavily, *Prochlorothrix* appears as the sister group of green chloroplasts (consistent with morphological and pigment evidence, Miller and Jacobs, 1989).

The greatest difficulties are caused by cases in which the likelihood of character-state change is itself subject to change during the evolution of a group. Especially troublesome are instances in which a character is independent when it first evolves but later becomes coupled with another character, or vice versa. Where there are such changes in the degree of independence during the evolution of a group, there does not appear to be any straightforward way to code or weight characters to reflect their changing evidential significance (Donoghue, 1989). We suspect that this is a very real problem in some morphological studies. In seed plants, for example, it appears that leaf and sporophyll evolution may have been strictly coupled in some parts of the phylogeny (simultaneous reduction in the evolution of coniferopsids) and uncoupled in other parts (e.g., within anthophytes) (Doyle and Donoghue, 1986). We are uncertain how often this problem will arise in molecular data, but there are ample opportunities for it to occur. For example, evolutionary changes in the secondary structure of a molecule could change pairing and functional relationships and hence the degree of independence.

Taken together, these observations indicate that the supposed differences between molecular and morphological data in terms of character independence and weighting are a matter of degree. In view of how little we still know about the relevant mechanisms of molecular evolution, the assumption that all sites are independent rests largely on faith. Just as genetic and developmental studies have revealed complex relationships among morphological traits, as we learn more about molecular mechanisms we are likely to discover many more (and even more subtle) forms of interdependence. For example, there is the possibility that some structural changes may have a "destabilizing" effect (Palmer et al., 1987), or that the transfer of a functional gene (Baldauf and Palmer, 1990) will result in a release from selective constraint and changes in substitution rate. In the meantime, it is brash to imply that weighting of morphological characters can never be justified or that the weighting of molecular characters is easily accomplished.

Independence of Data Sets

Another reason given for focusing exclusively on molecular characters is the desire to avoid circularity in studying morphological evolution (e.g., Olmstead, 1989; Sytsma, 1990). It would, of course, be circular to reach conclusions about the evolution of any trait in a phylogeny based exclusively on that character. But the desire to study the evolution of one or several morphological traits does not justify omitting all morphological characters. Surely there are other morphological traits that are independent and can help in establishing the phylogeny, and failure to consider such characters may yield an unresolved or inaccurate tree. Even the character of interest has some bearing on the inference of phylogeny. If the addition of this character to the analysis results in a change in topology, then it is not acceptable to leave it out and proceed to interpret its evolution. One simply needs more data.

This outlook may be clarified by turning the argument around. Suppose that one were interested in studying the evolution of a particular portion of the genome. An extreme form of the independence argument implies that molecular data should be eliminated altogether—that the phylogeny must be based solely on other data, perhaps from morphology. We assume that most readers will balk at this suggestion. Why, it will be asked, is it necessary to omit all molecular data in order to study some single aspect of molecular evolution? Furthermore, the tree based solely on morphology might be inaccurate—after all, it does not take into account the molecular data. These are precisely the points we made in the preceding paragraph with regard to morphology.

In order to obtain an accurate picture of the evolution of a given feature, the phylogeny should be based on all of the relevant evidence, rather than ignoring some (perhaps sizable) portion of the data on the grounds that it is similar in some respect to the trait of interest. The issue is the independence of the phylogeny from the character(s) of interest, not whether a trait happens to be classified as morphological or molecular. In any case, in view of the great concern that characters be independent (see above), it is ironic that the independence of data sets has been used as an argument for keeping them separate. If morphological and molecular characters really are independent of one another, this is a powerful argument for putting them together (Barrett et al., 1991). Exceptions arise when there are entities under consideration that have separate evolutionary histories, and the aim is to compare these histories. Thus, in studying hybridization, one might wish to construct separate trees based on cpDNA and on morphology or nuclear genes.

The Number of Characters

The beauty of molecular data is that there is potentially so much of it—and increased numbers add evidential weight and statistical power to phylogenetic inferences. In general, it appears that the more characters there are per taxon, the higher the level of confidence, at least as measured by the bootstrap (Sanderson, 1989). And this relationship is unaffected by the amount of homoplasy present, which tends to vary independently of the number of characters (Sanderson and Donoghue, 1989; but see Archie, 1989a).

In response to this argument, it should be remembered that it is possible to gather more and better morphological data than we have now, for example, through studies of development. Furthermore, morphological data can be gathered from more organisms/taxa. Even if DNA will soon be routinely obtained from herbarium specimens, it is unlikely that much molecular evidence will ever be obtained from fossils, notwithstanding the success of Golenberg et al. (1990) in sequencing the *rbc*L gene of a Miocene *Magnolia*. It may be that a clear picture of phylogeny will require a very good sample of taxa, including fossils, in which case there would be a significant advantage to morphological data (Donoghue et al., 1989). In this regard, the simulation studies of Wheeler (1991) are especially intriguing. Wheeler found that the number of characters used in the reconstruction accounted for most of the variation in cladogram resolution (although this also depended on whether evolution was assumed to be clocklike), but that cladogram accuracy depended largely on the number of taxa included in the analysis (either 4 or 12 in his study).

It is also important to note that increased clarity is by no means guaranteed to come cheaply. The ratio of phylogenetically uninformative (constant or autapomorphic) characters to potentially informative characters has typically been very high in molecular data, entailing a significant investment of time and money to obtain a moderate amount of relevant information. In the case of hominoid primates, only 54 potentially informative sites were discovered in a sample of 10,939 sites in three sequences from nuclear and mitochondrial DNA (Holmquist et al., 1988), and some of these potentially informative sites are homoplastic on the most parsimonious tree. It may be more efficient in many cases to concentrate on morphological data, especially considering the evident congruence of data sets discussed below.

For our purposes, the question is whether the number of molecular characters that might be amassed can justify ignoring whatever morphological data are available. In practice, there have seldom been so many more molecular characters that one would seriously entertain abandoning the morphological evidence. Thus, although Olmstead (1989) made a point of the large number of molecular characters that could be acquired, morphology still accounted for over one third of the (binary-coded) characters he analyzed in *Scutellaria*. Moreover, some morphological studies contain a very large number of characters, and it is unlikely that they will be outnumbered by informative molecular characters any time soon. The most impressive numbers, however, are found in vertebrates; for example, Gauthier et al. (1989) considered 972 potentially informative characters in analyzing relationships among 83 tetrapod taxa.

But even if many more molecular characters were available, should morphological evidence be abandoned? Doing so effectively assumes that the relatively few morphological traits would be overpowered by the molecular traits in such a way that they could have no effect on the outcome. In turn, this seems to imply that there is no significant variation in the level of support for different clades in molecular studies. If, on the contrary, some nodes happened to be supported by only a few character changes (perhaps as a function of the tempo of evolution rather than sampling error), then the addition of only a few morphological characters might tip the balance in favor of a new tree, as shown in Fig. 15.4. This effect would be even more pronounced if the few molecular characters in question showed some homoplasy.

In fact, in molecular studies conducted to date, there is often considerable variation in the level of support for different clades and considerable homoplasy. In some cases, the result is an almost complete lack of resolution, especially when trees in the neighborhood of the most parsimonious cladogram(s) are taken into consideration. This has been demonstrated by Bremer (1988) for amino acid data on angiosperm families (also see Archie, 1989b), and by Bremer and Bremer (1989) for rRNA oligonucleotide catalog data on blue-green algae. Lack of resolution of some clades is also apparent in many studies with cpDNA restriction site data, even when there is little homoplasy (Olmstead et al., 1990). The analysis of *Clarkia* by Sytsma et al. (1990) provides an example of this problem: although the traditional sections of the genus are well marked, resolution of relationships among the sections is completely lost in the consensus of the 25 trees within one step of the most parsimonious trees. Sytsma et al. (1990) postulate a rapid radiation early in the history of the genus to account for this pattern, as did Sytsma and Smith (1988) for sections of *Fuchsia,* and Chase and Palmer (in Palmer et al., 1988) for lack of resolution among several lineages within Oncidii-

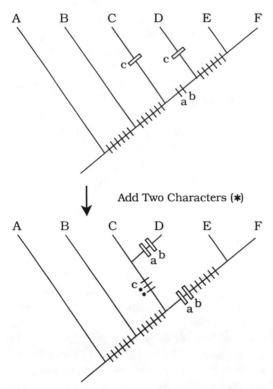

Add Two Characters (✱)

Figure 15.4. When character support is not uniformly distributed, it is possible for the addition of a small number of characters to change a topology based on a large number of characters. The upper cladogram (based on 21 characters) requires 22 steps, with homoplasy in character c. The addition of two new characters (✱), with derived states in taxa C and D, yields the lower cladogram, requiring 25 steps and homoplasy in characters a and b.

nae (Orchidaceae). In all of these cases, the addition of even a few morphological characters could help resolve relationships.

In general, there is nothing to be lost (and potentially much to be gained) by taking into account the morphological data. It may not have any effect on the outcome, either because it is congruent with the molecular data (but adds no resolution) or because whatever incongruence exists is resolved in favor of the molecular characters. On the other hand, the addition of morphological characters might help to resolve with greater confidence some unresolved portion of the molecular tree or might result in a change in topology in an area that is not strongly supported by molecular data. These effects are evident in the examples discussed by Miyamoto (1983), Hillis (1987), and Kluge (1989), and they are also apparent in a preliminary analysis of major seed plant groups based on a combination of rRNA sequence and morphological data (Donoghue, Zimmer, and Doyle, unpublished data). In the worst scenario, morphological data may be so noisy that they lower the resolution of an otherwise highly resolved molecular data set. We know of no cases in which this has occurred. Of course, the reverse might also occur, namely that noisy molecular data might obscure the signal in morphology. For example, it is possible that 5S rRNA data would obscure the picture of green plant relationships based on morphological characters (Bremer et al., 1987). If one suspected such a case, it would be valuable to explore whether the level of homoplasy in one or both of the separate data sets deviated from that seen in random data (cf., Archie, 1989b), and to examine confidence levels in the separate and combined data sets.

Patterns in Empirical Studies

Most of what we "know" about phylogeny is based on morphology, and most of that knowledge is probably not far off the mark. If nothing else, this has been confirmed time and again by molecular studies. Thus, it is not surprising that molecular data show tobacco and pea to be more closely related than either is to corn or to a liverwort (Ritland and Clegg, 1987). By now it is clear that congruence is the rule, not the exception, even in more controversial cases. Thus, Jansen and Palmer's cpDNA results for Asteraceae (Jansen and Palmer 1987; Palmer et al., 1988; Jansen et al., 1990; Jansen et al., Chapter 11, this volume) agree in many ways with Bremer's (1987) morphological evidence, especially as regards the monophyly of Asteroideae and the position of Barnadesiinae. We note that although Bremer (1987) included one molecular character in his analysis (a 22-kb inversion), the same result is obtained when this character is omitted (Bremer, personal communication). The monophyly of the Gnetales is also strongly supported both by morphological (Crane, 1985; Doyle and Donoghue, 1986) and rRNA sequence data (Zimmer et al., 1989; Hamby and Zimmer, Chapter 4, this volume). Many other examples of congruence are documented in this volume.

In other cases, the molecular and morphological data give consistent results, but one provides better resolution than the other. In the case of seed plant phylogeny, the morphological data provide rather strong support for the anthophyte clade, with Gnetales being more closely related to angiosperms than any other extant group (Crane, 1985; Doyle and Donoghue, 1986). In the rRNA sequence data, this relationship is less clear, with several alternative placements of Gnetales seen within one or two steps of the most parsimonious trees (Zimmer et al., 1989; Hamby and Zimmer, Chapter 4, this volume). In other instances, molecular data favor one of several alternatives based on morphological data; *Fuchsia* section *Skinnera* provides a good example (Sytsma et al., 1991; Sytsma and Smith, Chapter 13, this volume), and the question of the placement of the root of the angiosperms might also fall in this category (Donoghue and Doyle, 1989).

Cases of genuine disagreement are hard to find. In fact, we know of no case in which cladistic analysis of morphological data *strongly* supports a conclusion that is *strongly* contradicted by cladistic analysis of molecular data (but see Sytsma and Smith, 1988, for a different interpretation). Moreover, apparent discordance seems to be as common in comparing different morphological studies of the same group, or different molecular studies, as it is in comparing molecular versus morphological results (Wyss et al., 1987).

Reports of incongruence are largely based on comparing results obtained by different methods of analysis or, more often, one sort of data has been rigorously analyzed and the other has not. The *Clarkia-Heterogaura* example (Sytsma and Gottlieb, 1986; Sytsma and Smith, 1988, and Chapter 13, this volume) falls in this category: the morphological data have not yet been treated cladistically, and it is inappropriate to interpret the traditional classification as a phylogeny (Donoghue and Cantino, 1988; Doyle et al., 1990). We suspect that morphological results will be consistent with molecular results in this case but that the molecular data will provide a clearer resolution of the position of *Heterogaura* within *Clarkia* (but see Sytsma, 1990, for a different perspective). Other apparent instances of incongruence involve weak support for particular relationships in analyses based on one or both data sets; for example, the apparent difference between cpDNA and morphology regarding the status of the Lactucoideae (aside from Barnadesiinae). Although Jansen et al. (1990) favor the monophyly of Lactucoideae (based largely on Dollo parsimony), the Wagner parsimony analysis of their data set also produced trees in which Lactucoideae are paraphyletic, consistent with Bremer's (1987) result. Although Bremer's (1987) analysis did not strongly support the paraphyly of Lactucoideae (see Sanderson, 1989), additional morphological data have strengthened this hypothesis (Bremer, personal communication).

Other instances of apparent incongruence may be a function of the sample of taxa considered, which can certainly influence tree topology (Donoghue et al., 1989; Wheeler, 1991). For example, initial analyses of rRNA sequence data indicated that monocots were polyphyletic, with *Sagittaria* and *Potamogeton* arising within different dicot lines (Zimmer et al., 1989; Donoghue and Doyle, 1989). But these arrangements were only weakly supported, and with the addition of taxa to the analysis they are now seen to be less parsimonious (Hamby and Zimmer, Chapter 4, this volume). Similarly, apparent disagreement between Hamby and Zimmer's (1988) rRNA analysis of grasses and the morphological analysis of Kellogg and Campbell (1987) may be a function of the coverage of taxa in the two studies. Congruent results are obtained when the morphological data set is reduced to just those taxa included in the molecular study (Kellogg, personal communication), and the addition of taxa to the rRNA study might bring it in line with the larger morphological study (Hamby and Zimmer, Chapter 4, this volume).

The foregoing review of empirical studies, aside from confirming our expectations based on the theory of evolution, provides as strong an argument in favor of morphological data as it does in favor of molecular data. And if there is disagreement, it is not clear, without further evidence and/or analysis, which result (if any) is correct (Doyle, 1987). As Cracraft and Mindell (1989, p. 398) pointed out, "it is one of the ironies of our time" that another conclusion has been drawn by some molecular systematists: the reliability of molecular data are judged by congruence with morphological phylogenies, and then it is claimed that molecular data are more informative than morphology whenever there appears to be a conflict (e.g., Ahlquist et al., 1987). If morphology is so untrustworthy, why should it be used at all in evaluating molecular results? And if it is trustworthy enough to use in this capacity, why should we not trust it when conflicts arise? It is also ironic that even those who are most wary of morphological data nevertheless lean on it heavily in designing their own research, namely in choosing which groups to work on, which subgroups to sample, and so on. If it is admitted that morphological evidence has been useful in establishing phylogenetic relationships, then what excuse can there be for setting it aside?

Consequences of Ignoring Data

The temptation to ignore data is evidently great, but experience suggests that this should be resisted. Thus, we are suspicious of assertions to the effect that some forms of data are useless; for example, the claim that continuous variables cannot provide evidence on phylogeny (Pimentel and Riggins, 1987). In the same vein, molecular systematists will want to consider carefully the pronouncement by Wilson et al. (1989) that restriction site mapping is now outmoded and should be abandoned in favor of sequencing.

Here, the controversy over the use of fossils is especially instructive. Patterson (1981) and others have implied that it is acceptable to ignore fossils in reconstructing phylogeny. After all, fossils are generally incomplete and are therefore unlikely to have much effect on an analysis compared to extant organisms. As reasonable as this may seem on the surface, it is simply wrong. Inclusion of fossils can and does make a difference, both in terms of tree topology and the interpretation of character evolution (Donoghue et al., 1989). This is true even when data on fossils are limited. The fact that there are more data on extant organisms does not insure that this information will overwhelm the limited fossil evidence. Completeness and relevance to resolving a particular phylogenetic problem are not the same thing.

Although the case of fossils has to do with taxa, and only indirectly with characters, we trust that the parallel to the molecule versus morphology debate will be obvious. Morphological data, even when limited in comparison with molecular data, may be highly relevant. Just as with fossils, this may be especially true in the case of ancient, rapid radiations. As we noted in connection with Figs. 15.3 and 15.4, in such instances molecular data may be limited or very noisy, and the addition of even a few morphological characters—characters associated with early branching events that have since become fixed—could make a big difference (see above; Olmstead, 1989). What purpose is served in denying such possibilities at the outset? Why not try to make use of both morphological and molecular data?

Conclusions

The arguments presented above lead us to the conclusion that it is a mistake to set morphological data aside and base phylogeny reconstruction only on molecular evidence. As we have tried to show, arguments that appear to support this outlook are illogical, otherwise unconvincing, or rest largely on faith. It is obvious that there are highly desirable attributes associated with molecular data, especially the large number of characters that can potentially be obtained rather readily. But these attributes do not justify ignoring morphological characters, which have some fortunate properties of their own, such as their availability from a larger sample of organisms, including fossils. Of course, in most ways molecular and morphological data are quite similar—differences are mainly a matter of degree and do not map neatly onto the division of characters into molecular versus morphological.

Theoretically, there appear to be good reasons to retain both types of data, and empirical studies indicate not only that both types of data are relevant, but that failure to consider all of the data might lead to unresolved or even inaccurate results. Why, then, have we not analyzed all of the data? Undoubtedly, part of the problem is uncertainty over how best to proceed. In particular, it has been unclear whether to analyze data sets separately and find the consensus of the resulting trees, or to combine data sets and analyze them simultaneously. The pros and cons of these alternatives have been discussed at length elsewhere (see Miyamoto, 1985; Hillis, 1987; Cracraft and Mindell, 1989; Kluge, 1989; Barrett et al., 1991). In most cases we believe that it is best to combine data sets, rather than separate results. At least this is our recommendation when the same phylogenetic question is being addressed by the two data sets, which need not be the case if there are entities involved that have different modes of inheritance and hence more-or-less independent phylogenies (see Kawata, 1987; Avise, 1989). For example, as illustrated in several chapters in this volume, the study of hybridization is facilitated by tracing organelle and organism phylogenies separately. At the very least, it is important to recognize, as Barrett et al. (1991) have shown, that consensus trees (even strict consensus trees, which contain only those components that appear in all of the trees being compared) can be *positively incongruent* with the tree(s) based on an analysis of the pooled data. Thus, contrary to popular opinion, consensus techniques are not even a means of playing it safe, and it behooves the investigator to determine whether a consensus tree is really sanctioned by all of the data. This requires a combined analysis.

The prospect of combining molecular and morphological data sets raises a set of difficult issues, especially regarding character weighting and differences in the nature of the sampling of terminal taxa. Here, we can offer only a few comments on these issues. The feeling that it will be necessary to weight characters differentially in a combined analysis arises mainly from the worry that the data set with more characters (usually molecular) will overwhelm the data set with fewer characters (e.g., Kluge, 1983). As we stressed above, the sheer number of characters of a particular type is not as important as the nature of the character support and the distribution of homoplasy. Based on the overall congruence seen in empirical studies, complementarity may be the usual result.

The second problem—differences in the sampling of taxa—is not as widely appreciated. In morphological studies the character states assigned to a terminal taxon are usually based on information from many organisms, whereas in molecular studies there may be just a single representative of a taxon (which might be well nested within that taxon). Thus, the terminal taxon "conifers" in the Doyle and Donoghue (1986) analysis of seed plants is scored as a presumed basal state based on virtually all modern and fossil conifers, whereas in the rRNA sequence studies of Zimmer and colleagues (Zimmer et al., 1989; Hamby and Zimmer, Chapter 4, this volume) conifers are represented by sequences from just a few accessions. Some practical means of handling such discrepancies are explored elsewhere (Donoghue, Zimmer, and Doyle, unpublished data). Although it may seem that consensus techniques sidestep these issues, the problems are only hidden or arbitrarily resolved (Barrett et al., 1991).

In part, the reluctance to combine data sets may have a sociological basis. Enthusiasm over a new source of evidence is understandable, as are exaggerated claims on its behalf. But too often in the history of systematics the rising popularity of one sort of data takes place at the expense of another, which remains insufficiently explored. This is not just an incidental effect of limited resources—such replacement has often been actively pursued. That is, it is felt that the best way to promote the new data is to find fault with the old, and what could be better than to claim that the old data are worthless? But rhetoric of this sort, and the fads that it encourages, are unhealthy from the standpoint of our common goal, namely reconstructing the phylogeny of plants. Molecular data, when gathered carefully and analyzed in an appropriate manner, are obviously very useful in understanding evolutionary history, and the same can be said of morphological data. Both are, we believe, extremely promising avenues to pursue, and neither has come close to achieving its full potential. Ultimately, however, our efforts to reconstruct phylogeny will be judged by their success in integrating *all* of our observations, which means that more attention should be devoted to combining molecular and morphological evidence.

References

Ahlquist, J.E., Bledsoe, A.H., Sheldon, F.H., and Sibley, C.G. (1987) DNA hybridization and avian systematics. *Auk* **104**, 556–563.

Appels, R., and Honeycutt, R.L. (1986) rDNA: evolution over a billion years. In: *DNA Systematics Volume II: Plants* (ed. S.K. Dutta), CRC Press, Boca Raton, FL, pp. 81–135.

Archie, J.W. (1989a) A randomization test for phylogenetic information in systematic data. *Syst. Zool.* **38**, 239–252.

Archie, J.W. (1989b) Phylogenies of plant families: a demonstration of phylogenetic randomness in DNA sequence data derived from proteins. *Evolution* **43**, 1796–1800.

Avise, J.C. (1989) Gene trees and organismal histories: a phylogenetic approach to population biology. *Evolution* **43**, 1192–1208.

Baldauf, S.L., and Palmer, J.D. (1990) Evolutionary transfer of the chloroplast *tuf*A gene to the nucleus. *Nature* **334**, 262–265.

Barrett, M., Donoghue, M.J., and Sober, E. (1991) Against consensus. *Syst. Zool.*, in press.

Bremer, B., and Bremer, K. (1989) Cladistic analysis of blue-green procaryote interrelationships and chloroplast origin based on 16S rRNA oligonucleotide catalogues. *J. Evol. Biol.* **2**, 13–30.

Bremer, K. (1987) Tribal interrelationships of the Asteraceae. *Cladistics* **3**, 210–253.

Bremer, K. (1988) The limits of amino acid sequence data in angiosperm phylogenetic reconstruction. *Evolution* **42**, 795–803.

Bremer, K., Humphries, C.J., Mishler, B.D., and Churchill, S.P. (1987) On cladistic relationships in green plants. *Taxon* **36**, 339–349.

Britten, R.J. (1986) Rates of DNA sequence evolution differ between taxonomic groups. *Science* **231**, 1393–1398.

Brown, W.M., George, M., and Wilson, A.C. (1979) Rapid evolution of animal mitochondrial DNA. *Proc. Natl. Acad. Sci. USA* **76**, 1967–1971.

Carpenter, J.M. (1988) Choosing among multiple equally parsimonious cladograms. *Cladistics* **4**, 291–296.

Clegg, M.T. (1990) Dating the monocot-dicot divergence. *Trends Ecol. Evol.* **5**, 1–2.

Cracraft, J., and Mindell, D.P. (1989) The early history of modern birds: a comparison of molecular and morphological evidence. In: *The Hierarchy of Life* (eds. B. Fernholm, K. Bremer, and H. Jornvall), Elsevier, Amsterdam, pp. 389–403.

Crane, P.R. (1985) Phylogenetic analysis of seed plants and the origin of angiosperms. *Ann. Missouri Bot. Gard.* **72**, 716–793.

Davis, J.I. (1983) Phenotypic plasticity and the selection of taxonomic characters in *Puccinellia* (Poaceae). *Syst. Bot.* **8**, 341–353.

Davis, J.I. (1987) Genetic and environmental determination of leaf epidermal anatomy in *Puccinellia* (Poaceae). *Amer. J. Bot.* **74**, 1744–1749.

DeBry, R.W., and Slade, N.A. (1985) Cladistic analysis of restriction endonuclease cleavage maps within a maximum-likelihood framework. *Syst. Zool.* **34**, 21–34.

Donoghue, M.J. (1983) The phylogenetic relationships of *Viburnum*. In: *Advances in Cladistics*, Vol. 2 (eds. N.I. Platnick and V.A. Funk), Columbia University Press, New York, pp. 143–166.

Donoghue, M.J. (1989) Phylogenies and the analysis of evolutionary sequences, with examples from seed plants. *Evolution* **43**, 1137–1156.

Donoghue, M.J. (1991) The trouble with homology. In: *Keywords in Evolutionary Biology* (eds. E. Fox-Keller and E. Lloyd), Harvard University Press, Cambridge, in press.

Donoghue, M.J., and Cantino, P.D. (1988) Paraphyly, ancestors, and the goals of taxonomy: a botanical defense of cladism. *Bot. Rev.* **54**, 107–128.

Donoghue, M.J., and Doyle, J.A. (1989) Phylogenetic studies of seed plants and angiosperms based on morphological characters. In: *The Hierarchy of Life* (eds. B. Fernholm, K. Bremer, and H. Jornvall), Elsevier, Amsterdam, pp. 181–193.

Donoghue, M.J., Doyle, J.A., Gauthier, J., Kluge, A.G., and Rowe, T. (1989) The importance of fossils in phylogeny reconstruction. *Ann. Rev. Ecol. Syst.* **20**, 431–460.

Downie, S.R., Olmstead, R.G., Zurawski, G., Soltis, D.E., Soltis, P.S., Watson, J.C., and Palmer, J.D. (1991) Six independent losses of the chloroplast DNA *rpl2* intron in dicotyledons: molecular and phylogenetic implications. *Evolution*, in press.

Doyle, J.A., and Donoghue, M.J. (1986) Seed plant phylogeny and the origin of angiosperms: an experimental cladistic approach. *Bot. Rev.* **52**, 321–431.

Doyle, J.J. (1987) Plant systematics at the DNA level: promises and pitfalls. *Int. Org. Plant Biosyst. Newsl.* **8**, 3–7.

Doyle, J.J. (1991) Evolution of higher-plant glutamine synthetase genes: tissue specificity as a criterion for predicting orthology. *Mol. Biol. Evol.*, in press.

Doyle, J.J., Doyle, J.L., and Brown, A.H.D. (1990) A chloroplast DNA phylogeny of the wild perennial relatives of the soybean (*Glycine* subgenus *Glycine*): congruence with morphological and crossing groups. *Evolution* **44**, 371–389.

Farris, J.S. (1969) A successive approximations approach to character weighting. *Syst. Zool.* **18**, 374–385.

Felsenstein, J. (1978) Cases in which parsimony or compatibility methods will be positively misleading. *Syst. Zool.* **27**, 401–410.

Felsenstein, J. (1985) Confidence limits on phylogenies: an approach using the bootstrap. *Evolution* **39**, 783–791.

Felsenstein, J. (1988) Phylogenies and quantitative characters. *Ann. Rev. Ecol. Syst.* **19**, 445–471.

Fernholm, B., Bremer, K., and Jornvall, H. (eds.) (1989) *The Hierarchy of Life*, Elsevier, Amsterdam.

Gauthier, J., Cannatella, D., de Queiroz, K., Kluge, A.G., and Rowe, T. (1989) Tetrapod phylogeny. In: *The Hierarchy of Life* (eds. B. Fernholm, K. Bremer, and H. Jornvall), Elsevier, Amsterdam, pp. 337–353.

Gillespie, J.H. (1986) Rates of molecular evolution. *Ann. Rev. Ecol. Syst.* **17**, 637–665.

Golenberg, E. M., Giannasi, D.E., Clegg, M.T., Smiley, C.J., Durbin, M., Henderson, D., and Zurawski, G. (1990) Chloroplast DNA sequence from a Miocene *Magnolia* species. *Nature* **344**, 656–658.

Gottlieb, L.D. (1988) Towards molecular genetics in *Clarkia*: gene duplications and molecular characterization of PGI genes. *Ann. Missouri Bot. Gard.* **75**, 1169–1179.

Gould, S.J. (1985) A clock for evolution. *Nat. Hist.* **94**, 12–25.

Gould, S.J., and Lewontin, R.C. (1979) The spandrels of San Marco and the Panglossian paradigm: a critique of the adaptationist programme. *Proc. Roy. Soc. London [Biol.]* **205**, 581–598.

Gouy, M., and Li, W.-H. (1989) Phylogenetic analysis based on rRNA sequences supports the archaebacterial rather than the eocyte tree. *Nature* **339**, 145–147.

Hamby, R.K., and Zimmer, E.A. (1988) Ribosomal RNA sequences for inferring phylogeny within the grass family (Poaceae). *Plant Syst. Evol.* **34**, 393–400.

Hennig, W. (1966) *Phylogenetic Systematics,* University of Illinois Press, Urbana, IL.

Hillis, D.M. (1987) Molecular versus morphological approaches to systematics. *Ann. Rev. Ecol. Syst.* **18**, 23–42.

Hixson, J.E., and Brown, W.M. (1986) A comparison of the small ribosomal RNA genes from the mitochondrial DNA of the great apes and humans: sequence, structure, evolution, and phylogenetic implications. *Mol. Biol. Evol.* **3**, 1–18.

Holmquist, R., Miyamoto, M.M., and Goodman, M. (1988) Analysis of higher-primate phylogeny from transversion differences in nuclear and mitochondrial DNA by Lake's methods of evolutionary parsimony and operator metrics. *Mol. Biol. Evol.* **5**, 217–236.

Jansen, R.K., and Palmer, J.D. (1987) A chloroplast inversion marks an ancient evolutionary split in the sunflower family (Asteraceae). *Proc. Natl. Acad. Sci. USA* **84**, 5818–5822.

Jansen, R.K., Holsinger, K.E., Michaels, H.J., and Palmer, J.D. (1990) Phylogenetic analysis of chloroplast DNA restriction site data at higher taxonomic levels: an example from the Asteraceae. *Evolution* **44**, 2089–2105.

Kaplan, D.R. (1984) The concept of homology and its central role in the elucidation of plant systematic relationships. In: *Cladistics: Perspectives on the Reconstruction of Evolutionary History* (eds. T. Duncan and T. Stuessy), Columbia University Press, New York, pp. 51–69.

Kawata, M. (1987) Units and passages: a view for evolutionary biology and ecology. *Biol. Phil.* **2**, 415–434.

Kellogg, E.A., and Campbell, C.S. (1987) Phylogenetic analysis of the Gramineae. In: *Grass Systematics and Evolution* (eds T.R. Soderstrom, K.W. Hilu, C.S. Campbell, and M.E. Barkworth), Smithsonian Institution Press, Washington, DC, pp. 310–322.

Kimura, M. (1983) *The Neutral Theory of Molecular Evolution.* Cambridge University Press, Cambridge.

Kluge, A.G. (1983) Cladistics and the classification of the great apes. In: *New Interpretations of Ape and Human Ancestry* (eds. R.L. Ciochan and R.S. Corruccini), Plenum Press, New York, pp. 151–177.

Kluge, A.J. (1989) A concern for evidence and a phylogenetic hypothesis of relationships among *Epicrates* (Bovidae, Serpentes). *Syst. Zool.* **38**, 7–25.

Kluge, A.G., and Farris, J.S. (1969) Quantitative phyletics and the evolution of anurans. *Syst. Zool.* **18**, 1–32.

Kreitman, M.E., and Aguade, M. (1986) Excess polymorphism at the ADH locus in *Drosophila melanogaster. Genetics* **114**, 93–110.

Lake, J.A. (1987) A rate-independent technique for analysis of nucleic acid sequences: evolutionary parsimony. *Mol. Biol. Evol.* **4**, 167–191.

Lanyon, S.M. (1988) The stochastic mode of molecular evolution: what consequences for systematic investigations? *Auk* **105**, 565–573.

Lavin, M., Doyle, J.J., and Palmer, J.D. (1990) Evolutionary significance of the loss of the chloroplast-DNA inverted repeat in the Leguminosae subfamily Papilionoideae. *Evolution* **44**, 390–402.

Martin, P.G., Boulter, D., and Penny, D. (1985) Angiosperm phylogeny studied using sequences of five macromolecules. *Taxon* **34**, 393–400.

Martin, P.G., Dowd, J.M., and Stone, S.J.L. (1983) The study of plant phylogeny using amino acid sequences of ribulose-1,5-bisphosphate carboxylase. II. The analysis of small subunit data to form phylogenetic trees. *Austr. J. Bot.* **31**, 411–419.

Mayr, E. (1969) *Principles of Systematic Zoology*, McGraw-Hill, New York.

Melnick, D.J. (1990) Molecules, evolution and time. *Trends Ecol. Evol.* **5**, 172–173.

Mickevich, M.F., and Weller, S.J. (1990) Evolutionary character analysis: tracing characters on cladograms. *Cladistics* **6**, 137–170.

Miller, K.R., and Jacobs, J.S. (1989) On *Prochlorothrix*. *Nature* **338**, 303–304.

Mishler, B.D., Bremer, K., Humphries, C.J., and Churchill, S.P. (1988) The use of nucleic acid sequence data in phylogenetic reconstruction. *Taxon* **37**, 391–395.

Mishler, B.D., Thrall, P., Hopple, J.S., de Luna, E., and Vilgalys, R. (1990) The emperor's new clothes? A molecular approach to the phylogeny of bryophytes using chloroplast DNA. *Amer. J. Bot.* **77**, 146 (abstract).

Miyamoto, M.M. (1983) Frogs of the *Eleutherodactylus rugulosus* group: a cladistic study of allozyme, morphological, and karyological data. *Syst. Zool.* **32**, 109–124.

Miyamoto, M.M. (1985) Consensus cladograms and general classifications. *Cladistics* **1**, 186–189.

Morden, C.W., and Golden, S.S. (1989a) *psb*A genes indicate common ancestry of prochlorophytes and chloroplasts. *Nature* **337**, 382–385.

Morden, C.W., and Golden, S.S. (1989b) Corrigendum. *Nature* **339**, 400.

Olmstead, R.G. (1989) Phylogeny, phenotypic evolution, and biogeography of the *Scutellaria angustifolia* complex (Lamiaceae): inference from morphological and molecular data. *Syst. Bot.* **14**, 320–338.

Olmstead, R.G., Jansen, R.K., Michaels, H.J., Downie, S.R., and Palmer, J.D. (1990) Chloroplast DNA and phylogenetic studies in the Asteridae. In: *Biological Approaches and Evolutionary Trends in Plants* (ed. S. Kawano), Academic Press, San Diego, pp. 119–134.

Palmer, J.D. (1985) Comparative organization of chloroplast genomes. *Ann. Rev. Genet.* **19**, 325–354.

Palmer, J.D. (1987) Chloroplast DNA evolution and biosystematic uses of chloroplast DNA variation. *Amer. Natur.* **130**, S6–S29.

Palmer, J.D., Jansen, R.K., Michaels, H.J., Chase, M.W., and Manhart, J.R. (1988) Chloroplast DNA variation and plant phylogeny. *Ann. Missouri Bot. Gard.* **75**, 1180–1206.

Palmer, J.D., Osorio, B., Aldrich, J., and Thompson, W.F. (1987) Chloroplast DNA evolution among legumes: loss of a large inverted repeat occurred prior to other sequence rearrangements. *Curr. Genet.* **11**, 275–286.

Patterson, C. (1981) Significance of fossils in determining evolutionary relationships. *Ann. Rev. Ecol. Syst.* **12**, 195–223.

Patterson, C. (1982) Morphological characters and homology. In: *Problems of Phylogenetic Reconstruction* (eds. K.A. Joysey and A.E. Friday), Academic Press, London, pp. 21–74.

Patterson, C. (1987) Introduction. In: *Molecules and Morphology in Evolution: Conflict or Compromise?* (ed. C. Patterson), Cambridge Unversity Press, Cambridge, pp. 1–22.

Patterson, C. (1988) Homology in classical and molecular biology. *Mol. Biol. Evol.* **5**, 603–625.

Patterson, C. (1989) Phylogenetic relations of major groups: conclusions and prospects. In: *The Hierarchy of Life* (eds. B. Fernholm, K. Bremer, and H. Jornvall), Elsevier, Amsterdam, pp. 471–488.

Pimentel, R.A., and Riggins, R. (1987) The nature of cladistic data. *Cladistics* **3**, 201–209.

Remane, A. (1952) *Die Grundlagen des naturlichen Systems der vergleichenden Anatomie und der Phylogenetik,* Geest and Portig, Leipzig.

Riedl, R. (1978) *Order in Living Organisms,* Wiley, New York.

Ritland, K., and Clegg, M.T. (1987) Evolutionary analysis of plant DNA sequences. *Amer. Natur.* **130**, S74–S100.

Sanderson, M.J. (1989) Confidence limits in phylogenies: the bootstrap revisited. *Cladistics* **5**, 113–129.

Sanderson, M.J., and Donoghue, M.J. (1989) Patterns of variation in levels of homoplasy. *Evolution* **43**, 1781–1795.

Scherer, S. (1990) The protein molecular clock: time for a reevaluation. *Evol. Biol.* **24**, 83–106.

Schlichting, C.D. (1986) The evolution of phenotypic plasticity in plants. *Ann. Rev. Ecol. Syst.* **17**, 667–693.

Sibley, C.G., and Ahlquist, J.E. (1987) Avian phylogeny reconstructed from comparisons of the genetic material, DNA. In: *Molecules and Morphology in Evolution: Conflict or Compromise?* (ed. C. Patterson), Cambridge University Press, Cambridge, pp. 95–121.

Sidow, A., and Wilson, A.C. (1990) Compositional statistics: an improvement of evolutionary parsimony and its application to deep branches in the tree of life. *J. Mol. Evol.* **31**, 51–68.

Smith, A.B. (1989) RNA sequence data in phylogenetic reconstruction: testing the limits of its resolution. *Cladistics* **5**, 321–344.

Sober, E. (1988) *Reconstructing the Past: Parsimony, Evolution, and Inference,* MIT Press, Cambridge, MA.

Song, K.M., Osborne, T.C., and Williams, P.H. (1988) *Brassica* taxonomy based on nuclear restriction fragment length polymorphisms (RFLPs). *Theor. Appl. Genet.* **75**, 784–794.

Stebbins, G.L. (1950) *Variation and Evolution in Plants*, Columbia University Press, New York.

Steele, K.P., Holsinger, K.E., Jansen, R.K., and Taylor, D.W. (1988) Phylogenetic relationships in green plants—a comment on the use of 5S ribosomal RNA sequences by Bremer et al. *Taxon* **37**, 135–138.

Strauss, S.H., Palmer, J.D., Howe, G.T., and Doerksen, A.H. (1988) Chloroplast genomes of two conifers lack a large inverted repeat and are extensively rearranged. *Proc. Natl. Acad. Sci. USA* **85**, 3898–3902.

Sultan, S.E. (1987) Evolutionary implications of phenotypic plasticity in plants. *Evol. Biol.* **21**, 127–178.

Sytsma, K.J. (1990) DNA and morphology: inference of plant phylogeny. *Trends Ecol. Evol.* **5**, 104–110.

Sytsma, K.J., and Gottlieb, L.D. (1986) Chloroplast DNA evidence for the origin of the genus *Heterogaura* from a species of *Clarkia* (Onagraceae). *Proc. Natl. Acad. Sci. USA* **83**, 5554–5557.

Sytsma, K.J., and Smith, J.F. (1988) DNA and morphology: comparisons in Onagraceae. *Ann. Missouri Bot. Gard.* **75**, 1217–1237.

Sytsma, K.J., Smith, J.F., and Berry, P.E. (1991) Biogeography and evolution of morphology, breeding systems, flavonoids, and chloroplast DNA in the four Old World species of *Fuchsia* (Onagraceae). *Syst. Bot.* **16**, 257–269.

Sytsma, K.J., Smith, J.F., and Gottlieb, L.D. (1990) Phylogenetics in *Clarkia* (Onagraceae): restriction site mapping of chloroplast DNA. *Syst. Bot.* **15**, 280–295.

Syvanen, M., Hartman, H., and Stevens, P.F. (1989) Classical plant taxonomic ambiguities extend to the molecular level. *J. Mol. Evol.* **28**, 536–544.

Templeton, A.R. (1983) Convergent evolution and nonparametric inferences from restriction data and DNA sequences. In: *Statistical Analysis of DNA Sequence Data* (ed. B.S. Weir), Marcel Dekker, New York, pp. 151–179.

Turner, S., Burger-Wiersma, T., Giovannoni, S.J., Mur, L.R., and Pace, N.R. (1989) The relationship of a prochlorophyte *Prochlorothrix hollandica* to green chloroplasts. *Nature* **337**, 380–382.

Wagner, G.P. (1989) The origin of morphological characters and the biological basis of homology. *Evolution* **43**, 1157–1171.

Wheeler, W.C. (1990) Combinatorial weights in phylogenetic analysis: a statistical parsimony procedure. *Cladistics* **6**, 269–275.

Wheeler, W.C. (1991) Extinction, sampling, and molecular phylogenetics. In: *Extinction and Phylogeny* (ed. M. Novacek), Columbia Universty Press, New York, in press.

Wheeler, W.C., and Honeycutt, R.L. (1988) Paired sequence difference in ribosomal RNAs: evolutionary and phylogenetic implications. *Mol. Biol. Evol.* **5**, 90–96.

Wilson, A.C., Zimmer, E.A., Prager, E.M., and Kocher, T.D. (1989) Restriction mapping in the molecular systematics of mammals: a retrospective salute. In: *The Hierarchy of Life* (eds. B. Fernholm, K. Bremer, and H. Jornvall), Elsevier, Amsterdam, pp. 407–419.

Wolfe, K.H., Li, W.-H., and Sharp, P.M. (1987) Rates of nucleotide substitution vary greatly among plant mitochondrial, chloroplast, and nuclear DNAs. *Proc. Natl. Acad. Sci. USA* **84,** 9054–9058.

Wolters, J., and Erdmann, V.A. (1986) Cladistic analysis of 5S rRNA and 16S rRNA secondary and primary structure—the evolution of eukaryotes and their relation to Archaebacteria. *J. Mol. Evol.* **24,** 152–166.

Wu, C.-I., and Li, W.-H. (1985) Evidence for higher rates of nucleotide substitution in rodents than in man. *Proc. Natl. Acad. Sci. USA* **82,** 1741–1745.

Wyss, A.R., Novacek, M.J., and McKenna, M.J. (1987) Amino acid sequence versus morphological data and the interordinal relationships of mammals. *Mol. Biol. Evol.* **4,** 99–116.

Zimmer, E.A., Hamby, R.K., Arnold, M.L., LeBlanc, D.A., and Theriot, E.C. (1989) Ribosomal RNA phylogenies and flowering plant evolution. In: *The Hierarchy of Life* (eds. B. Fernholm, K. Bremer, and H. Jornvall), Elsevier, Amsterdam, pp. 205–214.

Zurawski, G., Bottomley, W., and Whitfeld, P.R. (1984) Junctions of the large single copy region and the inverted repeats in *Spinacia oleracea* and *Nicotiana debneyi* chloroplast DNA: sequence of the genes for tRNA[his] and the ribosomal proteins S19 and L2. *Nucleic Acids Res.* **12,** 6547–6558.

Zurawski, G., and Clegg, M.T. (1987) Evolution of higher plant chloroplast-encoded genes: implications for structure-function and phylogenetic studies. *Ann. Rev. Plant Physiol.* **38,** 391–418.

16

Character-State Weighting for Restriction Site Data in Phylogenetic Reconstruction, with an Example from Chloroplast DNA

Victor A. Albert, Brent D. Mishler, and
Mark W. Chase

Data derived from cleavage points of various restriction endonucleases in all three genomes present in eukaryotes—nuclear, mitochondrial, and chloroplast—have been used for phylogenetic reconstruction in diverse groups of organisms. Mapped restriction sites, which represent a sampling of a whole genome or of any specific sequence, can be considered estimates of homologous characters and their transformations. In this sense, restriction site data are like any other robustly derived systematic data (e.g., morphological). An important difference is that probabilities for character-state transformations within restriction site characters can now be formulated using hypothetical or empirical estimates of sequence evolution (DeBry and Slade, 1985). Such transformational probabilities can be incorporated into phylogenetic reconstruction using parsimony with the application of maximum likelihood character-state weights (Felsenstein, 1981a).

Organisms evolve by character-state transformations, and we believe that parsimony methods, weighted if possible, are well suited for phylogenetic estimation involving restriction site (or other) character transformations. No particular evolutionary processes, other than descent with modification and cladogenesis, are stipulated in cladistic analysis. Therefore, diverse evolutionary processes can be examined from the framework of a cladogram (despite Nei's worries, 1987, p. 326). Investigators not interested in exploring evolutionary mechanisms, specific homologies, and the sources and implications of homoplasy

We thank James S. Farris, Joseph Felsenstein, David L. Swofford, and Anita Gutierrez-Albert for comments and advice. All interpretations are, however, our own. We acknowledge John M. Mercer for modeling the branch-length inequality problem. We also thank Harold G. Hills for sequencing *rbc*L from *Lilium superbum* and *Colchicum speciosum* and our reviewers (Robert K. Jansen, Kent E. Holsinger, and one anonymous) for helpful comments. We acknowledge support from the NSF (grants BSR-8914635 to VAA, and BSR-8906496 to MWC), the North Carolina Board of Science and Technology (development award 89SE14 to BDM), and the American Orchid Society (to MWC).

might wish to use similarity measures for phenetic taxonomic purposes (after Nei, 1987).

Parsimony Methods in Phylogenetic Reconstruction

Thus far, all phylogenetic analyses of restriction site characters using parsimony have employed either the Wagner (Kluge and Farris, 1969; Farris, 1970) or Dollo (LeQuesne, 1974; Farris, 1977a,b) methods. Wagner parsimony permits a cleavage site to be gained or lost with equal weight. Dollo parsimony stipulates that a restriction site can evolve only once but that multiple losses are possible; the latter are then parsimoniously minimized. For a 6-bp recognition sequence, any one of 18 substitutions results in a loss, whereas only one substitution can cause a gain of a site from five nucleotides already in their proper order. Thus, although Dollo parsimony might seem superficially preferable for use with restriction site data, it is certainly too strict in allowing a restriction site to evolve only once.

The Inadequacy of Dollo Parsimony

Using a maximum-likelihood model similar to that of Felsenstein (1978), DeBry and Slade (1985) analyzed the efficacy of Wagner versus Dollo parsimony for phylogenetic reconstruction with restriction site data from animal mitochondrial DNA. Their conclusion, that Dollo parsimony was the most appropriate for reconstructing phylogeny, is only potentially relevant under the variables of their parametric model (i.e., those of mitochondrial DNA) and under their assumptions and methods of comparison. Several plant molecular systematists have invoked DeBry and Slade's study as justification for the use of Dollo parsimony in chloroplast DNA (cpDNA) restriction site studies (e.g., Sytsma and Gottlieb, 1986a,b; Jansen and Palmer, 1988; Palmer et al., 1988; Baldwin et al., 1990; Sytsma et al., 1990). Despite DeBry and Slade's (1985) conclusions, it is inappropriate to consider maximum-likelihood estimates of gain/loss probabilities calculated from one genome as support for using a particular parsimony method with data from a different genome. Furthermore, the comparison made by DeBry and Slade (1985) was only between Wagner and Dollo parsimony, without consideration of other alternatives.

In contrast to their own conclusions concerning Dollo parsimony, we present, from a formal extension of the probability models of DeBry and Slade (1985), a third alternative—weighting individual character-state transformations in DNA restriction sites. We demonstrate that neither strict Wagner nor Dollo parsimony is entirely appropriate for phylogenetic analysis of chloroplast, or other, DNA (after Jansen and Palmer, 1988), but Wagner produces more accurate topologies than Dollo. We also argue that Dollo parsimony is not an appropriate method of phylogenetic reconstruction because it effectively weights the independent gain of a character-state to infinity, which is biologically unrealistic. A parsimony model that permits, but appropriately biases against, known improbable events (such as parallel gains of a restriction site), will always be superior to either parsimony with unbiased reversibility (Wagner) or parsimony with evolutionary singularity (Dollo).

Parsimony with Maximum-Likelihood Transformational Weights

The character-state weighting alternative has only recently become practical. The latest versions of PAUP (3.0 for the Macintosh, designed by D. Swofford) and MacClade (W. Maddison and D. Maddison) contain an option for the construction and implementation of user-defined stepmatrices (see Sankoff and Cedergren, 1983; Williams and Fitch, 1989; Swofford and Olsen, 1990). Using stepmatrices, transformations between individual character states within a character can be funnelled through hypothetical constraints, which can be thought of as costs or weights of transformation. Weighting of entire characters, which has previously been available in the PAUP, Hennig86 (J.S. Farris), and PHYLIP (J. Felsenstein) computer packages, is rather different; all transformations within a highly weighted character are weighted equally. Whole-character weighting alone is inadequate for restriction site data, because we wish to reflect the known asymmetry in these data between forward and backward transformation.

We have developed a general model for weighted parsimony analyses of restriction site data that provides reasonable estimates for the relative weight or cost of state transformations within each character. The model presented was developed specifically for use with endonucleases with 6-bp recognition sites, but it could easily be altered for other types of endonucleases. Our weighting criteria are derived from transformational probabilities (after DeBry and Slade, 1985) and the conversion of these probabilities into maximum-likelihood weights (after Felsenstein, 1981a).

Chloroplast DNA and Character-state Weighting

The chloroplast genome is packed with genes, much like bacterial genomes (Palmer, 1985, 1991; Palmer et al., 1988). In fact, the sequence evolution of chloroplast genes may be under some kind of whole genome constraint. Clegg et al. (1986) found no statistical difference in nucleotide substitutions per codon site and similar transition/transversion ratios for the genes coding for ribulose-1,5-bisphosphate carboxylase/osygenase (*rbc*L) and those for the ATPase β- and β-subunits (*atp*B and *atp*E, respectively), although the latter evolves much more rapidly at the amino acid level (Palmer, 1985). Ritland and Clegg (1987) furthered the comparison between *rbc*L and *atp*B with several more taxa using a maximum-likelihood topology, and again, no significant evolutionary differences were found among the sequences by codon position. These results are surprising considering the functional and positional difference between *rbc*L and the genes coding for subunits of ATPase. We have used sequence data for *rbc*L to provide *a priori* estimates of substitution rates for our maximum-likelihood weighting model on the assumption that *rbc*L evolution sufficiently approximates that of the chloroplast genome as a whole, falling as it does between extreme substitution rate estimates, although nearer the more conservative (Palmer, 1991).

Determining *a priori* weights for character-state transformations over an approximately 150,000-bp genome from empirically observed substitutions in 1,428 homologous *rbc*L nucleotides (~1% sampling) is subject to error. Nevertheless, we contend that our estimates provide robust initial weighting criteria for protein-coding regions (the majority) of the chloroplast genome, needing perhaps minor corrections for the slowly evolving inverted repeats*, transfer RNA genes, and more rapidly evolving inter-and intragenic sequences when more is known about their evolution. Usage of lower or higher rates than that derived from *rbc*L, such as estimates for nonsynonymous substitutions in the cpDNA inverted repeats or synonymous substitutions in the single-copy regions (Palmer, 1991), affects only the correspondence between a given weight and estimates of segment lengths in years, and not the character-state weight formulations of our model (see Table 16.4 and pages 381–384).

In this chapter we consider first the primacy of homology and parsimony in phylogenetic reconstruction and the problems inherent to molecular phylogenetics, and then develop our character-state weighting scheme in detail and provide an example of its application to a real data set, namely, the sections of *Clarkia* (Onagraceae) (Sytsma et al., 1990). Finally, the effects of branch-length asymmetry are explored and shown to be unimportant for any restriction site data that could be expected to yield accurate estimates of phylogeny.

Phylogenetic Reconstruction:
Homology and Parsimony

All individual organisms that do or have existed are part of a true (but unknown) phylogeny, that is, they are connected by inherited, homologous character-state transformations in a topological relationship that can be estimated by sampling within characters and across taxa. All organisms, fossil or living, are similar to each other in some sense. However, similarity *per se* cannot be used as an accurate estimate of phylogeny because organisms do not evolve by similarity but by transformation. Phylogenetic and similarity (phenetic) trees can be similar

---

*These comprise approximately 9% (*Marchantia*) to 16% (*Nicotiana*) of the chloroplast genome and contain ribosomal RNA operons and some protein-coding genes (Palmer, 1985; Palmer et al., 1988), e.g., *rpl*2, with its average substitution rates for 1st and especially 3rd codon positions significantly lower than for *rbc*L and *atp*B (Zurawski and Clegg, 1987).

in overall topology (e.g., the molecular analyses of humans and anthropoid apes by Nei, 1987, and of *Clarkia* and *Heterogaura* in the Onagraceae by Sytsma and Gottlieb, 1986a, b). However, patterns of similarity can neither support nor create a phylogeny because their units of topological distance are at best only superficial mimics of the relevant units of phylogenetic comparison: homology and transformation within any character system (Mishler and DeLuna, 1991). Colless (1970) and Farris (1983) provided examples and discussion of incongruity between phenetic and phylogenetic trees.

If phylogeny represents a process of descent with modification, producing homologous character-state transformations, then a method that attempts to identify such characters and use them to reconstruct patterns of relative recency of common ancestry is to be preferred (Hennig, 1966). Cladistics is a transformational and genealogical approach to phylogeny; transformations between homologous character states are allowed (including reversals and convergences, i.e., homoplasies) but minimized subject to constraints imposed when all putatively homologous characters are summarized into a cladogram. Cladists wish to produce phylogenetic hypotheses based on the minimal necessary assumptions about evolutionary processes. These can then be used to erect evolutionary hypotheses based on the phylogenies.

Attempts at maximum-likelihood phylogenetic reconstructions (e.g., Felsenstein, 1981b) have alerted systematic theorists to the problem of highly unequal rates of evolution in different branches. Pairs of taxa with long branch lengths can be erroneously grouped by parsimony as sister taxa, due to accumulation of homoplasy (Felsenstein, 1978). The conditions under which parsimony can produce such a result (i.e., when internal branch lengths are overly short compared to external branch lengths) have sometimes been called the Felsenstein zone (e.g., Lake, 1987). Avoiding the Felsenstein zone means (1) including taxa sufficient to break up long external branches (Donoghue et al., 1989), or (2) using data with sufficiently low probabilities of character-state change (see below for a formalization of this concept).

Felsenstein (1981a) has recognized that for slowly evolving characters, for example, cpDNA sequences at least in theory (Palmer, 1985; Clegg et al., 1986; Ritland and Clegg, 1987), the parsimony estimate can also be a maximum-likelihood estimate (also discussed by Sober, 1988). This closeness implies that another criterion can be applied to phylogenies: character-state weighting using a maximum-likelihood framework for the calculation of character-transformation probabilities (e.g., DeBry and Slade, 1985) and their weights (Felsenstein, 1981a), which could then be applied to phylogeny reconstruction by simple ordered parsimony analysis (Farris, 1970). As pointed out by Felsenstein (1981a), when rates of change are sufficiently small, a weighted parsimony method becomes the proper maximum-likelihood method. We argue in the conclusion of this paper that phylogenetic characters (i.e., putatively homologous restriction sites in this example) are and should be selected on the basis of an expectation of sufficiently small rates of change. Which rates are "sufficiently" small will be seen to be a function of both differential rates and segment lengths in years.

Problems in Molecular Phylogenetics

Similarity and Phylogeny

With the increasing advancement of molecular technology over the past 15 years, much comparative DNA-based data have been generated from both restriction enzyme cleavage sites and whole nucleotide sequences. Many attempts to reconstruct phylogeny from such data have used similarity algorithms (e.g, UPGMA, see Sokal and Michener, 1958; Fitch and Margoliash, 1967; and distance Wagner, see Farris, 1972) in an inappropriate attempt to create phylogenetic tree topologies (one example illustrating all of these methods is presented by Nei, 1987). Theoretical objections to using distance data for phylogenetic reconstruction have been elaborated by several authors (e.g., Farris, 1983; Cracraft, 1987). There is also a clear, practical objection: why generate specific points of comparison (e.g., putative homologies) and then lose information by "boiling the comparison down" to a single number in a distance matrix?

Multiple-taxon comparisons with an underlying assumption of similarity have been used to construct topologies based on whole protein-coding genes (e.g., *rbc*L, see Palmer et al., 1988, who used only percent sequence divergence), and on partial sequences of whole genomes (e.g., mitochondrial DNAs from humans and anthropoids; see Nei, 1987, who used all the methods mentioned above on the data of Brown et al., 1982). For the latter data set, the maximum parsimony method of Fitch (1971, 1977) produced a most parsimonious tree (phylogeny) that was shorter by two steps than a tree agreeing with UPGMA, Fitch-Margoliash, and distance-Wagner topologies (Nei, 1987) and with the topology favored by Sibley and Ahlquist (1984) based on DNA–DNA hybridization (a measure of overall similarity). Specifically, maximum parsimony placed humans as sister to gorilla-chimpanzee, rather than gorillas as sister to human-chimpanzee. Nei's (1987, p. 319) caveats on parsimony methods seem directed to the data of Brown et al. (1982); we agree that when the nucleotide substitution rate is very large (as in the case of animal mitochondrial DNA) and if only a small number of nucleotides is being examined (Peacock and Boulter, 1975; Nei, 1987), parsimony can yield inconsistent phylogenies even if rates of substitution are constant among lineages—branch-length asymmetry may still be acute because of differential segment lengths in years. However, we prefer parsimony, no matter how large the homoplasy, to the output of similarity methods because only with an estimated phylogeny can the sources and evolutionary implications of character convergence be assessed, whether they represent substitution-rate differences between taxa or between portions of a DNA sequence.

DNA and Homology

Applying homology criteria to DNA can be difficult but theoretically no more so than any other character system. Restriction fragment length polymorphisms (RFLPs), which are characters with no mapped linkage to other such characters, make determination of homology impossible. Certainly, homology exists in some portions of RFLP data sets, but phylogenetic analyses will produce more homoplasy with these data than with mapped restriction site data simply because of the uncertainty of DNA fragment identity. In turn, mapped restriction sites have an intrinsic degree of uncertainty because of the resolution limits of Southern blot analyses; "homologous" cleavage positions may be off by a frame of 1 or even 50 bp. Only knowledge of the nucleotide sequence can insure restriction site homology. Thus, we consider cleavage sites that appear homologous via mappings to be better phylogenetic criteria than RFLPs but still only estimates of sequence homology.

Restriction site mapping amounts to only a sampling of the whole sequence being studied. For example, consider a hypothetical restriction site analysis of 10 taxa that detects 150 cleavage sites, 15 of which are phylogenetically informative. Thus, of 150,000 possible recognition sites (considering a circular genome of 150,000 bp; Nei, 1987), only 0.1% have been sampled. In turn, only 0.01% of all possible recognition sequences have been found to be phylogenetically informative. The effect of an arithmetic increase of the number of enzymes used on (1) the number of additional restriction sites recognized and (2) phylogenetic reconstruction of such data is not predictable *a priori*, but obviously more restriction sites would be found (both uninformative and informative) until the sampling reached the level of the entire sequence itself. Despite the "retrospective salute" to restriction site data provided by Wilson et al. (1989), this method of sequence sampling will undoubtedly continue in studies at the phylogenetic level where analysis of specific sequences has not yet been (but may soon be) found possible, for example, within some angiosperm families or familial subclades.*

Nucleotide sequences themselves may present problems with homology; for example, genes that code for ribosomal RNAs are notoriously difficult to align (even by computer algorithm) because of the mixture of evolutionarily (functionally) conserved and unconserved portions within the same sequence (Gray et al., 1984; Gerbi, 1985; Bremer et al., 1987; Mishler et al., 1988; Schleifer and Ludwig, 1989; Zimmer et al., 1989). Protein-coding genes are often easier to align; for example, *rbc*L sequences can be easily and accurately aligned by eye (as was done for this paper). Nevertheless, protein-coding genes are also subject

---

*For example, in the Asteraceae, whole familial (Jansen and Palmer, 1988) as well as intrasectional (Crawford et al., 1990) relationships have been examined with restriction site data. In the Orchidaceae, relationships within a subtribe (Chase and Palmer, this volume) and a clade within this subtribe (Chase and Palmer, 1989) have been investigated. See Palmer *et al.* (1988) for further examples.

to functional constraints. The arrangement of nucleotides into 3-bp codons automatically implies some character dependence (e.g., because there is no degeneracy of the genetic code in second positions, any second-position change in a codon will alter the amino acid), and some regions/domains of the protein might be more sensitive to amino acid change than others (which could produce dependence effects across several or more codons). Additionally, comparing DNA sequences involves comparing multistate characters with four potential identities at each position (A, T, G, or C) and no intrinsic ordering (i.e., only four states are available for each character, excluding methylated or otherwise altered bases). A bias in transformational probabilities (i.e., for a transition versus a transversion) would justify character-state weighting to optimize against rarer nucleotide transformations. Potential rate inequalities at specific codon positions or over few to many nucleotides (if they could be quantified) would justify the use of whole-character weights in addition to weighting transformations of nucleotide states within characters (Albert et al., unpublished data).

Nei (1987, p. 319) pointed out that, "Some numerical taxonomists are of the opinion that parsimony methods do not require the assumption of approximate constancy of nucleotide or amino acid substitution. This is true if the number of substitutions per site is very small." We agree but would point out that the latter condition must be true for sequences that could be expected to resolve cladistic relationships. We will demonstrate this assumption below both generally and as it applies to the evolution of the chloroplast genome, using the relatively sequence-conserved gene *rbc*L (Clegg et al., 1986; Ritland and Clegg, 1987) as representative.*

A Model for Weighting Character-State Transformations in Restriction Site Characters

DeBry and Slade (1985) derived a useful set of formulae concerning the probabilities of restriction site character-state transformations. However, their purpose was to evaluate the accuracy of Wagner versus Dollo parsimony for phylogenetic reconstruction from mitochondrial DNA restriction site data and not to use the probabilities to calculate and apply weights for transformations. For characters that have a low probability of change over time, Felsenstein (1981a) has suggested that probabilities of change could be logarithmically converted to weights for use with a maximum-likelihood parsimony analysis. The results of

---

*The previously unpublished *rbc*L sequences of *Lilium superbum* and *Colchicum speciosum* are presented in Appendix I. Other *rbc*L sequences used in our analyses were taken from published reports: (i) *Chlamydomonas moewusii* (Yang et al., 1986), (ii) *Marchantia polymorpha* (Fukuzawa et al., 1988), (iii) *Nicotiana tabacum* (Shinozaki and Sugiura, 1982), and (iv) *Petunia* (Aldrich et al., 1986).

DeBry and Slade (1985) and Felsenstein (1981a) can be combined to produce reasonable *a priori* character-state weights.

Templeton (1983a,b) has also produced a statistical method for weighting restriction site characters. We consider this algorithm inappropriate (after DeBry and Slade, 1985) because (1) it combines a probability model with both Wagner parsimony and compatibility methods to create a topology, (2) it groups restriction sites into classes based on the nonrandom distribution along sequences at which different enzymes cleave, and (3) it both implies evolutionary-rate differences between different recognition sequences and groups all recognition sites for each enzyme as single characters. Although we agree that restriction sites are nonrandomly distributed, we follow DeBry and Slade (1985) in treating all restriction sites as individual characters of equal *a priori* importance.

First, we present a formalization of the maximum-likelihood weighting scheme, and then we show how empirically derived data from chloroplast *rbcL* sequences can be used to construct a range of weights that vary according to expected substitutions per restriction site per segment of the phylogenetic tree.

Probabilities of Character-state Transformations in Restriction Site Data and their Maximum-likelihood Weights

After DeBry and Slade (1985), we define 6-bp recognition sequences as "words," presence of a complete recognition sequence as $+$, Templeton's (1983a,b) "one-off sites" as $0'$, and strings two or more nucleotides off from endonuclease recognition as 0. For purposes of assigning weights to word transformations, we need only to consider the probabilities of $0' \rightarrow +$ (gain) versus that of $+ \rightarrow 0$ (loss). We agree with DeBry and Slade (1985) that any 0 state has a negligible chance of becoming a word (approximately the probability of two substitutions times 18^{-2}, after DeBry and Slade), hence $0 \rightarrow +$ need not be considered in our model. Rather than $+ \rightarrow 0'$, $+ \rightarrow 0$ is the relevant "loss" probability because it allows for the possibility of word or $0'$ loss to a 0 state by two or more substitutions in a single segment.

The assumptions of DeBry and Slade's (1985) probabilities that we employ here are somewhat restrictive, but nonetheless provide a robust approximation:

1. The rate of substitution (branch length in the topological sense as a function of time) is assumed to be constant across lineages, which both we and DeBry and Slade recognize as unrealistic. However, DeBry and Slade note for mitochondrial DNA (and we would assert the same for chloroplast DNA) that the extent of rate variability within genomes and across taxa is unknown (for some estimates, see Sytsma and Gottlieb, 1986b, and Palmer et al., 1988). Nevertheless, we discuss below the robustness of this assumption with respect to the accommodation of rate asymmetries in a single phylogeny.

2. Ancestral populations are assumed to have been nonpolymorphic at the time of cladogenesis. DeBry and Slade cite problems with this assumption for mitochondrial DNA (e.g., Avise et al., 1979; Ferris et al., 1981). Similar ancestral polymorphisms have been identified in chloroplast DNA studies as well (e.g., Doyle et al., 1990a,b).

3. Homology of specific words is assumed. We have already discussed caveats to this assumption.

4. The base composition of the DNA sequence is assumed to be 1:1:1:1 for A, T, G, and C. Chloroplast DNA as a whole is A-T rich (relative to random expectation) by approximately 14 to 25% in *Chlamydomonas* and land plants, but other genomes exhibit a continuum to G-C richness (Brown, 1983; Palmer, 1985, 1991).

The following formulae are taken from DeBry and Slade (1985) with only minor alteration of some symbols. The expected number of changes in restriction site characters along a branch of phylogeny (i.e., a segment or lineage connecting two nodes on a tree or a node and an OTU) is

$$\lambda = \text{effective nucleotide substitutions/word} \cdot \text{segment} \tag{16.1}$$

where

$$\lambda = TS_e \tag{16.2}$$

if

$$T = \text{time interval in years per segment,}$$

and

$$S_e = \text{effective nucleotide substitutions/word} \cdot \text{year} \tag{16.3}$$

For phylogenetic purposes, we are interested in the number of effective nucleotide substitutions, that is, *realized* changes in a word during a segment due to point mutations that have been fixed in the lineage. Such estimates can best be derived from character optimization on *a priori* topologies (see the following section for justification). Note that this effective substitution rate is potentially different (and usually lower, see below) than one that would be preferred by a population geneticist (who would be concerned with *all* substitutions).

For the calculation of $0' \rightarrow +$ and $+ \rightarrow 0$ probabilities, two other probabilities must be incorporated: p_d, the probability of two consecutive substitutions at the same nucleotide position in a 6-bp word, and p_r, the probability of an exact reversal at a position.

$$p_d = \frac{\sum\limits_{i=1}^{6} r_i^2}{\sum\limits_{i=1}^{6} r_i} \qquad (16.4)$$

where r_i = the probability of substitution/year at codon position i, and $r_1 = r_4 =$ (1st position % • 6 • r_{ave}), $r_2 = r_5 =$ (2nd position % • 6 • r_{ave}), $r_3 = r_6 =$ (3rd position % • 6 • r_{ave}). If substitutions are random with respect to codon position, this calculation becomes

$$p_d = \frac{\sum\limits^{6} [(0.167)(6)(0.167)]^2}{\sum\limits^{6} [(0.167)(6)(0.167)]} = 0.167 \qquad (16.5)$$

because r_{ave} is always ⅙, or 0.167. If substitutions are nonrandom, then p_d is higher.

The calculation of p_r requires information on the number of transitions (Tr) versus transversions (Tv) found in pairwise or phylogenetic comparisons of nucleotide sequences. The general formula (counting all transversions as one class) is

$$p_r = (Tr)^2 + \left(\frac{Tv}{2}\right)^2 + \left(\frac{Tv}{2}\right)^2 \qquad (16.6)$$

where $Tr + Tv$ (as percentages) = 1. The expected transition:transversion ratio for comparisons of random nucleotide sequences is simply 1:2 because there are one transition and two transversions possible at any given nucleotide position. Thus, for this special case, $p_r = 0.333$ or ⅓. As the relative percentage of transitions increases, so does the probability of a reversal.

With p_d and p_r defined, the probabilities (as a function of λ) of $0' \rightarrow +$ and $+ \rightarrow 0$ can be specified (see DeBry and Slade, 1985, for a detailed justification):

$$P_{0' \rightarrow +}(\lambda) = \lambda e^{-\lambda}\left(\frac{1}{18}\right) + \frac{1}{2}\lambda^2 e^{-\lambda}\left(\frac{2}{18}\right)p_d\left(\frac{1}{3}\right) \qquad (16.7)$$

$$p_{+ \rightarrow 0}(\lambda) = 1 - \left[e^{-\lambda} + \frac{1}{2}\lambda^2 e^{-\lambda}p_d p_r\right] \qquad (16.8)$$

For example, in a simple scenario, considering λ = 0.1 (i.e., 1 substitution per 10 words per segment) and the random values for p_d and p_r, the probabilities for Eqs. (16.7) and (16.8) are approximately 0.005 and 0.100, respectively.

Table 16.1. Estimates of transition/transversion bias, codon position substitution percentages, and nucleotide substitutions for rbcL sequences based on phylogenetic inferences from two a priori cladograms.

	Tr/Tv	Codon position %			Substitutions*
		1	2	3	
"Green plant" cladogram (Fig. 16.1)	1.0	19.0	7.1	74.0	580
"Land plant" cladogram (Fig. 16.2)	1.5	16.8	7.2	76.1	447

*Numbers of total substitutions are taken from the ACCTRAN estimates provided by paup.

Hence, with λ at this value and with sequence randomness, the probability of word gain (Eq. 16.7) is approximately 20 times less than word loss (Eq. 16.8). The relative weights of these changes may be obtained by taking the negative natural logarithms of the probabilities (see Felsenstein, 1981a). These weights are approximately 5.30 (Eq. 16.7) and 2.30 (Eq. 16.8). Thus, a stepmatrix of word-transformation weights can be constructed from these values:

$$
\begin{array}{c c}
 & \text{(to)} \\
 & \begin{array}{c c} 0 & 1 \end{array} \\
\text{(from)} \quad \begin{array}{c} 0 \\ \\ 1 \end{array} & \begin{array}{|c c} 0 & 5.30 \\ \\ 2.30 & 0 \end{array}
\end{array}
$$

or, approximated to integers (suitable for PAUP 3.0),

$$
\begin{array}{c c}
 & \text{(to)} \\
 & \begin{array}{c c} 0 & 1 \end{array} \\
\text{(from)} \quad \begin{array}{c} 0 \\ \\ 1 \end{array} & \begin{array}{|c c} 0 & \mathbf{23} \\ \\ \mathbf{10} & 0 \end{array}
\end{array}
$$

which reflects a gain:loss character-state weight of 2.3:1. Thus, it can be seen that states with a low probability of transformation are those that have high weights relative to states with higher probabilities of transformation (after Felsenstein, 1981a).

The above weighting scheme requires that rates of change be small enough and the degree of branch-length asymmetries low enough such that the approximate consistency criterion for parsimony with binary characters (Felsenstein, personal communication) is not violated. This is specified in Felsenstein's (1981a, pp. 186–187) expression (3) and accompanying discussion, which will be treated in detail below.

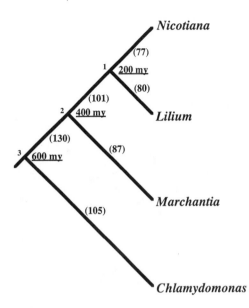

Figure 16.1. "Green plant" clado-gram. The topology was accepted *a priori* and is based on data presented by Bremer et al. (1987). Sequence data from *rbc*L were mapped passively onto the rooted cladogram using the CON-STRAINTS specification and the ACC-TRAN optimization of PAUP. Resulting branch lengths for the nucleotide data are shown in parentheses. Divergence times shown at nodes (numbered 1–3) are ap-proximate estimates from the fossil record.

Character-state Weighting Based on Empirical Estimates of rbcL Sequence Evolution

In an effort to establish transformational probabilities for restriction site data from the chloroplast genome, *rbc*L sequences from several taxa representing a wide range of evolutionary distances (in terms of years since cladogenesis) are compared in Table 16.1 for (1) codon position substitution percentages, (2) transition: transversion ratios, and (3) estimated numbers of nucleotide substitu-tions. Data for 1,428 homologous nucleotide positions representing the entire *rbc*L coding region from two taxa* were compared based on two *a priori* topolog-ies (from cladistic analysis of morphological, ultrastructural, and phytochemical data; as summarized by Bremer et al., 1987): (1) the "green plants" (Fig. 16.1), and (2) the "land plants" (Fig. 16.2). We took the accelerated transformation optimization (ACCTRAN; PAUP 3.0) estimates of branch lengths to compute patristic distances in terms of effective nucleotide substitutions for each segment in both phylogenies (Figs. 16.1 and 16.2). The ACCTRAN phylogenetic correc-tion was compared with two commonly used pairwise correction methods: Jukes-Cantor (1969) and Kimura 2-parameter (Kimura, 1980). The results of the three corrections were similar (Table 16.2). Nevertheless, where they differ, we favor the use of the phylogenetically derived measure, because it is a better estimate of the effective rate of substitutions actually fixed in lineages.

---

*Both *Chlamydomonas moewusii* and *Marchantia polymorpha* have *rbc*L coding regions of 1428 bp; if longer, other sequences were truncated to this length. For our purposes of estimating patristic distance, all variable sites were used, including autapomorphies.

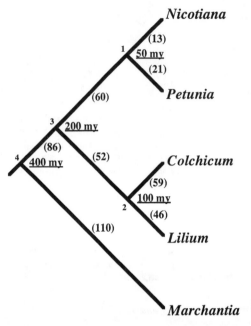

Figure 16.2. "Land plant" cladogram. The topology was accepted *a priori* and is based on data presented by Bremer et al. (1987). Sequence data from *rbc*L were mapped passively onto the rooted cladogram using the CONSTRAINTS specification and the ACCTRAN optimization of PAUP. Resulting branch lengths for the nucleotide data are shown in parentheses. Divergence times shown at nodes (numbered 1–4) are approximate estimates from the fossil record.

We estimated divergence times for all nodes of both trees using approximate dates from the fossil record (Taylor, 1988, on bryophytes; Cornet, 1986, on the putative angiosperm *Sanmiguelia*; Muller, 1981, 1984, and Walker and Walker, 1984, on pollen from Liliaceae and Asteridae; Figs. 16.1 and 16.2). We then estimated S_e (expression 16.3, above) for each segment (Table 16.3). The expression for our determination of S_e is

$$S_e = \frac{(\text{patristic distance in bp})(6 \text{ bp per word})}{(1{,}428 \text{ bp in } rbcL)(\text{divergence time in years})} \qquad (16.9)$$

From the PAUP-derived patristic distances and our estimates for divergence times, S_e varied by a factor of 3.7 in the "green plant" phylogeny and by a factor of 2.3 in the "land plant" phylogeny, with a mean of approximately 1.7×10^{-9} (Table 16.3). Using this estimate of S_e, we can now consider various values of λ, based on different segment lengths (T), and the weights derived from substitution of these λ values into Eqs. (16.7) and (16.8).

Continuing with our simple scenario, where p_d and p_r reflect random values, we can examine the proper weights for $0' \rightarrow +:+ \rightarrow 0$ for a range of values. For example, λs of 0.1, 0.01, and 0.001 reflect segment time intervals decreasing in order of magnitude from 6×10^7 to 6×10^6 and 6×10^5 years, based on *rbc*L (Table 16.4). At $\lambda = 0.1$ (segment length of approximately 60 myrs), the $0' \rightarrow +:+ \rightarrow 0$ weight ratio is 2.3:1. At $\lambda = 0.01$ (segment length of approximately

Table 16.2. *Estimated numbers of substitutions in* rbcL *between selected pairwise combinations of taxa.*

	Observed Substitutions	Estimated Substitutions				
		(Tr)	(Tv)	ACCTRAN	J-C	K2P
Nicotiana—Chlamydomonas	349	155	194	413^G	422.3	424.0
Nicotiana—Marchantia	247	132	115	$265^G, 269^L$	280.8	282.9
Nicotiana—Lilium	157	103	54	$157^G, 171^L$	169.8	171.6
Nicotiana—Petunia	34	16	18	34^L	34.6	34.6
Lilium—Colchicum	101	73	28	105^L	106.1	107.1

The observed substitutions are those found by simple pairwise comparison, and the subsets of these that were transitions (Tr) and transversions (Tv) are indicated. Three corrections for underestimated substitutions are compared: the "phylogenetic correction" of the ACCTRAN optimization in PAUP, the Jukes-Cantor method (J-C), and the Kimura 2-parameter method (K2P). Under the ACCTRAN column, the specific *a priori* topology from which the estimate came is indicated by superscript G (for the "green plant" cladogram, Fig. 16.1) or L (for the "land plant" cladogram, Fig. 16.2). Note the similarity of all of the corrections. The phylogenetic correction almost uniformly underestimates substitutions slightly compared to the other corrections, but see text for its justification.

6 myrs), the proper weight of word gain/loss becomes 1.6:1. Continuing this trend, at $\lambda = 0.001$ (segment length of approximately 600,000 yrs), the weight is 1.4:1 (these results and those for other values of λ are summarized in Table 16.4). The behavior of the maximum-likelihood weight as a function of λ is shown in Fig. 16.3. Only the weights lying in the relatively flat portion of the curve in Fig. 16.3 are biologically relevant, as will be described below.

The above segment length times are dependent upon the substitution rate estimated for *rbc*L. Alternative time intervals corresponding to the same λs are presented (Table 16.4) for lower and higher rates of sequence change (i.e., nonsynonymous substitutions in the inverted repeat and synonymous substitutions in the single-copy regions, respectively; Palmer, 1991). Thus the relationship between λ and segment length in years may change somewhat, but the weighting model and its limitations remain the same (Fig. 16.3).

To investigate the effects of transition:transversion ratios and codon position substitution percentages differing from the above random expectations, we examined the most extreme biases of both variables seen in our data (i.e., Tr:Tv = 2.6, 1st/13.9%—2nd/5.0%—3rd/81.2% for *Lilium-Colchicum* from direct pairwise comparison) and their effects on the likelihood weighting model. Considering an extremely high value of $\lambda = 1$ (estimated for *rbc*L as reflecting 600 myr segment length), p_d, calculated from Eq. (16.4), equals 0.34; p_r, calculated from Eq. (16.6), equals 0.56. When these probabilities are applied to Eqs. (16.7) and (16.8), the word gain/loss weight decreases from 8:1 (as calculated with random values for p_d and p_r) to 7.3:1. Thus, our prediction that increasing transition bias would increase the likelihood of reversal is true for enormous segment lengths. However, at a lower λ of 0.1 (corresponding to approximately 60-myr segment lengths, a more reasonable value) the $0' \rightarrow +:+ \rightarrow 0$ weight ratio is the same

Table 16.3. Estimated rates* of effective nucleotide substitutions per 6-bp restriction site per year (S_e) for rbcL data, based on patristic distance (in terms of ACCTRAN-estimated substitutions) and approximate divergence time estimates from the fossil record.

Segment	Estimate of S_e (times 10^{-9})
"Green plant" cladogram (Fig. 16.1)	
Nicotiana	1.6
Lilium	1.7
Marchantia	0.91
Chlamydomonas	0.73
Nodes 1–2	2.1
Nodes 2–3	2.7
Mean =	1.6
"Land plant" cladogram (Fig. 16.2)	
Nicotiana	1.1
Petunia	1.8
Lilium	1.9
Colchicum	2.5
Marchantia	1.2
Nodes 1–3	1.7
Nodes 2–3	2.2
Nodes 3–4	1.8
Mean =	1.8

*These rates are estimated for each segment on the *a priori* topologies shown in Figs. 16.1 and 16.2.

to two significant figures as it is when considering random sequence change: 2.3:1. This is because p_d and p_r enter the probability Eqs. (16.7 and 16.8) only in the second term, which specifies a second change in a word during a segment (see DeBry and Slade, 1985, for a more detailed justification). Thus, for small λs (i.e., all values reasonable for phylogenetic reconstruction), the effect of substitution biases is negligible. Figure 16.3 shows the relationship of weight-to-λ calculated without including any transition/transversion or codon position bias—a simplification that holds generally for all phylogenetically useful restriction site data from 6-bp recognition enzymes.

Implementation of Character-state Weighting

As with any form of phylogenetic analysis, proper understanding of the assumptions inherent to stepmatrix-weighted parsimony approaches is of the utmost importance in both producing and evaluating topological estimates. Swofford and Olsen (1990) have described briefly how several commonly employed parsimony criteria can be emulated using the "generalized parsimony" flexibility of stepmatrices. For example, stepmatrices are symmetrical in their transformational weights in an emulation of Wagner parsimony (e.g., Fig. 11a in Swofford and Olsen, 1990). As we have described above, they are asymmetrical in their transformational

Table 16.4. *Estimated relationship between* T *(segment length in years) and* λ *(expected number of nucleotide substitutions per 6-bp restriction site per segment) based on various* S_e *values.* Also shown are proper weights of restriction site gains over losses at each value of* λ.

T_{IR}	T_{rbcL}	T_{SC}	λ	Transformational weight
1.7×10^2	5.9×10^1	1.1×10^1	0.0000001	1.19
1.7×10^3	5.9×10^2	1.1×10^2	0.000001	1.20
1.7×10^4	5.9×10^3	1.1×10^3	0.00001	1.25
1.7×10^5	5.9×10^4	1.1×10^4	0.0001	1.30
1.7×10^6	5.9×10^5	1.1×10^5	0.001	1.40
1.7×10^7	5.9×10^6	1.1×10^6	0.01	1.60
1.7×10^8	5.9×10^7	1.1×10^7	0.1	2.30
1.7×10^9	5.9×10^8	1.1×10^8	1.0	8.0

*Estimates of S_e (effective substitution rate per 6-bp restriction site per year) range from the nonsynonymous substitution rate in the cpDNA inverted repeat (IR—0.6×10^{-9} substitutions per 6-bp restriction site per year), to that for *rbcL* (1.7×10^{-9}), to the synonymous substitution rate in the single-copy regions (SC—9.0×10^{-9}; calculated from Palmer, 1991).

weights for character-state weighting of restriction site data (also shown by Fig. 11d in Swofford and Olsen, 1990). The latter approach has been likened by Swofford and Olsen (1990, p. 463) to a "relaxed Dollo criterion." This comparison is misleading because the ability of such a stepmatrix to emulate the Dollo criterion is itself dependent upon the choice of ancestral-state assumptions.

Asymmetrical stepmatrices implemented with PAUP 3.0 require the specification of hypothetical ancestral states from which the initial costs of transformation are assessed. Should an investigator wish to specify the outgroup character states as the hypothetical ancestor, implementation of the character-state weighting model outlined above will indeed span the extremes between Wagner and Dollo analyses including that outgroup. The implied gain/loss weight of 1.0 for the Wagner criterion represents the bottommost limit, and some data-dependent weight (which should lie beyond the flat portion of the curve in Fig. 16.3 for most data)—along with any greater value approaching infinity—will yield the Dollo result. This occurs because as character-state gains are increasingly biased against with increasing weight, the parsimony algorithm will eventually force all gains to be singular, whereas multiple losses from ancestral states "present" are not prohibited because of their low relative weight. Specification of the outgroup as ancestor in an appropriately character-state weighted parsimony analysis can thus emulate unrooted applications of Dollo parsimony that include that outgroup (see also Swofford and Olsen, 1990, pp. 459–461).

An alternative approach is to make "no assumptions" about ancestral states and to produce topologies with forced ingroup monophyly. This can be implemented by (1) an ANCSTATES statement specifying all missing data, (2) constraining ingroup monophyly via a PAUP "constraints" statement in the PAUP BLOCK such that only trees meeting the requirement "((taxa 1 through n), outgroup)" are allowed, and (3) implementation of stepmatrix weighting at the

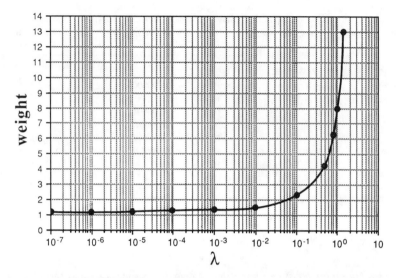

Figure 16.3. Proper weight of restriction site gains over losses, as a function of λ, which is genome independent. $\lambda = TS_e$, where T is time (i.e., length of a branch or segment of a cladogram) and S_e is the nucleotide substitution rate per restriction site per unit time. Note that the curve is nearly flat over the domain of λ for which restriction site data are likely to be phylogenetically useful. Thus, a general recommendation of character-state weight above 1.0 but below about 2.0 is justifiable for all restriction site data.

desired weight (Swofford, personal communication; see also the PAUP 3.0 interactive helpfile and user's manual). In contrast to the implementation discussed above, the Wagner criterion is still the bottommost weighting limit, but neither the Dollo criterion nor its topological solution can ever be achieved. As weight increases, a point will be reached at which no parallel gains remain; however, the topology is not a Dollo topology. This is because gains are being weighted against from a starting point of uncertain polarity (i.e., ANCSTATES "data missing"). Increasing weight will increasingly favor losses over gains until a data-dependent weight is reached (extending to infinity) at which "total loss" topologies are produced because the uncertain ancestor has been forced to a global "states present."

Total loss topologies are in fact the logical expectations of character-state weighting with values derived from enormous λs; such a λ implies an enormous rate of change, enormous segment lengths in years, or both. Restriction sites are not expected to show homology under such conditions because losses of any previously existing synapomorphies would be virtually assured because of the high relative probability of character-state loss (see above). Dollo parsimony is therefore a specialized subset of the total loss scenario; one initial gain is allowed, but a second gain in the same character is then weighted to infinity. The Dollo criterion requires a specification of polarity, whether executed rooted or unrooted (Swofford and Olsen, 1990). On the other hand, character-state weighting with the ancestor coded as "missing data" allows the differential weights of gain versus loss to set up polarities appropriate to the data being analyzed. We believe that the latter approach, although not assumption-free, is substantially less biased and is therefore preferable. Unlike Swofford and Olsen (1990), however, we would not term this character-state weighting scheme "relaxed Dollo," but rather "enhanced Wagner," because as mentioned above, Dollo topologies are not produced from this implementation at any step matrix weight, but subsets of equally parsimonious Wagner topologies can be, as will be demonstrated in the next section.

Application of Character-state Weighting to a Chloroplast DNA Restriction Site Data Set: The Sections of *Clarkia* (Onagraceae)

Sytsma and coworkers (Sytsma and Smith, 1988; Sytsma et al., 1990) have reviewed the literature pertaining to the wealth of morphological, hybridization, and isozymic data available on the genus *Clarkia*. Here, we are concerned with the phylogenetic treatment of the restriction site data set of representatives of *Clarkia* sections begun by Sytsma and Smith (1988) and more fully developed by Sytsma et al. (1990). Our first approach in reanalyzing these data was to check the Wagner and Dollo parsimony results presented by Sytsma et al. (1990). Of the 121 word gains/losses relative to the outgroup taxon (*Epilobium brachycarpum*) reported, 62 were synapomorphous and 59 autapomorphous as judged by Table 2 of Sytsma et al. (1990); this contrasts with their own report that "66 are phylogenetically informative and 56 are not."* Of the 18 restriction enzymes used in the analysis, two are unusual in that their recognition sequence involves more than six nucleotides (*Eco*NI recognizes CTT*NNNNN*AGG and *Bst*EII recognizes GGT*N*ACC, *N* being any nucleotide), yet they are still functionally 6-bp words under the above probability model. Wagner parsimony performed with the BRANCH AND BOUND option of PAUP 3.0 on only synapomorphous characters produced four equally parsimonious cladograms of 85 steps and with consistency indices (CI) of 0.729 (Fig. 16.4). These appear to encompass the same topologies shown by Sytsma et al. (1990, Fig. 2); our analysis includes an additional, fully resolved topology (Fig. 16.4B) if we are correct in interpreting their Figs. 2a and 2b (shown with one zero-length branch each) as equivalent to

---

*There are also discrepancies in their Table 2 (character 45 is absent) and in their Fig. 3 (characters 28, 53, and 73 are misplaced relative to their distribution in Table 2); the length of the tree in Fig. 3 should thus be 145 rather than the reported 146. We have used their data as in Table 2; fortunately, the above character discrepancies do not affect our results because they occur at nodes well defined by other restriction site states.

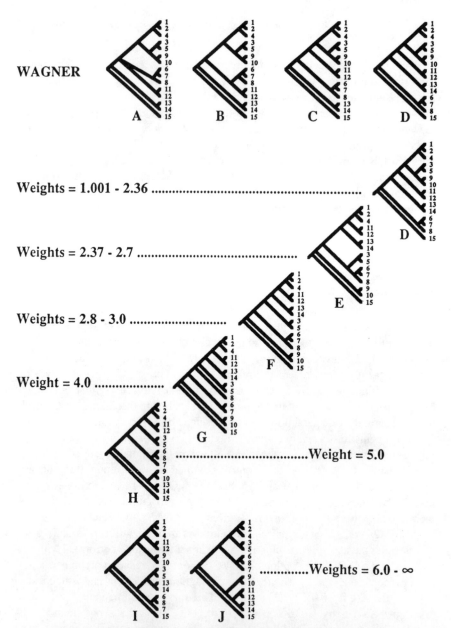

WAGNER

Weights = 1.001 - 2.36 ....

Weights = 2.37 - 2.7 ....

Weights = 2.8 - 3.0 ....

Weight = 4.0 ....

Weight = 5.0

Weights = 6.0 - ∞

Figure 16.4. Results of several analyses of the chloroplast DNA restriction site data set published by Sytsma et al. (1990) for *Clarkia* (Onagraceae). Shown are all equally most parsimonious cladograms generated using PAUP under Wagner parsimony (top) and character-state weighting of restriction site gains over losses at the stated values. Topologically identical cladograms are identified with the same letter. OTUs are the taxa numbered as in Table 16.5. Note the gradual topological transition evident from Wagner to the "total loss" scenario as the weight increases.

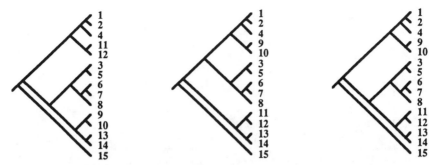

Figure 16.5. The Dollo parsimony topologies of Sytsma et al.'s (1990) *Clarkia* restriction site data set. Both rooted and unrooted Dollo approaches (see Swofford and Olsen, 1990) gave the same topologies (rooted trees are shown here). Note that character-state weighting can emulate the Dollo criterion and produce these particular topologies by specifying the outgroup states as the ancestral states and by implementing gain/loss weights of 4.0 to infinity.

our Fig. 16.4A (our analysis used the MULPARS and COLLAPSE options of PAUP). Sytsma et al. (1990, pp. 285–286) "converted" their four trees "to Dollo parsimony trees by changing the unlikely convergent gains or loss/gains to the more likely convergent losses or gain/losses," invoking the model of Templeton (1983a,b). From their "Dollo-izations," Sytsma et al. (1990) chose one of the four topologies (their Fig. 2a, our Fig. 16.4A) as the most likely, because it was also their single, most parsimonious Dollo tree obtained from PHYLIP. However, our own BRANCH AND BOUND rooted and unrooted Dollo parsimony analyses (performed with PAUP 3.0; see Swofford and Olsen, 1990, for further discussion) of the 62 synapomorphies produced three trees of 129 and 102 steps and CIs of 0.481 and 0.608, respectively (Fig. 16.5), all of which were different from the topologies obtained by Sytsma et al. (1990). We were unable to replicate Sytsma et al.'s (1990) Dollo parsimony results using any of the flexible options of PAUP 3.0 to emulate various appropriate, as well as inappropriate, applications of the Dollo criterion.

The *Clarkia* restriction site data set is of appropriately large size (62 synapomorphies) with relatively few taxa (15; Table 16.5); thus we considered it a good test for character-state weighting. Using the USERTYPE STEPMATRIX option of PAUP 3.0 and the "ancestral states missing" approach outlined above, we were able to examine the topological effects of different segment-length assumptions (in years), which are directly related to λ and hence to the proper character-state weight. Use of the BRANCH AND BOUND option of PAUP is not usually practical for data analysis with a stepmatrix because of the much greater time required for calculation (except when taxon number is low, e.g. 10 or fewer). Thus, for each weight, 10 replicates with random addition sequence were run using the HEURISTICS option with TREE BISECTION-RECONNECTION (TBR) branch swapping and ACCTRAN optimization.

Table 16.5.   Taxa examined in the chloroplast DNA restriction site analysis of Sytsma et al.
(1990).

Taxon number	Taxon
1	Clarkia biloba (Sympherica)
2	Clarkia rostrata (Sympherica)
3	Clarkia xantiana (Phaeostoma)
4	Clarkia unguiculata (Phaeostoma)
5	Clarkia bottae (Fibula)
6	Clarkia concinna (Eucharidium)
7	Clarkia breweri (Eucharidium)
8	Clarkia pulchella (Clarkia)
9	Clarkia imbricata (Godetia)
10	Clarkia williamsonii (Godetia)
11	Clarkia amoena subsp. huntiana (Rhodanthos)
12	Clarkia lassenensis (Rhodanthos)
13	Clarkia virgata (Myxocarpa)
14	Clarkia mildrediae (Myxocarpa)
15	Epilobium brachycarpum

Numerical abbreviations are supplied for interpretation of Fig. 16.4. Traditional sectional assign-
ments of the clarkias (after Sytsma et al., 1990) are indicated after each binomial.

At a weight of $0' \rightarrow +:+ \rightarrow 0$ (gain/loss) = 6:1 (segment lengths well over
500 million years for *rbc*L; see Table 16.4 and Fig. 16.3), the stepmatrix analysis
identified two "total loss" topologies that continue to be obtained as weight
approaches infinity (Fig. 16.4; note that because character-state weights are used,
consistency indices are not comparable to those produced by Wagner and Dollo
parsimony). At a gain/loss stepmatrix weight of 5:1, one tree, topologically
distinct from the two "total loss" trees, was found (Fig. 16.4). Unlike the "total
loss" trees, this topology included three character-state gains. Similarly, at stepmatrix weights 4:1 and 2.8–3.0:1, single, distinct topologies were produced with
the number of character-state gains increasing to 11 and 15, respectively (Fig.
16.4). At gain/loss weights between 2.37 and 2.7, an additional single topology
was produced with 25 character-state gains (including one parallel gain; Fig.
16.4). Stepmatrix gain/loss weights between a miniscule 1.001 and 2.36 produced
a single topology, identical to one of the Wagner trees (Fig. 16.4D). In our
opinion, the robustness of this topology over a broad range of all reasonable λs
makes it the optimal solution for these data.

Onagraceous pollen is known from the Maestrichtian of the late Cretaceous,
approximately 73 to 65 myrs ago (Raven, 1988). Hence, it is highly doubtful that
segment lengths in years (calculated from any particular rate of change; Table
16.4) among individual taxa of *Clarkia* could be anywhere near the upper bound
of the "optimal" weight range (i.e., those estimated from weight = 2.36). Even
if clarkias themselves date to the Maestrichtian, average *T*s over an entire *Clarkia*
phylogeny would be expected to be considerably lower. Likewise, the lower-
bound weight estimate (gain/loss weight = 1.001) is unreasonable in that it

would correspond to average segment lengths of less than a year! Therefore, all reasonable values of λ are encompassed by this range (Fig. 16.3). Although the topological resolution of the analysis increased through the use of *a priori* character-state weighting, the justification of the model lies in its increased accuracy in portraying transformational asymmetries in restriction site character-states, not in its ability to find fewer optimal trees.

The preferred character-state weighted estimate is a subset of the Wagner trees (Fig. 16.4), thus a comparison of its character-topology relationship to the other Wagner trees is revealing. All four trees require 23 homoplasies (as befits their equal parsimony under the Wagner criterion), but trees A and B require 13 homoplasious gains and 10 homoplasious losses, whereas trees C and D require only 10 homoplasious gains and 13 homoplasious losses. Thus, with even a slight optimization against parallel gains, either intuitive or mathematically formalized, trees C and D are highly favored over A and B (but note that tree A was favored by Sytsma et al., 1990). Although trees C and D are equivalent in the number of homoplasious gains with reference to the outgroup taxon, inclusion of the outgroup in a rooted, character-state weighted analysis with "ancestral states missing" (as described above) favors trees that can assign the greatest number of "states present" to the hypothetical ancestor. This is because transformational cost is minimized in segments leading to both in-group and out-group when the greatest number of character-state losses is allowed. Topology D does in fact produce the minimal number of weighted steps of the four topologies.

We will not discuss the systematic implications of our findings here. We only note that the potential exists for significantly improved phylogenetic estimation of *Clarkia* relationships through consideration of the probabilities and weights for transformations in the characters under study. It should be noted that the Dollo topologies (Fig. 16.5) are obtained from character-state weighting analyses only using the outgroup equals ancestor mode (described above) and only at gain/loss weights of 4 and above, which are well outside the flat region of the weight–λ curve (Fig. 16.3).

Segment Lengths and Phylogeny Reconstruction

The calculations made above assume a roughly constant λ across the phylogeny being reconstructed. However, individual λ values for particular segments may vary considerably, either due to variation in absolute times that particular lineages existed, or due to variation in substitution rates, or both. How similar must λs be such that our approximation of constancy is a robust assumption? An examination of the Felsenstein zone for binary systems, such as restriction site data, reveals that parsimony can accommodate a wide range of λs in the same analysis and remain consistent.

Felsenstein (1978) established exact conditions under which branch-length inequalities would make parsimony "positively misleading"; sufficiently long branches can tend to group together because of accumulated homoplasy rather than shared ancestry. The exact condition of consistency for the Wagner criterion, specified with the transformational probabilities p for two long branches and q for a short branch connecting them, is (after Felsenstein, 1978, 1983)

$$q(1 - q) > p^2 \qquad (16.10)$$

Felsenstein (1983, Fig. 3) provided a plot of this expression with q (X) and p (Y) axes both representing values from 0 to the limit 0.5 (see Felsenstein, 1983, for justification of this limit). Even with limiting values of q and p, the domain of consistency was found to be large.

More generally, an approximate consistency criterion can be established for parsimony analysis of binary characters that are not necessarily restricted to the unbiased reversibility criterion. This is simply

$$q > p^2 \qquad (16.11)$$

(Felsenstein, personal communication), a condition echoed in Felsenstein's paper that established consistency conditions for weighted parsimony analysis (1981a, Eq. 3). His 1981 inequality is more general in that it allows for asymmetrical p values, such that

$$q > (p)(\alpha p) \qquad (16.12)$$

where α is a modifier greater than 1.0. This expression reduces the approximate domain of consistency relative to the symmetry assumption in Eq. (16.11). Although Eq. (16.12) is a more biologically realistic criterion than Eq. (16.11), the latter is computationally simpler. Although we cannot provide a full mathematical treatment of the Felsenstein zone here, we can demonstrate a heuristic approximation of its boundaries based on Eq. (16.11) and λ-dependent probabilities.

From Eq. (16.11) and the discussion by Felsenstein (1981a, pp. 186–187), it is clear that for parsimony to be consistent, one change in the ith character in a short segment must be more probable than one change in that character in each of two long segments. This can be translated into Poisson-based probabilities of change, substituting $\lambda_1 = X$ (for short connecting branch) and $\lambda_2 = Y$ (for long branches):

$$Xe^{-X} > (Ye^{-Y})^2 \qquad (16.13)$$

or

$$Xe^{-X} - Y^2 e^{-2Y} > 0 \qquad (16.14)$$

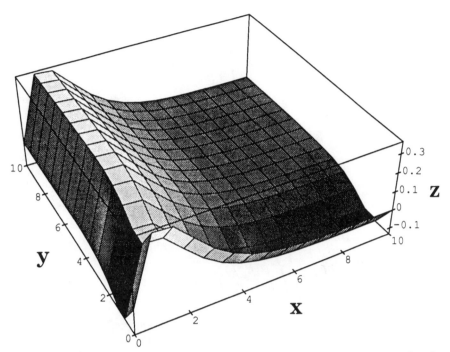

Figure 16.6. A three-dimensional surface representing Eq. (16.14), where X is λ for a short branch connecting two long branches each with λs of Y. Where the surface is positive on the Z axis, the approximate consistency solution for parsimony with binary characters holds. Negative regions of the surface indicate where branch-length inequality makes parsimony inconsistent, that is, the Felsenstein zone.

Thus, solutions for Eq. (16.14) show the approximate consistency of parsimony with positive values and inconsistency with negative ones. A three-dimensional plot of Eq. (16.14) is shown in Fig. 16.6, where λs from 0 to 10 are compared on the X and Y axes. Where the surface is positive as indicated on the Z axis, the approximate criterion for consistency is obtained. At Z values less than 0, the surface reflects the area of inconsistency, that is, the Felsenstein zone. The boundary between these two regions is defined by those values of X and Y for which Eq. (16.14) equals zero.

Contour plots of Eq. (16.14) are helpful for evaluating the consistency/inconsistency of combining particular X and Y λ-values. Considering the λ range examined in previous portions of this chapter, from near zero to 1.0, two contour plots are helpful: one expressing the full range from $\lambda = 0$ to $\lambda = 1.0$ (Fig. 16.7) and one expressing a more restricted but biologically more reasonable range from $\lambda = 0$ to $\lambda = 0.1$ (Fig. 16.8). In each plot, an approximate zero line delimits the boundary of the Felsenstein zone; the positive (consistent) region lies to the right of this line, and the negative Felsenstein zone lies to the left. Pairs of X and Y λ-

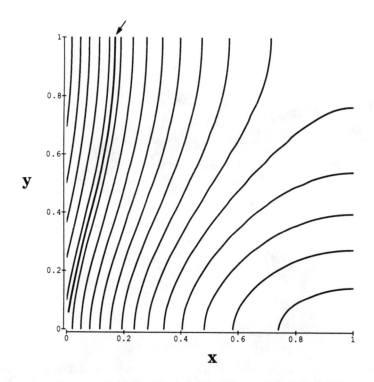

Figure 16.7. A contour plot representing Eq. (16.14) for λ values of X and Y between 0 and 1.0, where X is λ for a short branch connecting two long branches each with λs of Y. The line originating near $X, Y = 0$, indicated by arrow, is the approximate boundary of consistency (to the right) and inconsistency (to the left) of parsimony, that is, where the solution of Eq. (16.14) equals zero.

values for short and contiguous long branches (respectively) that graph to the right of the zero line can be present on a true tree and still permit consistent phylogenetic reconstruction using parsimony.

The net result of this heuristic investigation is that both of Felsenstein's conditions for validity of weighted parsimony methods using logarithmic transformations of probabilities of change (i.e., small absolute probabilities of change and relatively equal segment lengths) are obtained when λs are less than about 0.1. Thus, in the flat part of the curve shown in Fig. 16.3, λs for individual segments can differ greatly (e.g., $X = 0.02$, $Y = 0.08$; Fig. 16.8) with increasingly greater differentials "allowed" at lower probabilities of change (e.g., $X = 0.005$, $Y = 0.05$; Fig. 16.8). At λs nearing or in the vertical portion of the curve shown in Fig. 16.3, misleading transformations (homoplasies) increasingly tend to obscure the historically informative transformations. Our character-state weighting scheme is therefore justified over nearly the whole range of differential λs for which phylogenetic reconstruction is feasible.

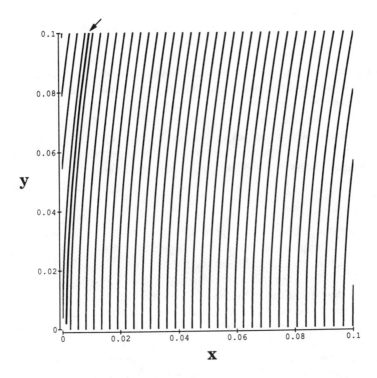

Figure 16.8. A contour plot of Eq. (16.14) for a more restricted range of λs than shown in Fig. 16.7, that is, X and Y from 0 to 0.1, where X is λ for a short branch connecting two long branches each with λs of Y. The line originating near X, $Y = 0$, indicated by arrow, is the approximate boundary of consistency (to the right) and inconsistency (to the left) of parsimony, that is, where the solution of Eq. (16.14) equals zero.

Practical Recommendations for Character-state Weighting of Restriction Site Data

The relative invariance in weight over a large range of values for λ (Fig. 16.3) allows some rules of thumb to be developed even with the usual uncertainty about divergence times and evolutionary rates. For mapped restriction site data produced from 6-bp recognition endonucleases, we suggest initial character-state weighting of gains over losses by a factor of approximately 1.3 for "low-level" analyses (e.g., species within a genus or even genera within a family), and by a factor of approximately 2.0 for "high-level" analyses (e.g., major groups of land plants, such as divisions—but note that restriction site synapomorphies are likely to be homoplasious at such a level). Even given these recommendations, we would suggest trying several other weights between 1.0 and 2.5. Performing an equally weighted Wagner analysis is also recommended, as this permits a comparative check for robustness of results to particular assumptions of λ.

We would never recommend the use of Dollo parsimony for phylogenetic analysis of restriction site data from any genome, despite its use in previous studies. First, its criterion of evolutionary singularity models the impossibility of parallel character-state gain, whereas such events are not impossible, but merely improbable with probabilities that can be estimated. Second, outgroup = ancestor implementation of character-state weighting yields Dollo topologies only at unreasonably high weights, which reflect either rapid molecular evolutionary rates, long times over which lineages existed, or both (their product being λ). If data must carry such assumptions, they are unlikely to provide reliable markers of phylogenetic history and should be rejected from the outset.

Coda

Our analysis provides a more general insight into the justification for Wagner parsimony. If the probability of change in a particular character in a given segment on a tree is low, then the unbiased reversibility (i.e., equal weights for forward and backward transformations) permitted by Wagner parsimony is approximated. We have shown that this is true even for a character system such as restriction site data, that is known to have a great asymmetry in transformation probabilities. Therefore, in character systems with lesser asymmetries, such as in much morphologic data, Wagner parsimony is even more closely approximated. It should also be noted that even for characters with great transformational asymmetries, a few characters undergoing multiple changes will cause an increase in homoplasy but will not necessarily affect the accuracy of phylogenetic reconstruction if similar homoplasies are not rife.

Felsenstein (e.g., 1978, 1981a) was the first to state clearly the general relationship between infrequent change and Wagner parsimony (see also Sober, 1988). Nevertheless, Felsenstein and many others (e.g., Sober) have failed to identify the necessary linkage between infrequent change and the phylogenetic utility of any putative homology (molecular or morphological). Various models of phylogenetic change (including Felsenstein's own; 1978) include rapid rates of transformation—often with restricted numbers of character states—making the problem with homoplasy acute. Such rates are unrealistic for any character that could be expected to be phylogenetically useful (i.e., a potential marker of shared cladistic history).

We are comfortable with an assumption of relatively infrequent change in the characters we select as potential phylogenetic markets (selected because of their constancy within operational toxonomic units [OTUs] or larger monophyletic groups). If we had evidence of rapid rates of change in certain characters relative to the phylogenetic level of an intended analysis, such characters would be rejected a priori. Note that this is not equivalent to an assumption that "nature is parsimonious" (low probability of change does not necessarily imply low homoplasy) or to an assumption of equality of evolutionary rates (such as is necessary to justify phenetic methods). The rules of thumb developed in this chapter for restriction site data have a general, intuitively pleasing application: a "good" data set for phylogenetic reconstruction is one in which less than about 1 out of 10 characters is expected to change on any particular branch (i.e., $\lambda <$ 0.1).

We advocate combining the information present in each individual hypothesis of homology (i.e., a character with two or more character states, such as a synapomorphous restriction site) by means of a maximum parsimony analysis. Such an analysis should include the application of character-state weights if sufficient *a priori* knowledge of relative transformation probabilities exists. Furthermore, such an analysis should also include whole-character weights if different sets of characters (e.g., from different genes or noncoding sequences, nucleotide strings within either type of sequence, or even codon positions) are known to be evolving at different rates. We are currently investigating various applications of both character and character-state weighting for DNA sequence data.

References

Aldrich, J., Cherney, B., Merlin, E., and Palmer, J.D. (1986) Sequence of the *rbc*L gene for the large subunit of ribulose bisphosphate carboxylase-oxygenase from petunia. *Nucleic Acids Res.* **14**, 9534.

Avise, J.C., Lansman, R.A., and Shade, R.O. (1979) The use of restriction endonucleases to measure mitochondrial DNA sequence relatedness in natural populations. I. Population structure and evolution in the genus *Peromyscus. Genetics* **92**, 279–295.

Baldwin, B.G., Kyhos, D.W., and Dvořák, J. (1990) Chloroplast DNA evolution and adaptive radiation in the Hawaiian silversword alliance (Asteraceae–Madiinae). *Ann. Missouri Bot. Gard.* **77**, 96–109.

Bremer, K., Humphries, C.J., Mishler, B.D., and Churchill, S.P. (1987) On cladistic relationships in green plants. *Taxon* **36**, 339–349.

Brown, W.M. (1983) Evolution of animal mitochondrial DNA. In: *Evolution of Genes and Proteins* (eds. M. Nei and R.K. Koehn), Sinauer Associates, Sunderland, MA., pp. 62–88.

Brown, W.M., Prager, E.M., Wang, A., and Wilson, A.C. (1982) Mitochondrial DNA sequences of primates: tempo and mode of evolution. *J. Mol. Evol.* **18**, 225–239.

Chase, M.W., and Palmer, J.D. (1989) Chloroplast DNA systematics of lilioid monocots: resources, feasibility, and an example from the Orchidaceae. *Amer. J. Bot.* **76**, 1720–1730.

Clegg, M.T., Ritland, K., and Zurawski, G. (1986) Processes of chloroplast DNA evolution. In *Evolutionary Processes and Theory* (eds. S. Karlin and E. Nevo), Academic Press, New York, pp. 275–294.

Colless, D.H. (1970) The phenogram as an estimate of phylogeny. *Syst. Zool.* **19**, 352–362.

Cornet, B. (1986) The leaf venation and reproductive structures of a late Triassic angiosperm, *Sanmiguelia lewisii*. *Evol. Theory* **7**, 231–309.

Cracraft, J. (1987) DNA hybridization and avian phylogenetics. *Evol. Biol.* **21**, 47–96.

Crawford, D.J., Palmer, J.D., and Kobayashi, M. (1990) Chloroplast DNA restriction site variation and the phylogeny of *Coreopsis* section *Coreopsis* (Asteraceae). *Amer. J. Bot.* **77**, 552–558.

DeBry, R.W., and Slade, N.A. (1985) Cladistic analysis of restriction endonuclease cleavage maps within a maximum-likelihood framework. *Syst. Zool.* **34**, 21–34.

Donoghue, M.J., Doyle, J.A., Gauthier, J., Kluge, A.G., and Rowe, T. (1989) The importance of fossils in phylogeny reconstruction. *Ann. Rev. Ecol. Syst.* **20**, 431–460.

Doyle, J.J., Doyle, J.L., and Brown, A.H.D. (1990a) Chloroplast DNA polymorphism and phylogeny in the B genome of *Glycine* subgenus *Glycine* (Leguminosae). *Amer. J. Bot.* **77**, 772–782.

Doyle, J.J., Doyle, J.L., Grace, J.P., and Brown, A.H.D. (1990b) Reproductively isolated polyploid races of *Glycine tabacina* (Leguminosae) had different chloroplast genome donors. *Syst. Bot.* **15**, 173–181.

Farris, J.S. (1970) Methods for computing Wagner trees. *Syst. Zool.* **19**, 83–92.

Farris, J.S. (1972) Estimating phylogenetic trees from distance matrices. *Amer. Natur.* **106**, 645–668.

Farris, J.S. (1977a) Phylogenetic analysis under Dollo's Law. *Syst. Zool.* **26**, 77–88.

Farris, J.S. (1977b) Some further comments on LeQuesne's methods. *Syst. Zool.* **26**, 220–223.

Farris, J.S. (1983) The logical basis of phylogenetic analysis. *Advances in Cladistics* **2**, 7–36.

Felsenstein, J. (1978) Cases in which parsimony or compatibility methods will be positively misleading. *Syst. Zool.* **27**, 401–410.

Felsenstein, J. (1981a) A likelihood approach to character weighting and what it tells us about parsimony and compatibility. *Biol. J. Linn. Soc.* **16**, 183–196.

Felsenstein, J. (1981b) Evolutionary trees from DNA sequences: a maximum likelihood approach. *J. Mol. Evol.* **17**, 368–376.

Felsenstein, J. (1983) Methods for inferring phylogenies: a statistical view. In: *Numerical Taxonomy* (ed. J. Felsenstein), Springer-Verlag, Berlin, pp. 315–334.

Ferris, S.D., Wilson, A.C., and Brown, W.M. (1981) Evolutionary tree for apes and humans based on cleavage maps of mitochondrial DNA. *Proc. Natl. Acad. Sci. USA* **78**, 2432–2436.

Fitch, W.M. (1971) Toward defining the course of evolution: minimum change for a specific tree topology. *Syst. Zool.* **20**, 406–416.

Fitch, W.M. (1977) On the problem of generating the most parsimonious tree. *Amer. Natur.* **111**, 223–257.

Fitch, W.M., and Margoliash, E. (1967) Construction of phylogenetic trees. *Science* **155**, 279–284.

Fukuzawa, H., Kohchi, T., Sano, T., Shirai, H., Umesono, K., Inokuchi, H., Ozeki, H., and Ohyama, K. (1988) Structure and organization of *Marchantia polymorpha* chloroplast genome. III. Gene organization of the large single copy region from *rbc*L to *trn*I(CAU). *J. Mol. Biol.* **203**, 333–351.

Gerbi, S.A. (1985) Evolution of ribosomal DNA. In: *Molecular Evolutionary Genetics* (ed. R.J. MacIntyre), Plenum Press, New York, pp. 419–517.

Gray, M.W., Sankoff, D., and Cedergren, R.J. (1984) On the evolutionary descent of organisms and organelles: a global phylogeny based on a highly conserved structural core in small subunit RNA. *Nucleic Acids Res.* **12**, 5837–5852.

Hennig, W. (1966) *Phylogenetic Systematics* (transl. D. Davis and R. Zangerl), University of Illinois Press, Urbana, IL.

Jansen, R.K., and Palmer, J.D. (1988) Phylogenetic implications of chloroplast DNA restriction site variation in the Mutisieae (Asteraceae). *Amer. J. Bot.* **75**, 751–764.

Jukes, T.H., and Cantor, C.R. (1969) Evolution of protein molecules. In: *Mammalian Protein Metabolism* (ed. H.N. Munro), Academic Press, New York, pp. 21–132.

Kimura, M. (1980) A simple method for estimating evolutionary rate of base substitutions through comparative studies of nucleotide sequences. *J. Mol. Evol.* **16**, 111–120.

Kluge, A.G., and Farris, J.S. (1969) Quantitative phyletics and the evolution of anurans. *Syst. Zool.* **18**, 1–32.

Lake, J.A. (1987) A rate-independent technique for analysis of nucleic acid sequences: evolutionary parsimony. *Mol. Biol. Evol.* **4**, 167–191.

LeQuesne, W.J. (1974) The uniquely evolved character concept and its cladistic application. *Syst. Zool.* **18**, 201–205.

Mishler, B.D., Bremer, K., Humphries, C.J., and Churchill, S.P. (1988) The use of nucleic acid sequence data in phylogenetic reconstruction. *Taxon* **37**, 391–395.

Mishler, B.D., and DeLuna, E. (1991) The use of ontogenetic data in bryophyte systematics. In: *Advances in Bryology* Vol. 4 (ed. N.S. Miller), J. Cramer, New York, in press.

Muller, J. (1981) Fossil pollen records of extant angiosperms. *Bot. Rev.* **47**, 1–142.

Muller, J. (1984) Significance of fossil pollen for angiosperm history. *Ann. Missouri Bot. Gard.* **71**, 419–443.

Nei, M. (1987) *Molecular Evolutionary Genetics*. Columbia University Press, New York.

Palmer, J.D. (1985) Evolution of chloroplast and mitochondrial DNA in plants and algae. In: *Molecular Evolutionary Genetics* (ed. R.J. MacIntyre), Plenum Press, New York, pp. 131–240.

Palmer, J.D. (1991) Plastid chromosomes: structure and evolution. In: *Cell Culture and Somatic Cell Genetics in Plants*, Vol. 7, (eds. L. Bogorad and I.K. Vasil), in press.

Palmer, J.D., Jansen, R.K., Michaels, H.J., Chase, M.W., and Manhart, J.R. (1988) Chloroplast DNA variation and plant phylogeny. *Ann. Missouri Bot. Gard.* **75**, 1180–1206.

Peacock, D., and Boulter, D. (1975) Use of amino acid sequence data in phylogeny and evaluation of methods using computer simulation. *J. Mol. Biol.* **95**, 513–527.

Raven, P.H. (1988) Onagraceae as a model of plant evolution. In: *Plant Evolutionary Biology* (eds. L.D. Gottlieb and S.D. Jain), Chapman and Hall, New York, pp. 85–107.

Ritland, K., and Clegg, M.T. (1987) Evolutionary analysis of plant DNA sequences. *Amer. Natur.* **130**, S74–S100.

Sankoff, D., and Cedergren, R.J. (1983) Simultaneous comparison of three or more sequences related by a tree. In: *Time Warps, String Edits and Macromolecules: The Theory and Practice of Sequence Comparison* (eds. D. Sankoff and J.B. Kruskal), Addison-Wesley, London, pp. 253–263.

Schleifer, K.H., and Ludwig, W. (1989) Phylogenetic relationships among bacteria. In: *The Hierarchy of Life* (eds. B. Fernholm, K. Bremer, and H. Jörnvall), Elsevier, Amsterdam, pp. 103–117.

Shinozaki, K., and Sugiura, M. (1982) The nucleotide sequence of the tobacco chloroplast gene for the large subunit of ribulose-1,5-bisphosphate carboxylase/oxygenase. *Gene* **20**, 91–102.

Sibley, C.G., and Ahlquist, J.E. (1984) The phylogeny of the hominoid primates, as indicated by DNA–DNA hybridization. *J. Mol. Evol.* **20**, 2–15.

Sober, E. (1988) *Reconstructing the Past: Parsimony, Evolution, and Inference*, MIT Press, Cambridge, MA.

Sokal, R.R., and Michener, C.D. (1958) A statistical method for evaluating systematic relationships. *University of Kansas Sci. Bull.* **28**, 1409–1438.

Swofford, D.L., and Olsen, G.J. (1990) Phylogeny reconstruction. In: *Molecular Systematics* (eds. D.M. Hillis and C. Moritz), Sinauer Associates, Sunderland, MA., pp. 411–501.

Sytsma, K.J., and Gottlieb, L.D. (1986a) Chloroplast DNA evidence for the origin of the genus *Heterogaura* from a species of *Clarkia* (Onagraceae). *Proc. Natl. Acad. Sci. USA* **83**, 5554–5557.

Sytsma, K.J., and Gottlieb, L.D. (1986b) Chloroplast DNA evolution and phylogenetic relationships in *Clarkia* sect. *Peripetasma* (Onagraceae). *Evolution* **40**, 1248–1261.

Sytsma, K.J., and Smith, J.F. (1988) DNA and morphology: comparisons in the Onagraceae. *Ann. Missouri Bot. Gard.* **75**, 1217–1237.

Sytsma, K.J., Smith, J.F., and Gottlieb, L.D. (1990) Phylogenetics in *Clarkia* (Onagraceae): restriction site mapping of chloroplast DNA. *Syst. Bot.* **15**, 280–295.

Taylor, T.N. (1988) The origin of land plants: some answers, more questions. *Taxon* **37**, 805–833.

Templeton, A.R. (1983a) Phylogenetic inference from restriction endonuclease cleavage maps with particular reference to the evolution of humans and the apes. *Evolution* **37**, 221–244.

Templeton, A.R. (1983b) Convergent evolution and nonparametric inferences from restriction data and DNA sequences. In: *Statistical Analysis of DNA Sequence Data* (ed. B.W. Weir), Marcel Dekker, New York, pp. 151–179.

Walker, J.W., and Walker, A.G. (1984) Ultrastructure of lower Cretaceous angiosperm pollen and the origin and early evolution of flowering plants. *Ann. Missouri Bot. Gard.* **71**, 464–521.

Williams, P.L., and Fitch, W.M. (1989) Finding the minimal changes in a given tree. In: *The Hierarchy of Life* (eds. B. Fernholm, K. Bremer, and H. Jörnvall), Elsevier, Amsterdam, pp. 453–470.

Wilson, A.C., Zimmer, E.A., Prager, E.M., and Kocher, T.D. (1989) Restriction mapping in the molecular systematics of mammals: a retrospective salute. In: *The Hierarchy of Life* (eds. B. Fernholm, K. Bremer, and H. Jörnvall), Elsevier, Amsterdam, pp. 407–419.

Yang, R.C.A., Dove, M., Seligy, V.L., Lemieux, C., Turmel, M., and Narang, S.A. (1986) Complete nucleotide sequence and mRNA-mapping of the large subunit of ribulose-1,5-bisphosphate carboxylase/oxygenase (Rubisco) from *Chlamydomonas moewusii. Gene* **50**, 259–270.

Zimmer, E.A., Hamby, R.K., Arnold, M.L., Leblanc, D.A., and Theriot, E.C. (1989) Ribosomal RNA phylogenies and flowering plant evolution. In: *The Hierarchy of Life* (eds. B. Fernholm, K. Bremer, and H. Jörnvall), Elsevier, Amsterdam, pp. 205–214.

Zurawski, G., and Clegg, M.T. (1987) Evolution of higher plant chloroplast DNA-encoded genes: implications for structure-function and phylogenetic studies. *Ann. Rev. Plant Physiol.* **38**, 391–418.

Appendix I. Sequences of *rbc*L from *Lilium superbum* (*M. W. Chase 112*) and *Colchicum* sequences are part of an ongoing survey focusing on the lilioid monocots. Periods in the *Lilium*

Colchicum	atg	tca	cca	caa	aca	gaa	act	aaa	gca	agt	gct	gga	ttc	aaa	gcc	ggt
Lilium	...	...	...	...	...	...	...	...	...	...	...	...	...	...	..t	...
Colchicum	act	gat	atc	ttg	gca	gca	ttc	cga	gta	act	ccg	caa	ccg	ggg	gtt	ccg
Lilium	...	...	..t	...	...	...	...	...	...	...	..t	...	..c	..a	...	..a
Colchicum	aca	act	gtg	tgg	act	gat	gga	ctt	acc	agt	ctt	gat	cgt	tat	aaa	gga
Lilium	...	...	...	...	...	...	...	...	...	...	...	...	...	..c	...	..g
Colchicum	tat	gta	gct	tat	cct	tta	gac	ctt	ttc	gaa	gaa	ggt	tct	gtt	act	aac
Lilium	...	...	...	...	...	...	...	...	..t	...	...	...	...	...	...	...
Colchicum	cta	cgt	cta	gag	gat	cta	cga	att	ccc	cct	gct	tat	tcc	aaa	act	ttc
Lilium	...	...	..g	...	...	..g	...	...	..t	a..	t.	...	...	...	...	..t
Colchicum	ggt	cgt	ccc	cta	ttg	gga	tgt	acc	att	aaa	cca	aaa	ttg	gga	tta	tcc
Lilium	...	...	...	...	..a	...	...	...	...	...	...	...	...	...	...	..t
Colchicum	ttt	acc	aag	gat	gat	gaa	aac	gtg	aat	tcc	caa	cca	ttt	atg	cgt	tgg
Lilium	...	...	...	...	...	...	...	...	..c	..a	...	...	...	...	...	...
Colchicum	acg	ggc	gaa	atc	aaa	gga	cat	tac	ttg	aat	gcg	act	gcg	ggt	acg	tgt
Lilium	...	..t	...	...	...	...	...	...	...	...	..t	...	...	...	..a	...
Colchicum	gta	atg	cat	gac	tat	tta	acc	ggg	ggc	ttc	acc	gca	aat	act	agc	ttg
Lilium	...	...	...	...	..c	...	..g	...	..a	...	..t	...	...	...	...	...
Colchicum	cat	gca	gtt	att	gat	aga	cag	aaa	aat	cat	ggt	atg	cat	ttt	cgt	gta
Lilium	...	...	...	...	...	...	...	...	...	...	...	...	...	...	..c	...
Colchicum	gta	gta	ggt	aaa	ctg	gaa	ggg	gaa	cgt	gag	atg	act	tta	ggt	ttt	gtc
Lilium	...	...	...	...	...	...	..a	...	..c	...	...	...	...	...	...	..t
Colchicum	ttc	acc	caa	gat	tgg	gtc	tct	atg	cca	ggt	gtt	ttg	ccc	gtg	gct	tcc
Lilium	...	..t	...	...	...	...	...	...	...	...	...	c..	...	...	...	..g
Colchicum	gat	tct	gtg	cta	cag	ttc	ggc	ggg	gga	act	tta	gga	cat	cct	tgg	ggg
Lilium	...	...	..a	...	...	...	...	..a	...	...	...	..g	..c	...	...	..a
Colchicum	gct	cgt	aat	gag	gga	cgt	gat	ctt	gct	agt	gaa	ggt	aat	gaa	att	atc
Lilium	...	...	...	..a	..g	...	...	...	...	c..	..g	...	...	...	...	...
Colchicum	aag	gag	atc	aaa	ttc	gag	ttc	gaa	ccg	gta	gat	aag	cta	gat	gtc	aaa
Lilium	...	..c	..a	...	...	.t.	..g	c..	..a	...	...	...	...	...	aca	g.g

speciosum (*M. W. Chase 109*; vouchers deposited at UNC). These previously unpublished sequence indicate agreement with the *Colchicum* sequence.

```
gtt  aaa  gat  tac  aaa  ttg  act  tat  tat  act  cct  gaa  tac  caa  acc  aag  gat      [99]
...  ...  ...  ...  ...  ...  ...  ...  ...  ...  ...  ..c  ..t  g..  ...  ..a  ...

cct  gaa  gaa  gca  ggg  gct  gca  gta  gct  gcc  gag  tct  tct  act  ggt  aca  tgg      [198]
..c  ...  ..g  ...  ...  ..a  ..g  ...  ..c  ...  ..a  ...  ..c  ...  ...  ...  ...

cga  tgc  tac  ggc  atc  gag  aaa  gtt  att  ggg  gaa  gat  aat  caa  tat  att  gct      [297]
...  ...  ...  ca.  ...  ...  .gc  ...  g..  ...  ..g  ..a  ...  ...  ...  ..c  ...

atg  ttt  act  tcc  att  gtg  ggt  aat  gta  ttc  ggt  ttt  aaa  gcc  tta  cga  gct      [396]
...  ..c  ...  ...  ...  ...  ...  ...  ...  ..t  ...  ..c  ...  ...  c..  ...  ...

caa  ggc  ccg  ccc  cac  ggc  atc  caa  gtt  gaa  cga  gat  aaa  ⁺tg  aac  aag  tat      [495]
...  ...  ...  ..t  ..t  ...  ...  ...  ...  ...  a.g  ...  ...  ...  ...  ...  ...

gca  aag  aac  tac  ggt  agg  gct  gtt  tat  gaa  tgt  cta  cgt  ggt  gga  ctt  gat      [594]
...  ...  ...  ..t  ...  ..a  ...  ...  ...  ...  ...  ..g  ..c  ...  ...  ...  ...

aga  gat  cgt  ttc  tta  ttt  tgt  gcc  gaa  gca  att  tat  aaa  gcg  caa  gcc  gaa      [693]
...  ..c  ...  ...  ...  ...  ...  ...  ...  ...  ...  a..  ...  ...  ...  ...  ...

gaa  gaa  atg  atg  aaa  aga  gcc  gta  ttt  gct  aga  gaa  ttg  gga  gtt  cct  atc      [792]
...  ...  ...  ...  ...  ..g  ...  ...  ...  ...  ...  ...  ...  ...  ...  ...  ...

gct  cat  tat  tgc  cgc  gac  aac  ggc  cta  ctt  ctt  cac  att  cac  cgc  gca  atg      [891]
t..  ...  ...  ...  ..a  ...  ...  ...  ...  ...  ...  ...  ...  ..t  ..t  ...  ...

cta  gcg  aaa  gca  tta  cgt  atg  tct  ggt  gga  gat  cat  att  cac  gct  ggt  aca      [990]
...  ..t  ...  ...  ...  ...  ...  ..c  ...  ..c  ...  ...  ...  ...  ...  ...  ...

gat  tta  tta  cgt  gat  gat  ttt  att  gaa  aaa  gac  cga  agt  cgt  ggt  att  tct      [1089]
...  ...  c.g  ..c  ...  ...  ...  g..  ...  ...  ...  ...  ...  ...  ...  ...  .t.

ggg  ggt  att  cat  gtt  tgg  cat  atg  cct  gcc  ttg  act  gaa  atc  ttc  ggg  gat      [1188]
...  ...  ...  ...  ...  ...  ...  ...  ...  ..t  c.a  ..c  ...  ...  ..t  ...  ...

aat  gcg  ccg  ggt  gca  gta  gct  aat  cga  gtg  gct  tta  gaa  gca  tgt  gta  caa      [1287]
...  ..a  ..a  ...  ...  ...  ...  ...  ...  ...  ...  ...  ..g  ...  ...  ...  ...

cgt  gaa  gct  tgc  aaa  tgg  agt  cca  gag  cta  gct  gcc  gct  tgt  gaa  gta  tgg      [1386]
...  ...  ...  ...  ...  ...  ...  ..t  ..a  ...  ...  ..t  ...  ...  ...  ...  ...

aaa  taa  [1440]
..g  a..  taa  [1443]
```

17

Polymorphism, Hybridization, and Variable Evolutionary Rate in Molecular Phylogenies

Kermit Ritland and *James E. Eckenwalder*

The ideal data for molecular systematics are DNA sequences that are homogeneous along their length for evolutionary rate and represent a taxon by a single sample. Ideally, the taxa studied will have undergone complete genetic isolation following divergence from ancestral taxa. These ideal features often represent assumptions in the reconstruction of phylogeny. Evolutionary processes that cause their violation can either be regarded as a source of statistical bias of potential significance, as another type of evolutionary change that should be incorporated into the procedure of phylogenetic inference, or as features of molecular evolution worthy of study. In this chapter, we consider how polymorphism, hybridization, and variable evolutionary rate can influence our inferences about phylogeny. We also turn the problem around and occasionally consider how phylogenetic analyses can characterize these as processes of evolution.

We believe that the characterization of these factors involves a balance between the simplicities of modeling and the complexities of reality and that, quite importantly, the adequacy of this compromise can be judged by the statistical testing of such models. In the phylogenetic analysis of molecular data, systematists have predominantly used the cladistic methods of parsimony or compatibility, which build phylogenetic trees minimizing the number of character changes in a data set. These approaches are not model-based in the formal sense and are not explicit in their assumptions; hence they can be difficult to place into a statistical framework. Consequently, the testing of assumptions, and the characterization of processes that lead to violations of assumptions, have been uncommon in molecular systematics. This chapter is written in the spirit of encouraging workers to discuss further the properties of their methods and to explore new avenues of data analysis.

As a quantity about which we develop our models and center our discussion,

---

Acknowledgments: We thank the reviewers for their comments. This research is supported by grants from the Natural Sciences and Engineering Research Council of Canada.

we use the "evolutionary distance" between DNA sequences. This distance is defined as the number of substitutions separating two or more sequences, or in cladistic terminology, the number of character-state changes. It is commonly depicted as the branch length in a cladogram. This focus upon "patristic" distance may seem irrelevant to cladists, who consider branch order in a cladogram paramount, but it is relevant to questions both about evolutionary taxonomy and about relative rates of evolution.

Because we consider changes at only a single representative nucleotide site, we cannot consider other genetic changes such as deletions, inversions, and changes of secondary structure. At a single nucleotide site, there are also many processes that influence the evolutionary distance that are beyond the scope of our consideration, such as population size, generation time, and selection intensity. Herein, we consider only the processes of time and mutation rate, and subsequently consider how variation of rate, polymorphism, and hybridization alter our inferences about these quantities. Although molecular phylogenies may be constructed from many types of data, including isozymes, amino-acid sequences, RFLPs, and RNA and DNA sequences, we also restrict our discussion to the last category, DNA sequences. We acknowledge that restriction site approaches still predominate in plant systematics, and probably will into the near future. However, our treatment of sequence-based analyses more clearly illustrates such problems as variation of evolutionary rate, whereas restriction-site-based analyses either ignore or cannot deal with these problems. DNA data also provide superior information, and, with the advent of the PCR method, may eventually become the predominant class of data.

We also restrict our discussion to simple phylogenies, usually involving just three sequences, as they are sufficient to illustrate the problem. Four sequences introduce the problem of uncertainty of node placement (as opposed to branch length), which is an intriguing and largely uninvestigated class of statistical problems. The small parameter space of three-sequence phylogenies allows us to specify exact models, and in some cases, to derive statistics to characterize those evolutionary processes that violate the assumptions of the phylogenetic model.

Evolutionary Distance

Fundamental to the questions we consider is the idea of evolutionary distance between homologous DNA sequences. The quantity of distance is important because of the three categories of phylogenetic inference methods given in 1988 by Felsenstein (parsimony and compatibility, distance matrix, and likelihood), two (distance matrix and likelihood) use a distance statistic. In the distance matrix method, distance is calculated between contemporary sequences. In the likelihood method, distance is calculated between contemporary sequences and their nearest node within the tree topology, as well as between all nodes. Thus in this chapter, we are not specifically concerned with the distance-matrix method to construct trees, but are more generally concerned with methods that incorporate branch distances in the phylogeny-building algorithm.

Some workers have pointed out problems with distance-based analyses. For example, in response to Felsenstein (1984), Farris (1985) gives an example (his Eq. 2) involving three amino acid sequences, wherein the number of differences between the triplet of amino acids at a site is allowed to have an arbitrary distribution. Therefore, sequences may be derived from a process that includes a variable evolutionary rate, which is one of the problems we discuss in this chapter. Some objections about distance-based analyses, and generally "model-based" analyses, seem to be due to a misunderstanding about the meaning and use of statistical expectation and about how much statistical models can tell us regarding changes at a single site versus parsimony inferences (e.g., no nucleotide site ever has 0.5 substitutions, cf. Farris, 1985, p. 65, just as no family ever has 2.1 children). However, we will not debate this issue in this chapter.

From the standpoint of either phenetic similarity or genetic relationship (the identity-by-descent of nucleotide sites), the measure of choice for evolutionary distance has been the average number of nucleotide substitutions at a site (Jukes and Cantor, 1969; Li et al., 1985; Felsenstein, 1988). This number does not equal the fraction of sites that differ between two sequences, unless sequences are closely related, with a sequence divergence of 10% or less, so that repeated changes at the same site are unlikely.

When sequences diverge such that multiple substitutions at the same site are frequent, models for the process of base substitution are needed to estimate the actual number of substitutions that have taken place during the history of the site. Just as many methods have been proposed to measure genetic distance from gene frequencies, several models for finding evolutionary distance between DNA sequences have been proposed over the past two decades. Most models consider the nucleotide site as the unit of evolution (as opposed to the codon), and, of these, most assume that new, mutant nucleotides are drawn from a pool of nucleotides of fixed frequency (e.g., 0.25; c.f. Jukes and Cantor, 1969; Holmquist and Pearl, 1980; Kimura, 1981; Kaplan and Risko, 1982). A consequence of this assumption, not always stated, is that the nucleotide frequencies of sequences are likewise constrained (usually to be equally frequent, to ensure stationarity of the Markovian substitution process). These assumptions allow rather elaborate, but solvable, models allowing up to six different mutation rates.

For estimating evolutionary distance, most workers use either the Jukes-Cantor estimator, Tajima-Nei's modification of the Jukes-Cantor estimator, or Kimura's 3ST estimator. The Jukes-Cantor estimator is

$$\hat{D}_{12} = -(\tfrac{3}{4}) \ln [1 - 4S_{12}/3]$$

where S_{12} is the fraction of sites sharing the same bases between two sequences S_1 and S_2, and the hat denotes the estimated quantity; this assumes a single mutation rate and bases equally frequent. Tajima and Nei's (1984) modification of Jukes and Cantor's estimator is

$$\hat{D}_{12} = -b \ln [1 - S_{12}/b],$$

for

$$b = 1 - \pi_t^2 + \pi_c^2 + \pi_a^2 + \pi_g^2$$

and π the base frequencies. Kimura's (1981) 3ST estimator is

$$\hat{D}_{12} = -(\tfrac{1}{4}) \ln [(1 - 2P - 2Q)(1 - 2P - 2R)(1 - 2Q - 2R)]$$

where P, Q, and R are the proportions of transition, transversion I, and transversion II events, respectively; this allows three mutation rates but assumes equally frequent nucleotide frequencies.

By contrast, a second class of models (cf. Felsenstein, 1981; Hasegawa et al., 1985; Ritland, unpublished), places fewer constraints on base composition by assuming that new, mutant nucleotides are drawn from a pool with frequencies equal to those in the contemporary sequences. Although analytic solutions for these distance measures can be more complex, these models are more appealing, both because they allow arbitrary base composition and because their structure is analogous to population genetic models for measuring relatedness between individuals within populations.

To see this latter connection, consider Felsenstein's (1981) model. At a site, let the probability that base $i(i = $ T, C, A, and G) changes to base $j(j = $ T, C, A, or G) in a small interval of time be $u\pi_j dt$, where u is a constant mutation rate and $j = t$, c, a, or g, where t, c, a, and g correspond to bases T, C, A, and G. These π are assumed to be simply the frequencies of T, C, A, and G in the contemporary DNA sequences. By solving a differential equation, after an interval of time t the frequencies of the 16 possible combinations of bases between sequences S_1 and S_2, arranged in a matrix denoted $\mathbf{P}_{12}$, is

$$\mathbf{P}_{12} = e^{-2ut}\begin{bmatrix} \pi_t & 0 & 0 & 0 \\ 0 & \pi_c & 0 & 0 \\ 0 & 0 & \pi_a & 0 \\ 0 & 0 & 0 & \pi_g \end{bmatrix} + (1 - e^{-2ut})\begin{bmatrix} \pi_t^2 & \pi_c\pi_t & \pi_a\pi_t & \pi_g\pi_t \\ \pi_t\pi_c & \pi_c^2 & \pi_a\pi_c & \pi_g\pi_c \\ \pi_t\pi_a & \pi_c\pi_a & \pi_a^2 & \pi_g\pi_a \\ \pi_t\pi_g & \pi_c\pi_g & \pi_a\pi_g & \pi_g^2 \end{bmatrix} \quad (17.1)$$

where the columns of $\mathbf{P}_{12}$ index bases T, C, A, and G of the first sequence, and the rows index these bases for the second sequence. This equation graphically shows how a fraction e^{-2ut} of sites is identical-by-descent, and a remaining fraction $(1 - e^{-2ut})$ of sites is not-identical-by-descent, consisting of a random association of bases. Note that if two sequences share the same base at a site, these bases may be the same because of parallel mutations and not shared ancestry. The fraction e^{-2ut} is the "relatedness" (the fraction of sites identical-by-descent), but the *number* $-\ln[e^{-2ut}] = 2ut$ is the evolutionary distance (the average number of substitutions separating S_1 and S_2).

A formula to estimate evolutionary distance under this model can be derived by equating the observed similarity of two sequences S_1 and S_2, denoted S_{12}, to the sum of the diagonal elements of $\mathbf{P}_{12}$, and solving for ut. This gives the estimator for evolutionary distance, denoted D_{12}, as

$$\hat{D}_{12} = -\ln\left[\frac{S_{12} - H_2}{1 - H_2}\right] \tag{17.2}$$

where

$$H_2 = \pi_t^2 + \pi_c^2 + \pi_a^2 + \pi_g^2.$$

H_2 is analogous to the expected homozygosity of gene frequency.

One can argue that specific models of base change, such as discussed above, are not really needed, because for most sequence comparisons any statistical bias introduced by various assumptions about homogeneity of substitutional rates is often small compared to the statistical error of estimates, at least for sequences that are either short in length (ca. 1 kb) or not very divergent. Regarding divergence, simulation studies that evaluate the bias of distance measures (Takahata and Kimura, 1981; Tajima and Nei, 1985; Li et al., 1985) generally find that for evolutionary distances less than about 0.5, any model will give suitable estimates of distance, but for distances greater than 1.0, more elaborate models may be required, but used with caution and tested for their fit to the data. To test the fit, the estimated parameter values (e.g., evolutionary distance, nucleotide frequencies) are substituted into the model (e.g., Eq. 17.1), and this matrix of expectations is compared to the observed matrix of base substitutions with goodness-of-fit statistics such as the chi-square test or the likelihood ratio test (Ritland and Clegg, 1987).

Others will argue that even more elaborate models do not incorporate the true complexity of the molecular evolutionary process. How can the rarer events of sequence deletion, duplication, or rearrangement enter into the above measures of evolutionary distance? What about the pool of transfer RNAs that place constraints on acceptable mutations? Most recent attention on this issue has concentrated upon models that distinguish synonymous from nonsynonymous (amino-acid changing and hence selectively constrained) substitutions, and also distinguish different types of nonsynonymity (Miyata and Yasunaga, 1980; Li et al., 1985; Lewontin, 1989). However, models of base change with many parameters are not necessarily better for two reasons: (1) often, compensating assumptions must be made to ensure mathematical tractability, and (2) the information about existing parameters can be diluted, resulting in larger sampling errors, and outweighing any reduction in bias provided by the more complex model. For example, a complete codon-level model would require a 61 × 61 matrix analogous to Eq. (17.1) but with many more parameters. Thus we can take comfort in the likelihood that most other models or theories would make indistinguishably different predictions, when we remember that the quantities we estimate have a statistical uncertainty.

Polymorphism

The possession of two or more variants of a given sequence within a single species (or, in the extreme, within a single individual) creates several difficulties for phylogenetic reconstruction. However, because many recent studies have assayed just one individual or population for each taxon, the extent of sequence polymorphism is largely unknown, at least in plants. In the absence of natural selection, we expect within-population polymorphism for sequences to be greater for those sequences that show higher between-population evolutionary distances (Hudson et al., 1987). Thus, polymorphism is likely to be a problem with rapidly evolving, noncoding DNA and less of a problem with slowly evolving coding DNA, as well as with chloroplast DNA. The probability of discovering sequence polymorphism when two or more individuals (or populations) of a species are examined would be proportional to the Shannon and Weaver (1949) information measure (H) among sequences in the population. Polymorphism is more likely to be discovered if several variants are equally common than if one sequence is predominant and the other is rare. However, one cannot ever rule out polymorphism, because no matter how many individuals are discovered with identical sequences, there may be a variant just around the corner.

The Consequences of Ignoring Polymorphism

Because the extent of sequence polymorphism in natural plant populations is largely unknown, it follows that the consequences of ignoring polymorphism when estimating evolutionary distances are largely speculative at this point. Regardless, two examples of a simple, three-taxon problem, with some simplifying assumptions, will represent instructive examples from the continuum of possible consequences. Because each pair of species has a single time depth to divergence, for the purposes of the examples we assume that there is some kind of taxon-based evolutionary distance that is estimated by a particular sequence distance or combination of sequence distances (cf. Avise, 1989).

Example 1. The three taxa T_1, T_2, and T_3 share the same two variants S_1 and S_2 in the same (equal) frequencies (Fig. 17.1A). The true evolutionary distances among the three taxa are thus all equal to zero. The estimated evolutionary distances will also equal zero 25% of the time, when the same single variant is

(a)   (b)   (c)

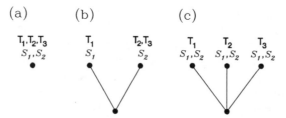

Figure 17.1. A. The "true" cladogram (a trichotomy with zero branch length) for three polymorphic taxa T_1, T_2, and T_3, each with the same pair of variant sequences, S_1 and S_2. B. A false topology (no trichotomy) is inferred when only one variant is assumed, but different variants are encountered in different taxa. C. Biased branch lengths are inferred even when all sequences are found in all taxa.

discovered in each of the three species. In 75% of investigations, two taxa will be found to have the same variant, distinct from that in the third taxon. The evolutionary distance from the third taxon to the two that share the same sequence is estimated as being greater than zero. The overall impact of the polymorphism is thus an overestimation of the true evolutionary distance based on the sequence of interest and an essentially coincidental resolution of the apparent trichotomy (Figs. 17.1B and 17.1C) among the three taxa when both sequences are known. Increasing the frequency of one of the variants will decrease the probability of overestimation, but not its magnitude. Conversely, divergently increasing the frequencies of two of the variants will increase the probability of overestimating the true evolutionary distance.

Example 2. Each of the three taxa has two distinct variants in equal frequencies (Fig. 17.2A), descended from the same original pair of variants (primes denote homologous sequences). We can define the true evolutionary distance for each pair of species as the average of the two distances based on the most closely homologous variants. If the substitution rates in these variants are truly clocklike, then the two distances that contribute to each interspecies evolutionary distance would be the same. However, even if molecular evolution is clocklike overall, it is unlikely that it will be so in detail, and so the constituent distance pairs are likely to be different. In any event, the "true" evolutionary distance derived in this way will be the smallest distance that can be constructed using all four variants present in each pair of species. When only a single variant is discovered in each species, the estimated evolutionary distances will always differ from the true values, except in the unlikely circumstance of a perfect molecular clock. As with the first example, in 25% of investigations, the three sequences discovered will all represent one of the two sets of most closely homologous variants. In this case, the correct cladistic structure will be recovered, but the calculated distances will vary unpredictably from the "true" values. The variance in estimating the

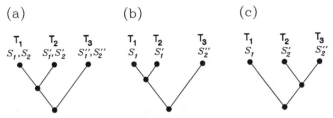

Figure 17.2. A. The "true" cladogram for three polymorphic taxa, each with two variant sequences descended from the same original pair of sequences. B. Biased branch lengths caused by the assay of only one sequence of each taxon. C. Incorrect topological inference also caused by the assay of only one sequence per taxon.

distance arises, as usual, from a violation of the assumption of homogeneity of substitution rates, in this case not through time or in different parts of the sequence, but between co-occurring copies of the same sequence. This variance is the most commonly occurring consequence of polymorphism. In another 25% of cases (Fig. 17.2B), the true cladistic structure will also be recovered, because most closely homologous variants are found in T_1 and T_2, whereas the variant found in T_3 comes from the other set. In these cases D_{12} varies unpredictably, while D_{13} and D_{23} are overestimated. Finally, 50% of the time, even the true cladistic structure will not be recovered (Fig. 17.2C). In each case of this type, as in the preceding quarter of cases, one distance will vary unpredictably from the true distance, while the other two distances will be overestimated.

Incorporating Polymorphism into the Phylogeny

In both of the above examples, the only systematic bias arising from undiscovered polymorphism is to overestimate two of the three molecular evolutionary distances in the three-taxon problem. This bias holds whether polymorphism occurs within individuals (= heterozygosity or gene duplication) or only among individuals. Only in the most extreme type of polymorphism, when the variants of each of the species are not close homologs (as in Fig. 17.3, where they arise by differential loss from an even more polymorphic ancestor), does systematic overestimation of distances disappear completely, and the variance generated by among-copy differences in substitution rates becomes the sole source of the error due to undetected polymorphism.

Even when an investigation discovers polymorphism, there are still problems and questions for phylogenetic reconstruction. The consequences will vary considerably depending upon the distributions of the perceived and actual polymorphisms within and among individuals and populations. Generally speaking, the degree of polymorphism present will always be underestimated, with fewer species apparently than actually polymorphic, and each apparently polymorphic species with fewer apparent variants than it actually has. Under these conditions, evolutionary distances will again be overestimated, but other effects will vary. Once again, it is instructive to set up a number of examples to show the range of other effects on phylogenetic analyses.

Polymorphism within individuals will usually arise as a result of sexual reproduction. At the diploid level, it is simply an example of heterozygosity (or less commonly a result of some form of gene duplication) and implies the polymorphism of the population containing the sampled individual. The variant sequences are likely to express variance in evolutionary rate between the copies and hence to contribute to increased variance in the estimates of evolutionary distances based on them, thus broadening confidence intervals in resultant phylogenies. In polyploid individuals, the variant sequences will usually have been imported via hybridization after divergence of parental sequences. Under these conditions, it does not seem appropriate to combine estimates of evolutionary distance based on the co-occurring sequences, but rather to calculate separate distances for each sequence, which can then be used to postulate reticulations in the resulting cladograms. Further implications of hybridization will be considered under that heading.

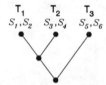

T_1 T_2 T_3
S_1,S_2 S_3,S_4 S_5,S_6

Figure 17.3. A three-taxa cladogram, where variants of each taxon have no particular relationship to one another.

The recognition of polymorphism among individuals within populations parallels isozyme variation, so a frequency-based evolutionary distance may then be appropriate for between-population comparisons, when different populations share the same known variants, so that the sequence-based distance would be zero for all populations. The situation is more complicated when different monomorphic populations of a species have distinct variants, so that polymorphism is primarily among populations. This might well suggest that a reassessment of the conspecificity of the population is in order. They could represent cryptic sibling species, but, considering the probability and consequences of undetected polymorphism discussed above, such a conclusion should be entertained with due caution. The situation would merit a specific investigation, involving more intensive molecular sampling coupled with biosystematic approaches. The errors in distance estimates, from either falsely treating a polymorphic species as several monomorphic species or of treating a series of cryptic species as one polymorphic species, can each alter other relationships in a reconstructed phylogeny.

The Phylogeny of Sequences versus the Phylogeny of Taxa

One final approach to known polymorphism is to reconstruct a phylogeny of all the individual variant sequences and only later map taxa to that phylogeny (Pamilo and Nei, 1988). This approach should successfully distinguish three sources of polymorphism.

1. If taxa map directly to monophyletic groups of sequence variants, then those variants will have arisen by diversification within the taxa in question.

2. If two or more taxa map in parallel to two or more monophyletic sets of variants, then they will have inherited the polymorphism from a common ancestor.

3. If a single taxon maps to two unrelated (that is nonmonophyletic) variants, there are three main explanations. It may be a pseudo-taxon with multiple origins, it may be the sole member of a clade whose other survivors are secondarily monomorphic, or it may have arisen by hybridization. Independent lines of evidence should help select among these possibilities.

Hybridization and Introgression

Taxa are normally assumed to have evolved from a common ancestor by a process of branching followed by complete genetic isolation (Cavalli-Sforza and Edwards, 1967). A period of hybridization shortly following speciation is not thought to weaken this assumption provided the period is short relative to the periods between successive speciation events. However, in plants, species (e.g., species of *Quercus*) may hybridize long after speciation, and the formation of polyploid species often involves hybridization.

These two basic hybrid situations, diploid hybridization and alloploid speciation, have different analytical consequences, and these are not the same for nuclear as for organellar sequences. Although we briefly discuss the detection and analysis of hybridization, this is the subject of Chapter 7 by Rieseberg and Brunsfeld (this volume), and our main goal will be to explore the effects of undetected hybridization on phylogenetic reconstruction via sequence evolutionary distance. In keeping with the theme of this chapter, this represents a departure from the assumptions of the distance model, in this case the assumption of uniform divergence.

Undetected Hybridization

The uniparental inheritance of organellar DNA is the same for both diploid and polyploid hybrids. It acts like a weighting function that assigns a weight of 1.0 to the maternal sequence in most angiosperms and 0.0 to the paternal sequence. The sequence evolutionary distance of a new hybrid itself (but, of course, not its organellar genome) to the pollen parent is thus overestimated as equaling the distance between the pollen and ovulate parents, whereas the other two distances are unaltered. Overestimation of the distance to the pollen parent will remain even when the sequences representing all three taxa have changed since the time of hybridization, as might be the case with an old alloploid. Posthybridization divergence will also eliminate the zero distance between the hybrid and the ovulate parent. Although the identity of a diploid entity to the mother species might raise some question about its status, if it becomes stabilized at the diploid level by some means and divergence occurs, then, in a routine survey of the organellar genome alone, hybridization would never be suspected and the hybrid would simply be recognized as the sister-group of the ovulate parent. For alloploids, in contrast, the presumption of hybridization will always be there, whatever the relationship of the organellar genome. Thus, organellar sequences, by themselves, always distort organismal relationships by overestimating one distance in the standard three-taxon problem.

Nuclear sequences in hybrids pose different problems. The basic effect of hybridization on the nuclear genome is to introduce intranuclear polymorphism, so many of the conclusions in the section on polymorphism are applicable to hybridization. For diploid hybrids, this polymorphism is equivalent to heterozygosity, whereas for polyploids it corresponds to gene duplication.

One approach for detecting admixture (in our case, hybridization) was developed by Thompson (1975) and elaborated upon by Lathrop (1982). Their method involved fitting a model of evolutionary divergence and admixture to the data, and estimating the divergence times and admixture. They noted that in many cases, phylogeny and admixture cannot be separated. In general, any pattern of genetic relationships depicted by an evolutionary tree can be mimicked by an appropriately chosen stationary migration process (Felsenstein, 1982).

Thus, hybridization is most likely to remain undetected at the diploid level, although the equivalent hybridization between two polyploids at the same level may also escape discovery. Hybridization will remain hidden when the resulting polymorphism is undiscovered, because only a single sequence is assigned to the hybrid. Analytically, this is equivalent to the uniparental inheritance of organellar sequences with the exception that, instead of favoring the maternal or paternal genome, each parental genome has a 50% chance of contributing the detected sequence.

Irrespective of which sequence is detected, the analytical consequence is still to overestimate one of the three distances. If we follow the suggestion in the section on "pure" polymorphism that the "true" sequence distances derived from polymorphic taxa should be averages of the distances based on each of the variants for each taxon, then the distance to the other parent would be underestimated. There would be no bias in the overestimation versus the underestimation and so the net effect would be zero. Instead, ignorance of the hybrid-based polymorphism would act as a switch that either overestimated or underestimated the distances to the maternal and paternal genomes. Of course, if the parents are themselves polymorphic, then the particular variants found in the hybrid appear with a probability equal to their frequency in the parental species. In these circumstances, undetected hybridization (and polymorphism) still lead to either biased overestimation or compensating over- and underestimation, depending on the method of calculating evolutionary distance.

Detected Hybridization

Since polyploids will always be suspected of being of hybrid origin, their polymorphism is likely to be detected and can be used, in conjunction with organellar genomes, to help analyze their origins. The information derived can include identifying the parent species, which one served as the ovulate parent, and whether the polyploid arose more than once independently (e.g., Soltis and Soltis, 1989). While these topics are covered elsewhere in this volume, we would like to discuss three interesting issues concerning polyploids. These depend on detecting polymorphism in the polyploid and recognizing the specific homologies of its variants to those of the other two species (its parents or their descendants) in the problem. This may be difficult if the time since hybridization is much longer than the time from separation of the parents to the hybridization. This will be a problem primarily with paleopolyploids, whose diploid relatives are probably extinct and which may themselves be diploidized and monomorphic for the sequence of interest.

Presupposing that our data meet the conditions outlined, it is possible to distinguish alloploidy from autoploidy in certain cases. Let us suppose that we have two diploid taxa T_1 and T_3, monomorphic for distinct sequences, and a tetraploid, T_2, with one sequence related to that in T_1 and a second related to that in T_3. If T_2 is an autoploid, it implies that the common ancestor of T_1 and T_3 was itself polymorphic for the sequence. Let us assume that T_2 is cladistically closer to T_1 (Fig. 17.4A). On the other hand, if T_2 is an alloploid derived from hybridization between ancestors of T_1 and T_3, then the most parsimonious interpretation is that their common ancestor was monomorphic for a sequence ancestral to both sequences (Fig. 17.4B). If the distances from T_2 to T_1 and T_3 are each based on its most nearly homologous sequence, and, if sequence divergence has been approximately clocklike, then $D_{12} < (D_{13} \cong D_{23})$ would imply autoploidy, whereas $D_{12} > (D_{13} \cong D_{23})$ would imply alloploidy.

Given that we know the progenitor species, we can also estimate the relative

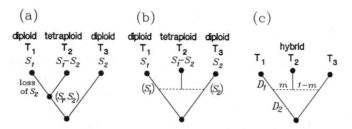

Figure 17.4. Autoploidy versus alloploidy in three-taxa cladograms. A. Autoploidy. B. Alloploidy. C. In some cases, the relative time since hybridization, $D_1/(D_1 + D_2)$, and the admixture proportion m, can be estimated from the three pairwise distances.

time since hybridization and any admixture (introgression) when the evolutionary distance is based upon either several sequences per taxon or several isozyme loci per taxon. Suppose the hybrid taxon T_2 was formed at a time in the past when nonhybrid taxa were a distance D_1 apart. Following hybridization, the nonhybrid taxa diverge an additional distance D_2, where $D_1 + D_2 = D_{12}$ (Fig. 17.4C). To add generality, suppose the hybrid taxon ultimately inherits a proportion m of genes from parent T_1 (if $m \neq \frac{1}{2}$, introgression or backcrossing occurs after the initial hybridization event). From Fig. 17.4C, under rate constancy, the three pairwise distances between the taxa are expected to be

$$D_{12} = D_1 + D_2$$
$$D_{13} = mD_1 + D_2$$
$$D_{23} = (1 - m)D_1 + D_2$$

Note that at the time of hybridization, the hybrid taxon has an evolutionary distance of $0.5D_1$ to either parental taxon. Solving for D_1, D_2, and m gives

$$D_1 = 2D_{12} - D_{13} - D_{23}$$
$$D_2 = D_{13} - D_{23} - D_{12}$$
$$m = \frac{(D_{12} - D_{23})}{(2D_{12} - D_{13} - D_{23})} \tag{17.3}$$

(Ritland, unpublished). To construct a dendrogram involving both nonhybrid and hybrid taxa, a two-step procedure can be followed. First, a dendrogram of distances among nonhybrid taxa is constructed. Then, hybrid taxa are placed in the dendrogram.

Allotetraploids are also excellent candidates for testing the accuracy of a molecular clock (cf. following sections), as long as the parental lineages are also still available for testing. Since the sequences from the parent species are brought into association in the hybrid at the same time (Fig. 17.4B), under the molecular clock hypothesis, $D_{12} = D_{13}$. As long as the sequences in T_1 and T_3 are direct descendants of the sequences in the species that gave rise to T_2, any deviation from equality implies nonhomogeneous substitution rates for the sequence in question. Once one or more sets of appropriate test organisms were discovered, it would be possible to test many different genes or other sequences using them.

Finally, if hybridization is suspected, either because of polyploidy, or for other reasons, the best approach to its study is to investigate and compare both organellar and nuclear sequences. These comparisons can be used to reveal a great deal about the nature of the hybridization. Despite this analytical tractability when hybridity is known or suspected, unsuspected hybridization can easily remain undetected if the attending polymorphism is overlooked. Even if polymorphism is discovered, it is not always easy to separate hybridization from other possible causes.

Variation of Evolutionary Rate

The fundamental feature of evolution is change, and yet inferences about phylogeny often assume that substitution or mutation rates remain constant throughout a phylogeny and through branches within a phylogeny. Surely rates do vary, and the question arises as to their extent and effect upon estimates when ignored. The issue of rate constancy, or "molecular clocks," started with Zuckerkandl and Pauling (1962) and Margoliash (1963), who noted that rates of amino acid substitution in hemoglobin and cytochrome C were approximately the same among lineages. Subsequent studies utilizing the more informative DNA sequence data have indicated that rates do vary, but the extent is subject to dispute (e.g., Langley and Fitch, 1974; Kimura, 1983; Gillespie, 1986a,b; Bulmer, 1989). In this section, we discuss two types of rate variation: "through time," or within and among branches of a phylogeny, and "through space," or among sites within a sequence.

Variation of Rate Among Lineages

Phylogenetic reconstruction can be difficult when rates of evolution differ greatly among lineages and when evolutionary distances between sequences are large. In a four-sequence, unrooted tree, the problem becomes especially difficult when the central branch is short and the tips are of long and unequal lengths, the general problem being that "long branches attract," causing misinference of topology (Li et al., 1987). Certain methods, such as maximum likelihood and Lake's (1987) "evolutionary parsimony," are more appropriate when evolutionary rates differ among lineages (Felsenstein, 1978; Saitou and Imanishi, 1989). The latter method has generated recent interest because it is also free of many assumptions about the base substitution process, including constancy of rate among sites, although some assumptions are made about symmetry of base

(a)   (b)

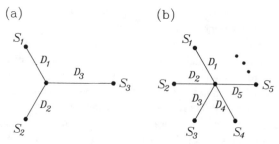

Figure 17.5. Two situations that allow testing for a molecular clock. A. A three-sequence comparison (the root lies on the branch leading to S_3, but its position is irrelevant. B. A star phylogeny.

substitutions (Felsenstein, 1988), including (contrary to Lake's assertions) that the two types of transversion rates must be equal (Jin and Nei, 1990).

The existence of rate differences among lineages thus is an important issue. As a means to detect differences of evolutionary rate between lineages, the "relative rate test" was proposed by Wilson et al. (1977) for cases when divergence times are unknown. Generally, this is a test for differences of branch length in a dendrogram. Suppose we have three sequences S_1, S_2, and S_3 related by the unrooted dendrogram of Fig. 17.5A. Under an "additive" model of divergence (cf. Olsen, 1987), which is expected when rates of evolution do not covary among branches (i.e., rates do not vary along the sequence), the expected evolutionary distances between sequences are

$$D_{12} = D_1 + D_2$$
$$D_{13} = D_1 + D_3$$
$$D_{23} = D_2 + D_3 \qquad\qquad (17.4)$$

If the sequences are clocklike in their evolution, the two shortest branches of Fig. 17.5A should be of identical length, that is, $D_1 = D_2$. Thus, the null hypothesis to be tested is $(D_1 - D_2) = 0$, or equivalently $(D_{13} - D_{23}) = 0$. If sequences are closely related, the variance of $(D_{13} - D_{23})$ is the variance of the sum of two independent Poisson variables. However, if sequences are distantly related, the estimates of D_1 and D_2 are not independent, but rather negatively correlated, and the variance of $(D_{13} - D_{23})$ is inflated to a value that is difficult to determine, but could be found via the bootstrap method (see Felsenstein, 1985).

A second relative-rate test was proposed by Kimura (1983) specifically for data on several sequences, which all diverged from a common ancestor at nearly the same time, forming a "star phylogeny" (Fig. 17.5B). Under the null hypothesis, the branches are all of equal length. The ratio of the variance of branch distances to the mean branch distance (the "variance/mean ratio") is the test statistic, with

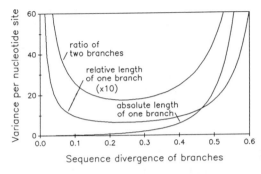

Figure 17.6. Optimal sequence divergence for estimating relative branch lengths (evolutionary distance) in a phylogeny. Their statistical variances are at minimum over a broad range of 10% to 40% nucleotide divergence between branches. The statistical variance of absolute branch length always increases with branch length.

values significantly greater than unity indicating a variable clock (Gillespie, 1986a).

In this test, the statistical covariance of distances between pairs of sequences with a lineage in common is always positive (others are zero). This results in overestimation of the variance/mean ratio, particularly when distances are great, unless the appropriate corrections are made (Gillespie, 1986b; Bulmer, 1989). An additional component of bias not considered by many workers is whether the star phylogeny is indeed a star phylogeny; if the phylogeny is not star, and a star is assumed, the variance/mean ratio will also be overestimated (Gillespie, 1989). Thus, accurate paleontologic evidence of a star phylogeny is required for this test, which may be difficult to obtain for many plant taxa.

Optimal Sequence Divergence for Phylogenetic Hypotheses

An alternative form of the relative rate test is that the ratio of branch lengths should equal 1.0 under the clock hypothesis. When considering this aspect of the phylogeny (the ratio of branch lengths, which can be considered tree "shape"), Ritland and Clegg (1990) found an "optimal" sequence divergence for testing phylogenetic hypotheses, at which estimation variances were at a minimum (Fig. 17.6). In addition to an optimum for the ratio of branch lengths, an optimum was also found for the ratio of one branch length to all branch lengths. No optimum was found for branch length by itself, as expected.

Figure 17.6 shows both optima occur at approximately 25% divergence (relative to a maximum of 75%), although it effectively spans the range from 10% to 40% divergence. These results assume all four bases are equally frequent and assume homogeneous mutation rates; such optima undoubtedly exist in more complicated situations, but probably at lower divergence. With larger divergence, there are problems of sequence alignment and the choice of an appropriate model of base change is more important. Taking these problems into consideration, the true optimal sequence divergence is probably lower than 25%, perhaps at approximately 10% divergence.

An optimal sequence divergence occurs because, as sequences diverge, the proportion of highly informative "single-hit" sites reaches a maximum, then decreases upon further sequence divergence, whereupon the less informative, "multiple-hit" sites become predominant. This idea of an optimal divergence places importance upon the rate of divergence as a criterion for choice of genes to use for molecular systematics. The optimum is generally larger than desired for parsimony analysis, as at even 10% divergence there would be significant homoplasy in a tree of sequences.

Variation of Rate Within a Lineage

If variation of rate is observed among branches or lineages of a phylogeny, it is likely that variation of rate occurred through time *within* branches. What is the effect of such variation upon the expected estimate of branch length, or evolutionary distance? To find this, suppose that during the divergence of a sequence from its ancestor, the mutation rate was a random variable U with mean u and variance v. The recursion for the fraction of sites "F" identical-by-descent between the sequence and its ancestor is $F(t + 1) = (1 - U)F(t)$. After t generations, the expected gene-identity is $F(t) = e^{E[t\ln(1-U)]} \cong e^{-t(u + v/2)}$, where E denotes expectation and $F(0) = 1$. The estimate of evolutionary distance obtained by all distance measures is $-\ln [F(t)] = t(u + v/2)$. Thus, temporal variation of rate increases the rate of divergence above that expected from the mean rate of divergence, and results in overestimation of average evolutionary change by approximately $vt/2$.

Methods to estimate temporal variation of rate are needed, but such variation is probably impossible to estimate with models with just one rate of mutation (e.g., transition and transversion rates assumed equal). Interestingly, a two-rate model seems to allow estimation of the temporal variance of the *difference* between transition and transversion rates, using just two homologous sequences (Ritland, unpublished), although the power to detect such variation of rate is probably low.

Variation of Rate among Known
Regions of a Sequence

An almost ubiquitous assumption of phylogenetic sequence analysis is that the process of base change is homogeneous among nucleotide sites, i.e., variation of substitutional rate does not occur. Clearly, many types of genes or sequences violate this assumption. For example, the large rRNA has conservative domains, which can fold into a universal core of secondary structure, as well as more rapidly evolving domains. Also, most coding sequences show differences of substitutional rate between codon positions, with second, first, and third positions showing successively larger rates of change.

Generally, if one knows *a priori* about regions with different evolutionary rates, then evolutionary distances and phylogenies should be found separately for each rate class. For example, coding sequences should be partitioned into 1st, 2nd, and 3rd codon positions. When results from different rate classes are combined in such ways, statistical tests are still possible (Li, 1989). However, the problem of how phylogenies are best combined is complex, as the best phylogeny is not necessarily the average phylogeny (implying equal weight placed upon each class of data). This is because, firstly, different types of data may provide information about qualitatively different aspects of a phylogeny: conservative regions seem best for finding branch order, whereas variable regions seem to provide more information about branch length (Ritland and Clegg, 1987). Secondly, a region of lower evolutionary rate is more informative about branch order in a cluster of closely related sequences than in a cluster of distantly related sequences in the same phylogeny (Qu et al., 1988). Thus it seems these weights would vary among regions of the gene, as well as among regions of the phylogeny! Although such a weighting scheme is complex, the likelihood approach will provide such weights automatically when the total likelihood of all data is maximized, wherein the total likelihood is the product of likelihood for each class of data, assuming the same branch-order for each class of data.

Variation of Rate among Unknown Regions of a Sequence

If variation of rate among regions of a sequence is suspected, but the boundaries of these regions are unknown, then statistical methods for detecting and characterizing variation of rate within a sequence are warranted. Furthermore, it would be useful to incorporate any rate variation into our phylogeny-building algorithms, both to remove bias from our estimate of average rate and to characterize further the evolutionary relationships among the sequences. There are at least two approaches to characterizing within-sequence variation of evolutionary rate. The first involves calculating the number of changes at each site (across sequences), then computing statistics based upon this inferred distribution of changes along the sequence. The second involves direct computation of statistics to characterize within-sequence variation.

The Distribution of Changes along a Sequence

The first approach infers the distribution of substitutions along the sequence. This inference is straightforward for two sequences sufficiently similar such that multiple substitutions are rare, as then there is a direct correspondence between observed changes and actual changes. From a pairwise comparison, variation of rate can be detected *if* it is spatially clustered: under the null hypothesis of no clustering the distribution of the number of nucleotides separating substitutions follows a geometrical distribution (Brown and Clegg, 1983), which is the basis for a statistical test (Zurawski et al., 1984). A sequential clustering method may also detect regions of increased or decreased variation (Clegg et al., 1986).

When more than two sequences are simultaneously compared, the power to detect variation of rate is increased but the common ancestry of sequences complicates the analysis. At the simplest, if the total evolution among sequences is low, the tree constructed by a parsimony method can map the substitutional changes to particular branches in the tree, and the total number of changes at a site during the history of the molecule can be counted. Parsimony, however, does not consider these changes as uncertain inferences with a certain statistical variance, as evidenced by the often occurrence of alternative, equally parsimonious mapping of characters on the tree, especially for more diverged sequences that show homoplasy.

The probabilistic nature of base changes within a phylogeny can be described by the model upon which the maximum-likelihood method for estimating phylogenies is based (e.g., Felsenstein, 1981). This method can be extended to give the expected number of changes for each branch in the phylogeny, conditioned upon the estimates obtained from the entire sequence and conditioned upon the data observed at the particular nucleotide site (Ritland, unpublished). A computer program, NSEQS, was written by Ritland to find the maximum-likelihood phylogeny from a set of n sequences. This program separately estimates the transition and transversion substitution rates and is available upon receipt of a DOS formatted floppy disk. At the conclusion of the sequence tree estimates, the program computes the expected number of changes at each site for each branch; the sum of changes over branches gives the total number for the site over the phylogeny.

With the program, the maximum-likelihood tree of 10 rbcL sequences spanning blue-greens to higher plants was found (the taxa are listed in Ritland and Clegg, 1987). Figure 17.7 shows the expected number of substitutions at each codon site, by codon position. At the slowly evolving 2nd codon position of rbcL, at most three substitutions were observed at any one site, whereas up to six changes were inferred at the 3rd position. At the 3rd position, the number of changes is not integer-valued because of probabilistic corrections made for the possibility of multiple, unseen substitution, whereas at the 2nd position, changes are very close (but not completely) integer-valued because the slow rate of evolution at this position makes multiple substitutions very unlikely.

These data suggest several analyses. The correlation of changes between positions of a codon all have highly significant values (0.385, 0.184, and 0.137 for pairs 1-2, 1-3, and 2-3, respectively). The autocorrelation of changes between successive codon positions was significantly positive for only the second position ($r = 0.10$, $p < 0.001$), indicating clustering of changes occurs at sites under selective constraints. Other methods developed for spatial analysis of ecologic patterns should be fruitful to apply here. There are patterns to the variation of rate, but these are rather weak and probably do not substantially violate the

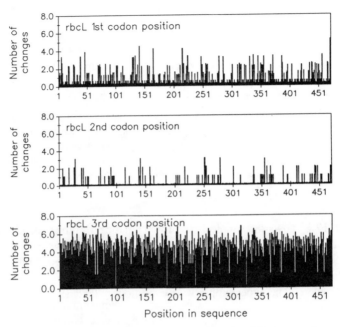

Figure 17.7. The distribution of changes along the *rbc*L sequence in a comparison of 10 sequences, by codon position.

assumptions of independent changes among sites made by the maximum-likelihood method.

Characterizing Within-sequence Variance of Rate

Another approach to studying within-sequence variation of rate involves the fitting of parametric models to the data. A general method, applicable to three or more DNA sequences, involves using the estimated parameters (the branch lengths) of the phylogeny to generate an expected distribution of data, which represents the null hypothesis of no variation of rate. If there is an excess of sequences with all sites identical (detected by testing the fit of data to expectations), then variation of rate must exist (Ritland and Clegg, 1987).

An alternative, model-based approach for directly estimating within-sequence variation of rate can be developed by the simultaneous comparison of three sequences (Ritland, unpublished). Information about variation of rate exists because a complete description of the genetic relationship between three sequences includes not only their pairwise evolutionary distances, but also a "third-order" evolutionary distance. This additional statistic provides information about variation of rate.

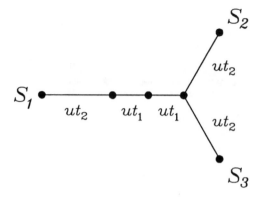

Figure 17.8. An unrooted, three-sequence cladogram for finding within-sequence variation of evolutionary rate. A time t_1 precedes the divergence of sequences S_2 and S_3, whereupon all three sequences evolve for a time t_2. Along each sequence, the mean evolutionary rate is u and the variance of evolutionary rate is v.

Figure 17.8 shows the simplest case for estimating within-sequence variation of evolutionary rate. In this figure, a time t_1 precedes the divergence of sequences S_2 and S_3, whereupon all three sequences evolve for a time t_2. We assume a molecular clock with a rate of substitution u, which varies along the sequence with variance v. The expected pairwise evolutionary distances are

$$D_{23} = (2u - v)t_2$$
$$D_{12} = D_{13} = (2u - v)(t_1 + t_2) \qquad (17.5)$$

These distances D_{12}, D_{13}, and D_{23} can be estimated from observed DNA sequences via Eq. (17.2).

The estimate of branch length to sequence S_1 equals $2u(t_1 + \frac{1}{2}t_2)$, and would normally be found as $\hat{D}_1 = \frac{1}{2}(D_{12} + D_{13} - D_{23}) = (2u-v)(t_1 + \frac{1}{2}t_2)$. Thus, within-sequence variation of rate causes evolutionary distance to be downwardly biased by $v(t_1 + \frac{1}{2}t_2)$. Note this is opposite from the *upward* bias caused by a temporal variation of rate, as discussed in a previous section.

To estimate v separately and remove this bias, we introduce a three-sequence evolutionary distance with expectation

$$D_{123} = (2u - v)t_1 + (3u - 3v)t_2, \qquad (17.6)$$

which is related to the number of independent substitutions occurring during the evolution of all three sequences [$\exp(-D_{123})$ is the probability that the triplet of bases at a site are all identical-by-descent]. Using the same arguments as for the derivation of Eq. (17.2), an estimator for the three-sequence distance is

$$\hat{D}_{123} = -\ln\left[\frac{(S_{123} - H_3) + (H_3 - H_2)(S_{12} + S_{13} + S_{23} - 3H_2)/(1 - H_2)}{1 - 3H_2 + 2H_3}\right] \qquad (17.7)$$

where S_{123} is the fraction of the three DNA sequence sites sharing the same base and $H_3 = \pi_t^3 + \pi_c^3 + \pi_a^3 + \pi_g^3$ (other terms were defined previously). With this three-sequence distance, we can solve for the variance/mean ratio, v/u, as

$$\left(\frac{\hat{v}}{u}\right) = \left[\frac{2D_{12} + 2D_{13} + 2D_{23} - 4D_{123}}{D_{12} + D_{13} + 4D_{23} - 2D_{123}}\right] \tag{17.8}$$

This is an example of a statistic describing an evolutionary process that violates the assumption of the phylogenetic inference procedure.

In this formula, t_2 canceled out. This means that evolution only during the period t_2 contributed information about v/u and thus that the precision of estimates is greater with star phylogenies. Finally, this equation allows us to obtain an unbiased estimate of branch length as

$$\hat{D}_l^u = \frac{2\hat{D}_1}{2 - \left(\dfrac{\hat{v}}{u}\right)} \tag{17.9}$$

This is an example of how statistics that characterize violations of the assumptions can, in turn, be used to refine our original estimates. The statistical properties of the above estimators, extensions to nonclock models, and examples will be presented elsewhere.

Conclusions

In this chapter, we have addressed some of the statistical issues that arise in the phylogenetic analysis of sequence data. We have intentionally not dealt with the most important issue, the uncertainty or variance of the "point" estimate of phylogeny reviewed by Felsenstein (1988), as well as other issues in a broader context (Swofford and Olsen, 1990). Instead we have considered some of the assumptions and potentials of phylogenetic analysis, with an emphasis on plant-oriented problems. Although our simplistic models may seem to grind up character information and spit out single numbers, the "bottom-up" approach of starting with the simple model, and adding parameters to create more complex models, allows a careful analysis of factors that may bias our estimates in more complex data analyses.

If one has several sequences to examine for rate variation, the bottom-up approach would involve first testing the three most distantly related sequences. One could then adopt a hierarchic strategy, wherein a second, nearly independent analysis can then be performed upon two other sequences more closely related to one of the three original sequences, and so on. Although it may be possible to formulate four-sequence and higher-order models for hybridization and variation of rate, such models would become very complex very quickly.

One can raise the disconcerting possibility that the statistics for estimating variation of rate are too model-dependent for general application to all types of sequence data. Although we have noted that one can test for the fit of an appropriate base-substitution model to the data, which would seem to validate the subsequent estimate of evolutionary distance, goodness-of-fit does not guarantee that the model is correct: the true model can never be proven, only rejected. The only truly robust estimate of evolutionary distance would be for closely related sequences on the order of 10% or less divergence, which do not show significant levels of multiple substitutions and hence really do not need any model of base substitution to describe their evolution. At least for sequences at this level of relationship, the various statistics on hybridization and variation of rate would be robust. This level of divergence includes chloroplast sequences at the first and second codon positions among all higher plants, so that even quite wide comparisons can be robust if the appropriate genes are studied.

References

Avise, J.C. (1989) Gene trees and organismal histories: a phylogenetic approach to population biology. *Evolution* **43**, 1192–1208.

Bulmer, M. (1989) Estimating the variability of substitution rates. *Genetics* **123**, 615–619.

Brown, A.H.D., and Clegg, M.T. (1983) Analysis of variation in related DNA sequences. In: *Statistical Analysis of DNA Sequence Data* (ed. B.W. Weir), Marcel Dekker, Inc., New York, pp. 107–132.

Cavalli-Sforza, L.L., and Edwards, A.W.F. (1967) Phylogenetic analysis. Models and estimation procedures. *Amer. J. Hum. Genet.* **19**, 233–257.

Clegg, M.T., Ritland, K., and Zurawski, G. (1986) Processes of chloroplast DNA evolution. In: *Evolutionary Processes and Theory* (eds. S. Karlin and E. Nevo), Academic Press, New York, pp. 275–294.

Farris, J.S. (1985) Distance data revisited. *Cladistics* **1**, 67–85.

Felsenstein, J. (1978) Cases in which parsimony or compatibility will be positively misleading. *Syst. Zool.* **27**, 401–410.

Felsenstein, J. (1981) Evolutionary trees from DNA sequences: a maximum likelihood approach. *J. Mol. Evol.* **17**, 368–376.

Felsenstein, J. (1982) How can we infer geography and history from gene frequencies? *J. Theor. Biol.* **96**, 9–20.

Felsenstein, J. (1984) Distance methods for inferring phylogenies: a justification. *Evolution* **38**, 16–24.

Felsenstein, J. (1985) Confidence limits on phylogenies: an approach using the bootstrap. *Evolution* **39**, 783–791.

Felsenstein, J. (1988) Phylogenies from molecular sequences: inference and reliability. *Ann. Rev. Genet.* **22**, 521–565.

Gillespie, J.H. (1986a) Rates of molecular evolution. *Ann. Rev. Ecol. Syst.* **17**, 637–655.

Gillespie, J.H. (1986b) Natural selection and the molecular clock. *Mol. Biol. Evol.* **3**, 138–155.

Gillespie, J.H. (1989) Lineage effects and the index of dispersion of molecular evolution. *Mol. Biol. Evol.* **6**, 636–647.

Hasegawa, M., Kishino, H., and Yano, T. (1985) Dating of the human-ape splitting by a molecular clock of mitochondrial DNA. *J. Mol. Evol.* **22**, 160–174.

Holmquist, R., and Pearl, D. (1980) Theoretical foundations for quantitative paleogenetics. III. The molecular divergence of nucleic acids and proteins for the case of genetic events of unequal probability. *J. Mol. Evol.* **16**, 211–267.

Hudson, R.R., Kreitman, M., and Aguade, M. (1987) A test of neutral molecular evolution based on nucleotide data. *Genetics* **116**, 153–159.

Jin, L., and Nei, M. (1990) Limitations of the evolutionary parsimony method of phylogenetic analysis. *Mol. Biol. Evol.* **7**, 82–102.

Jukes, T.H., and Cantor, C.R. (1969) Evolution of protein molecules. In: *Mammalian Protein Metabolism* (ed. H.N. Munro), Academic Press, New York, pp. 21–132.

Kaplan, N., and Risko, K. (1982) A method for estimating rates of nucleotide substitution using DNA sequence data. *Theor. Pop. Biol.* **21**, 318–328.

Kimura, M. (1981) Estimation of evolutionary distances between homologous nucleotide sequences. *Proc. Natl. Acad. Sci. USA* **78**, 454–458.

Kimura, M. (1983) *The Neutral Theory of Molecular Evolution,* Cambridge University Press, Cambridge.

Lake, J.A. (1987) A rate-independent technique for analysis of nucleic acid sequences: evolutionary parsimony. *Mol. Biol. Evol.* **4**, 167–191.

Langley, C.H., and Fitch, W.M. (1974) An examination of the constancy of the rate of molecular evolution. *J. Mol. Evol.* **3**, 161–177.

Lathrop, G.M. (1982) Evolutionary trees and admixture: phylogenetic inference when some populations are hybridized. *Ann. Hum. Genet.* **46**, 245–255.

Lewontin, R.C. (1989) Inferring the number of evolutionary events from DNA coding sequence differences. *Mol. Biol. Evol.* **6**, 15–32.

Li, W.-H. (1989) A statistical test of phylogenies estimated from sequence data. *Mol. Biol. Evol.* **6**, 424–435.

Li, W.-H., Luo, C.C., and Wu, C.I. (1985) Evolution of DNA sequences. In: *Molecular Evolutionary Genetics* (ed. R.J. MacIntyre), Plenum Press, New York, pp. 1–94.

Li, W.-H., Wolfe, K.H., Sourdis, J., and Sharp, P.H. (1987) Reconstruction of phylogenetic trees and estimation of divergence times under nonconstant rates of evolution. *Cold Spring Harbor Symp. Quant. Biol.* **52**, 847–856.

Margoliash, E. (1963) Primary structure and evolution of cytochrome C. *Proc. Natl. Acad. Sci. USA* **50**, 672–679.

Miyata, T., and Yasunaga, T. (1981) Rapidly evolving mouse α-globin related pseudogenes and its evolutionary history. *Proc. Natl. Acad. Sci. USA* **78**, 450–453.

Olsen, G.J. (1987) Earliest phylogenetic branchings: comparing rRNA-based evolutionary trees inferred with various techniques. *Cold Spring Harbor Symp. Quant. Biol.* **52**, 825–837.

Pamilo, P., and Nei, M. (1988) Relationships between gene trees and species trees. *Mol. Biol. Evol.* **5**, 568–583.

Qu, L.H., Nicoloso, M., and Bachelleria, J.P. (1988) Phylogenetic calibration of the 5' terminal domain of large rRNA achieved by determining twenty eukaryotic sequences. *J. Mol. Evol.* **28**, 113–124.

Ritland, K., and Clegg, M.T. (1987) Evolutionary analysis of plant DNA sequences. *Amer. Natur.* **130**, S74–S100.

Ritland, K., and Clegg, M.T. (1990) Optimal DNA sequence divergence for testing phylogenetic hypotheses. In: *Molecular Evolution* (eds. S.J. O'Brien and M.T. Clegg), UCLA Symposia on Molecular and Cellular Biology, New Series, Vol. 122, Alan R. Liss, New York, pp. 289–296.

Saitou, N., and Imanishi, T. (1989) Relative efficiencies of the Fitch-Margoliash, maximum-parsimony, maximum-likelihood, minimum-evolution and neighbor-joining methods of phylogenetic tree construction in obtaining the correct tree. *Mol. Biol. Evol.* **6**, 514–525.

Shannon, C.E., and Weaver, W. (1949) *The Mathematical Theory of Communication*, University of Illinois Press, Urbana, IL.

Soltis, D.E., and Soltis, P.S. (1989) Allopolyploid speciation in *Tragopogon*: insights from chloroplast DNA. *Amer. J. Bot.* **76**, 1119–1124.

Swofford, D.L., and Olsen, G.J. (1990) Phylogeny reconstruction. In: *Molecular Systematics* (eds. D.M. Hillis and C. Moritz), Sinauer Associates, Inc., Sunderland, MA, pp. 411–501.

Tajima, F., and Nei, M. (1984) Estimation of evolutionary distance between nucleotide sequences. *Mol. Biol. Evol.* **1**, 269–285.

Takahata, N., and Kimura, M. (1981) A model of evolutionary base substitutions and its application with special reference to rapid change of pseudogenes. *Genetics* **98**, 641–657.

Thompson, E.A. (1975) *Human Evolutionary Trees*, Cambridge University Press, Cambridge.

Wilson, A.C., Carlson, S.S., and White, T.J. (1977) Biochemical evolution. *Ann. Rev. Biochem.* **46**, 573–639.

Zuckerkandl, E., and Pauling, L. (1962) Molecular disease, evolution and genic heterogeneity. In: *Evolving Genes and Proteins* (eds. V. Bryson and H.J. Vogel) Academic Press, New York, pp. 97–116.

Zurawski, G., Clegg, M.T., and Brown, A.H.D. (1984) The nature of nucleotide sequence divergence between barley and maize chloroplast DNA. *Genetics* **106**, 735–749.

Index

DATE DUE

MAR 1 9 1994			
APR 3 0 1995			
APR 1 9 1999			
APR 0 9 2000			
MAY 1 0 2000			
4/6/08			